NOVEL ECOSYSTEMS

Hoh Rainforest, Olympic National Park, Washington, USA (Photo: RJ Hobbs)

*"Do not let spacious plans for a new world divert your energies
from saving what is left of the old."* Winston Churchill

Pinus radiata invading grassland in the Mackenzie Basin, New Zealand (Photo: RJ Hobbs)

*"As I did stand my watch upon the hill,
I look'd toward Birnam, and anon, methought,
The wood began to move."*
Shakespeare, Macbeth, Act 5 Scene 5

NOVEL ECOSYSTEMS

INTERVENING IN THE NEW ECOLOGICAL WORLD ORDER

Edited by

Richard J. Hobbs
University of Western Australia, Australia

Eric S. Higgs
University of Victoria, Canada

Carol M. Hall
University of Victoria, Canada

⟨W⟩WILEY-BLACKWELL

A John Wiley & Sons, Ltd., Publication

This edition first published 2013 © 2013 by John Wiley & Sons, Ltd
Chapter 29 'Coming of Age in a Trash Forest' © 2011 by Emma Marris

Wiley-Blackwell is an imprint of John Wiley & Sons, formed by the merger of Wiley's global Scientific, Technical and Medical business with Blackwell Publishing.

Registered office: John Wiley & Sons, Ltd, The Atrium, Southern Gate, Chichester, West Sussex, PO19 8SQ, UK

Editorial offices: 9600 Garsington Road, Oxford, OX4 2DQ, UK
                        The Atrium, Southern Gate, Chichester, West Sussex, PO19 8SQ, UK
                        111 River Street, Hoboken, NJ 07030-5774, USA

For details of our global editorial offices, for customer services and for information about how to apply for permission to reuse the copyright material in this book please see our website at www.wiley.com/wiley-blackwell.

Library of Congress Cataloging-in-Publication Data
Novel ecosystems : intervening in the new ecological world order / edited by Richard J. Hobbs, Eric S. Higgs, and Carol M. Hall.
       p. cm.
    Includes bibliographical references and index.
    ISBN 978-1-118-35422-3 (cloth)
    1. Ecosystem management.   2. Ecosystem health.   3. Ecological disturbances.   4. Nature–Effect of human beings on.
I. Hobbs, R. J. (Richard J.)   II. Higgs, Eric, 1958–   III. Hall, Carol M.
    QH75.N68 2013
    333.72–dc23
                                2012031506

A catalogue record for this book is available from the British Library.

Wiley also publishes its books in a variety of electronic formats. Some content that appears in print may not be available in electronic books.

Cover image by Richard Hobbs. Plant colonization on Mayan ruins at Uxmal, Yucatan Peninsula, Mexico.
Cover design by Steve Thompson

Set in 9/11 pt PhotinaMT by Toppan Best-set Premedia Limited

Printed in the UK

CONTENTS

v

Companion website

This book is accompanied by a companion website:

www.wiley.com/go/hobbs/ecosystems

The website includes:

Figures and Tables from the book for downloading

CONTRIBUTORS

Katy Beaver   Plant Conservation Action group (PCA), Victoria, Mahé, Seychelles

Thomas J. Brandeis   Southern Research Station, USDA Forest Service, Knoxville, Tennessee, USA

Peter Bridgewater   Global Garden Consulting, UK

Wylie Carr   Society and Conservation, University of Montana, USA

F. Stuart Chapin III   Institute of Arctic Biology, University of Alaska Fairbanks, Fairbanks, USA

Erle C. Ellis   Geography & Environmental Systems, University of Maryland, USA

John J. Ewel   Department of Biology, University of Florida, USA

Mirijam Gaertner   Centre for Invasion Biology, Department of Botany and Zoology, Stellenbosch University, South Africa

Mark R. Gardener   Charles Darwin Foundation, Galapagos Islands, Ecuador, and School of Plant Biology, University of Western Australia, Australia

Carol M. Hall   School of Environmental Studies, University of Victoria, Canada

Lauren M. Hallett   Department of Environmental Science, Policy & Management, University of California, Berkeley, USA

James A. Harris   Environmental Science and Technology Department, Cranfield University, UK

Laurel M. Hartley   Department of Integrative Biology, University of Colorado, USA

Barbra A. Harvie   School of GeoSciences, Institute of Geography and the Lived Environment, University of Edinburgh, UK

Eileen H. Helmer   International Institute of Tropical Forestry USDA Forest Service, Río Piedras, Puerto Rico

Eric S. Higgs   School of Environmental Studies, University of Victoria, Canada

Richard J. Hobbs   Ecosystem Restoration and Intervention Ecology (ERIE) Research Group, School of Plant Biology, University of Western Australia, Australia

Kristin B. Hulvey   Ecosystem Restoration and Intervention Ecology (ERIE) Research Group, School of Plant Biology, University of Western Australia, Australia

Stephen T. Jackson   Department of Botany and Program in Ecology, University of Wyoming, USA and Southwest Climate Science Center, US Geological Survey, Arizona, USA

Jill F. Johnstone   Department of Biology, University of Saskatchewan, Canada

Thomas A. Jones   USDA Agricultural Research Service, Logan, Utah, USA

Patricia L. Kennedy   Department of Fisheries and Wildlife & Eastern Oregon Agriculture & Natural Resource Program, Oregon State University, USA

Steven V. Kokelj   Cumulative Impact Monitoring Program, Aboriginal Affairs and Northern Development Canada, Yellowknife, Northwest Territories, Canada

Christoph Kueffer   Plant Ecology, Institute of Integrative Biology, Swiss Federal Institute of Technology (ETH), Zurich, Switzerland

Lori Lach   Ecosystem Restoration and Intervention Ecology (ERIE) Research Group, School of Plant Biology and the Centre for Integrative Bee Research, University of Western Australia, Australia

Trevor C. Lantz   School of Environmental Studies, University of Victoria, Canada

Andrew Light   Institute for Philosophy & Public Policy, George Mason University, USA and Center for American Progress, Washington, D.C., USA

Ariel E. Lugo   International Institute of Tropical Forestry, USDA Forest Service, Rio Pedras, Puerto Rico

Pete Manning   School of Agriculture, Food and Rural Development, University of Newcastle, UK

Sebastián Martinuzzi   Department of Forest and Wildlife Ecology, University of Wisconsin–Madison, USA

Emma Marris   Columbia, Missouri, USA

Joseph Mascaro   Department of Global Ecology, Carnegie Institution for Science, Stanford, California, USA

James Mougal   National Park Authority, Victoria, Mahé, Seychelles

Peter J. Mumby   Marine Spatial Ecology Lab, School of Biological Sciences, University of Queensland, Australia

Stephen D. Murphy   Department of Environment and Resource Studies, University of Waterloo, Canada

Cara R. Nelson   Department of Ecosystem and Conservation Sciences, College of Forestry and Conservation, University of Montana, USA

Jesse B. Nippert   Division of Biology, Kansas State University, USA

Michael P. Perring   Ecosystem Restoration and Intervention Ecology (ERIE) Research Group, School of Plant Biology, University of Western Australia, Australia

Cristina E. Ramalho   Ecosystem Restoration and Intervention Ecology (ERIE) Research Group, School of Plant Biology, University of Western Australia, Australia

David M. Richardson   Centre for Invasion Biology, Department of Botany and Zoology, Stellenbosch University, South Africa

Martin D. Robards   Arctic Beringia Program Director, Wildlife Conservation Society, New York, USA

Steve Schwarze   Communication Studies, University of Montana, USA

Timothy R. Seastedt   Department of Ecology and Evolutionary Biology, University of Colorado, USA

Rachel J. Standish   Ecosystem Restoration and Intervention Ecology (ERIE) Research Group, School of Plant Biology, University of Western Australia, Australia

Brian M. Starzomski   School of Environmental Studies, University of Victoria, Canada

Katharine N. Suding   Department of Environmental Science, Policy & Management, University of California, Berkeley, USA

Allen Thompson   School of History, Philosophy, and Religion, Oregon State University, USA

Pedro M. Tognetti   Departamento de Métodos Cuantitativos y Sistemas de Información, Facultad de Agronomia, University of Buenos Aires, Argentina

Laith Yakob   School of Population Health, University of Queensland, Australia

Laurie Yung   Resource Conservation Program, College of Forestry and Conservation, University of Montana, USA

ACKNOWLEDGEMENTS

Are any ecosystems still mostly historical?
Is novelty a continuum or categorical?
Why are we concerned?
And what have we learned?
Are novel ecosystems real or just metaphorical?

–Richard Hobbs (with homage to all the fine limericks penned at and
after the Pender Island workshop)

The workshop and subsequent activities leading to the production of this book was funded by the Australian Research Council, via an Australian Laureate Fellowship to RJH and the Centre of Excellence for Environmental Decisions, and by the Restoration Institute, University of Victoria, Canada. We thank all the participants in the workshop at Poet's Cove, Pender Island, Canada, dedicated staff at Poet's Cove, Heather Gordon and volunteers who helped with transportation. To the authors, we offer a toast to your energy, enthusiasm and collegiality during the writing process. All chapters were peer reviewed, and we thank all reviewers for their contribution to ensuring that the material is both sound and, hopefully, accessible. We also thank our partners Gillian, Stephanie and David, children Katie, Hamish and Logan and other family for their support and patience, especially when editing took us away from other activities for long periods. Finally, thanks to Logan for coining the term 'freakosystem'.

Part I

Introduction

Chapter 1

INTRODUCTION: WHY NOVEL ECOSYSTEMS?

Richard J. Hobbs[1], Eric S. Higgs[2] and Carol M. Hall[2]

[1]Ecosystem Restoration and Intervention Ecology (ERIE) Research Group, School of Plant Biology, University of Western Australia, Australia

[2]School of Environmental Studies, University of Victoria, Canada

"Certain people always say we should go back to nature. I notice they never say we should go forward to nature. It seems to me they are more concerned that we should go back, than about nature."

Gottlieb 1947

On Santa Cruz island in the Galapagos Islands, management of the humid highlands has taken a different path from what we might imagine for ecosystems celebrated for their role in bringing awareness of evolutionary processes. What began as a conventional restoration exercise of returning ecosystems to their historical conditions (characterized as treeless and dominated by the endemic shrub *Miconia robinsoniana*) became, over two decades, something quite different (see Chapter 22). Pervasive ecosystem change driven by an invasion of non-native species such as the red quinine tree (*Cinchona pubescens*), black rats (*Rattus rattus*) and others, means that achieving the original goals is increasingly unrealistic. It is simply impossible, or at least practically impossible, to recover historical ecosystems. The focus of goals now rests on key species of conservation interest and involves a constantly adaptive approach to control invasive species without any realistic intention of eliminating them. Here is a novel ecosystem. It defies conventional management approaches, and demands a new way of thinking

about our interventions in and responsibilities toward ecosystems. Indeed, we have entered an era characterized more and more by novel ecosystems. The Galapagos case is one of many treated in this book, and from these we draw broader lessons and approaches.

This book presents a challenge to the conservation and environmental management communities, but not in the way the reader may expect. Let's take an unusual tack and state clearly what this book is *not* about. It is not a polemic against traditional conservation approaches that recognize the value of protecting places and ecosystems that retain their original biota and historical character. It is not a suggestion that traditional restoration approaches aiming to restore historical ecosystem structures and functions are no longer relevant. It is not an argument for giving up on attempts to keep non-native species out of countries, regions or parks and to effectively control them where they are causing significant problems. Finally, it is not an argument that novelty *per se* is a good thing and should be encouraged. Rather, this book asks us to

---

be open to new goals and approaches (or traditional approaches applied in new ways), and to help shape the development of a management framework to address rapidly changing ecosystems in a way that benefits the well-being of both humans and other species.

By raising the issue of novel ecosystems, some fear that we are simply paving the way for a more *laissez-faire* attitude to conservation, admitting defeat with regard to traditional conservation and restoration goals and more generally moving over to the 'dark side'. These are legitimate concerns that will be discussed in subsequent chapters. However, our motivation for pulling together a book that examines novel ecosystems from a variety of perspectives is less sinister and more pragmatic. Regardless of the range of views and perceptions about novel ecosystems, their existence is becoming ever-more obvious and prevalent in today's rapidly changing world. We can choose to ignore these systems as being unworthy of conservation concern or we can accept (even if reluctantly) the importance of figuring out what we know about novel ecosystems and what to do about them.

What exactly are novel ecosystems? As illustrated earlier, these are systems that differ in composition and/or function from present and past systems as a consequence of changing species distributions, environmental alteration through climate and land use change and shifting values about nature and ecosystems (Harris et al. 2006; Root and Schneider 2006; Ricciardi 2007, see Chapter 6). Such systems can arise either from the cessation of past management practices or because of changes in mostly unmanaged systems. These new systems have been termed 'novel', 'emerging' or 'no-analog' (Milton 2003; Hobbs et al. 2006; Williams and Jackson 2007) and have been considered primarily in relation to invasive species or climate change. The realization of the importance of this type of system has been arrived at from differing perspectives and directions: this includes the insights that ecosystems are always changing (see Chapter 7), but that the current interest stems from the rate and pervasiveness of change resulting from multiple environmental trends. Until recently, the types of ecosystem resulting from these trends have largely been ignored both in ecological theory and in practical management, and yet they now loom large as a growing part of the world in which we live. This is rendering it increasingly critical that we carefully consider novel ecosystems in detail: what are they, what do we know

about them, how do we manage them and how do we tackle the important ecological, social, ethical and policy issues they raise?

The idea of novel ecosystems is not itself new, but how far back it can be traced depends on how broadly it is interpreted. We know, for instance, that the ancient Greeks contemplated constant change as a reality of life (Plato quotes Heraclitus as saying "Everything changes and nothing remains still . . . and . . . you cannot step twice into the same stream", *Cratylus* 402a). Chapter 5 explores the more recent ecological antecedents of the idea, and several of the book's authors have thought extensively about the topic over the past few decades (Bridgewater 1990; Ewel et al. 1991; Chapin and Starfield 1997; Seastadt et al. 2008), while examples of novel ecosystems are being documented in the literature (Lugo 2004; Lugo and Helmer 2004; Wilkinson 2004; Lindenmayer et al. 2008; Mascaro et al. 2008).

In this book, we use the framework presented by Hobbs et al. (2009) as a starting point in which varying degrees of alteration of abiotic and/or biotic components result in systems that move away from their historical configuration and dynamics into different configurations (Fig. 3.2). This framework identifies a gradation in level of change, with moderately changed systems forming a hybrid state and more extensively changed systems forming a novel state. Inherent in this formulation is the idea that there may be thresholds in play, both ecological and social, that effectively prevent the return of the system from a novel state to a less altered state. These ideas are elaborated upon in Chapter 3.

Examples of novel ecosystems include disused shale dumps in Scotland (Chapter 35), places transformed by rock-plowing and subsequently overtaken by non-native plant species in the Everglades in Florida (Chapter 2) and many island systems such as in the Seychelles (Chapter 27) and Puerto Rico (Chapter 9) where non-native species have formed apparently persistent alternative biotic communities. The Scotland and Everglades examples represent places where abiotic conditions have been dramatically altered, following which a different biological community establishes and persists. The Seychelles and Puerto Rico examples present cases where biotic change has been initiated by human activities such as vegetation clearing and use for production, but has then been hastened by the presence of particular plant species that thrive in the newly created or abandoned systems. In contrast, on Christmas Island, Australia, it is the invasion of an animal species (the yellow crazy ant) rather than a plant species that, through its direct and indirect effects, has dramatically altered the food web dynamics and functioning of the forest ecosystem (see Chapter 14).

In addition to novel ecosystems, which can be thought of as displaying an entirely novel biotic and abiotic configuration, there are also many cases where novel elements have become integral parts of the system. This can range from the prevalence of a single species that was not present before to completely transformed systems such as plantations becoming important resources for native species. For example, in the first category, in Nothofagus forests in the foothills of the Andes in Argentina, the non-native European bumble bee *Bombus ruderatus* has all but replaced the native bumble bee *Bombus dahlbomii* as the main pollinator of the major understory plant species *Alstroemeria aurea* (Madjidian et al. 2008). In the second category, extensive plantations of northern hemisphere pines bordering the city of Perth in Western Australia now provide important habitat and food resources for the endangered Carnaby's black-cockatoo *Calyptorhynchus latirostris* (Valentine and Stock 2008). These and numerous other examples illustrate how, whether we like it or not, species are re-assorting or taking advantage of new situations. In the Nothofagus forests, it seems highly unlikely that the non-native bumble bee could ever be eradicated and it would probably be foolish to attempt this in any case. For the cockatoos in Western Australia, it would also appear foolish to consider removing all the non-native pines, or at least to consider doing so without providing alternative food and shelter options for the cockatoos.

These examples also illustrate the need for careful consideration of traditional conservation and restoration practices and norms. Novel ecosystems may appear to exist outside the standard conservation paradigms of protecting special places and species. However, what happens when special places start to change because of invasive species and climate change? What happens when special species start depending on novel ecosystems and assemblages for their persistence? In addition, what happens when we begin to piece together increasing amounts of information that suggest that simply protecting protected areas is unlikely to achieve conservation goals but rather that the overall landscape – including protected, managed and altered systems – is important? In addition to the biophysical aspects inherent in these discussions, there are important social, ethical and policy questions surrounding

how humans perceive and cope with, or adapt to, the changing situation. Although novel ecosystems may be increasingly pervasive across the globe, people's responses to these systems are likely to vary greatly and will include everything from gleeful acceptance through continued denial to outright hostility.

The book has its origins in a workshop held on Pender Island, British Columbia, Canada in May 2011 (Fig. 1.1). We recognized that, while the idea of novel ecosystems has been widely discussed in the ecological literature, such systems pose immense challenges scientifically, ethically and also from a practical and policy perspective. While there has been considerable discussion there has been, to date, little concrete advice to give to managers and policy makers on how to deal with these systems. The workshop therefore brought together selected researchers from a variety of disciplines and also managers and policy makers who are confronting the issues surrounding novel ecosystems and asking how we intervene in such ecosystems in meaningful and effective ways.

The workshop consisted of short presentations from each participant followed by group discussions on various issues surrounding novel ecosystems. Discussions were lively and brought to light both common perspectives and, at times, strongly contested differing points of view. Distilling the initial outcomes of the discussion sessions led to the original formulation of the outline of the book, together with a set of commitments from participants to lead or participate in particular chapters. Other outputs were also discussed, and a 'Call for Action' was generated (http://www.restorationinstitute.ca/projects) and reported by Bridgewater et al. (2011). After the workshop, we utilized the web-based file sharing system 'Dropbox' to initiate and develop the book structure and content. This proved a fascinating and fruitful choice as participants eagerly continued with discussions and debates initiated at the workshop. Many chapters have multiple authors, and some fundamental discussions have continued until the final stages of editing. As editors, rather than trying to impose a received wisdom on authors, we have tried to allow these discussions to run their course and intervened only where we perceived impasses that needed to be resolved. We have also intervened in order to ensure a degree of flow and consistency through the book. However, that task was significantly minimized by the file-sharing process that allowed general access to all chapters as they were being drafted. Constructing the book has therefore

Figure 1.1 Participants in the workshop at Poet's Cove, Pender Island, May 2011. Left to right, standing: Pat Kennedy, Cara Nelson, Tim Seastedt, Jim Harris, Peter Bridgewater, Erle Ellis, Karen Keenleyside, Carol Hall, Keith Bowers, Jack Ewel, Tom Jones, Lori Lach, Mark Gardener, Eric Higgs, Dave Richardson, Richard Hobbs (plus time-keeping parrot), Brian Starzomski, Joe Mascaro, Emma Marris, Mike Perring, Ariel Lugo, Allen Thompson, Steve Jackson. Kneeling: Steve Murphy, Katie Suding, Pedro Tognetti, Rachel Standish, Christoph Kueffer, Lauren Hallett, Laurie Young, Kris Hulvey. (Absent from photo: Andrew Light, Trevor Lantz, Terry Chapin.)

been a highly interactive process with authors from throughout the book chipping in, asking awkward questions and seeking clarifications on chapters throughout the process.

Although we have at all times encouraged clarity and accessibility, we have also allowed authors to maintain their distinctive voices. The issues covered in many chapters are difficult and at times contentious, and agreement was not always easy. Despite a thread of consistency through the book, incongruities remain and not all ends are nicely tied up. To arrive at any other outcome would, we feel, not accurately represent the current state of discussion on novel ecosystems. While chapters in this book contribute to and develop ideas around the topic, they are certainly not the last word; we anticipate that they will open up further discussion rather than closing it off.

The book consists of various different types of contribution. There are the *main chapters* that aim to explore current understanding of novel ecosystems, their biophysical, social and ethical dimensions and

how we might go about managing them effectively. These are complemented by a set of *case studies* that collectively illustrate in concrete terms the ideas and questions raised throughout the book. These case studies vary from detailed and lengthy analyses to short descriptive illustrations. Finally, a number of *perspectives* from authors are included that illustrate how different people perceive or came to know or appreciate novel systems, or give local anecdotes of how the issues surrounding novel ecosystems are being played out in different settings.

The book is organized in seven parts that collect together chapters within broad themes, starting with this introduction and overview of the book (Part 1). Part 2 (Chapters 2–6) provides a foundation for the rest of the book by introducing a range of concepts and ideas about what novel ecosystems are and how they can be considered from both theoretical and practical perspectives, particularly in relation to deciding when and how to intervene in such systems. A conceptual framework is presented in Chapter 3, followed by an examination of islands where novel ecosystems are increasingly the norm and a testing ground for considering some of these concepts to frame the broader topics examined throughout the book (Chapter 4). A closer look at the origins of the concept of novel ecosystems (Chapter 5) leads to a summary of our current understanding of what these systems are, and how they might be broadly characterized and defined (Chapter 6).

Part 3 (Chapters 7–16) examines the key characteristics of novel ecosystems in more detail, starting with a reminder that ecosystems have always been in flux and hence that novelty is nothing new (Chapter 7). Moving to the present, Chapter 8 considers whether it is possible to estimate the overall extent of novel systems in terrestrial and marine environments at a global scale, while Chapter 9 presents an example of mapping novel ecosystems at a more regional scale in Puerto Rico. Subsequent chapters present the current understanding of different aspects of novel ecosystems. Chapter 10 offers insights into understanding climate change impacts on species, communities and ecosystems, while Chapter 11 describes how issues and concepts of alien plant invasions relate to novel ecosystems. Subsequent chapters cover aspects of novel ecosystems that are often overlooked. Chapter 12 uses two case studies to describe how infectious diseases are emerging with rapidly changing ecosystems. Fauna also play a key role in formation and management of novel ecosystems and Chapter 14 provides an overview of relevant issues from this perspective, drawing on many rich examples in the process.

Part 4 (Chapter 17–28) then brings us to one of the central themes of the book: when and how to intervene in novel ecosystems. Building from the conceptual framework described in Chapter 3, this part presents a framework for making decisions regarding the type of ecosystem being managed and the options available for management (Chapter 18). This is complemented by a collection of case studies which illustrate various aspects involved in making these decisions such as identifying whether an ecosystem is hybrid or novel, choosing references and addressing social and ecological barriers (Chapters 19–23). An important aspect of management is the ability to assess the degree of novelty of any given system, and Chapters 24 and 25 discuss aspects of how to measure and recognize novel systems. On making decisions regarding what plant material to use when considering management or restoration of novel ecosystems, Chapter 26 illustrates the hierarchical nature of decision making required to

implement interventions. In other words, decisions at the broad level of the framework presented earlier in this part will generally have many sets of more detailed decisions nested within them.

In the discussions of management and intervention of novel ecosystems, questions of values and perceptions repeatedly surface and Part 5 (Chapters 29–35) examines these questions in more detail. Chapter 30 considers how engagement of people with novel ecosystems is best accomplished through a variety of participatory and decision-making tools. Chapter 31 considers values of novel ecosystems from an ethical perspective, while Chapter 33 suggests that policy needs to recognize and incorporate novel ecosystems.

Part 6 (Chapters 36–41) starts a discussion of what lies ahead for novel ecosystems and how people interact with them. This starts with an articulation of the many concerns that can be raised with regards to novel ecosystems and their increased recognition and acceptance (Chapter 37). As a counterpoint, Chapter 38 discusses novel ecosystems in urban areas and points to their likely contribution to human well-being through the provision of essential ecosystem services. The likely meshing of considerations on managing novel ecosystems with broader ideas relating to ecosystem stewardship then provides a potential roadmap for how things might profitably progress in the future both for human well-being and conservation goals (Chapters 39 and 40). Case studies and perspectives are distributed throughout the book, and this last part is bookended by two contrasting perspectives: one articulating at a personal level the concerns surrounding novel ecosystems (Chapter 36) and the other presenting a view that novel ecosystems are here to stay and hence need to be embraced more fully by society (Chapter 41).

In the concluding part (Chapter 42) we, as editors, provide a synthesis of the key lines of thought running through the book and pull out some of the main issues that arose, together with some of the important aspects that require ongoing discussion and thought.

The book is organized, we hope, in a logical way that leads the reader through a progression of topics. However, we also recognize that many readers will not wish to read the material sequentially and would rather aim for the parts of most relevance to their particular interests. Chapter 3 will provide a broad overview of many of the main concepts covered in the book. Thereafter, someone with more interest in management issues is likely to find Part 4 of most value, while a person with a broad interest in our current understanding of the ecology of novel ecosystems will be drawn to Part 3. On the other hand, someone interested in the more philosophical and social aspects of the topic will concentrate on Parts 4 and 5. This characterization is, however, too simplistic, and many parallel themes occur throughout the book and are also highlighted in the case studies and perspectives. Whatever course you chart through the book, we hope our endeavors will lead to a clearer picture of the current state of play regarding how we might understand, manage and interact with novel ecosystems.

REFERENCES

Bridgewater, P.B. (1990) The role of synthetic vegetation in present and future landscapes of Australia. *Proceedings of the Ecological Society of Australia*, **16**,129–134.

Bridgewater, P., Higgs, E.S., Hobbs, R.J. and Jackson, S.T. (2011) Engaging with novel ecosystems. *Frontiers in Ecology and Environment*, **9**, 423.

Chapin, F.S. and Starfield, A.M. (1997) Time lags and novel ecosystems in response to transient climatic change in Alaska. *Climate Change*, **35**, 449–461.

Ewel, J.J., Mazzarino, M.J. and Berish, C.W. (1991) Tropical soil fertility changes under monocultures and successional communities of different structure. *Ecological Applications*, **1**, 289–302.

Gottlieb, A. (1947) The Ides of Art: The Attitudes of Ten Artists on Their Art and Contemporaneousness. *The Tiger's Eye*, **1**, 42–52.

Harris, J.A., Hobbs, R.J., Higgs, E. and Aronson, J. (2006) Ecological restoration and global climate change. *Restoration Ecology*, **14**, 170–176.

Hobbs, R.J., Arico, S., Aronson, J., Baron, J.S., Bridgewater, P., Cramer, V.A., Epstein, P.R., Ewel, J.J., Klink, C.A., Lugo, A.E., Norton, D., Ojima, D., Richardson, D.M., Sanderson, E.W., Valladares, F., Vilà, M., Zamora, R. and Zobel, M. (2006) Novel ecosystems: Theoretical and management aspects of the new ecological world order. *Global Ecology and Biogeography*, **15**, 1–7.

Hobbs, R.J., Higgs, E. and Harris, J.A. (2009) Novel ecosystems: implications for conservation and restoration. *Trends in Ecology and Evolution*, **24**, 599–605.

Lindenmayer, D.B., Fischer, J., Felton, A., Crane, M., Michael, D., Macgregor, C., Montague-Drake, R., Manning, A. and Hobbs, R.J. (2008) Novel ecosystems resulting from landscape transformation create dilemmas for modern conservation practice. *Conservation Letters*, **1**, 129–135.

Lugo, A.E. (2004) The outcome of alien tree invasions in Puerto Rico. *Frontiers in Ecology and Environment*, **2**, 265–273.

Lugo, A.E. and Helmer, E. (2004) Emerging forests on abandoned land: Puerto Rico's new forests. *Forest Ecology and Management*, **190**, 145–161.

Madjidian, J., Morales, C. and Smith, H. (2008) Displacement of a native by an alien bumblebee: lower pollinator efficiency overcome by overwhelmingly higher visitation frequency. *Oecologia*, **156**, 835–845.

Mascaro, J., Becklund, K.K., Hughes, R.F. and Schnitzer, S.A. (2008) Limited native plant regeneration in novel, exotic-dominated forests on Hawai'i. *Forest Ecology and Management*, **256**, 593–606.

Milton, S.J. (2003) 'Emerging ecosystems': a washing-stone for ecologists, economists and sociologists? *South African Journal of Science*, **99**, 404–406.

Ricciardi, A. (2007) Are modern biological invasions an unprecedented form of global change? *Conservation Biology*, **21**, 329–336.

Root, T.L. and Schneider, S.H. (2006) Conservation and climate change: the challenges ahead. *Conservation Biology*, **20**, 706–708.

Seastadt, T.R., Hobbs, R.J. and Suding, K.N. (2008) Management of novel ecosystems: Are novel approaches required? *Frontiers in Ecology and the Environment*, **6**, 547–553.

Valentine, L.E. and Stock, W. (2008) Food Resources of Carnaby's Black-Cockatoo (*Calyptorhynchus latirostris*) in the Gnangara Sustainability Strategy study area. http://ro.ecu.edu.au/ecuworks/6147. Gnangara Sustainability Strategy Taskforce, Perth.

Wilkinson, D.M. (2004) The parable of Green Mountain: Ascension Island, ecosystem conservation and ecological fitting. *Journal of Biogeography*, **31**, 1–4.

Williams, J.W. and Jackson, S.T. (2007) Novel climates, no-analog communities, and ecological surprises. *Frontiers in Ecology and the Environment*, **5**, 475–482.

Part II

What are Novel Ecosystems?

Part II

What are Novel Ecosystems?

Chapter 2

CASE STUDY: HOLE-IN-THE-DONUT, EVERGLADES

John J. Ewel

Department of Biology, University of Florida, USA

In the early 1900s, farmers in South Florida, USA, found an isolated area of wetlands (Fig. 2.1, upper panel) that had a sufficiently long dry season and deep enough soils to make vegetable farming worthwhile. Farmers liked the isolation of the site, removed as it was from the pest loads of surrounding fields. Furthermore, the risk of crop lost to frost was very much reduced in this warmest of continental US climates. This area, which later came to be known as the Hole-in-the-Donut, covered several thousand hectares. Farming there greatly expanded in two bursts following initial colonization. One expansion was triggered by the completion of a road to the nearest town in 1915; the second came after World War II when crawler tractors and heavy plows capable of crushing limestone and mixing it with the thin layer of marl soil on its surface created a deeper, better-aerated soil. This rock plowing of nearly 2400 hectares, plus the original area of deeper soils, together made commercial agriculture a lucrative enterprise on some 4000 hectares.

In 1947 Everglades National Park was inaugurated, and today more than 370,000 hectares of land lie within its boundaries. The park completely surrounded the agricultural lands, giving rise to the sobriquet Hole-in-the-Donut: the doughnut was the park and the hole the farmland. Not surprisingly, the accouterments of intensive agriculture (fertilizers, aerial application of pesticides, heavy equipment) were not welcomed by the park. Under threat of condemnation, the farmers sold their lands to the government in the early 1970s. At that point it was still a donut, but the hole was now abandoned soil, much of it rock-plowed, to which fertilizer had been applied for decades. Every weed in South Florida found it to be a great place to grow.

Schinus terebinthifolius (Brazilian pepper or Christmas berry) is an alien tree species that has been present in South Florida as an ornamental since the mid-1800s, but was so inconspicuous in the wild that it went unmentioned in major ecological surveys conducted in the 1940s and 1950s. It proved, however, to be exceptionally well suited to the former agricultural lands in the Hole-in-the-Donut, forming long-lived thickets of tangled stems that were almost impenetrable (Fig. 2.1, lower panel). By the late 1970s, park biologists realized that in buying out the farmers, they had traded one problem for another. *Schinus* not only transformed the viewscape from wet prairie to woodland, but it began to invade adjacent unplowed ecosystems (especially pine-dominated rocklands) despite attempts to exclude it with prescribed fire.

At that time I was invited to conduct research in the Hole-in-the-Donut with the objective of helping the park staff contain *Schinus* and consider ways to restore the Hole-in-the-Donut to some semblance of its former self. My colleagues and I worked in the Hole-in-the-Donut for four years, and the more we learned about the environment and about *Schinus*, the more intractable the problem seemed. *Schinus* is obligately

Novel Ecosystems: Intervening in the New Ecological World Order, First Edition. Edited by Richard J. Hobbs, Eric S. Higgs, and Carol M. Hall.
© 2013 John Wiley & Sons, Ltd. Published 2013 by John Wiley & Sons, Ltd.

Figure 2.1 Upper panel: native wet prairie vegetation growing on shallow marl overlying pitted limestone (inset). Lower panel: following rock plowing and subsequent cessation of agriculture, a novel forest dominated by the invasive introduced tree (*Schinus terebinthefolius*) but containing many other species, both native and introduced, developed on the rock-plowed soils (inset). This anthrosol is well aerated, supports mycorrhizae, contains residual phosphorus and has developed a surface horizon high in organic matter.

mycorrhizal, thus able to take advantage of the deep well-aerated anthrosol; it was pollinated by native insects and dispersed by native birds and mammals; and its southern hemisphere reproductive phenology landed it in a regeneration niche six months out of phase with native competitors. In short, *Schinus* was right at home in the Hole-in-the-Donut and, unlike farmers, the threat of condemnation did not faze it.

Post-agriculture vegetation in the Hole-in-the-Donut was undeniably lush, and the plant life fueled substantial animal life. Clouds of tree swallows gorged on vast quantities of fruits off the native wax myrtle bushes that formed monospecific stands in some places. Raccoons were extremely abundant and, in fact, were important long-distance dispersers of *Schinus* seed. Deer were sighted frequently. Although this was at a time when the number of surviving Florida panthers was at its nadir (in the low 20s), panther sightings were not uncommon in the Hole-in-the-Donut. Whether this was due to the abundance of food (raccoons, deer, etc.), as was common lore around the Research Center (but unsubstantiated by data), or whether it was due to great visibility on two long lightly traveled, elevated farm roads that ran east–west for miles and were the highest ground around, is unknown. But there were frequent panther sightings: that is certain. The Hole-in-the-Donut now supported a truly novel ecosystem composed of a mix of plant species including many aliens; it housed abundant bird and mammal life, not to mention mosquitoes in clouds so thick that we bought military-strength DEET by the case; and it occupied an anthrosol later conferred its own Soil Series name by the US Department of Agriculture.

Toward the end of my research tenure in the park, I was asked to participate in a meeting convened among park biologists, scientists, resource managers and administrators. We were joined by administrative staff from higher in the park service bureaucracy who flew in from Atlanta. I was asked: "Jack, what should we do about this *Schinus* problem?" I briefly related how our recent soil studies had revealed high concentrations of residual phosphorus (presumably from fertilizer) and described graduate student Chip Meador's findings on the positive mycorrhizal status of all those weeds, including *Schinus*, in what had been essentially a non-mycorrhizal ecosystem prior to rock plowing. Tongue-in-cheek, I made the remark that "The only way to restore the native wet prairies would be to come in here

with bulldozers and cart that anthrosol out of the Hole-in-the-Donut." I then turned back to business and went on at length to summarize what we had learned about the ecology and life history of *Schinus* that made it so well suited to these former farmlands. Recognizing the futility of past blunt-force efforts (Fig. 2.2), I described what I thought at the time might be a promising approach to convert *Schinus* forests to native-species dominance. The recommended tactic consisted of killing all the females of this dioecious tree species and leaving the males in place to act as a nurse crop for native species, which we had observed reproducing in the understory. (Our t-shirts were to have said "Kill the Mothers!") With time, presumably the *Schinus* would die out – the males of old age and the females from triclopyr toxicity – and native species would take over.

Slogging around in a mosquito-filled *Schinus* tangle with a backpack sprayer and trying to get a ring of herbicide around each stem of a ten-trunked tree is slow-going work, even if you can find someone willing to do it. It is to the park's credit that they tried single-tree herbicide treatments, and it is not surprising that it was a flop. In subsequent years a number of approaches to restoration were undertaken, but only one of them gave promise of success at large scale. In 1997, 17 years after that meeting in the Research Center, word reached me that the park was undertaking Hole-in-the-Donut restoration by removing the anthrosol. I was incredulous.

That dramatic action reflected the thinking of the time: invasive exotics were to be controlled at all costs. In the Hole-in-the-Donut, this continues to be the story today. Heavy machinery is used to scrape up the rock-plowed soil, which is loaded onto big dump trucks and stockpiled in low mounds in the Hole-in-the-Donut (Fig. 2.3). The operation is clearly visible on Google Earth (longitude 80° 40′ 30″ W; latitude 25° 22′ 40″ N) and work has been completed on more than 1700 hectares (about two-thirds of the land authorized for treatment). More than $3 \times 10^6 \, \mathrm{m}^3$ of soil have been moved; if piled on an American football field, the resulting mound would be a rectangular column reaching 628 m (about six times taller than the highest hill in the state). Native plants are again reclaiming the Hole-in-the-Donut, and the prairie viewscape is being restored (Fig. 2.4). The new substrate is not a twin of the original; it is relatively smooth and solid whereas the original limestone was rugged and pitted, containing pockets of marl and organic matter.

Figure 2.2 In the mid-1970s some of the *Schinus*-dominated vegetation was bulldozed into windrows, but this did not affect restoration on most sites. The novel ecosystem rebounded quickly, as the anthrosol remained even though the plants were piled and burned.

Figure 2.3 In the mid-1990s the decision was made to remove the anthrosol. The process, still underway, consists of: (a) cutting the vegetation; (b) loading the debris and soil; (c) hauling the material to local repositories within the Hole-in-the-Donut; and finally (d) scraping the remaining soil off the underlying limestone base rock. Photographs courtesy of Everglades National Park.

Figure 2.4 Vegetation that develops after soil removal resembles that of the original wetland and is maintained by prescribed fire. Photograph by Todd Osborne, courtesy of Everglades National Park and the photographer.

And what of the elusive panther? It would be a tragedy indeed if I reported that the last of Florida's panthers was scared out of the Hole-in-the-Donut by earth movers, to die of starvation, mosquito bites and collisions with automobiles. Happily, that is not the case. After much study, consultation and public debate, authorities made the decision in the mid-1990s to give the handful of remaining panthers a genetic boost by introducing eight Texan females. The genotype was sacrificed for the phenotype, but there are now more than 120 panthers in Florida. Our State Mammal is a novel hybrid.

ACKNOWLEDGEMENTS

I thank staff of Everglades National Park for conversations, data, reports, comments on the draft manuscript and photographs. JR Snyder of the US Geological Survey directed me to information sources and provided feedback. Responsibility for the perspectives expressed in this case study lies solely with the author.

FURTHER READING

Aziz, T., Sylvia, D.M. and Doren, R.F. (1995) Activity and species composition of arbuscular mycorrhizal fungi following soil removal. *Ecological Applications*, **5**, 776–784.

Dalrymple, G.H., Doren, R.F., O'Hare, N.K. Norland, M.R. and Armentano, T.V. (2003) Plant colonization after complete and partial removal of disturbed soils for wetland restoration of former agricultural fields in Everglades National Park. *Wetlands*, **23**, 1015–1029.

Doren, R.F., Whiteaker, L.D., Molnar, G. and Sylvia, D. (1990) Restoration of former wetlands within the Hole-in-the-Donut in Everglades National Park. *Proceedings 17th Annual Conference on Wetlands Restoration and Creation*, Tampa, FL.

Ewel, J. 1986. Invasibility: Lessons from South Florida, in *Ecology of Biological Invasions of North America and Hawaii* (eds H. Mooney and J. Drake). Springer-Verlag, NY, 214–230.

Ewel, J.J., Ojima, D.S. Karl, D.S. and DeBusk, W.F. (1982) *Schinus* in successional ecosystems of Everglades National Park. South Florida Research Center Report T-676. National Park Service, Everglades National Park, Homestead FL.

Krauss, P. 1987. Old field succession in Everglades National Park. South Florida Research Center Report. SFRC-87/0. National Park Service, Everglades National Park, Homestead FL.

Li, Y.C. and Norland, M. (2001) The role of soil fertility in invasion of Brazilian Pepper (*Schinus terebinthifolius*) in Everglades National Park, Florida. *Soil Science*, **166**, 400–405.

Loope, L.L. and Dunevitz, V.L. (1981) Impact of fire exclusion and invasion of *Schinus terebinthifolius* on limestone rockland pine forests of southeastern Florida. South Florida Research Center Report T-645. National Park Service, Everglades National Park, Homestead FL.

Loope, L.L. and Dunevitz, V.L. (1981) Investigations of early plant succession on abandoned farmland in Everglades National Park. South Florida Research Center Report T-644. National Park Service, Everglades National Park, Homestead FL.

Meador, R.E. II. (1977) The role of mycorrhizae in influencing succession on abandoned Everglades farmland. MSc thesis. University of Florida, Gainesville.

Orth, P.G. and Conover, R.A. (1975) Changes in nutrients resulting from farming in the Hole-in-the-Doughnut, Everglades National Park. *Proceedings of the Florida State Horticultural Society*, 221–225.

Roman, J. (2011) *Listed: Dispatches from America's Endangered Species Act*. Harvard University Press, Cambridge MA.

Smith, C.S., Serra, L., Li, Y., Inglett, P. and Inglett, K. (2011) Restoration of disturbed lands: the Hole-in-the-Donut restoration in the Everglades. *Critical Reviews in Environmental Science and Technology*, **41**(S1), 723–739.

Chapter 3

TOWARDS A CONCEPTUAL FRAMEWORK FOR NOVEL ECOSYSTEMS

Lauren M. Hallett[1], Rachel J. Standish[2], Kristin B. Hulvey[2], Mark R. Gardener[3], Katharine N. Suding[1], Brian M. Starzomski[4], Stephen D. Murphy[5] and James A. Harris[6]

[1]Department of Environmental Science, Policy & Management, University of California, Berkeley, USA

[2]Ecosystem Restoration and Intervention Ecology (ERIE) Research Group, School of Plant Biology, University of Western Australia, Australia

[3]Charles Darwin Foundation, Galapagos Islands, Ecuador, and School of Plant Biology, University of Western Australia, Australia

[4]School of Environmental Studies, University of Victoria, Canada

[5]Department of Environment and Resource Studies, University of Waterloo, Canada

[6]Environmental Science and Technology Department, Cranfield University, UK

3.1 INTRODUCTION

Endangered birds often garner conservation action and the Rodrigues fody (*Foudia flavicans*) is no exception. Dependent on mature-stand forests on the smallest of the Mascarene Islands, the Rodrigues fody (Fig. 3.1) experienced a population crash when the majority of its habitat was converted for agriculture in the 1960s. What was exceptional about the fody was its manner of recovery. Before its population could be completely decimated, it was saved in part by the

expansion of fast-growing non-native trees that quickly fulfilled the mature-stand habitat requirement of the bird (Impey et al. 2002; popular coverage and interpretation by Fox 2003 and Marris 2011). Its story highlights three key points we explore throughout this chapter. First, it indicates that novel species interactions should be considered in conservation efforts. Second, it demonstrates that novel ecosystems can provide some of the same functions as their historical counterparts. Lastly, it serves as a cautionary tale: the fody nearly went extinct due to anthropogenic land

Figure 3.1 A male Rodrigues fody (*Foudia flavicans*) displaying. Photograph courtesy of Dubi Shapiro.

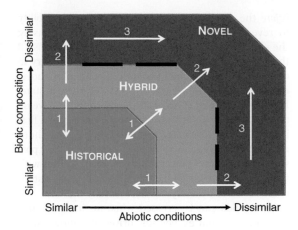

Restoration threshold

Figure 3.2 Types of ecosystems under varying levels of biotic and abiotic change. A historical ecosystem remains within its historical range of variability; a hybrid ecosystem is biotically and/or abiotically dissimilar to its historical ecosystem but is capable of returning to the historical state; novel ecosystems are biotically and/or abiotically dissimilar to the historical state and have passed a threshold such that they cannot be returned to the historical state. Pathways represent possible directions of change: (1) shifts from historical to hybrid ecosystems that are reversible; (2) non-reversible shifts from historical or hybrid ecosystems to novel ecosystems; and (3) further biotic and abiotic shifts are possible within novel ecosystems. From Hobbs et al. (2009). Reproduced with permission of Elsevier.

change and native forests are recovering slowly, if at all (Impey et al. 2002). Slowing anthropogenic drivers of ecosystem change, such as land conversion and climate change, is the primary way to reduce the frequency of these types of conservation challenges. Novel ecosystems and associated species interactions may be a significant secondary tool in conservation situations.

This chapter is about when and how to intervene in novel ecosystems. It provides a brief introduction to many of the ideas and concepts addressed in greater detail in later chapters. It is not an argument for the virtue of novel ecosystems *per se*; given the choice, most of us would opt to mitigate many of the processes driving ecosystem change. In a world of rapid human-induced change however, the power of the novel ecosystem concept is its pragmatism. Novel ecosystems can serve conservation aims, whether by maintaining species diversity or providing ecosystem services. Here we develop a framework to aid in evaluation of such benefits. We first describe approaches to identify thresholds shifts into novel territory. Second, we consider how functional similarities between novel and historical ecosystems can inform decisions about when and how to intervene in novel ecosystems. We conclude with a discussion of practical considerations and methods for managing these systems.

3.2 THRESHOLDS AND ANTICIPATING DRAMATIC ECOSYSTEM SHIFTS

Novel ecosystems are composed of non-historical species configurations that arise due to anthropogenic environmental change, land conversion, species invasions or a combination of the three. They result as a consequence of human activity but do not depend on human intervention for their maintenance (Hobbs et al. 2006; Chapter 5). Because it can be easier to reverse the effects of some drivers of ecological change but not others, a useful distinction is between hybrid and novel ecosystems. While both hybrid and novel ecosystems are composed of new species combinations and/or abiotic conditions, hybrid ecosystems can more readily be returned to their historical states whereas significant barriers prevent novel ecosystems from returning to their historical states (Fig. 3.2).

This distinction is important for two reasons. First, a primary management goal is often to prevent threshold shifts that result in novel ecosystems. This requires the ability to differentiate a hybrid from a novel ecosystem *before* the shift occurs, and also requires the ability to reverse or control the effects of drivers causing the shift. Second, when a threshold has been crossed that in practice is irreversible, it becomes necessary to develop new management goals for the resulting novel ecosystem other than returning it to its historical state. Later in this chapter we will discuss how to develop management goals for novel ecosystems, but it is first important to characterize possible thresholds. Without a robust consideration of threshold dynamics, it is possible that the novel label becomes over-prescribed with the unintended consequence that the conservation potential of some ecosystems is not fully realized.

Thresholds can be crossed when an increase in a continuous, often exogenous, driver of change, such as nitrogen deposition or climate change, accumulates to a point at which the system can no longer absorb the change and instead shifts into a different state. Identifying these tipping points can help managers prioritize their efforts in hybrid ecosystems (Suding and Hobbs 2009). For example, known thresholds tipped by rising salinity in the Wheatbelt of south-western Australia (Cramer and Hobbs 2002) helped managers decide when to intervene to prevent a large freshwater lake from becoming saline (Froend et al. 1997; Wallace 2003). Managers with extensive ecological knowledge and access to long-term datasets may be better placed than most to make use of the threshold approach in a predictive manner (Bestelmeyer 2006; Bestelmeyer et al. 2011). There will always be uncertainty about the exact location of thresholds however, and for systems characterized by complex dynamics it may not be possible to develop indicators of an impending regime shift (Hastings and Wysham 2010). Consequently, a combination of a threshold approach with risk assessment – what are the consequences if a threshold is crossed? – is needed to help guide decision making (Polasky et al. 2011; Chapter 18).

Species invasions can similarly drive ecosystems across thresholds. Species invasions are often dynamic and difficult to anticipate, and invasion-driven thresholds may be passed before they have even been noticed (Box 3.1). For example, relatively little research has been conducted to show quantitative thresholds of biotic and abiotic impacts of biological invasions on native assemblages (Gaertner et al. 2009). Understanding invasion-related thresholds is further confounded by the time lags between invasion and impact (Sax and Gaines 2008). For example, the non-native quinine tree (*Cinchona pubescens*) now comprises 20% of the cover in shrub lands in the highlands of the Galapagos Islands. This invasion has resulted in a reduction in the abundance of most native plant species but, as yet, no local extinctions (Jäger and Kowarik 2010). Difficulty in projecting the future trajectory of the quinine invasion leaves managers uncertain as to whether the native and non-native species will continue to co-exist or whether the tree invasion will eventually result in extinctions. Species extinctions are clear examples of irreversible change but, in practice, changes in species abundances can also be irreversible if such changes entail additional effects on ecosystem composition or structure (Box 3.1). Such examples underscore the need to couple uncertainty with risk assessment when designing interventions to prevent threshold shifts (Chapter 18).

When the driver of change is external to the site, managers may be able to predict but unable to prevent the system crossing a threshold. For example, nitrogen deposition on nutrient-poor soils can have widespread effects on ecosystem composition and function but, despite known threshold dynamics, managers may be unable to reverse the driver (Bobbink et al. 2010). Consequently, these ecosystems are likely to become novel and management efforts may be better served by curbing the effects of threshold shifts rather than attempting to control the underlying driver (this approach is discussed further in Section 3.6.2 and in Chapter 18). In contrast, when a driver occurs at the site level, such as in the salinity example earlier, managers may aim to intervene in the hybrid system before the threshold is crossed.

Altered fire regimes are a common anthropogenic change that can result in both hybrid and novel systems. In Illinois barrens (a woodland-prairie ecotone), for example, fire suppression shifted plant community composition from prairie to woodland species (Anderson et al. 2000). Reintroduction of fire was associated with an increase in prairie species abundances; after long periods of fire-suppression however, the system passed a threshold such that fire reintroduction was not sufficient to restore historical community composition (Anderson et al. 2000). Although this threshold may still be reversible, the additional costs to restore the historical community are greater if fire reintroductions occur after the system has passed a definitively hybrid state. Sometimes altered fire regime can be the result of other thresholds being crossed; these are very difficult to reverse. For example, invasion of the introduced pasture grass *Andropogon gayanus* created vastly hotter

Box 3.1 Pragmatic management of remnants of the humid highlands of the inhabited islands of the Galapagos: Are thresholds useful and can we prevent shifts into novel states?

Much of the humid highlands and transition zones of the inhabited islands of Galapagos have been transformed to a novel state by land use and biological invasions (Watson et al. 2009). On the island of Santa Cruz, which was permanently settled in the 1920s, change has been rapid. Although the island has experienced longer-term disturbance from wild stock, the most significant anthropogenic change occurred when much of its native forests were cleared between 1960 and 1980. In the south-eastern part, however, which has less attractive land for agriculture (lower rainfall and more rocky), there remains a small patch of transitional zone forest which could be considered to be hybrid, that is, it still has its original composition and structure. It has a canopy of native trees and intact shrub layers, whose species are still actively recruiting. This forest has small patches of invasive elephant grass (*Pennisetum purpureum*), *Lantana camara* and low densities of Cuban cedar (*Cedrela odorata*) and guava (*Psidium guayaba*). The most insipid and widespread species is the ground cover, *Tradescantia fluminensis*, which has a cover of greater than 50%.

Are thresholds a good measure to guide intervention? Has this system shifted from hybrid to novel? The first barrier to answering these questions is that we have very little data on the historical state.

Moreover, Galapagos vegetation, especially in drier areas, is inherently variable: species abundances wax and wane with patterns of wet and dry years (Hamann 1975, 1985). Because the invasion process is just beginning it is highly dynamic; consequently, the eventual state of the system is difficult to envision. If we assume the historical state to be something structurally similar to its current state but with all native plant diversity, chemical control could reduce most invasive species to low densities and return the system to its historical state. However, *Tradescantia fluminensis* invasion has probably crossed a threshold of impact because it competes with and prevents the recruitment of native herbaceous species (M. Gardener, Charles Darwin Foundation, Galapagos, personal observation). This threshold has never been quantified in the Galapagos but has been quantified in New Zealand (Standish et al. 2001). In hybrid ecosystems thresholds are still reversible. However, the disturbance created by trying to remove this species with chemical methods could be highly perverse: it may damage the biodiversity and ecosystem process and could potentially facilitate further invasion by other species. In short, although this system is relatively pristine and can be maintained in its current state, it is novel and not hybrid because it cannot go back to its historical state.

---

fires than normal, killing native savanna species in northern Australia and resulting in a near monoculture (Rossiter et al. 2003).

Human activity at the site level, such as land conversion and subsequent abandonment, can rapidly push ecosystems past thresholds. It is easier to identify a threshold once it has been passed and, while the specific barriers to recovery can be hard to identify, the fact that a system has crossed some sort of threshold may be obvious. For example, if overgrazed vegetation does not recover after the removal of livestock, managers can probably assume that a threshold is preventing its recovery (Westoby et al. 1989). In this case, the more pertinent question is whether or not the threshold effects are *reversible*. This question is central to restora-

tion ecology. Within that context, descriptions of ecological filters (Hobbs and Norton 2004; Funk et al. 2008) and state-and-transition models (Wilkinson et al. 2005; Rumpff et al. 2011) are common frameworks to test and characterize the presence of thresholds at a site. Site-level experimental tests and adaptive management are tools to identify specific barriers to ecosystem recovery and to decide if management interventions can reverse their effects (Chapter 18). Two additional considerations are important here. First, it is possible that a system is so altered that the totality of thresholds acting at the site cannot be easily identified. For these highly degraded landscapes, bet-hedging management that employs an array of approaches may be the most effective strategy. Second, the social and economic costs as well as the ecological consequences of intervention can determine whether in practice the threshold is actually reversible, or whether the ecosystem is or should be managed as a novel ecosystem (further described in Section 3.6.1 and Chapter 18).

Ecological and social barriers can interact in a variety of ways. First, social and ecological drivers can combine to create novel ecosystems. Earlier we described suppressed fire frequency as a site-level driver that could be reversed to prevent a hybrid ecosystem from becoming novel. Fire suppression, however, is often due to social pressure to prevent fires near human habitation; housing growth around natural areas may in practice turn ecologically reversible factors into irreversible drivers of ecosystem change (Radeloff et al. 2010). Importantly, as described in Chapter 5, ecosystems are composed of individuals that move independently of one another and in response to their environment. Humans are no exception to this individualistic concept; rather, people and societies are capable of adapting to and valuing aspects of ecosystem change. The value humans derive from some aspects of novel ecosystems may form a social threshold that cannot, and possibly should not, be reversed (Marris 2011).

The notion of an irreversible threshold is therefore a multidimensional one that includes ecological and social components. In reality, it may often be theoretically possible to reverse many thresholds but not practical due to knowledge, social or resource constraints. At times, new methods or approaches may shift perceptions of whether the same threshold is reversible or not. These themes are expanded on in Section 3.6.2 and in Chapter 18. Once an irreversible threshold has been identified, however, managers know they have a novel ecosystem and are faced with decisions of how to manage it. If there is no going back, what is next? We suggest that a consideration of function in novel ecosystems can aid in setting goals related to both biodiversity conservation and ecosystem services.

3.3 FUNCTION AS A MANAGEMENT GOAL

An underlying assumption in much of environmental management is that maintaining or restoring a historical species assemblage is the best approach to achieve a suite of other common goals from biodiversity conservation to ecosystem service provisioning. As anthropogenic change increases, this assumption becomes less defensible (Hobbs et al. 2011). As described earlier, escalating change may push an increasing number of ecosystems past thresholds such that they cannot be returned to their historical states. Rather than abandoning restoration or persisting with futile efforts, these systems may require a shift in evaluation metrics. Are there specific conservation goals or ecosystem services that can be provided through further management?

The idea of 'function' in ecology is used in three main ways that are relevant to the management of novel ecosystems (Jax 2005). First, function can refer to interactions between species or between a species and its environment. To understand function at the species level we ask, how does a species affect its environment, and how is it affected by its environment (Naeem 2002)? Second, function can refer to the collective effect of a complex set of interactions on the processes that sustain the functioning of the whole ecosystem. This meaning of ecosystem function is broad in scope and so prompts a different set of questions, for example, what do individual species or groups of species contribute to particular ecosystem functions? How do individual functions sum to affect the functioning of the whole ecosystem (Grime 1998)? Third, when ecosystem functions are considered in relation to human well-being, they become ecosystem services. Ecosystem services can take a variety of forms from regulatory (e.g. climate regulation and pollination) and supporting (e.g. nutrient cycling) services to provisioning (e.g. food and water) and cultural services (Millennium Ecosystem Assessment 2005).

An explicit focus on function provides metrics to set management goals in novel ecosystems. Novel ecosystems by definition differ from historical ecosystems in their biotic and/or abiotic characteristics, but functional similarities between past and present species can potentially mitigate the effects these changes have on ecosystem functions (Benayas et al. 2009). Environmental filtering can cause trait compositions to converge even while species compositions diverge (Fukami et al. 2005), and consequently novel ecosystems with altered biotic composition may still function like their historical ecosystems.

Figure 3.3 illustrates the space of possible relationships between historical, hybrid and novel ecosystems and their functional similarity to the historical system. For both historical systems and functionally similar hybrid systems, interventions may be a low priority

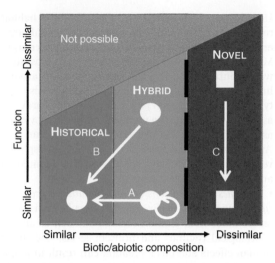

Figure 3.3 A state-space of functional similarity to the historical ecosystem in relation to abiotic and biotic novelty. Depending on management goals, functional similarity in this context may refer to habitat provision, ecosystem service provision or diversity maintenance. Compositionally similar but functionally dissimilar ecosystems are unlikely to occur and are labeled 'not possible'. Circles represent ecosystems that have not crossed a threshold into a novel state; squares represent ecosystems that have. Three pathways are considered in order of management preference: (A) when possible, functionally similar historical or hybrid ecosystems should be managed to prevent threshold shifts into a novel state; (B) functionally dissimilar hybrid ecosystems should be prioritized for restoration to their historical state; and (C) novel ecosystems should be managed to maintain or restore functional similarity to the historical state.

with one exception: managers may aim to understand possible threshold points in the system and at times intervene to prevent an irreversible threshold from being crossed (Fig. 3.3, pathway A). For novel ecosystems with high functional similarity to historical systems, intervention may similarly be a lower priority. Alternatively, for both hybrid and novel ecosystems in which key functions are lost, interventions to restore those functions may be a priority. For hybrid systems this can be achieved by returning the ecosystem to its historical state (Fig. 3.3, pathway B). For novel ecosys-

tems, however, interventions to restore functions may be more successful if they are not restricted to promoting the historical species pool (Fig. 3.3, pathway C).

3.4 SCALES OF ECOSYSTEM FUNCTIONING

Conservation of single species, as in the case of rare and endangered species, requires consideration of the functional traits and relationships of the species of concern. Managers should aim for a novel ecosystem to provide the focal species with functionally similar habitat to the historical ecosystem. As in the case of the Rodrigues fody, these relationships are often capitalized on for bird conservation. In the British Isles, for example, native blue tits began to feed upon non-native gall wasps hosted by the non-native Turkey oak after a decline in native oak and caterpillar populations (Stone et al. 2008; Hobbs et al. 2009). Similarly, removal of non-native pines outside of Perth, Australia was recently slowed when it was recognized that pine seeds had become an important food source of the endangered Carnaby's black cockatoo (Valentine and Stock 2008). Other examples of novel ecosystems providing habitat for bird species (Gleditsch and Carlo 2011) led one editorialist to answer the question 'do native birds care whether their berries are native or exotic?' with a simple 'no' (Davis 2011). While birds are a well-studied taxa for these questions, there is evidence the same principles hold true more broadly from the seed-disperser requirements of trees (Dungan et al. 2002) to the habitat requirements of beetles (Pawson et al. 2008).

To achieve other goals, a focus on ecosystem-level functions and services may become important. Increasingly, managers look to enhance ecosystem services as well as maintain key species. For some regulatory and supporting services, the origin of the species may matter very little; rather, they may depend on whether specific functional groups are present (Mascaro et al. 2012). Although an extreme example of ecosystem change, mine-site reclamation illustrates this well. Vegetation removal prior to mining results in a need to quickly re-stabilize soils once mining is complete. As a consequence, managers often plant fast-growing or deep-rooted plants regardless of their origin to ensure that this supporting service is restored (Richardson et al. 2010). At the landscape level, the well-reputed Working for Water program in South Africa partners with local communities to remove invasive species based not on the origin of species, but rather on how species affect ecosystem function. In working landscapes in the United States, range managers employ a range-assessment protocol that evaluates soil stability, hydrology and biotic integrity without considering species identity (Pyke et al. 2002). Finally, cultural ecosystem services are not necessarily tied to historical species assemblage. This is particularly evident in urban landscapes, in which novel ecosystems can reflect people's preference for nature that may not include historical ecosystems (Chapter 38).

3.5 PUTTING IT TOGETHER: MULTIPLE FUNCTIONS AND FUNCTIONAL TRADE-OFFS

Although specific functions may be similar between novel and historical ecosystems, no two species are exactly alike or functionally redundant. As a consequence, it is unlikely that a novel ecosystem will be similar to its historical analog for *all* functions. Rather, goals based on function require managers to choose specific functional relationships, conservation priorities and ecosystem services to focus on. Literature is rapidly growing on when and where ecosystem services can be bundled versus when there are trade-offs between services, but our understanding is still limited as to how services interact (Nelson et al. 2009; Lavorel et al. 2011). Similarly, there is often uncertainty about whether specific functions in a novel ecosystem will be retained over time. Ecosystem functions can shift with environmental variability, and there is growing evidence that multiple species are needed to maintain the same ecosystem function over time (Isbell et al. 2011). As a consequence, in many cases managers will be willing to consider ecosystem processes in tandem but not in place of biodiversity or native species diversity (Thompson and Starzomski 2007; Duffy 2009). Threshold effects may limit the feasibility of restoring native diversity to novel ecosystems but, to the extent that restoring native species is possible, synergies and trade-offs may similarly exist between this goal and goals based on species conservation and ecosystem services.

These decisions are context-dependent, but the following examples illustrate the range of trade-offs and choices faced by managers.
• *Management for species conservation may not translate to ecosystem conservation.* Kirtland's warbler (*Setophaga*

kirtlandii) is a rare bird with very specific habitat requirements, spending its winters in the Bahamas and summers in jack pine (*Pinus banksiana*) barrens in Michigan. Historically the jack pine barrens ecosystem was maintained by fire, but fire suppression around human population centers has reduced this ecosystem type. Strategic logging in jack pine plantations provides a habitat analog for the warbler and has been viewed as a good choice for conservation (Houseman and Anderson 2002). However, logging fails to replicate vegetation diversity and stand structure (Spaulding and Rothstein 2009).
• *Synergies and trade-offs between cultural services, ecosystem function and biodiversity.* Indirect human effects on wetlands near human habitation, such as altered hydrological regimes and nutrient levels, and direct human effects such as recreation can result in major compositional and structural change. In urban wetlands in New Jersey, Ehrenfeld (2004) found that increased human use of wetlands resulted in both emerging ecosystem functions as well as trade-offs among ecological and social functions. For example, increased potential to store flood waters was associated with an increase in plant diversity but a decrease in the presence of vertebrates. Low water levels were associated with more vertebrates but also with increased disturbance from humans, such as trash dumping. Different again, areas with a lot of human recreational use also had low plant diversity (Ehrenfeld 2004).

These examples highlight that understanding the synergies and trade-offs in achieving different conservation goals and ecosystem services is a key aspect of novel ecosystem management. At times this may entail tough value judgments to set realistic goals for ecosystem management. In other cases, such as the example of urban wetlands, a landscape perspective may be required to achieve multiple goals across several sites. Often, complementarity between species that provide different ecosystem functions may allow managers to augment management interventions to achieve a core goal while also achieving additional functions and services. For example, to guide restoration decisions aimed at controlling post-fire invasion of cheatgrass (*Bromus tectorum*), Wainger et al. (2010) developed an optimization model that focused on several ecosystem services (antelope hunting, property protection from fire, sage-grouse habitat and forage production) and incorporated cost-effectiveness ratios of restoration options. They found that if managers selected sites to optimize multiple services and utilized treatments with the greatest cost-effectiveness ratios (often the lowest intensity treatment), service benefits would increase three-fold. In this example and in general, practical and cost considerations shape ecosystem management. These constraints are expanded upon in the following sections.

3.6 FROM GOALS TO IMPLEMENTATION: PRACTICAL CONSIDERATIONS AND NOVEL METHODS

3.6.1 Practical considerations

Considering the functional relationships of species and ecosystem services helps provide goals for interventions in novel ecosystems. Practical constraints may however have an effect on when it is appropriate to intervene in novel ecosystems.

First, consideration should be given to temporal factors that may affect how long interventions are likely to persist in the system. In ecosystems characterized by frequent or intense disturbances, the effects of management may be superseded by subsequent disturbance. In tropical forests, for example, secondary succession following anthropogenic disturbance and natural disturbances such as hurricanes can routinely result in the assembly of new species combinations (Chazdon 2003). As a consequence, attempts to manage for specific species assemblages in these ecosystems are often futile. Over longer timescales, management with disturbance may also become unnecessary due to the self-organizing capacity of ecosystems. For example, Thompson et al. (2007) suggest that hurricane disturbance in the Luquillo Mountains of Puerto Rico serve as a check on species unable to tolerate infrequent but intense disturbance. In other situations, however, reducing the frequency of a disturbance may be the chief intervention necessary to achieve management goals. Increased fire frequency in the Amazon, for example, can result in large shifts in forest structure. Policies to control fire disturbance are therefore an important management response to prevent large shifts in ecosystem function (Nepstad et al. 2001).

Second, spatial factors will matter for the scale of management action required and its likelihood of success. For example, the location of a site in relation to source populations can largely influence the long-term success of interventions. Working along the Sacramento River, Holl and Crone (2004) found that plant restoration was more successful near remnant forests. Limited evidence exists on how local and landscape factors dictate restoration and management success, but the need to consider these factors is of growing research and management interest (Brudvig 2011). Further, while novel species assemblages can provide site-level conservation benefits and ecosystem services, they may also have wider landscape-level effects. Novel ecosystems near sites with a more exclusive management focus on historical assemblages, such as national parks or 'museum-style' conservation sites, might undermine those management efforts. Having local management projects out of sync with other patches at the regional or landscape scale may lead to the failure of restoration projects at one or both sites; at the very least, it may alter important local-regional relationships that structure local diversity (Ricklefs 1987; Starzomski et al. 2008). Thus, the net value of novel ecosystems may matter in relation to landscape factors.

Third, intervention in novel ecosystems should be based on an understanding of costs as well as benefits. Miller and Hobbs (2007) emphasize that in ecosystem management costs generally do not scale linearly with benefits. For some management objectives, initial conservation goals may be achieved with a minimal initial cost while achieving subsequent improvements becomes disproportionately costly. For example, when assessing habitat restoration options for grassland birds in suburban Chicago, Snyder et al. (2007) found that large areas distant from urban development could be restored relatively inexpensively whereas the costs of both restoration and land acquisition increased greatly for land parcels nearer to urban areas. On the other hand, large initial costs may be necessary to achieve initial benefits. For example, native plant restoration efforts in the Gulf Islands of British Columbia, Canada, were stymied by herbivores including introduced fallow deer (*Dama dama*) and Canada geese (*Branta canadensis*). These herbivores have negative effects on local vegetation because they facilitate competitively superior non-native grasses (Best and Arcese 2008) by increasing nutrient loading in conjunction with non-native species propagule supply and herbivory (Best 2008; Gonzales and Arcese 2008). Consequently, the Parks Canada Agency has found it necessary to invest additional funds to fence an entire island within the park (Eagle Island) to achieve their goal of restoring native plants.

Further, and especially in relation to novel ecosystems, it is important to consider the inadvertent impacts of management interventions to functions already maintained within an ecosystem. Herbicide application, for example, can deleteriously affect desired species populations as well as curb populations of non-desired species. Removal of non-native species may create 'weed-shaped holes' (Buckley et al. 2007) that without additional management expense will simply be refilled, sometimes with invasive species that can cause even greater changes to the desired species composition (Jäger et al. 2009). Lastly, when non-native species naturalize in an ecosystem they can form facilitative relationships with existing species and provide valued (though not necessarily historical) ecosystem functions. In Puerto Rico, for example, the invasive tree *Leucaena leucocephala* serves as a nurse plant for native species and also reduces risks of fire damage by decreasing fuel loads (Santiago-Garcia et al. 2008). Decisions on whether to remove a species should therefore include consideration of the costs required to replace the ecosystem functions it provides.

While many of these constraints would suggest higher tolerance of novel assemblages and non-native species, the greater uncertainty associated with novel assemblages also suggests that precaution is an important element of novel ecosystem management. Many invasive species go through low-abundance lapse phases before becoming highly abundant. In Germany, for example, 51% of the 184 woody weed species had a lapse phase longer than 200 years before they became invasive (Kowarik 1995). In southeastern USA, Kudzu (*Puerarua montana*) was planted widely in the early 1900s because it reduced soil erosion in drought years. Since that time it has become highly invasive, covering over 3 million hectares in the eastern USA (Forseth and Innis 2004). Humans have long managed ecosystems for specific functions and the approach we advocate here is no different, but the history of Kudzu in the USA emphasizes the need to temper a focus on ecosystem function with caution for an uncertain future.

3.6.2 Novel methods

In the framework we have presented, a novel ecosystem paradigm shifts management concerns from the specific goal of maintaining historical ecosystems toward an admittedly more qualitative consideration of how the ecosystem functions to provide species habitat and ecosystem services. As a consequence, many management decisions focus on *what* to value and *when* to intervene in these ecosystems. Valuing some novel species interactions, however, also introduces new approaches for *how* to manage ecosystems (Seastedt et al. 2008). Traditional management responses to species invasions facilitated by novel drivers often include removing or reversing the driver and targeting the invasive species to restore native assemblages. These approaches implicitly assume that successional trajectories can be predicted and that removing drivers of change is sufficient to reverse change (Hobbs and Norton 1996). A growing body of research on priority effects (Belyea and Lancaster 1999; Starzomski et al. 2008) and alternative stable states (Froend et al. 1997; Suding et al. 2004) suggests that these assumptions are not always true (Firn et al. 2010). Further, in a world of global climate change and shifting nutrient cycles, many drivers of ecosystem change are playing out at a scale beyond the control of site managers. To achieve many conservation goals, new approaches become necessary.

These new approaches tend to take two forms. First, new species interactions can mitigate the effects of novel drivers, even if those drivers cannot be reversed. Grazing as a management tool is often used in this context, from mimicking the effects of now-suppressed fire regimes (Seastedt et al. 2008) to mitigating the effects of shifting nutrient cycles (Weiss 1999; Box 3.2) and even curbing the dominance of species whose initial invasion was facilitated by grazing (Firn et al. 2010).

Second, when a system has been degraded through multiple pathways, it can be difficult to isolate and account for all possible thresholds that pose barriers to restoration. Uncertainty about the identity of all thresholds need not be an excuse for inaction, however. For plant restoration, Seastedt et al. (2008) suggest seeding species with a wider range of functional traits and environmental tolerances than present in the historical community. In areas where seed source is not limited, this provides a cost-effective way to increase the likelihood that some desired species establish at the site. For example, in an attempt to restore a mined gravel pit that was once tallgrass prairie, Cherwin et al. (2009) seeded grasses with moisture requirements spanning a wide (500 mm) rainfall gradient. This resulted in a mixed-grass community that, while different from the historical community, was able to persist and provided many of the same ecosystem functions (Seastedt et al. 2008; Cherwin et al. 2009).

Box 3.2 A novel management technique to restore species diversity in California serpentine grasslands

Covering only 1% of California's landscapes, serpentine grasslands contain 10% of California's endemic plant species (Safford et al. 2005). These endemics include *Lasthenia californica*, which in the spring lives up to its common name of goldfields, and *Plantago erecta*, which serves as the key host plant for the well-studied highly endangered Bay Area checkerspot butterfly. In short, California serpentine grasslands are prime for management focused on native species diversity. Historically, little intervention was necessary to achieve this goal. The high level of endemic serpentine plant species is due their adaptation to harsh low-nutrient soil conditions that characterize serpentine soils. These same soil characteristics have also historically restricted the establishment of non-native Mediterranean grasses that have successfully invaded most other California grasslands. More recently, however, nitrogen deposition from automobiles has provided a release which allows non-native annual grasses to invade serpentine systems (Weiss 1999). When these tall thatch-forming annual grasses become dominant in serpentine systems, they dramatically reduce many native species abundances (Weiss 1999). Removing nitrogen from the system or slowing rates of deposition are outside of the control of site managers. Cattle grazing however serves to remove much of the biomass of non-native species from the serpentine, and cattle trampling reduces the accumulation of non-native thatch that suppresses native species (Weiss 1999). Consequently, cattle grazing constitutes a novel management technique to mitigate the effects of a novel and, from a species conservation perspective, deleterious driver of change.

3.7 WHEN DOES A NOVEL STATE BECOME THE REFERENCE?

Throughout this chapter we have retained the assumption that the historical ecosystem is an appropriate ref-

erence (if not always for species composition, then for key species interactions and ecosystem function). It is worth tempering this assumption with a consideration of ecological history. First, ecosystems are constantly in flux and the time point which we designate 'historical' will be different relative to past time points. Natural climate change coupled with other environmental changes and contingencies have generated no-analog species combinations throughout ecological history (Jackson et al. 2009). Second, a growing body of literature indicates that human societies have long altered ecosystems to suit their needs (Mann 2005; Gammage 2011). This literature both challenges notions of what is natural and suggests that long-past human actions may leave legacies that continue to shape ecosystems. Consequently, a consideration of history provides justification for modern societies to accept some ecosystem change and intervene in ways that promote biodiversity and ecosystem services (Jackson and Hobbs 2009). As species assemblages shift and adapt to anthropogenic change, there may be instances in which novel ecosystems are preferable to any historical ecosystem. This is either because they provide functions that would be lost in the attempt at traditional restoration, or because emergent assemblages are better able to respond to ongoing environmental change. This is the rationale behind efforts to increase connectivity for species to adapt to climate change, and also behind more extreme management suggestions such as assisted colonization (Loss et al. 2011). While these management questions are largely outside of the scope of this chapter, the reasons they are considered highlight the importance of assessing what species and functions an ecosystem currently maintains before attempting to alter it.

3.8 CONCLUDING NOTES

We live in a world shaped by widespread and escalating human activity. As long as anthropogenic change exists, other species will continue to respond to it. This basic premise of the novel ecosystem framework is the foundation for both realistic and optimistic conservation actions. Attempts to restore historical assemblages that do not consider the costs, long-term probabilities of success and ecological consequences of these actions are likely to have unexpected and often unwanted results. Trying to understand when to value emerging species assemblages and interactions, in contrast, provides new opportunities for biodiversity conservation. Acknowledging the dynamism inherent in ecosystems should underpin research and theory as we move toward a robust framework for managing ecosystems in a world of rapid change.

ACKNOWLEDGEMENTS

We thank Keith Bowers, Jack Ewel, Tom Jones, Karen Keenleyside and Tim Seastedt for discussions that developed ideas in this chapter and Steve Jackson, Ariel Lugo and Cara Nelson for helpful comments on earlier drafts.

REFERENCES

Anderson, R.C., Schwegman, J.E. and Anderson, M.R. (2000) Micro-scale restoration: A 25-year history of a southern Illinois barrens. *Restoration Ecology*, **8**, 296–306.

Belyea, L.R. and Lancaster, J. (1999) Assembly rules within a contingent ecology. *Oikos*, **86**, 402–416.

Benayas, J.M.R., Newton, A.C., Diaz, A. and Bullock, J.M. (2009) Enhancement of biodiversity and ecosystem services by ecological restoration: a meta-analysis. *Science* (Washington), **325**, 1121–1124.

Best, R.J. (2008) Exotic grasses and feces deposition by an exotic herbivore combine to reduce the relative abundance of native forbs. *Oecologia*, **158**, 319–327.

Best, R.J. and Arcese, P. (2008) Exotic herbivores directly facilitate the exotic grasses they graze: mechanisms for an unexpected positive feedback between invaders. *Oecologia*, **159**, 139–150.

Bestelmeyer, B.T. (2006) Threshold concepts and their use in rangeland management and restoration: The good, the bad, and the insidious. *Restoration Ecology*, **14**, 325–329.

Bestelmeyer, B.T., Goolsby, D.P. and Archer, S.R. (2011) Spatial perspectives in state–and–transition models: a missing link to land management? *Journal of Applied Ecology*, **48**, 746–757.

Bobbink, R., Hicks, K., Galloway, J., Spranger, T., Alkemade, R., Ashmore, M., Bustamante, M., Cinderby, S., Davidson, E., Dentener, F., Emmett, B., Erisman, J.W., Fenn, M., Gilliam, F., Nordin, A., Pardo, L. and De Vries, W. (2010) Global assessment of nitrogen deposition effects on terrestrial plant diversity: a synthesis. *Ecological Applications*, **20**, 30–59.

Brudvig, L.A. (2011) The restoration of biodiversity: where has research been and where does it need to go? *American Journal of Botany*, **98**, 549–558.

Buckley, Y.M., Bolker, B.M. and Rees, M. (2007) Disturbance, invasion and re-invasion: managing the weed-shaped hole in disturbed ecosystems. *Ecology Letters*, **10**, 809–817.

Chazdon, R.L. (2003) Tropical forest recovery: legacies of human impact and natural disturbances. *Perspectives in Plant Ecology Evolution and Systematics*, **6**, 51–71.

Cherwin, K.L., Seastedt, T.R. and Suding, K.N. (2009) Effects of nutrient manipulations and grass removal on cover, species composition, and invasibility of a novel grassland in Colorado. *Restoration Ecology*, **17**, 818–826.

Cramer, V.A. and Hobbs, R.J. (2002) Ecological consequences of altered hydrological regimes in fragmented ecosystems in southern Australia: Impacts and possible management responses. *Austral Ecology*, **27**, 546–564.

Davis, M. (2011) Do native birds care whether their berries are native or exotic? No. *Bioscience*, **61**, 501–502.

Duffy, J.E. (2009) Why biodiversity is important to the functioning of real-world ecosystems. *Frontiers in Ecology and the Environment*, **7**, 437–444.

Dungan, R.J., O'Cain, M.J., Lopez, M.L. and Norton, D.A. (2002) Contribution by possums to seed rain and subsequent seed germination in successional vegetation, Canterbury, New Zealand. *New Zealand Journal of Ecology*, **26**, 121–127.

Ehrenfeld, J.G. (2004) The expression of multiple functions in urban forested wetlands. *Wetlands*, **24**, 719–733.

Firn, J., House, A.P.N. and Buckley, Y.M. (2010) Alternative states models provide an effective framework for invasive species control and restoration of native communities. *Journal of Applied Ecology*, **47**, 96–105.

Forseth, I.N. and Innis, A.F. (2004) Kudzu (Pueraria montana): History, physiology, and ecology combine to make a major ecosystem threat. *Critical Reviews in Plant Sciences*, **23**, 401–413.

Fox, D. (2003) Using exotics as temporary habitat: an accidental experiment on Rodrigues Island. *Conservation*, **4**, 32–37.

Froend, R.H., Halse, S.A. and Storey, A.W. (1997) Planning for the recovery of Lake Toolibin, Western Australia. *Wetlands Ecology and Management*, **5**, 73–85.

Fukami, T., Bezemer, T.M., Mortimer, S.R. and Van Der Putten, W.H. (2005) Species divergence and trait convergence in experimental plant community assembly. *Ecology Letters*, **8**, 1283–1290.

Funk, J.L., Cleland, E.E., Suding, K.N. and Zavaleta, E.S. (2008) Restoration through reassembly: plant traits and invasion resistance. *Trends in Ecology & Evolution*, **23**, 695–703.

Gaertner, M., Den Breeyen, A., Hui, C. and Richardson, D.M. (2009) Impacts of alien plant invasions on species richness in Mediterranean-type ecosystems: a meta-analysis. *Progress in Physical Geography*, **33**, 319–338.

Gammage, B. (2011) *The Biggest Estate on Earth: How Aborigines Made Australia*. Allen & Unwin, Crows Nest, NSW.

Gleditsch, J.M. and Carlo, T.A. (2011) Fruit quantity of invasive shrubs predicts the abundance of common native avian frugivores in central Pennsylvania. *Diversity and Distributions*, **17**, 244–253.

Gonzales, E.K. and Arcese, P. (2008) Herbivory more limiting than competition on early and established native plants in an invaded meadow. *Ecology*, **89**, 3282–3289.

Grime, J.P. (1998) Benefits of plant diversity to ecosystems: immediate, filter and founder effects. *Journal of Ecology*, **86**, 902–910.

Hamann, O. (1975) Vegetational changes in the Galapagos Islands during the period 1966–1973. *Biological Conservation*, **7**, 37–59.

Hamann, O. (1985) The El Niño influence on the Galápagos vegetation, in *El Niño in the Galápagos Islands: The 1982–1983 Event* (eds G. Robinson and E. Del Pino), Charles Darwin Foundation, Quito, Ecuador, 299–330.

Hastings, A. and Wysham, D.B. (2010) Regime shifts in ecological systems can occur with no warning. *Ecology Letters*, **13**, 464–472.

Hobbs, R.J. and Norton, D.A. (1996) Towards a conceptual framework for restoration ecology. *Restoration Ecology*, **4**, 93–110.

Hobbs, R.J. and Norton, D.A. (2004) Ecological filters, thresholds and gradients in resistance to ecosystem reassembly, in *Assembly Rules and Restoration Ecology: Bridging the Gap Between Theory and Practice* (eds V.M. Temperton, R.J. Hobbs, T. Nuttle and S. Halle), Island Press, Washington, DC, 72–95.

Hobbs, R.J., Arico, S., Aronson, J., Baron, J.S., Bridgewater, P., Cramer, V.A., Epstein, P.R., Ewel, J.J., Klink, C.A., Lugo, A.E., Norton, D., Ojima, D., Richardson, D.M., Sanderson, E.W., Valladares, F., Vila, M., Zamora, R. and Zobel, M. (2006) Novel ecosystems: theoretical and management aspects of the new ecological world order. *Global Ecology and Biogeography*, **15**, 1–7.

Hobbs, R.J., Higgs, E. and Harris, J.A. (2009) Novel ecosystems: implications for conservation and restoration. *Trends in Ecology & Evolution*, **24**, 599–605.

Hobbs, R.J., Hallett, L.M., Ehrlich, P.R. and Mooney, H.A. (2011) Intervention ecology: Applying ecological science in the twenty-first century. *Bioscience*, **61**, 442–450.

Holl, K.D. and Crone, E.E. (2004) Applicability of landscape and island biogeography theory to restoration of riparian understorey plants. *Journal of Applied Ecology*, **41**, 922–933.

Houseman, G.R. and Anderson, R.C. (2002) Effects of jack pine plantation management on barrens flora and potential Kirtland's warbler nest habitat. *Restoration Ecology*, **10**, 27–36.

Impey, A.J., Cote, I.M. and Jones, C.G. (2002) Population recovery of the threatened endemic Rodrigues fody (Foudia flavicans) (Aves, Ploceidae) following reforestation. *Biological Conservation*, **107**, 299–305.

Isbell, F., Calcagno, V., Hector, A., Connolly, J., Harpole, W.S., Reich, P.B., Scherer–Lorenzen, M., Schmid, B., Tilman, D., Van Ruijven, J., Weigelt, A., Wilsey, B.J., Zavaleta, E.S. and Loreau, M. (2011) High plant diversity is needed to maintain ecosystem services. *Nature*, **477**, 199–202.

Jäger, H. and Kowarik, I. (2010) Resilience of native plant community following manual control of invasive Cinchona pubescens in Galapagos. *Restoration Ecology*, **18**, 103–112.

Jäger, H., Kowarik, I. and Tye, A. (2009) Destruction without extinction: long-term impacts of an invasive tree species on Galápagos highland vegetation. *Journal of Ecology*, **97**, 1252–1263.

Jackson, S.T. and Hobbs, R.J. (2009) Ecological restoration in the light of ecological history. *Science*, **325**, 567–569.

Jackson, S.T., Betancourt, J.L., Booth, R.K. and Gray, S.T. (2009) Ecology and the ratchet of events: Climate variability, niche dimensions, and species distributions. *Proceedings of the National Academy of Sciences of the United States of America*, **106**, 19685–19692.

Jax, K. (2005) Function and 'functioning' in ecology: what does it mean? *Oikos*, **111**, 641–648.

Kowarik, I. (1995) Time lags in biological invasions with regard to the success and failure of alien species, in *Plant Invasions: General Aspects and Special Problems* (P. Pysek, K. Prach, M. Rejmanke and M. Wade, eds), SPB Academic Publishing, Amsterdam, pp. 15–38.

Lavorel, S., Grigulis, K., Lamarque, P., Colace, M.–P., Garden, D., Girel, J., Pellet, G. and Douzet, R. (2011) Using plant functional traits to understand the landscape distribution of multiple ecosystem services. *Journal of Ecology*, **99**, 135–147.

Loss, S.R., Terwilliger, L.A. and Peterson, A.C. (2011) Assisted colonization: Integrating conservation strategies in the face of climate change. *Biological Conservation*, **144**, 92–100.

Mann, C.C. (2005) *1491: New Revelations of the Americas before Columbus*. Alfred A. Knopf, New York.

Marris, E. (2011) *Rambunctious Garden: Saving Nature in a Post-wild World*. Bloomsbury, New York, NY.

Mascaro, J., Hughes, R.F. and Schnitzer, S.A. (2012) Novel forests maintain ecosystem processes after the decline of native tree species. *Ecological Monographs*, **82**, 221–238.

Millennium Ecosystem Assessment. (2005) *Ecosystems and Human Well-being*. Synthesis. Island Press, Washington, DC.

Miller, J.R. and Hobbs, R.J. (2007) Habitat restoration: Do we know what we're doing? *Restoration Ecology*, **15**, 382–390.

Naeem, S. (2002) Ecosystem consequences of biodiversity loss: The evolution of a paradigm. *Ecology*, **83**, 1537–1552.

Nelson, E., Mendoza, G., Regetz, J., Polasky, S., Tallis, H., Cameron, D.R., Chan, K.M.A., Daily, G.C., Goldstein, J., Kareiva, P.M., Lonsdorf, E., Naidoo, R., Ricketts, T.H. and Shaw, M.R. (2009) Modeling multiple ecosystem services, biodiversity conservation, commodity production, and trade-offs at landscape scales. *Frontiers in Ecology and the Environment*, **7**, 4–11.

Nepstad, D., Carvalho, G., Barros, A.C., Alencar, A., Capobianco, J.P., Bishop, J., Moutinho, P., Lefebvre, P., Silva, U.L. and Prins, E. (2001) Road paving, fire regime feedbacks, and the future of Amazon forests. *Forest Ecology and Management*, **154**, 395–407.

Pawson, S.M., Brockerhoff, E.G., Meenken, E.D. and Didham, R.K. (2008) Non–native plantation forests as alternative habitat for native forest beetles in a heavily modified landscape. *Biodiversity and Conservation*, **17**, 1127–1148.

Polasky, S., Carpenter, S.R., Folke, C. and Keeler, B. (2011) Decision–making under great uncertainty: environmental management in an era of global change. *Trends in Ecology & Evolution*, **26**, 398–404.

Pyke, D.A., Herrick, J.E., Shaver, P. and Pellant, M. (2002) Rangeland health attributes and indicators for qualitative assessment. *Journal of Range Management*, **55**, 584–597.

Radeloff, V.C., Stewart, S.I., Hawbaker, T.J., Gimmi, U., Pidgeon, A.M., Flather, C.H., Hammer, R.B. and Helmers, D.P. (2010) Housing growth in and near United States protected areas limits their conservation value. *Proceedings of the National Academy of Sciences of the United States of America*, **107**, 940–945.

Richardson, P.J., Lundholm, J.T. and Larson, D.W. (2010) Natural analogues of degraded ecosystems enhance conservation and reconstruction in extreme environments. *Ecological Applications*, **20**, 728–740.

Ricklefs, R.E. (1987) Community diversity: Relative roles of local and regional processes. *Science*, **235**, 167–171.

Rossiter, N.A., Setterfield, S.A., Douglas, M.M. and Hutley, L.B. (2003) Testing the grass–fire cycle: alien grass invasions in the tropical savannas of Northern Australia. *Diversity and Distributions*, **9**, 169–176.

Rumpff, L., Duncan, D.H., Vesk, P.A., Keith, D.A. and Wintle, B.A. (2011) State-and-transition modelling for Adaptive Management of native woodlands. *Biological Conservation*, **144**, 1224–1236.

Safford, H.D., Viers, J.H. and Harrison, S.P. (2005) Serpentine endemism in the California flora: a database of serpentine affinity. *Madroño*, **52**, 222–257.

Santiago-Garcia, R.J., Colon, S.M., Sollins, P. and Van Bloem, S.J. (2008) The role of nurse trees in mitigating fire effects on tropical dry forest restoration: A case study. *Ambio*, **37**, 604–608.

Sax, D.F. and Gaines, S.D. (2008) Species invasions and extinction: The future of native biodiversity on islands. *Proceedings of the National Academy of Sciences of the United States of America*, **105**, 11490–11497.

Seastedt, T.R., Hobbs, R.J. and Suding, K.N. (2008) Management of novel ecosystems: are novel approaches required? *Frontiers in Ecology and the Environment*, **6**, 547–553.

Snyder, S.A., Miller, J.R., Skibbe, A.M. and Haight, R.G. (2007) Habitat acquisition strategies for grassland birds in an urbanizing landscape. *Environmental Management*, **40**, 981–992.

Spaulding, S.E. and Rothstein, D.E. (2009) How well does Kirtland's warbler management emulate the effects of

natural disturbance on stand structure in Michigan jack pine forests? *Forest Ecology and Management*, **258**, 2609–2618.

Standish, R.J., Robertson, A.W. and Williams, P.A. (2001) The impact of an invasive weed Tradescantia fluminensis on native forest regeneration. *Journal of Applied Ecology*, **38**, 1253–1263.

Starzomski, B.M., Parker, R.L. and Srivastava, D.S. (2008) On the relationship between regional and local species richness: a test of saturation theory. *Ecology*, **89**, 1921–1930.

Stone, G.N., Van Der Ham, R.W.J.M. and Brewer, J.G. (2008) Fossil oak galls preserve ancient multitrophic interactions. *Proceedings of the Royal Society B: Biological Sciences*, **275**, 2213–2219.

Suding, K.N. and Hobbs, R.J. (2009) Threshold models in restoration and conservation: a developing framework. *Trends in Ecology & Evolution*, **24**, 271–279.

Suding, K.N., Gross, K.L. and Houseman, G.R. (2004) Alternative states and positive feedbacks in restoration ecology. *Trends in Ecology & Evolution*, **19**, 46–53.

Thompson, J., Lugo, A.E. and Thomlinson, J. (2007) Land use history, hurricane disturbance, and the fate of introduced species in a subtropical wet forest in Puerto Rico. *Plant Ecology*, **192**, 289–301.

Thompson, R. and Starzomski, B.M. (2007) What does biodiversity actually do? A review for managers and policy makers. *Biodiversity and Conservation*, **16**, 1359–1378.

Valentine, L.E. and Stock, W. (2008) Food resources of Carnaby's black cockatoo (*Calyptorhynchus latirostris*) in the Gnangara sustainability strategy study area. Report to Forest Products Commission, Perth, Australia.

Wainger, L.A., King, D.M., Mack, R.N., Price, E.W. and Maslin, T. (2010) Can the concept of ecosystem services be practically applied to improve natural resource management decisions? *Ecological Economics*, **69**, 978–987.

Wallace, K.J. (2003) Lake Toolibin: working together. *Pacific Conservation Biology*, **9**, 51–57.

Watson, J., Trueman, M., Tufet, M., Henderson, S. and Atkinson, R. (2009) Mapping terrestrial anthropogenic degradation on the inhabited islands of the Galápagos archipelago. *Oryx*, **44**, 79–82.

Weiss, S.B. (1999) Cars, cows, and checkerspot butterflies: Nitrogen deposition and management of nutrient-poor grasslands for a threatened species. *Conservation Biology*, **13**, 1476–1486.

Westoby, M., Walker, B. and Noymeir, I. (1989) Opportunistic management for rangelands not at equilibrium. *Journal of Range Management*, **42**, 266–274.

Wilkinson, S.R., Naeth, M.A. and Schmiegelow, F.K.A. (2005) Tropical forest restoration within Galapagos National Park: Application of a state-transition model. *Ecology and Society*, **10**, 28–43.

Chapter 4

ISLANDS: WHERE NOVELTY IS THE NORM

John J. Ewel[1], Joseph Mascaro[2], Christoph Kueffer[3], Ariel E. Lugo[4], Lori Lach[5] and Mark R. Gardener[6]

[1]Department of Biology, University of Florida, USA

[2]Department of Global Ecology, Carnegie Institution for Science, Stanford, California, USA

[3]Plant Ecology, Institute of Integrative Biology, Swiss Federal Institute of Technology (ETH), Zurich, Switzerland

[4]International Institute of Tropical Forestry, USDA Forest Service, Río Piedras, Puerto Rico

[5]Ecosystem Restoration and Intervention Ecology (ERIE) Research Group, School of Plant Biology and the Centre for Integrative Bee Research, University of Western Australia, Australia

[6]Charles Darwin Foundation, Galapagos Islands, Ecuador, and School of Plant Biology, University of Western Australia, Australia

4.1 INTRODUCTION

Islands have much to offer our understanding of novel ecosystems and many of the concepts presented in the previous chapter (Chapter 3), for it is there that novel assemblages of species have developed most rapidly and dramatically. Start with a depauperate biota, wait millennia and add people (with their unique capacity to break down biogeographic barriers), and novelty is the inevitable outcome. Islands lack the biological buffering capacity of continents, with their vast areas and rich biota. That is not to say that all island ecosystems are novel assemblages; many islands, including some that are densely populated, still harbor ecosystems composed almost exclusively of native species growing on land unchanged by human activity. But the vulnerability of island ecosystems to change has long been recognized, and it is a well-documented generality.

In Section 4.2 we briefly discuss the three main factors that influence island ecosystem novelty: physical geography, biogeography and human ecology. There is a massive literature on each of these themes, and we make no attempt to review it here. Instead, our objective is to provide some context for development of the main topic of interest in keeping with the principal subject matter of this book: when and how to intervene in novel ecosystems. Intervention is covered in Section 4.3, where it is discussed in relation to need, barriers and feasibility.

A chapter on islands could include all manner of bounded, isolated environments, from ponds to mountain peaks. We restrict our purview to non-continental land masses currently surrounded by ocean, whether that has been true throughout their geological history or not.

Where we make comparative statements our reference is always the continental land masses, but to be

concise we do not always state this explicitly. The geographic expertise of the authors is concentrated primarily in the tropics and subtropics, and this is reflected in most of the examples cited. Why include a separate chapter that deals specifically with islands? In our view, what we see on islands today may indicate what can be expected on continents tomorrow. In many ways, islands are the window to the future.

4.2   INSULAR TRAITS THAT FOSTER NOVELTY

A number of island traits influence ecosystem novelty. Some of those described here are unique to islands while others pertain equally well to continents; inevitably, there are many exceptions to statements cast here as generalizations. Nevertheless, most traits described fit most islands and, more importantly, their influence in determining when and how to intervene in novel ecosystems has broad applicability.

4.2.1   Physical geography

The relatively small area of islands facilitates change in the wake of outside forces. Small land areas mean short internal distances, which in turn mean rapid expansion of outside agents. For example, an agriculture-based human society quickly exploits suitable land (Rolett and Diamond 2004) or an introduced species is soon dispersed into all suitable habitats (Whittaker and Fernández-Palacios 2007).

Given similar latitudes and physiography, an island and continent would be expected to have similar numbers of bioclimatic zones and be subject to similar kinds and frequencies of disturbance. An important difference arises because continents typically have some habitats that cover huge areas, portions of which are likely to escape agents of change through vastness alone, whereas each habitat type on an island is relatively small and susceptible to change in its entirety. Consequently, population sizes and areas of occupancy of many island species, especially habitat specialists, are naturally small which makes them particularly vulnerable to extinction (Caujapé-Castells et al. 2010).

A second way that islands and continents frequently differ is in uniformity of parent material. On many islands the entire land mass derives from a single, relatively uniform substrate, whether from continental

fragments, uplift, volcanism or biotic activity (in the case of atolls). Continents, in comparison, typically contain a greater variety of parent materials, so a single climate regime is likely to cover soils that differ in their developmental history. This observation does not ignore the great environmental heterogeneity of some (particularly large) islands often caused by steep climate gradients. Even on islands possessing such heterogeneity however, the sameness of parent material may yet be observed. This can contribute to the feeling of sameness in ecological communities, even when moving across steep climatic gradients (e.g. Vitousek 2004).

4.2.2   Biogeography

Isolation by water confers islands with famously unique biogeographic attributes. Natural colonization rates decline with distance from donor continents and diminishing island size, such that the smallest, most remote islands tend to have fewest species. As a result, entire functional groups and life forms may be missing; large mammals, a life form limited by both oversea dispersal ability and area of habitat available on islands, are the most conspicuous absentees. Other functional groups become bottled as anachronistic 'Lazarus' taxa now extirpated from continents. Plants occasionally become relicts, thriving in the absence of competitors that long ago vanquished them on the continents. As these filters alternatively limit or enhance the relative abundances of various groups, they yield disharmonic flora and fauna, not reflecting the diversity at higher levels of taxonomic classification that characterize their donor continents (Carlquist 1966). A spider might raft to a remote island on an airborne strand of web, but a frog is unlikely to survive the swim; a coconut might tolerate the float, while an orange would not. Disharmony and this loss of functional groups decrease with proximity to continents, and these in turn confer islands with uniquely high rates of human-mediated species invasion and subsequent species replacement.

Landscape age, like island size and distance from donor continents, leads to biogeographic differences among islands (e.g. Parent et al. 2008). Time and genetic isolation in an environment without finely divided niche foster evolutionary radiation of endemics, a process that has been recognized since Bates and Darwin and is now well documented in many groups of plants and animals (Whittaker and Fernández-Palacios 2007). Evolution of defenses against enemies that would be present on larger continental landscapes is relaxed in endemics that evolve in enemy-free habitats. We therefore find the textbook examples: flightless birds, thornless raspberries and scent-free mints. When enemies do arrive, the defenseless endemics are faced with challenges for which they are ill-equipped, and extinction often follows.

Sometimes, however, evolving in isolation and in small habitats provides survival benefits. The small populations that typify island endemics have likely survived genetic bottlenecks, conferring them with an attribute of great value when outside changes reduce population sizes below those that might not be tolerated by continental species (Adsersen 1989). Furthermore, a high degree of plasticity in island biota may bode well for adaptation to climate change. One recent analysis of 150 datasets encompassing both continents and islands indicated that organisms that evolved with opportunity to expand into a broad range of habitats that were not previously saturated with habitat specialists may be pre-adapted to change (Laurence et al. 2011).

An impoverished, disharmonic biota is almost certain to assemble into a community having a less complex structure than one with a full complement of functional groups. In forests at latitudes where the biogeographer might expect to find trees of many height classes, a few species sometimes dominate structure; these are often shorter than typical of the same climatic zone on continental land masses. In tropical and subtropical latitudes, the shorter stature may be due to wind storm frequency or it may simply reflect the genetic potential for height growth of the species that colonized. Furthermore, with relatively few competing species of similar stature, selection pressure for increased height is likely to be low even if the genetic potential is present. This was the case in the Galapagos for example, where a *Miconia* shrubland was readily invaded by the *Cinchona* tree (Jäger et al. 2007; Fig. 4.1).

Broad realized niches are typical on remote islands, where relatively few species occupy many environments and perform many roles. Plant-animal interactions, for instance, are often highly generalized (Kaiser-Bunbury et al. 2010). Likewise, the few species present might fill many roles in stand development from primary succession colonist to undisturbed ecosystem dominant (Mueller-Dombois 2008). A prime

Figure 4.1 Trees in a formerly treeless ecosystem. *Cinchona pubescens* forest in what was previously *Miconia robinsoniana* shrubland, Santa Cruz, Galápagos. One outcome of this invasion is increased substrate for epiphytes. Photograph courtesy of Mandy Trueman.

example is *Metrosideros polymorpha*, a Hawaiian endemic that previously dominated nearly all forests in the archipelago, thriving under rainfall regimes from 200 to more than 11,000 mm per year, substrates 0–4.6 million years old and from sea level to the tree line (Vitousek 2004). Today, its niche space is collapsing as introduced species permeate its range. It is out-competed at the onset of primary succession by nitrogen-fixing trees (Hughes and Denslow 2005), after fires fueled by grasses (Hughes et al. 1991), under high light by gap-filling pioneer trees (Mascaro et al. 2012) and in dense shade by a clonal understory tree (Zimmerman et al. 2008).

Islands are also famous for adaptive radiations, which can yield high species diversity and rather small realized niches. In these cases, however, the resulting species diversity may have the same vulnerability (as a group) to introduced organisms as a single wide-ranging species. For example, the lobeliad radiation in Hawai'i resulted in spectacular endemic diversity, but the resulting species have similar reproductive and morphological traits. Most Hawaiian lobeliads depend on native honeycreepers (another famous radiation) for pollination (Givnish et al. 2009), and in this way the whole group became vulnerable to a single event: the colonization of Hawai'i by avian malaria, which decimated honeycreeper populations. Similarly, the spread of feral ungulates throughout Hawaiian forests threatens all lobeliads, whose stems are notoriously weak and easily crushed.

4.2.3 Human ecology

People are the agents of change that lead to novelty everywhere. Recognition of the unique aspects of insular human ecology is useful when attempting to understand the ubiquitous and rapid development of novel ecosystems on islands. The history of some islands has been one of human population turnover, sometimes motivated by resource scarcity, sometimes by cultural clashes and sometimes by shifts in global power. The resulting cultural changes have typically led to more marked change than on islands colonized and utilized by a single cultural group over a long period of time. Each time the culture changes, the new culture brings with it new species and new approaches to land use, both of which facilitate novelty.

Even where cultural continuity is sustained, the sea is the inescapable boundary, and limits to agricultural expansion – fertile soil and fresh water – are quickly reached. It becomes expensive to further intensify agriculture, requiring labor or fossil energy to provide the requisite water and nutrients. At this stage, some combination of three pathways is likely: sustain the status quo through labor-intensive agriculture and fisheries (rarely a stand-alone outcome); degrade land, potentially followed by population collapse (e.g. Rapa Nui); or substitute imports for indigenous resources (nowadays, most islands). Imports track the ever-extending reach of global trade and it is that switch – from resource autonomy to import dependence – that has led to widespread novelty of island ecosystems.

It is no surprise that the physical and biogeographic features of islands, coupled with human behavior, lead to uniquely high rates of species invasion and subsequent species replacement. Islands are notorious for their high abundance and diversity of introduced species (Lonsdale 1999), and it is those newcomers growing together that form the novel ecological communities so widespread on islands today. Furthermore, in some places much of the resident non-native flora has not yet naturalized suggesting that further expansion of novel ecosystems is in the offing (e.g. Galápagos; Trueman et al. 2010).

Species introductions often begin with those that provide goods and services not available from the local flora and fauna. At one time the human-mediated flow of species around the world was largely intentional with ship captains, horticulturists, bird fanciers and forage scientists ranking high among the vectors. With increased sensitivity to movement of non-native species, intentional introductions are now secondary sources and stowaways have taken the lead. Because islands are so import-dependent, they receive a significant share of unintentional introductions, especially in cargo containing imported foodstuffs. For example, a recent study in Galápagos found that two cargo ships bringing produce from mainland Ecuador unintentionally carried an incredible 179 invertebrate taxa that were not native to the islands (Herrera 2011). Further exacerbating the risk of unwanted entry, few islands have adequate staff to thoroughly screen cargo (e.g. Zapata 2007).

Not infrequently, an island society that becomes import-dependent relaxes pressure on the land. In Puerto Rico, for example, forest cover increased from the 1960s to the present as a result of rural-to-urban migration (Lugo 2004). But what of the composition of those new forests? The abundance of non-native species characteristic of islands typically leads to novel ecosystems. In the absence of pressure on the land for agriculture, these mature to their logical conclusion: ecosystems that differ substantially in composition from those that dominated historically. These novel ecosystems sometimes harbor native species as subordinate members of the community (Lugo and Brandeis 2005; Kueffer et al. 2007; Mascaro 2011). Nevertheless, successful invasion elsewhere is a strong predictor of invasion in a new locale, resulting in a familiar sameness in species composition among novel ecosystems on islands of comparable latitude and climate (Mueller-Dombois and Fosberg 1998; Castro et al. 2010).

4.3 INTERVENTION

In keeping with sound ecosystem management, intervention in novel ecosystems is called for only after clear goals have been set, the resources needed to accomplish the objectives are available and there is a reasonable expectation of success. Managers often focus first on the scarcity of resources with which to do the job, but defining the goals and objectives (while not always easy) must be the first priority. Conservation value is a key and obvious goal, but intervention to accomplish it might have unpredictable implications for other goods and services; these must also be considered before attempting intervention. For example, the removal of introduced mangrove trees may satisfy a conservation objective but have a negative effect on productivity of native fish (Fig. 4.2). Sometimes enhancement of those non-conservation goods and services is the objective itself: increase water yield, augment the rate of carbon sequestration or enhance images in the viewshed. Goal-setting for intervention efforts is complicated by the fact that it is not uncommon for transformations (or the species that lead to them) to be valued as positive by some sectors of society and negative by others. Should a minority value judgment ever override that of the majority? Where unique products of evolution are threatened and where intervention might sustain them, the morally responsible answer is 'yes'.

The two attributes of novel ecological communities that typically motivate intervention on islands are species composition and ecosystem functioning. By definition, novelty derives from species composition and the effects of this composition on the abiotic environment (see Chapter 6); the high endemism, small populations and species loss that characterize many islands combine to lower the threshold for conversion to a novel condition (see Chapter 3). Species-focused preservation or restoration is therefore a common objective of intervention on islands, as it is on continents. Where intervention is aimed at species composition, the target is reasonably well defined (even though the pre-human-contact flora and fauna may not be completely known) and success or failure is readily measured.

In the case of ecosystem functioning, there is unlikely to be a reference point so the target is much more diffuse. Nevertheless, there is considerable overlap between intervention to change composition and intervention to achieve changes in processes; intervention in a novel ecosystem to change functioning invariably involves manipulation of species composition. Typically, this means reducing the abundance of one or more introduced species that affect particular ecosystem processes. In the absence of historical ecosystems where processes are known however, success or failure of intervention aimed at ecosystem functioning must be measured against the performance of the novel ecosystem not subjected to management.

The high degree of endemism commonly encountered on islands, coupled with small population sizes of those endemics and the consequent high risk of species loss, can motivate a strong desire to restore an ecosystem to its historical analogue if it is even theoretically possible. The desire to save endemic plant and animal species often becomes a matter of local pride and, when coupled with popular recognition of high vulnerability to invasions of non-native species, makes intervention a prominent component of resource management on many islands. Those experiences may offer insights to those who work on larger land masses and are faced with decisions on assessment of the *need* to intervene, recognition of potential *barriers* to intervention and assessing the *feasibility* of intervention.

4.3.1   Need

Under what circumstances is intervention in novel ecosystems on islands essential or even warranted? The need for intervention ranges across a gradient from conserving or restoring a nearly intact original system to managing a novel system that provides valued goods or services. The combination of need and opportunity to restore nearly intact ecosystems to their historical state often occurs at upper elevations, especially on tropical and subtropical islands. There are two reasons for this: (1) lowland ecosystems were typically (but not inevitably; e.g. Galápagos) the first to be obliterated following human settlement of islands; and (2) the richness of introduced species diminishes with elevation (Fig. 4.3). Hence, the ecosystems that retain native species as well as a reasonable semblance of pre-colonization structure are typically at higher elevation, and these tend to be the sites where intervention is focused on species composition. Classic historical restorations in lowland ecosystems often prove impossible or limited to educational exercises, except where species traits and treatment effects are well known.

Although species-driven intervention is typically done on behalf of one or more highly valued species, the opposite is often true as well: the need for intervention arises when one component, and not the entire ecosystem, is characterized as a bad actor. Sometimes the undesirable element is a thorn-covered plant or a toxic herp, but more often than not it is a species that poses a threat outside the novel ecosystem. It might be a potentially invasive weedy species that would have an economic impact on agriculture, a predator that would pose a threat to potential prey elsewhere or an arthropod that is a potential vector of pathogenic organisms. The key element in this kind of intervention

Figure 4.2 Stand of non-native mangrove, *Rhizophora mangle*, Hawai'i Island (a) before and (b) after chemical control using herbicide. The structure is a fyke net, used to trap fish during an ebbing tide. There were higher densities of native than non-native fish in the mangroves, and the killing of the mangroves did not reduce densities of non-native fish. Photographs and findings by Richard A. MacKenzie.

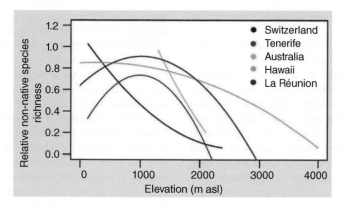

Figure 4.3 Changes in non-native plant species richness with altitude from three islands: Tenerife, Hawaii and La Réunion. Continental data from Switzerland and Australia are included for comparison. The relative richness of non-native species diminishes at high elevation, whether peak richness occurs at low or mid-elevation. Republished with permission of Ecological Society of America, from Pauchard et al (2009); permission conveyed through © Clearance Center, Inc.

is protection of something valued outside of the novel ecosystem housing the problematic species. The novel ecosystem itself might be tolerable, or even highly valued, but a particular component might merit efforts at containment or (rarely) eradication.

Deviations from historical functioning can be especially marked when new functional groups become part of the mix. If the resulting changes have undesirable and broad scale effects the need to intervene is typically perceived to be high, despite the limited prospects of success. There are many examples of such process-changing naturalizations covering a broad range of functional groups: nitrogen-fixing trees in Hawai'i (Vitousek and Walker 1989; Hughes and Denslow 2005); ants on Christmas Island (O'Dowd et al. 2003; see also Chapter 14); foxes on the Aleutian Islands (Croll et al. 2005); rabbits on sub-Antarctic islands (Bergstrom et al. 2009); tree snakes (now sustained by non-native lizards) on Guam (Fritts and Rodda 1998); a parasitic fly causing high mortality of Darwin's finches in Galápagos and thereby affecting seed dispersal and pollination (Koop et al. 2011); and fire-prone grasses in many places (D'Antonio and Vitousek 1992). However, enhanced richness of an existing functional group is not typically regarded as disruptive. In Puerto Rico, for example, African grasses in pastures and unmown urban lots have influenced the composition of the bird community. From a forested island having two native, granivorous bird

species, the island now has 18 granivores (the two natives plus 16 non-natives) all associated with novel grass communities (Raffaele 1989).

Water is a resource in short supply on many islands, and the need to intervene in a novel ecosystem might be motivated by a desire to augment water yield. Nevertheless, it is difficult to know whether or not intervention (manipulation of plant species or life form composition, for example) will have the desired effect. Transpiration is driven by a combination of physical processes, plant morphology and anatomy and phenology and when these change, water relations change. There is some evidence that those introduced plant species that tend to dominate novel ecosystems have greater leaf area per unit ground area, and therefore potentially higher rates of transpiration, than their predecessors (van Kleunen et al. 2010). Although the difference in water use between historical and novel might be only a few percent, this can be a crucial amount in dry climates (e.g. Thaxton et al. 2011). Evapotranspiration from tree plantations is sometimes high enough to reduce water yield (e.g. pines in Fiji; Waterloo et al. 1999), and this might also be the case from watersheds covered with novel ecosystems. Nevertheless, there are no published examples of water flows from paired watersheds (historical and novel) with which to substantiate or refute this possibility. Those who would intervene in a novel ecosystem with the expectation of increasing water yield are therefore advised to proceed only when data indicate a reasonable likelihood of success.

There is a perception among some members of the conservation community that ecosystem novelty is 'bad'. Nevertheless, novel ecosystems characterize much of today's world (see Chapters 8 and 9). They often provide goods and services valued by society, and many provide habitat for native and even endemic species (e.g. Lugo 2004). But even when those novel ecosystems do not harbor native plant and animal species in need of protection, there is no reason they should not be managed to optimize ecosystem services. Novel ecosystems afford unbridled opportunities for intervention: species re-introductions for conservation, sequestration of carbon, watershed protection, recreation, timber and non-timber forest products, and more. To perceive a need to intervene only for purposes of species preservation is to miss opportunities.

4.3.2 Barriers

Just as determining when intervention is needed, knowing when intervention may not be called for is important. With their histories of ecosystem change and close relationships between people and environment, islands offer some lessons in making that determination. Some barriers to intervention are obstacles to be overcome before management actions can proceed; other barriers are counter-indications, which signal a need to carefully assess the full range of consequences before moving forward (see Chapter 18).

One barrier to intervention – landscape size – is more formidable on islands, where it is an insurmountable physical obstacle, than on continents. The limited area of island landscapes effectively reduces the options available for response to allochthonous agents of change such as climate and sea level. On a continental landscape, there is potential for developing corridors that provide opportunities for species redistribution in response to environmental change; in the smaller landscapes that characterize islands however, long-distance corridors are seldom possible.

The first potentially surmountable barrier encountered by almost every ecosystem manager who proposes to intervene in a novel ecosystem is that of public opinion. Just like natural systems, novel assemblages have their proponents. Proposed interventions that do not adequately address the concerns of public opinion

supportive of the status quo are unlikely to be implemented. The necessity to engage public opinion early in the process of setting goals and objectives is essential (see Chapter 30), and the process requires the use of three tools that tend to be minor (or missing) components of the biologist's toolkit: education, mediation and conflict resolution. Gaining public buy-in is often the most difficult part of intervention.

Intervention is counter-indicated where it is likely to reduce habitat suitability for a threatened species that uses the novel ecosystem. Sometimes the role of novelty in supporting a threatened species is subtle. For example, it is not uncommon for native species to become at least partially dependent upon non-natives that provide food, pollinate flowers or disperse seeds. Management actions that disrupt those inter-specific linkages, even when inadvertent, would be ill-advised; avoiding them requires knowledge of the interactions within the novel system.

The risk of unintended consequences may be greater on islands than on continents. This may be because the comparatively low species-richness of island communities is only modestly buffered against change, such that a shift in environmental conditions resulting from management can lead to conditions that favor potentially explosive growth of non-target species. For example, six of the nine endemic ant species in Mauritius are restricted to less than one hectare occupied by a dark, cool, dense thicket of native and invasive plants near the summit of Le Pouce mountain (Alonso 2010; Fig. 4.4). Removal of the non-native invasive plants would likely change the abiotic conditions and would almost certainly facilitate invasion by the ant *Pheidole megacephala*, which has probably played a role in the extirpation of native ants at lower altitudes in Mauritius and already occurs on the border of this patch (L. Lach, unpublished data, 2011).

The consequences of intervention for species composition are not the only barriers to action: sometimes an abiotic service is the key. Water yield on high islands is a good example of the value of novelty, and intervention in novel systems that augment water yield is risky. Cloud-water interception by tall vegetation in the wet montane tropics, for example, typically augments rainfall by at least 5–20%. During the dry season, canopy drip can exceed rainfall in drier climates (Bruijnzeel 2004). Mountainous islands in the trade wind latitudes almost invariably exhibit this phenomenon and, on islands where the historical vegetation was scant or short stature (e.g. Ascension Island; Lanai in the

Figure 4.4 Dense patch of native and non-native plants that is the last remaining habitat of six endemic ants species, Le Pouce, Mauritius. Intervention would likely lead to invasion by *Pheidole megacephala* which occurs on the edges of the patch and has been implicated in the local extirpation of native ants elsewhere. Photograph: Lori Lach.

Hawaiian Islands), novel ecosystems containing tall trees were intentionally created to augment the capture of dry-season water.

Even more striking than the differences in water yield due to changes in species composition or vegetation stature are those between green and brown ecosystems: de-vegetated landscapes may not transpire, but they do not store much water either and they are subject to high rates of soil erosion. Re-vegetation, even by a novel assemblage, is invariably a more desirable watershed management alternative than bare soil. Unless there are compelling reasons to do so, drastic intervention in successfully revegetated watersheds (novel or not) is seldom warranted. Those beautiful, green mountains on many islands are blanketed with novel ecosystems, established to protect watersheds against erosion and now valued for their aesthetic value by tourists and locals alike. Intervention without

substantial assurance of successful restoration to historical conditions is unlikely to be warranted.

Sometimes there is a tendency to intervene in a novel system by attacking species based solely on their identity and reputations elsewhere. This is a common mistake and one that might result in expenditure of resources that would be better deployed on other management actions. For example, rats and mongooses have been part of Puerto Rico's wet forests for so long that they constitute a small part of the local food web (Willig and Gannon 1996) and no longer pose a threat to those native species that have survived thus far. Given their relatively low abundance and wide distribution, attempts to eradicate these small mammals would likely be futile. Another example is that the goats introduced five centuries ago onto nearby Mona Island continue to feed on rare and endangered plants as part of their diet that comprises 20% of the flora (Meléndez

Ackerman et al. 2008); there are no records of goat-driven extinctions, however. Predictions of no success, or no impact if successful, are money-saving counter-indicators of intervention.

4.3.3   Feasibility

Intervention is not to be undertaken lightly anywhere, but islands – heavily modified by people and of modest geographic complexity – are often excellent candidates for active management. Furthermore, the high ende-mism and small population sizes of native biota dis-cussed earlier are strong motivations for intervention.

Not infrequently, the proper course of intervention calls for attempts to contain or, where possible, eradi-cate one or more non-native species whose effects are deemed harmful (Figs 4.2, 4.5). This is particularly pertinent on islands because of their highly simplified food webs dominated by few, often non-native, species (e.g. Bergstrom et al. 2009). Dealing with such species on continents is very difficult, especially when the target species is well established and widespread. The likelihood of successful eradication is greater on islands however because the area, number of habitats and number of individuals to be covered by the effort are land-limited. Furthermore, efficacy of control is more easily assessed when area is limited and refugia are few.

There is a substantial record of successful eradica-tions from islands, and the procedures are becoming more successful with experience and the availability of better tools (see many examples in Veitch and Clout 2002; Veitch et al. 2011). Goats, for example, have been eliminated from more than 120 islands (Campbell and Donlan 2005). A particularly ambitious example was Project Isabela in the Galápagos archipelago, which removed more than 140,000 goats from 500,000 ha. The main factors that led to success were sufficient resources, technical know-how, mitigation of non-target impacts and support from the local community (Carrion et al. 2011). Substantial success has also been achieved with other mammals including rats, cats, pigs and sheep. Plants often prove more difficult however, especially if populations cover sizable areas. Managers considering eradication of non-native plant species would do well to note the words of Mack and Lonsdale (2002): *"The record of eradicating invasive plants . . . con-sists of few clear victories, some stalemates, and many defeats."* Likewise, attempts to eradicate ants, even on islands, almost always fail.

In addition to simplifying eradications, the compara-tively small and isolated land area of islands facilitates species introductions and re-introductions. A small area can however be a two-edged sword with respect to species introductions, for undesired species invasions can cover all suitable habitats in a very short time. A stunning example is that of the galling wasp (*Quadrasticus erythrinae*) which was detected in early 2005 on native Hawaiian Wiliwili trees (*Erythrina sandwicensis*) as well as introduced congeners; within six weeks it had spread across all of the high islands (Rubinoff et al. 2010).

Social acceptability sometimes makes intervention more feasible on islands than on continents. On islands, the degree of novelty already present opens the door to tinkering with introduced-species composition to achieve specific objectives. Members of the conserva-tion community who would be very skeptical of any proposal to modify the composition of an ecosystem composed of native species are often neutral, or even welcoming, toward active management that affects structure and composition of novel ecosystems. Never-theless, even on islands with their relatively small and accessible human populations, insufficient involvement of the community can compromise long-term success (Gardener et al. 2010).

Sometimes the novel ecosystem itself can provide the matrix into which endangered species can be reintroduced and thereby saved. This is the case when endangerment was caused by over-exploitation or by land-cover conversion to agriculture. An example from Puerto Rico is typical: more than a century ago, sug-arcane supplanted floodplain forests believed to have been dominated by a variety of native tree species including *Manilkara bidentata* and *Pterocarpus officinalis* (Abelleira-Martínez and Lugo 2008). Although both of these tree species were almost extirpated from the coastal plain, they can still be found on remnant allu-vial sites. After the abandonment of sugarcane cultiva-tion in the 1970s, forests dominated by the introduced tree *Spathodea campanulata* developed on the flood-plains and continue to occupy them 40 years later. More than half of the species in these novel forests are indigenous (including two endemics), but *Manilkara* and *Pterocarpus* are absent. The extirpation of a forest type and its eventual replacement by a novel forest that does not include the former dominants, even where soil fertility and hydrology are intact, raises questions about the reversibility of novel forests. Will *Pterocarpus* (which is wind-dispersed) or *Manilkara* (animal-dispersed

Figure 4.5 Intervention targeting plant species capable of transforming ecosystem structure or functioning. (a) Girdling the N-fixing tree, *Falcataria moluccana*, using bark spuds. National Park of American Samoa. Photograph courtesy of Tavita Togia. (b) Precision application of purple-dyed herbicide to individual *Miconia calvescens* growing in remote native-dominated vegetation. Orange hose connects herbicide reservoir on helicopter to pilot-controlled spray ball. Photograph courtesy of Maui Invasive Species Committee. (c) The killed trees are *Castilla elastica*, originally introduced to Samoa as a source of raw material for cricket balls. Cricket is all but gone; *Castilla* remains. This ecosystem grades from actively managed agroforest on the lower slope to nearly native forest on the upper slope. Tutuila, American Samoa. Photograph courtesy of Katie Friday.

and requiring specific microsites; O'Farrill et al. 2011) ever re-colonize? Deliberate seed introductions and seedling transplants might speed the process provided that these two species prove competitive in the new biotic milieu. A comparable situation has been analyzed in Seychelles, where novel *Cinnamomum verum* dominated forests are managed as habitat for rare native plants and animals (Kueffer et al. 2010; Seychelles case study in Chapter 27).

Propagule availability is not the only factor that leads to rarity of former dominants in novel ecosystems. In many instances a species' absence is due to change in competitive balances among species; introduced diseases, herbivores or predators; sustained exploitation; or changes in abiotic conditions that make it impossible for the endangered species to thrive (Caujapé-Castells et al. 2010). In those situations the agents of change must be dealt with directly, or reintroduction will fail for the same reasons that led to the original decline of the species.

Extinction rates on islands can be high, thus increasing the likelihood the novel ecosystem may be missing some of its structural/functional 'parts'. One logical approach is reconstruction or intervention that chooses species for inclusion or elimination based on the traits of species in the historical community. This approach has been used as a basis for designing sustainable agroecosystems, and it is likely to serve equally well when selecting species for design of less-utilitarian ecological systems, especially where invasion resistance is a criterion for restoration (Funk et al. 2008). What better place to experiment with, and eventually implement, designer ecosystems than on islands that have already suffered species losses and undergone human-mediated invasions?

Some of the most innovative and boldest efforts at restoring functionality involve rewilding on islands. Rewilding with non-native taxon substitutes – replacing recently extinct species with extant analogues from other geographic areas to resurrect lost ecosystem functioning – is controversial, primarily because the focus has been on replacing continental megafauna species that went extinct in the late Pleistocene (Donlan et al. 2006). In contrast, rewilding on islands offers much less controversial restoration scenarios. Insular mega-vertebrates weighed hundreds rather than thousands of kilograms, and many of them went extinct only a few hundred years ago compared with 12,000–40,000 years ago for continental Pleistocene megafauna. Consequently, practitioners working on degraded

islands around the world are spearheading the use of taxon substitutes involving birds, fruit bats, lizards and other taxonomic groups (Hansen 2010). The Mascarene Islands in the Western Indian Ocean offer an excellent example. By replacing recently extinct endemic giant tortoises with extant tortoises from other islands in the region, herbivory and seed dispersal likely to benefit native plants and control invasive non-native plants have been reinstated (Hansen 2010; Griffiths et al. 2011; Fig. 4.6).

Figure 4.6 Aldabra giant tortoise, *Aldabrachelys gigantea*, introduced on Rodrigues as herbivores and seed dispersers to replace two endemic *Cylindraspis* giant tortoise species, both recently extinct. Many endemic plant species have anti-tortoise-herbivory traits so the introduced tortoises, here and in Mauritius, seem to prefer invasive species such as this non-native *Leucaena leucocephala*. Photograph and text courtesy of Dennis Hansen.

4.4 WINDOW TO THE FUTURE?

At first glance, it may seem counter-intuitive to itemize the differences between islands and continents while also arguing that islands are a window into a future of novelty for all of Earth's ecosystems. If novelty arises on islands due to a lack of biogeographic buffering capacity, for instance, why should we predict the same outcome on continents? Buffering capacity is not the only factor determining the amount of change in an ecosystem, however: the strength of the *driver* is also paramount. While islands appear far more vulnerable to the biotic and abiotic forces that cause novel ecosystems to emerge, novel ecosystems are the likely results as these forces grow stronger and more widespread on continents.

Consider anthropogenic climate change. Although affecting all ecosystems on Earth, climate change is already altering island ecosystems. Limited area, lack of corridors and lower habitat heterogeneity (in many cases) imply high vulnerability of islands to climate change, and recent evidence confirms this. Coastline erosion due to sea-level rise is shrinking the land area of some Pacific Island nations made vulnerable by their small size and may literally wipe them off the map. Warming also allows the spread of disease vectors that directly cause avian extinctions (Benning et al. 2002). On continents, ecosystems may have higher buffering capacity, but this will mean little when climate change leads to wholesale biome shifts as predicted within one century by most global circulation models (Bergengren et al. 2011). High species diversity and a range of parent materials in an African savanna will not halt desertification, nor will the vast range of black spruce (*Picea mariana*) in North America shield it from thawing permafrost.

The global human footprint is also tracking what has already occurred on islands (see Chapters 8 and 9). For instance, islands have been referred to as 'paradise' for introduced species (Denslow 2003); indeed, their contribution to novelty is most apparent there. But comparative studies suggest that the fundamental ecological controls underlying species introductions are part of a global pattern of pervasive and widespread changes in ecological communities that are mechanistically consistent between islands and continents. For example, Sax and Gaines (2003) found that plant species richness at the regional scale is increasing on both islands and continents because species introductions far exceed extinctions. Similarly, novel ecosystems

completely dominated by introduced species emerge on islands and continents alike when the introduced species in question add functional diversity not present in the historical ecosystem.

Novel ecosystems are so ubiquitous on islands that it is not surprising that there have been many attempts to manipulate them to achieve particular ends. Some of those interventions have been successful and some have failed, but we can learn from all of them. Most of them are applicable to continents. Five lessons from islands and elsewhere have particularly widespread applicability.

1. *Recognize and embrace the need for social success.* Ecologists, land managers and conservation biologists tend to be strong in the biological and physical sciences, but weak in the social sciences. It is not unheard of for intervention to begin and be halted soon after because a broad range of community members objected. Lack of social acceptance usually leads to no long-term biological or physical success. It is essential to identify stakeholders and involve them in planning from the start.

2. *Assess the odds of success or failure prior to intervening.* Funds are scarce for resource managers everywhere, so it is astonishing that investments are made that have little chance of success. Value judgments are inevitably involved, and some circumstances call for desperate measures. Futile expenditures are not in the best interest of resource managers, however.

3. *Evaluate the full range of values involved.* Intervention motivated by a single factor without considering ancillary impacts is ill advised. It is better to assess the full range of values provided by a novel ecosystem than to act based exclusively on one value, only to learn belatedly that intervention was the wrong course of action.

4. *Guard against unintended consequences.* These occur within and outside of novel ecosystems subjected to intervention. It is a general tenet of system science that to understand any unit, one must understand the next largest system of which it is a part as well as the next smallest unit that is a part of the system of interest. Avoid those unintended outcomes by understanding the target system and its parts as well as the larger landscape in which the intervention is to occur.

5. *Recognize local constraints and capitalize on local opportunities.* Resource managers would do well to act cautiously, paying due attention to unique local conditions. Nevertheless, intervention is hampered when each situation is regarded as so completely unique that it must be fully understood before any action whatsoever can be taken. Balanced judgment is called for.

Adhering to those five guidelines does not guarantee success, but not paying attention to any one of them greatly increases the likelihood of failure. Resource managers on islands have demonstrated that they can sometimes do it well. They have learned from their mistakes and others are now in a good position to benefit from their experiences.

ACKNOWLEDGMENTS

We thank Dennis Hansen (Institute of Evolutionary Biology and Environmental Studies, University of Zurich, Switzerland) who generously wrote the paragraph on rewilding, provided Figure 4.6, and constructively reviewed the entire chapter. Katie Friday, Rich MacKenzie, Flint Hughes and Tracy Johnson, all with the US Forest Service in Hilo, Hawai'i, provided photographs or put us on the track of the right images. The involvement of JJE in this effort was supported in part by the US National Science Foundation.

REFERENCES

Abelleira-Martínez, O. and Lugo, A.E. (2008) Post sugar cane succession in moist alluvial sites in Puerto Rico, in *Post-agricultural succession in the Neotropics* (ed. R.W. Myster), Springer, New York, 73–92.

Adsersen, H. (1989) The rare plants of the Galápagos islands and their conservation. *Biological Conservation*, **47**, 49–77.

Alonso, L. (2010) Ant conservation: Current status and a call to action, in *Ant Ecology* (eds L. Lach, C.L. Parr and K.L. Abbott), Oxford University Press, Oxford UK, 59–74.

Benning, T.L., LaPointe, D., Atkinson, C.T. and Vitousek, P.M. (2002) Interactions of climate change with biological invasions and land use in the Hawaiian Islands: Modeling the fate of endemic birds using a geographic information system. *PNAS*, **99**, 14246–14249.

Bergengren, J.C., Waliser, D.E. and Yung, Y.L. (2011) Ecological sensitivity: a biospheric view of climate change. *Climatic Change*, doi: 10.1007/s10584-10011-10065-10581.

Bergstrom, D.M., Lucieer, A., Kiefer, K., Wasley, J., Belbin, L., Pedersen, T.K. and Chown, S.L. (2009) Indirect effects of invasive species removal devastate World Heritage Island. *Journal of Applied Ecology*, **46**, 73–81.

Bruijnzeel, L.A. (2004) Hydrological functions of tropical forests: not seeing the soil for the trees? *Agriculture, Ecosystems and Environment*, **104**, 185–228.

Campbell, K. and Donlan, C.J. (2005) Goat eradications on islands. *Conservation Biology*, **19**, 1362–1374.

Carlquist, S. (1966) The biota of long-distance dispersal. I. Principles of dispersal and evolution. *The Quarterly Review of Biology*, **41**, 241–270.

Carrion, V., Donlan, C.J., Campbell, K.J., Lavoie, C. and Cruz, F. (2011) Archipelago-wide island restoration in the Galapagos Islands: reducing costs of invasive mammal eradication programs and reinvasion risk. *PLoS ONE* **6**:e18835. doi: 18810.11371/journal.pone.0018835.

Castro, S.A., Daehler, C.C., Silva, L., Torres-Santana, C.W., Reyes-Betancort, J.A., Atkinson, R., Jaramillo, P., Guezou, A. and Jaksic, F.M. (2010) Floristic homogenization as a teleconnected trend in oceanic islands. *Diversity and Distributions*, **16**, 902–910.

Caujapé-Castells, J., Tye, A., Crawford, D.J., Santos-Guerra, A., Sakai, A., Beaver, K., Lobin, W., Florens, F.B.V., Moura, M., Jardim, R., Gómes, I. and Kueffer, C. (2010) Conservation of oceanic island floras: present and future global challenges. *Perspectives in Plant Ecology, Evolution and Systematics*, **12**, 107–130.

Croll, D.A., Maron, J.L., Estes, J.A., Danner, E.M. and Byrd, G.V. (2005) Introduced predators transform subarctic islands from grassland to tundra. *Science*, **307**, 1959–1961.

D'Antonio, C.M. and Vitousek, P.M. (1992) Biological invasions by exotic grasses, the grass fire cycle, and global change. *Annual Review of Ecology and Systematics*, **23**, 63–87.

Denslow, J.S. (2003) Weeds in paradise: thoughts on the invasibility of tropical islands. *Annals of the Missouri Botanical Garden*, **90**, 119–127.

Donlan, C.J., Berger, J., Bock, C.E., Bock, J.H., Burney, D.A., Estes, J.A., Foreman, D., Martin, P.S., Roemer, G.W., Smith, F.A., Soule, M.E. and Greene, H.W. (2006) Pleistocene rewilding: An optimistic agenda for twenty-first century conservation. *American Naturalist*, **168**, 660–681.

Fritts, T.H. and Rodda, G.H. (1998) The role of introduced species in the degradation of island ecosystems: a case history of Guam. *Annual Review of Ecology and Systematics*, **29**, 113–140.

Funk, J.L., Cleland, E.E., Suding, K.N. and Zavaleta, E.S. (2008) Restoration through reassembly: plant traits and invasion resistance. *Trends in Ecology and Evolution*, **23**, 695–703.

Gardener, M.R., Atkinson, R. and Renteria, J.L. (2010) Eradications and people: lessons from the plant eradication program in Galapagos. *Restoration Ecology*, **18**, 20–29.

Givnish, T.J., Millam, K.C., Mast, A.R., Paterson, T.B., Theim, T.J., Hipp, A.L., Henss, J.M., Smith, J.F., Wood, K.R. and Sytsma, K.J. (2009) Origin, adaptive radiation and diversification of the Hawaiian lobeliads (Asterales: Campanulaceae). *Proceedings of the Royal Society B*, **276**, 407–416.

Griffiths, C.J., Hansen, D.M., Zuël, N., Jones, C.G. and Harris, S. (2011) Resurrecting extinct interactions with extant substitutes. *Current Biology*, **21**, 762–765.

Hansen, D.M. (2010) On the use of taxon substitutes in rewilding projects on islands, in *Islands and Evolution* (eds V. Pérez-Mellado and C. Ramon), Institut Menorquí d'Estudis, Maó, Menorca, Spain, 111–146.

Herrera, H.W. (2011) Monitoreo de invertebrados terrestres en barcos de carga desde Guayaquil a Galápagos. Report for the Ecuadorian Ministry of Environment, Puerto Ayora, Galápagos.

Hughes, F., Vitousek, P.M. and Tunison, T. (1991) Alien grass invasion and fire in the seasonal submontane zone of Hawaii. *Ecology*, **72**, 743–746.

Hughes, R.F. and Denslow, J.S. (2005) Invasion by a N_2-fixing tree alters function and structure in wet lowland forests of Hawai'i. *Ecological Applications*, **15**, 1615–1628.

Jäger, H., Tye, A. and Kowarik, I. (2007) Tree invasion in naturally treeless environments: Impacts of quinine (*Cinchona pubescens*) trees on native vegetation in Galápagos. *Biological Conservation*, **140**, 297–307.

Kaiser-Bunbury, C.N., Traveset, A. and Hansen, D.M. (2010) Conservation and restoration of plant-animal mutualisms on oceanic islands. *Perspectives in Plant Ecology Evolution and Systematics*, **12**, 131–143.

van Kleunen, M., Webber, E. and Fischer, M. (2010) A meta-analysis of trait differences between invasive and non-invasive plant species. *Ecology Letters*, **13**, 235–245.

Koop, J., Huber, S., Laverty, S. and Clayton, D. (2011) Experimental demonstration of the fitness consequences of an introduced parasite of Darwin's finches. *PLoS ONE* **6**:e19706. doi: 19710.11371/journal.pone.0019706.

Kueffer, C., Schumacher, E., Fleischmann, K., Edwards, P.J. and Dietz, H. (2007) Strong belowground competition shapes tree regeneration in invasive *Cinnamomum verum* forests. *Journal of Ecology*, **95**, 273–282.

Kueffer, C., Schumacher, E., Dietz, H., Fleischmann, K. and Edwards, P.J. (2010) Managing successional trajectories in alien-dominated, novel ecosystems by facilitating seedling regeneration: a case study. *Biological Conservation*, **143**, 1792–1802.

Laurence, W.F., Useche, D.C., Shoo, L.P., Herzog, S.K., Kessler, M. and 48 others. (2011) Global warming, elevational ranges and the vulnerability of tropical biota. *Biological Conservation*, **144**, 548–557.

Lonsdale, W.M. (1999) Global patterns of plant invasions and the concept of invasibility. *Ecology*, **80**, 1522–1536.

Lugo, A.E. (2004) The outcome of alien tree invasions in Puerto Rico. *Frontiers in Ecology and the Environment*, **2**, 265–273.

Lugo, A.E. and Brandeis, T.J. (2005) New mix of alien and native species coexists in Puerto Rico's landscapes, in: *Biotic Interactions in the Tropics: Their Role in the Maintenance of Species Diversity* (eds D.F.R.P. Bursalem, M.A. Pinard and S.E. Hartley), Cambridge University Press, Cambridge.

Mack, R.M. and Lonsdale, W.R. (2002) Eradicating invasive plants: Hard-won lessons for islands, in *Turning the Tide: Eradication of Invasive Species* (eds C.R. Veitch and M.N. Clout), IUCN SSC Invasive Species Specialist Group, Gland, Switzerland and Cambridge, UK, 164–172.

Mascaro, J. (2011) Eighty years of succession in a non-commercial plantation on Hawai'i Island: Are native species returning? *Pacific Science*, **65**, 1–15.

Mascaro, J., Hughes, F.R. and Schnitzer, S.A. (2012) Novel forests maintain ecosystem processes after the decline of native tree species. *Ecological Monographs*, doi: 10.1890/11-1014.1.

Meléndez Ackerman, E.J., Cortés, C., Sustache, J., Aragón, S., Morales Vargas, M., García Bermúdez, M. and Fernández, D.S. (2008) Diet of feral goats in Mona Island Reserve, Puerto Rico. *Caribbean Journal of Science*, **44**, 199–205.

Mueller-Dombois, D. (2008) Pacific island forests: successionally impoverished and now threatened to be overgrown by aliens? *Pacific Science*, **62**, 303–308.

Mueller-Dombois, D. and Fosberg, F.R. (1998) *Vegetation of the Tropical Pacific Islands*. Springer, NY.

O'Dowd, D.J., Green, P.T. and Lake, P.S. (2003) Invasion 'meltdown' on an oceanic island. *Ecology Letters*, **6**, 812–817.

O'Farrill, G., Chapman, C.A. and Gonzalez, A. (2011) Origin and deposition sites influence seed germination and seedling survival of *Manilkara zapota*: implications for long-distance, animal-mediated seed dispersal. *Seed Science Research*, doi:10.1017.

Parent, C.E., Caccone, A. and Petren, K. (2008) Colonization and diversification of Galápagos terrestrial fauna: a phylogenetic and biogeographical synthesis. *Philosophical Transactions of the Royal Society B*, **363**, 3347–3361.

Pauchard, A., Kueffer, C., Dietz, H., Daehler, C.C., Alexander, J., Edwards, P.J., Arévalo, J.R., Cavieres, L., Guisan, A., Haider, S., Jakobs, G., McDougall, K., Millar, C.I., Naylor, B.J., Parks, C.G., Rew, L.J. and Seipel, T. (2009) Ain't no mountain high enough: Plant invasions reaching new elevations. *Frontiers in Ecology and the Environment*, **7**, 479–486.

Raffaele, A.H. (1989) The ecology of native and introduced granivorous birds in Puerto Rico, in *Biogeography in the West Indies: Past, Present, and Future* (ed. C.A. Woods), Sandhill Crane Press, Gainesville, FL, 541–566.

Rolett, B. and Diamond, J. (2004) Environmental predictors of pre-European deforestation on Pacific islands. *Nature*, **431**, 443–446.

Rubinoff, D., Holland, B.S., Shibata, A., Messing, R.H. and Wright, M.G. (2010) Rapid invasion despite lack of genetic variation in the erythrina gall wasp (*Quadrastichus erythrinae* Kim). *Pacific Science*, **64**, 23–31.

Sax, D.F. and Gaines, S.D. (2003) Species diversity: from global decreases to local increases. *Trends in Ecology and Evolution*, **18**, 561–566.

Thaxton, J.M., Cordell, S., Cabin, R.J. and Sandquist, D.R. (2011) Non-native grass removal and shade increase soil moisture and seedling performance during Hawaiian dry forest restoration. *Restoration Ecology*, doi: 10.1111/j.1526-1100X.2011.00793.x.

Trueman, M., Atkinson, R., Guézou, A.P. and Wurm, P. (2010) Residence time and human-induced propagule pressure at work in the alien flora of Galapagos. *Biological Invasions*, **12**, 3949–3960.

Veitch, C.R. and Clout, M.N. (eds) (2002) Turning the tide: the eradication of invasive species. *Proceedings of the International Conference on the Eradication of Island Invasives*. Gland, Switzerland, IUCN.

Veitch, C.R., Clout, M.N. and Towns, D.R. (eds) (2011) Island invasives: eradication and management. *Proceedings of the International Conference on Island Invasives*. Gland, Switzerland: IUCN and Auckland, New Zealand.

Vitousek, P.M. (2004) *Nutrient Cycling and Limitation: Hawai'i as a Model System*. Princeton University Press, Princeton, NJ.

Vitousek, P.M. and Walker, L.R. (1989) Biological invasion by *Myrica faya* in Hawai'i: plant demography, nitrogen fixation, ecosystem effects. *Ecological Monographs*, **59**, 247–265.

Waterloo, M.J., Bruijnzeel, L.A., Vugts, H.F. and Rawaqa, T.T. (1999) Evaporation from *Pinus caribea* plantations on former grassland soils under maritime tropical conditions. *Water Resources Research*, **35**, 2133–2144.

Whittaker, R.J. and Fernández-Palacios, J.M. (2007) *Island Biogeography: Ecology, Evolution, and Conservation*. 2nd edition. Oxford University Press, Oxford, UK.

Willig, M.R. and Gannon, M.R. (1996) Mammals, in *The Food Web of a Tropical Rain Forest* (eds D.P. Reagan and R.B. Waide), The University of Chicago Press, Chicago, IL, 399–431.

Zapata, C.E. (2007) Evaluation of the quarantine and inspection system for Galapagos (SICGAL) after seven years, in *Galapagos Report 2006–2007*. Puerto Ayora, Galapagos, Ecuador: Charles Darwin Foundation, Galapagos National Park & INGALA.

Zimmerman, N., Hughes, R.F., Cordell, S., Hart, P., Chang, H.K., Perez, D., Like, R.K. and Ostertag, R. (2008) Patterns of primary succession of native and introduced plants in lowland wet forests in Eastern Hawai'i. *Biotropica*, **40**, 277–284.

Chapter 5

ORIGINS OF THE NOVEL ECOSYSTEMS CONCEPT

Joseph Mascaro[1], James A. Harris[2], Lori Lach[3], Allen Thompson[4], Michael P. Perring[5], David M. Richardson[6] and Erle C. Ellis[7]

[1]Department of Global Ecology, Carnegie Institution for Science, Stanford, California, USA

[2]Environmental Science and Technology Department, Cranfield University, UK

[3]Ecosystem Restoration and Intervention Ecology (ERIE) Research Group, School of Plant Biology and the Centre for Integrative Bee Research, University of Western Australia, Australia

[4]School of History, Philosophy, and Religion, Oregon State University, USA

[5]Ecosystem Restoration and Intervention Ecology (ERIE) Research Group, School of Plant Biology, University of Western Australia, Australia

[6]Centre for Invasion Biology, Department of Botany and Zoology, Stellenbosch University, South Africa

[7]Geography & Environmental Systems, University of Maryland, USA

"In such ways *anthropogenic ecosystems* differ from those developed independently of man. But the essential formative processes . . . are the same."

Arthur G. Tansley, 1935

5.1 INTRODUCTION

Reading Tansley's (1935) opus, combatively titled 'The use and abuse of vegetational concepts and terms' nearly a century on, is illuminating in many ways. For one, the prose leaps off the page in a manner rarely seen in ecology journals today: he moves through emotional states no modern editor would allow, gleefully deconstructing the exactitude of his 'old friend' Fredrick Clements' proposition of a 'climatic climax', a single terminus of succession for a whole biome. He betrays his academic nature, variously accepting and rejecting terms that appear throughout the literature. He accepts 'climax' (within which he notably allows slow gradual change), coins the term 'ecosystem' (surely his greatest gift to the field), then roundly rejects the idea that ecosystems are a kind of 'complex organism', a phrase he finds too inexact to be useful. He italicizes profusely (evidenced in the opening quotation, for which the emphasis is original), an effort to convey final authority; like Clements, Tansley exhibits ecology's roots in botany, where taxonomy and strict classification reigned.

Novel Ecosystems: Intervening in the New Ecological World Order, First Edition. Edited by Richard J. Hobbs, Eric S. Higgs, and Carol M. Hall.
© 2013 John Wiley & Sons, Ltd. Published 2013 by John Wiley & Sons, Ltd.

The lasting value of Tansley's paper comes not in his concluding litany of terminology (most of which is now altered or abandoned), but in his flashes of uncertainty about nature. He raises questions to which he does not know the answer: "Is man part of 'nature' or not?" He frames the quintessential debate in early 20th century ecology without making a ruling: "Many ecologists hold that all vegetation is always changing. It may be so: we do not know enough either to affirm or to deny so sweeping a statement." With statements like these, Tansley hints at an important insight: strict definitions of ecosystem types may never be tractable because there is too much spatiotemporal variation to categorize.

The organizing objective of this chapter was to categorically define a *novel ecosystem*. Doing so at the outset would imply that we were already convinced of their existence as something separate from other ecosystems, a notion upon which our read of Tansley has cast doubt. We therefore begin not by declaring what novel ecosystems are or are not (a task which we shall come to presently) but by reviewing the foundational principles that point to the existence and importance of novel ecosystems. From these, and a brief review of previous formulations of the concept, we hope the idea will emerge for the reader as it has for the authors. We ultimately step into synthesis, and present a new framework for the novel ecosystems concept.

5.2 FOUNDATIONS OF THE NOVEL ECOSYSTEMS CONCEPT

The novel ecosystem concept rests on three foundational principles, the first of which follows from Gleason (1926): "It may be said that every species of plant is a law unto itself, the distribution of which in space depends on its individual peculiarities of migration and environmental requirements." In Gleason's time, ecology was populated largely by naturalists and botanists with an instinctual urge to classify. It is remarkable that he was able to look at the same forests and grasslands as his predecessors and devise so different a hypothesis to explain them, one that was abjectly allergic to classification. Gleason viewed communities as whisperers of organization: the ephemeral overlapping of ranges combined with a faint alignment. Decades went by before the individualistic concept became more widely accepted, but Gleason did live to see it embraced by giants such as John Curtis and Robert

Whittaker. When ecology mourned Gleason in 1975, Robert McIntosh wrote for *Torrey* that "Gleason's then heretical idea is now widely recognized as part of the conventional wisdom in ecology" (McIntosh 1975). Yet controversy remained.

Had he lived another decade, Gleason would have received a gift reserved for all but a few theoreticians: irrefutable proof. In the form of sediment from North American lakes, the past was being steadily disrobed by palynological studies (Davis 1981, 1983; Pielou 1991; Delcourt 2002). These samples, which were largely from a period of glacial retreat at the close of the Wisconsinan 8,000 to 14,000 years ago, painted a clear picture: in times of environmental change, tree species move as individuals and not as part of discrete communities or organisms (Fig. 5.1). While students from Madison to Cornell can rattle off tree and shrub species that 'characterize' the beech-maple forest association today, students visiting the same habitat in a previous interglacial would find mesic forests of quite different composition. Such patterns inspired one paleoecologist contributing to this volume to call himself a 'radical Gleasonian' (Jackson 2006). From the tropics, Janzen (1985) concluded that the seduction of species *belonging* to communities was better thought of as species *fitting* into a habitat.

The individualistic concept provides the raw biological material from which the novel ecosystems concept is made. Because the flora and fauna move independently of one another in response to their environment, communities are constantly in flux at some temporal resolution. Without this property of the biosphere, novel ecosystems would not be possible. In some areas, such as those undergoing secondary succession, the fluctuations may be rapid and repeatable; in others, such as those undergoing changes in temperatures or carbon dioxide concentrations, the 'fluctuations' may be directional and permanent. Temperature drove tree species to move during glacial advances and retreats (Davis 1981), while CO_2 concentrations in the ocean caused shifts in foraminifera communities (Pagani et al. 1999). Where the ranges of species expanded or contracted (whatever the cause), communities were altered. Most recently, a massive increase in the rate of species introductions has unfolded: prior to human arrival, Hawaii lacked reptiles, amphibians, flightless mammals and ants, yet now it has all in abundance (Ziegler 2002).

The earlier examples highlight community changes caused both by biotic (i.e. range shifts) and abiotic

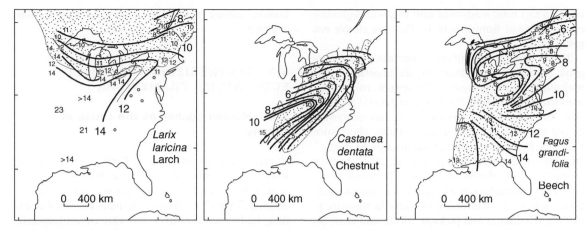

Figure 5.1 The Gleasonian framework of ecological communities. In the North American temperate forests shown, black bands indicate the southernmost extent for each species in time (1000 years). The varying trajectories of these ranges reflect the independent movement of each species as they coped with warming temperatures at the close of the Wisconsinan glaciation. From Davis (1983). Reproduced by permission of Missouri Botanical Garden Press.

forcing (i.e. responses to environmental change) and this leads to the second foundational concept upon which novel ecosystems rest: the biotic and abiotic characteristics of an ecosystem are tethered. They interact, following from one another and feeding back on one another (Tansley 1935; Jenny 1941; Odum 1969; Naeem 2002). Jenny (1980) made perhaps the most humorous declaration, calling the biotic factor ". . . a real bugbear. Like everybody else I could see that the vegetation affects the soil and the soil affects the vegetation, the very *circulus vitiosus* I was trying to avoid." Twenty years later, Naeem (2002) came to a similarly baffling (but editorially sanitized) conclusion: ". . . biodiversity is a product of its environment, and the antithesis, that the environment is, in part, a product of the organisms within it, is also correct." Empirical studies spanning the globe confirm strong interactions between the biotic and abiotic parts of ecosystems, from tubeworms mollifying habitat around oceanic vents (Cordes et al. 2003) to warming temperatures roasting tropical trees (Clark et al. 2003). Abiotic change leads to biotic change, and *vice versa* (this is, incidentally, why this book isn't called 'Novel Communities').

A third concept completes the groundwork for novel ecosystems: humans, like previously massive perturbations to the biosphere (Behrensmeyer et al. 1992; Vermeij 2005), cause changes in ecosystems that are directional and permanent. Indeed, Chapin and Starfield (1997) first invoked the term 'novel ecosystem' to characterize what they believed would be the ultimate outcome of anthropogenic changes to climate, disturbance regimes and species composition in boreal latitudes. Old views in terrestrial plant succession referred to human-modified ecosystems such as farms as "dis-climax" (Clements 1916). Note the root "dis" (having a negative or reversing force) and the similarity to the more modern term for the same systems: 'degraded' (reduced in rank). Humans have unequivocally had a negative impact on myriad species and ecosystems, particularly by co-opting land for the production of resources (Vitousek et al. 1997). However, when we are left to qualify a river that has increased in trophic complexity due to non-native fishes on Hawaii (Ziegler 2002), or an old Puerto Rican baseball field that is thick with pantropical tree species whose leaves have never before competed for sunlight (Lugo and Helmer 2004), we crumple at the term 'degraded'. To us, this term feels flawed and inexact. Clearly, neither 'degraded' nor its predecessor has sufficient descriptive power for ecosystems that are diverse and (in some cases) have undergone centuries of succession, or whose component species are experiencing unrestricted natural selection in their novel environs. Both terms imply change in a negative direction than can only be corrected by change in a positive direction, when in fact *novel ecosystems* exist as a non-regressive consequence of human activity. They are simply not going back.

The widening recognition of pervasive anthropogenic change in ecosystems was the impetus for formalizing the novel ecosystems framework. Following a workshop in Granada, 2002, Milton (2003) reported that an 'emerging ecosystem' was: "an ecosystem whose species composition and relative abundance have not previously occurred within a given biome." At a subsequent workshop in Brasilia, 2003, various examples of novel ecosystems were discussed. Some centered on disturbances wrought by human intervention, such as the conversion of Cerrado savanna to pasture (Hoffmann and Jackson 2000), and some on indirect consequences of human activities such as increases in non-native species with consequent collapse of 'historical' ecosystems. These examples were numerous, spanned continents and oceans and demonstrated a contemporary outcome of Gleason's individualistic concept as well as a strong linkage between the abiotic and biotic. For instance, Lugo (2004) reported on the growth of alien forests on deforested lands in Puerto Rico; they were eventually re-colonized by native species, but remained dominated by non-natives. These 'new forests' provided important ecological functions, such as the repair of soil structure and fertility and the restoration of forest cover and biodiversity at sites formerly used for production. The disparate threads ultimately coalesced in a seminal paper by Hobbs and colleagues (2006). They defined novel ecosystems as those that exhibit: "(1) novelty: new species combinations with the potential for changes in ecosystem functioning; and (2) human agency: ecosystems that are the result of deliberate or inadvertent human action, but do not depend on continued human intervention for their maintenance". The authors went on to suggest that that it was becoming clear that there were more 'novel' than 'historical' systems, and it was high time that we devoted ecological thought to them.

In subsequent sections we advance from these foundational concepts (i.e. the individualistic concept, the abiotic–biotic tether and pervasive human-caused change) to synthesize the novel ecosystems framework. Using the original Hobbs et al. (2006) components of *novelty* and *human agency* as a starting point, we develop this synthesis by first considering (1) where and how human agency leads to novelty; (2) what level of novelty constitutes a novel ecosystem; and (3) how human agency acts after a novel ecosystem has emerged.

5.3 SYNTHESIZING THE NOVEL ECOSYSTEMS FRAMEWORK

5.3.1 Human agency as the cause of novelty

Humans now influence all of Earth's ecosystems to some degree, either directly or indirectly (Vitousek et al. 1997; Ellis et al. 2010), and we might therefore conclude that all modern ecosystems are novel (with their higher temperatures and CO_2 concentrations). However, this is not a particularly useful exercise either theoretically or for management (but see Chapter 41 by Marris et al.), and a careful look at the pathways that lead to novelty in ecosystems reveals variation both geographically and with respect to human decision making.

Consider that human effects differ in space. For instance, 'onsite' influences such as development and plantings contrast with 'offsite' influences such as climate change, anthropogenic N deposition or non-native species colonization. In practice, nearly all of Earth's spaces have both onsite and offsite influences occurring simultaneously at some level (Ellis et al. 2010). However, there are examples in which offsite drivers can cause novel ecosystem emergence without onsite activity by humans. For instance, the increasing abundance of lianas in remote areas of tropical forest is one of many compositional changes for which atmospheric CO_2 increases may be a cause (Schnitzer and Bongers 2011). Similarly, climate change is presently altering cryptic Antarctic ecosystems never before visited by humans (Convey and Smith 2006).

Human agency is also expressed through different types of human intent, that is, deliberate versus inadvertent actions. Hawaii was deliberately populated with introduced tree species in the early 20th century in an attempt to foster ecosystem services in heavily logged and burned watersheds (Woodcock 2003). This contrasts with the many inadvertent introductions that afflicted the islands' biodiversity, from avian malaria to argentine ants (Ziegler 2002). Note that most human activity that has unintentionally resulted in the creation of novel ecosystems (as an accidental side-effect) was or is nonetheless intentional activity, undertaken for some other purpose. These actions are deliberate but the consequent novel ecosystems are an inadvertent result. For instance, Ascension Island is situated in the mid-Atlantic and provided a sanctuary for explorers in the 1500s; they deliberately populated it with goats and sheep and were inadvertently left with an island almost devoid of vegetation (Wilkinson 2004). Additionally, deliberate creation of new ecosystems may be undertaken for various purposes (e.g. 'designer ecosystems', Kueffer and Daehler 2009). This was the remedy selected by botanist Joseph Hooker after stopping off at a barren, goat-nibbled Ascension in 1843, whereupon he began to populate it with foreign plants (Wilkinson 2004).

Hobbs and colleagues (2006) focused on human agency in their effort to differentiate novel ecosystems from other types of ecosystems. They arranged novel ecosystems in the middle of a single axis with 'wild' (or 'historical') ecosystems at one end and intensive agriculture (or other sites intensely managed for human use) at the other (Fig. 5.2), a gradient originally described by Sanderson et al. (2002). Historical ecosystems became novel when subject to invasion (e.g. Weiss 1999), an offsite and inadvertent result of human-caused expansion of species ranges. Similarly, agricultural systems became novel when abandoned (e.g. Lugo 2004), an onsite and inadvertent result of human land use. Importantly, as noted by Kueffer and Daehler (2009), this typology suggests that novel ecosystems are not under direct onsite human control. Put another

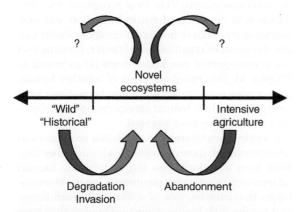

Figure 5.2 Hobbs et al. (2006) made the first attempt at formally defining novel ecosystems, arguing that they arise from either the (1) degradation and invasion of 'wild' or natural/seminatural systems or (2) the abandonment of intensively managed systems. From Hobbs et al. (2006). With permission from John Wiley & Sons.

way, humans do not *prescribe* the abiotic and biotic characteristics of novel ecosystems as they do in agricultural fields, for example. Rather, novel ecosystems are the *response* of the biosphere to human influence.

5.3.2 Novelty versus novel ecosystems

Hobbs et al. (2009) outlined a scheme for identifying historical and novel ecosystems as well as hybrid mixtures of the two, depending on quantifying biotic and abiotic differences from historical ecosystems, namely, novelty (Fig. 3.2). Hobbs et al. (2009) emphasize the second foundational concept behind novel ecosystems, noting that "biotic and abiotic factors often change simultaneously and act synergistically". Indeed, as human influences on biotic or abiotic ecosystem components have increased exponentially, the literature has become festooned with examples of this linkage. Introduced species, for example, are not static additions to the roster of ecological communities; they bring their particular physiologies and life histories to bear on their environment, modifying productivity, soil nutrients and development and physical ecosystem structure among others (e.g. Ehrenfeld 2003; Asner et al. 2008). The door swings both ways, with abiotic changes often facilitating new species colonization or species loss (e.g. MacDougall and Turkington 2005; Wardle et al. 2011).

After how much biotic or abiotic change does something *categorically* become a novel ecosystem? This might be approached in the same way that doctors handle diagnoses. For example, a patient may be diagnosed with a condition if he or she presents with a certain number of symptoms that characterize the condition. Ecosystems might similarly be arrayed according to their degree of novelty in abiotic or biotic factors, and might be deemed 'novel ecosystems' if they exist at sufficient *distance* from either the abiotic or biotic attributes of the historical system (Fig. 5.3). How far a distance? A logical distinction is that the system has crossed a threshold past which any return possibility is a problem of overcoming a strong tendency of the system to remain in its novel state. Consider hysteresis, in which a change in an environmental condition, usually accompanied by positive feedback, leads to the movement of a system to an alternative state (Suding et al. 2004). Even with the reversal of the environmental change, the altered state is maintained well beyond the condition at which the

(a) Theoretical axes of novelty

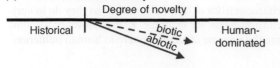

(b) Applied axes of novelty

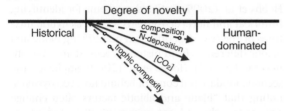

Figure 5.3 A more nuanced view of novelty envisions not a Boolean state, but a quantifiable axis of disparity from the historical condition. (a) Theoretically, this condition can be summarized by biotic and abiotic parts (as depicted in Fig. 3.2) but, (b) in an applied context, either for determination of ecosystem status or informing management, novelty would be measured as the real difference in various abiotic and biotic parameters.

original existed due to negative feedback (e.g. Simenstad et al. 1978). To be a true alternative stable state, exactly the same environmental conditions should lead to alternative states (depending upon initial conditions; Clark et al. 2005) but alternative states may also be observed without hysteresis (Suding and Hobbs 2009). Consider rapid changes in species composition at a tipping point along an environmental gradient, such as alterations to atmospheric N inputs or hydrologic regimes. In such cases, the new community is not stable at the same environmental condition of the old, yet overcoming the massive system-wide inertia of prolonged N deposition or hydrologic change may be impossible.

Hobbs et al. (2009) make a practical distinction between hybrid ecosystems, for which novelty might still practically be reversed (perhaps with great difficulty), and novel ecosystems, for which such reverses are impossible because the ecosystem has crossed an irreversible threshold. The placement of such thresholds with respect to quantitative levels of abiotic or biotic differences from the historical ecosystem might be expected to differ greatly among climates and biomes. For instance, small temperature changes in

boreal latitudes may contribute to melting permafrost (and subsequent ecosystem consequences), whereas the same temperature change may have a more limited influence in temperate latitudes.

Is desirability a factor in differentiating novel ecosystems? We suggest that it should not be. Desirability is a value-based assessment often linked to ecosystems services, and not everyone will place the same value on those services. For example, the Louisiana red-swamp crayfish has been introduced to aquatic systems widely outside its native range. In Kenya it has been implicated in the complete disappearance of native floating and submerged vegetation in Lake Naivasha with significant consequences for the ecosystem (Smart et al. 2002). However, because it decreases the habitat for and preys on native snails that host schistosome flukes, its presence in waterways also results in decreased prevalence of schistosomiasis, an intestinal or urinary disease to which over 12 million Kenyans are susceptible (Mkoji et al. 1999). While admirers of native aquatic plants in Kenya may consider the novelty of crayfish-invaded ecosystems to be highly undesirable, those who are at risk of schistosomiasis may consider it an improvement over the historical state.

5.3.3 Human agency after novelty

Hobbs et al. (2006) argue that human agency must operate as an originating (but not a sustaining) cause of novel ecosystems. This view recognizes the contribution of unrestricted natural selection and new species interactions to novel community assembly and novel ecosystem organization. However, humans also act as managers of novel ecosystems (as addressed in Chapter 3). This raises questions of whether human agency may actually prevent a novel ecosystem from emerging, and how human agency changes novel ecosystems after they have emerged.

Consider agricultural and plantation ecosystems in this context. Are these ecosystems only novel once they have been abandoned, no longer receiving human intervention? Exotic pine plantations are novel assemblages that exist because of deliberate human action and require little human intervention once they have been established. Although they are planted for eventual harvesting, decades will pass during which they may provide novel habitat for native species (Quine and Humphrey 2010). Some may never be harvested (Mascaro 2011). Do we therefore consider them to be novel ecosystems? If so, can we say the same of an apple orchard or a canola field, which may also provide novel habitat for pollinators and herbivores, but differ in that they have a much shorter time frame over which they do not require maintenance? Where do we draw the line? Invoking Gleason's allergy to classification, we point out that many ecosystems clearly share characteristics of human design and unmanaged novelty.

What occurs when humans undertake restorative action? Consider the framework proposed by Hobbs et al. (2009) in which restoration is directed toward the 'real' historical state (Fig. 3.2), versus that of Kueffer and Daehler (2009) in which restoration is seen as directed toward a historical *analogue* that is perpetuated only by continued human intervention. Importantly, both views agree that classic restoration efforts are directionally opposed to novelty and that beyond a threshold (i.e. in a novel ecosystem) these classic efforts are akin to trying to put the toothpaste back in the tube (see Chapter 3). The views also agree that the restorative action itself constitutes human agency (e.g. by deliberately altering abiotic and biotic attributes). In some instances the level of human agency may be relatively low, particularly when historical ecosystems are well defined and the target ecosystem is not very distant from the historical condition, but there remains a measurable effort needed.

In many systems however, truly recreating a historical analogue (if such a thing can even be defined) is often practically impossible (e.g. riparian ecosystems). At best it would require expensive long-term interventions, very often including engineering (Richardson et al. 2007). Restorations based on this proposition might be viewed as directing ecosystem management toward a fake or inauthentic ecosystem if it is clear that the thing being 'restored' is no more than a list of species that happened to prevail historically on that site (Callicott 2002). Management of such systems for exhibition of the historical parameters may be seen as 'fossilizing' an ecosystem in which the manager is essentially married to an infinite resource commitment, without which the novelty of the system would return. This view suggests that you cannot restore an ecosystem in the same way you can restore a painting (Gunn 1991), and recognizes that self-organization in novel ecosystems may be better tuned to future environmental conditions (Williams and Jackson 2007). Practical considerations are further addressed in other chapters (Chapters 3 and 18), including novel man-

agement strategies that approach novel ecosystems with the intent to improve biodiversity, ecosystem function and human welfare, but relax or abandon the notion of returning them to a historical condition (e.g. Hughes et al. 2007; Hobbs et al. 2009).

5.3.4 A novel ecosystems framework

We have argued that the novel ecosystems concept is grounded first in Gleason's individualistic concept that species respond differently when faced with environmental change; ecosystems are therefore not the discrete units they may appear to be. Following this, we devised a *continuous* space in which to ordinate novel ecosystems versus other ecosystems by quantifying the essential components of *novelty* and *human agency* (Fig. 5.4).

In considering novelty, we synthesize two previous iterations: (1) a single axis in which novelty is defined as the abiotic or biotic distance from a historical condition (Fig. 5.4); and (2) two axes, abiotic and biotic, where novel ecosystems occupy the space having high abiotic or biotic distance (or both) from historical ecosystems (Fig. 3.2). Given the second foundational concept behind novel ecosystems (that the biotic and abiotic properties of ecosystems are tethered), we note that the ordination of abiotic *and* biotic axes is not strictly necessary for the purposes of *defining* novel ecosystems (although this retains importance in outlining management approaches and predicting ecosystem function; see Chapter 3). This is implicit in Figure 3.2 (from Hobbs et al. 2009), which shows that novel ecosystems can result from abiotic or biotic novelty or any combination thereof (i.e. they occupy a space equidistant from the origin).

For human agency, we synthesize a gradient of human influence presented by Hobbs et al. 2006 (Fig. 5.2). We follow Kueffer and Daehler (2009) in placing this human agency axis orthogonal to compositional novelty. Here, we specify agency that is 'human design' (deliberate, onsite human activity) where we mean any action by humans to *prescribe* the abiotic or biotic properties of ecosystems. Novelty can therefore occur through human design by way of deliberate, onsite human agency (e.g. land use for agriculture followed by release), inadvertently and away from human design (e.g. the spreading of an introduced species, the uptake of anthropogenic nitrogen) or various combinations of these.

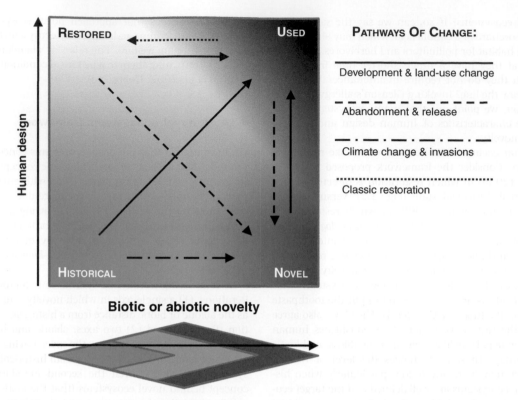

Figure 5.4 The novel ecosystems framework emerges by intersecting quantitative axes for *novelty* and *human design*. As outlined by Hobbs et al. (2009), novelty is the distance in biotic or abiotic properties from the historical state (i.e. note the *x*-axis placement overtop of the Hobbs et al. framework, see Fig. 3.2). Following Kueffer and Daehler (2009), human design represents the onsite, human prescription of abiotic and biotic properties that characterize human-used ecosystems (e.g. lands managed for production) and classic restorations (i.e. to resemble historical composition). The four corners represent absolute states that are rarely (if ever) observed, with the vast majority of real ecosystems contained within. Vectors indicate the 'motion' of a given ecosystem across the state space, and represent various drivers that cause ecosystem change over time. For example, climate change will cause ecosystems to gradually change their abiotic or biotic components, becoming more novel even though human design is absent. By contrast, a corn field is a human-used system which, upon release, may be colonized by new species and become a novel ecosystem. Additional specific examples are shown in Figure 5.5 and described in the text. Reproduced from Hobbs et al. (2009) and Kueffer and Daehler (2009).

The intersection of these two axes mandates four states that are analogous to 'absolute zero'; they are measurable, knowable conditions even though they may be rarely (if ever) observed. Most obviously, historical ecosystems are strictly those never altered by human agency. At the highest level of novelty, we find human-used ecosystems with prescribed abiotic and biotic characteristics that differ from the historical state (e.g. agriculture) as well as novel ecosystems sharing no compositional elements with the historical state, but without onsite human agency to influence their processes.

At the highest levels of human design, both classic restorations and human-used ecosystems share a high level of human intent to control their properties and processes. In the case of agriculture this control is intended to provide for basic human subsistence, while in classic restorations this control is intended to emulate the historical abiotic or biotic composition of an ecosystem. As examples here and in other chapters demonstrate, this level of control reduces the extent to which natural processes may operate (e.g. species sorting, natural selection).

Virtually all real ecosystems will be contained within the state space, often exhibiting elements of multiple ecosystem types. Where the level of compositional novelty and human design can be quantified (e.g. % of introduced species, persons per square mile), an ecosystem's change over time will be depicted by its migration across the space (as a vector). Thresholds within this framework would be annotated along the x axis of novelty, although they would differ according to context. A threshold would therefore keep a given system in a novel state despite attempts to drive its abiotic or biotic characteristics in the direction of a historical ecosystem (negative x direction).

5.3.5 Examples

Lugo (2004) detailed the emergence of novel forests following agricultural abandonment on Puerto Rico (Fig. 5.5a).

1. *Historical tropical moist forest.* Observed only by the first colonists of Puerto Rico, the historical ecosystem includes only indigenous species, many of them endemic to Puerto Rico. Remnants can be seen today within the Luquillio National Forest, although even these forests are no longer strictly historical.

2. *Agricultural development.* Banana, sugar cane and other crops usurp the historical ecosystem, representing the prescription by humans of species composition, ecosystem structure and – to the extent that we can control it – ecosystem function.

3. *Abandonment and colonization by introduced grasses.* A gradual human migration to urban centers results in the abandonment of agricultural fields, releasing these fields from former human prescription of their biotic and abiotic properties. The legacy of human influence remains in the form of altered soil nutrients (Silver et al. 2000) as well as matrix landscapes with agricultural and urban land uses (Lugo and Helmer 2004), and these vacant lands are colonized rapidly by wind-dispersed light-demanding grass species. This initial colonization represents the first appearance of a distinctly novel ecosystem on the site. The composition no longer has any resemblance to the historical tropical moist forest (where grass species were absent).

4. *Succession to introduced trees.* The physical conditions that pervade the grasslands are hostile to the indigenous tree flora but non-native tree species readily colonize and thrive in the open sites, exhibiting novel growth strategies not found in the native flora (e.g. flood and fire tolerance). The sites lack native species and remain compositionally novel, but lose structural novelty as they become forested.

5. *Colonization by native trees in the understory.* The novel forests partially revert in species composition as they are colonized by native trees in the understory. Lugo (2004) observed that after approximately 40 years, basal area is roughly half native and half introduced. Endemic tree species can also be found. Structurally, these forests greatly resemble the historical ecosystem.

Weiss (1999) reviewed the outcome of nitrogen enrichment in serpentine grasslands in California (Fig. 5.5b).

1. *Historical serpentine grasslands.* These grasslands support more than 100 grass and forb species, having low soil N and Ca availability, as well as high heavy metal concentrations. Relic versions remain at the Jasper Ridge Biological Reserve.

2. *Anthropogenic N deposition.* Predominately dry N deposition increases N content by as much as 10–30 times historical levels.

3. *Non-native grass colonization.* Introduced grass species (including *Bromus* spp.) thrive under the high N levels and increase in abundance at the expense of native vegetation. In many areas these grass species can attain 100% cover, forming a distinctly novel ecosystem.

4. *Novel management.* Grazing by cattle considerably reduces introduced grass species abundances, because cattle preferentially select grasses over forbs. This limited intervention by humans amounts to a minor prescription of species composition (i.e. cattle introduction) that results in a substantial recovery of historical grass and forb composition, even while N levels may remain high.

Other chapters in this volume expand on these examples and consider more, but even these simple examples highlight the complexities of novel ecosystem emergence. In the example of Puerto Rico, it is human land use that is the originating force for novel ecosystem emergence (Fig. 5.5a, position 2). This is a deliberate, onsite human action, after which the novel ecosystem is an inadvertent result of abandonment (Fig. 5.5a, position 3). Here there is a distinct human design pathway, but in California the opposite is the case. Human-caused N deposition is an inadvertent consequence of heavy industry, with offsite effects. Subsequent intervention (Fig. 5.5b, position 4) is a

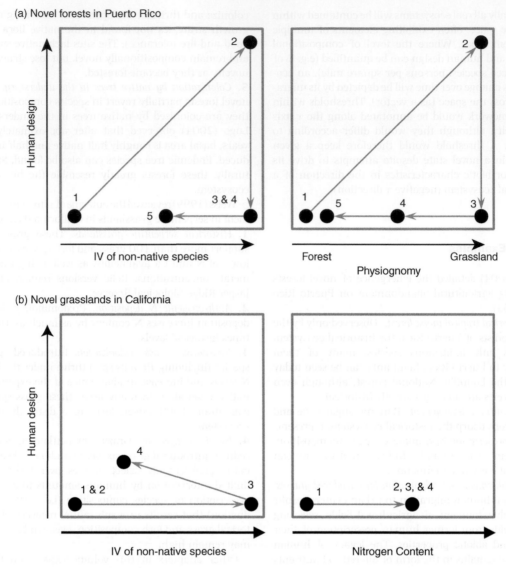

Figure 5.5 Conceptual examples of the emergence of novel ecosystems in (a) Puerto Rico and (b) California. In each case, the left panels highlight changes in biotic properties from historical states (IV is importance value, an index of dominance and density), the right panels highlight an abiotic property (i.e. physical structure in the case of Puerto Rico and N content in the case of California) and the numbers represent different states in time for the same ecosystem in space, as outlined in the text.

design element – albeit a low-impact element of grazing – that creates a system resembling, but not the same as, that which prevailed historically. The conceptual framework (Fig. 5.4) suggests that a classic restoration which would disallow cattle (a non-native species) would require a greater human design effort.

The tether between abiotic and biotic factors is also revealed. In Puerto Rico, introduced grass species result in a high level of structural novelty, forming a grassland where one did not exist (Fig. 5.5a, position 3). Further, while the penultimate novel forest maintains a high level of compositional novelty, it loses structural novelty as it is colonized by introduced tree species and becomes a tropical forest again (Fig. 5.5a, position 4).

5.4 CONCLUSIONS

Earth's communities and ecosystems have disassembled and emerged anew in the wake of environmental upheavals throughout the planet's history. In the past, these changes were driven by asteroids, volcanoes, glaciers and the rise and fall of oceans and mountains. Today, humans are driving extensive and pervasive change, and novel ecosystems are the unambiguous response.

The debate elucidated in this chapter highlights the perils of a one-sentence definition for novel ecosystems. Indeed, we have shown that even as the formal concept of a novel ecosystem has developed over time (Milton 2003; Lugo 2004; Hobbs et al. 2006; Hobbs et al. 2009; Kueffer and Daehler 2009), a precise declaration has not been forthcoming. On theoretical grounds (vis-à-vis Gleason), we advocate strongly for a continuous understanding (Fig. 5.4). Nevertheless, it is important that we appreciate that we are living in a changing world and that societal restoration norms of systems with historical species composition may not be a suitable or even possible future intervention target. In these landscapes, the existence of novel ecosystems must be recognized and their future management debated. Here, a categorical definition can have considerable utility.

Based on the framework proposed, for a given space on the Earth's surface, a *novel ecosystem* is not: (1) a system that would have occupied that space in the past (i.e. part of a historical range of variability); (2) managed intensively for specific production or built over; or (3) managed with the purpose of reproducing the historical ecosystem (i.e. classic restoration).

A novel ecosystem is one of abiotic, biotic and social components (and their interactions) that, by virtue of human influence, differ from those that prevailed historically, having a tendency to self-organize and manifest novel qualities without intensive human management. Novel ecosystems are distinguished from hybrid ecosystems by practical limitations (a combination of ecological, environmental and social thresholds) on the recovery of historical qualities.

This definition is considered further in the following chapter, and we reflect on how it adds to the ecological, environmental and management lexicon moving forward.

REFERENCES

Asner, G.P., Hughes, R.F., Vitousek, P.M., Knapp, D.E., Kennedy-Bowdoin, T., Boardman, J., Martin, R.E., Eastwood, M. and Green, R.O. (2008) Invasive plants transform the three-dimensional structure of rain forests. *Proceedings of the National Academy of Sciences of the United States of America*, **105**, 4519–4523.

Behrensmeyer, A.K., Damuth, J.D., Dimichele, W.A., Potts, R., Sues, H.-D. and Wing, S.L. (1992) *Terrestrial Ecosystems Through Time: Evolutionary Paleoecology of Terrestrial Plants and Animals*. University of Chicago Press, Chicago, IL.

Callicott, J.B. (2002) Choosing appropriate temporal and spatial scales for ecological restoration. *Journal of Bioscience*, **27**, 409–420.

Chapin, F.S. III and Starfield, A.M. (1997) Time lags and novel ecosystems in response to transient climatic change in artic Alaska. *Climatic Change*, **35**, 449–461.

Clark, B.R., Hartley, S.E., Suding, K.N. and De Mazancourt, C. (2005) The effect of recycling on plant competitive hierarchies. *American Naturalist*, **165**, 609–622.

Clark, D.A., Piper, S.C., Keeling, C.D. and Clark, D.B. (2003) Tropical rain forest tree growth and atmospheric carbon dynamics linked to interannual temperature variation during 1984–2000. *Proceedings of the National Academy of Sciences of the United States of America*, **100**, 5852–5857.

Clements, F.E. (1916) *Plant Succession: An Analysis of the Development of Vegetation*. Carnegie Institute of Washington, Washington, D.C.

Convey, P. and Smith, R.I.L. (2006) Responses of terrestrial Antarctic ecosystems to climate change. *Plant Ecology*, **182**, 1–10.

Cordes, E.E., Bergquist, D.C., Shea, K. and Fisher, C.R. (2003) Hydrogen sulphide demand of long-lived vestimentiferan tube worm aggregations modifies the chemical environment at deep-sea hydrocarbon seeps. *Ecology Letters*, **6**, 212–219.

Davis, M.B. (1981) Quaternary history and the stability of forest communities, in *Forest Succession: Concepts and Applications* (eds D.C. West, H.H. Shugart and D.B. Botkin), Springer-Verlag, New York, NY, USA, 134–153.

Davis, M.B. (1983) Quaternary history of deciduous forests of Eastern North America and Europe. *Annals of the Missouri Botanical Garden*, **70**, 550–563.

Delcourt, H.R. (2002) *Forests in Peril: Tracking Deciduous Trees from Ice-Age Refuges into the Greenhouse World*. The McDonald and Woodward Publishing Company, Blacksburg, VA.

Ehrenfeld, J.G. (2003) Effects of exotic plant invasions on soil nutrient cycling processes. *Ecosystems*, **6**, 503–523.

Ellis, E.C., Goldewijk, K.K., Siebert, S., Lightman, D. and Ramankutty, N. (2010) Anthropogenic transformation of the biomes, 1700 to 2000. *Global Ecology and Biogeography*, **19**, 586–606.

Gleason, H.A. (1926) The individualistic concept of the plant association. *Bulletin of the Torrey Botanical Club*, **53**, 7–27.

Gunn, A.S. (1991) The restoration of species and natural environments. *Environmental Ethics*, **13**, 291–310.

Hobbs, R.J., Arico, S., Aronson, J., Brown, J.S., Bridgewater, P., Cramer, V.A., Epstein, P.R., Ewel, J.J., Klink, C.A., Lugo, A.E., Norton, D., Ojima, D., Richadson, D.M., Sanderson, E.W., Valladares, F., Vila, M., Zamora, R. and Zobel, M. (2006) Novel ecosystems: theoretical and management aspects of the new ecological world order. *Global Ecology & Biogeography*, **15**, 1–7.

Hobbs, R.J., Higgs, E. and Harris, J.A. (2009) Novel ecosystems: implications for conservation and restoration. *Trends in Ecology & Evolution*, **24**, 599–605.

Hoffmann, W.A. and Jackson, R.B. (2000) Vegetation-climate feedbacks in the conversion of tropical savanna to grassland. *Journal of Climate*, **13**, 1593–1602.

Hughes, T.R., Rodriques, M.J., Bellwood, D.R., Ceccarelli, D., Hoegh-Guldberg, O., Mccook, L., Moltschaniwskyj, N., Pratchett, M.S., Steneck, R.S. and Willis, B. (2007) Phase shifts, herbivory, and the resilience of coral reefs to climate change. *Current Biology*, **17**, 360–365.

Jackson, S.T. (2006) Vegetation, environment, and time: the origination and termination of ecosystems. *Journal of Vegetation Science*, **17**, 549–557.

Janzen, D.H. (1985) On ecological fitting. *Oikos*, **45**(3), 308–310.

Jenny, H. (1941) *Factors of Soil Formation: A System of Quantitative Pedology*. McGraw-Hill, New York, NY, USA.

Jenny, H. (1980) *The Soil Resource: Origin and Behavior*. Springer-Verlag, New York, NY, USA.

Kueffer, C. and Daehler, C.C. (2009) A habitat-classification framework and typology for understanding, valuing, and managing invasive species impacts, in *Management of Invasive Weeds* (ed. Inderjit), Springer Science and Business Media BV, 77–101.

Lugo, A.E. (2004) The outcome of alien tree invasions in Puerto Rico. *Frontiers in Ecology and the Environment*, **2**, 265–273.

Lugo, A.E. and Helmer, E. (2004) Emerging forests on abandoned land: Puerto Rico's new forests. *Forest Ecology and Management*, **190**, 145–161.

MacDougall, A.S. and Turkington, R. (2005) Are invasive species the drivers or passengers of change in degraded ecosystems? *Ecology*, **86**, 42–55.

Mascaro, J. (2011) Eighty years of succession in a non-commercial plantation on Hawai'i Island: are native species returning? *Pacific Science*, **65**, 1–15.

McIntosh, R.P. (1975) H.A. Gleason: 'Individualistic Ecologist' 1882–1975: His contributions to ecological theory. *Bulletin of the Torrey Botanical Club*, **102**, 253–273.

Milton, S.J. (2003) 'Emerging ecosystems': a washing-stone for ecologists, economists and sociologists? *South African Journal of Science*, **99**, 404–406.

Mkoji, G.M., Hofkin, B.V., Kuris, A.M., Stewart-Oaten, A., Mungai, B.N., Kihara, J.H., Mungai, F., Yundu, J., Mbui, J., Rashid, J.R., Kariuki, C.H., Ouma, J.H., Koech, D.K. and Loker, E.S. (1999) Impact of the crayfish *Procambarus clarkii* on *Schistosoma haematobium* transmission in Kenya. *American Journal of Tropical Medicine and Hygine*, **61**, 751–759.

Naeem, S. (2002) Ecosystem consequences of biodiversity loss: The evolution of a paradigm. *Ecology*, **83**, 1537–1552.

Odum, E.P. (1969) The strategy of ecosystem development. *Science*, **164**, 262–270.

Pagani, M., Arthur, M.A. and Freeman, K.H. (1999) Miocene evolution of atmospheric carbon dioxide. *Paleoceanography*, **14**, 273–292.

Pielou, E.C. (1991) *After the Ice Age: The Return of Life to Glaciated North America*. The University of Chicago Press, Chicago, IL, USA.

Quine, C.P. and Humphrey, J.W. (2010) Plantations of exotic tree species in Britain: irrelevant for biodiversity or novel habitat for native species? *Biodiversity and Conservation*, **19**, 1503–1512.

Richardson, D.M., Holmes, P.M., Esler, K.J., Galatowitsch, S.M., Stromber, J.C., Kirkman, S.P., Pysek, P. and Hobbs, R.J. (2007) Riparian vegetation: degradation, alien plant invasions, and restoration projects. *Diversity and Distributions*, **13**, 126–139.

Sanderson, E.W., Jaiteh, M., Levy, M.A., Redfrod, K.H., Wannebo, A.V. and Woolmer, G. (2002) The human footprint and the last of the wild. *BioScience*, **52**, 891–904.

Schnitzer, S.A. and Bongers, F. (2011) Increasing liana abundance and biomass in tropical forests: emerging patterns and putative mechanisms. *Ecology Letters*, **14**, 397–406.

Silver, W.L., Ostertag, R. and Lugo, A.E. (2000) The potential for carbon sequestration through reforestation of abandoned tropical agricultural and pasture lands. *Restoration Ecology*, **8**, 394–407.

Simenstad, C.A., Estes, J.A. and Kenyon, K.W. (1978) Aleuts, sea otters, and alternative stable-state communities. *Science*, **200**, 403–411.

Smart, A.C., Harper, D.M., Malaisse, F., Schmitz, S., Coley, S. and Gouder De Beauregard, A.-C. (2002) Feeding of the exotic Louisiana red swamp crayfish, Procambarus clarkii (Custacea, Decapoda), in an African tropical lake: Lake Naivasha, Kenya. *Hydrobiologia*, **488**, 129–142.

Suding, K.N. and Hobbs, R.J. (2009) Threshold models in restoration and conservation: a developing framework. *Trends in Ecology & Evolution*, **24**, 271–279.

Suding, K.N., Gross, K.L. and Houseman, G.R. (2004) Alternative states and positive feedbacks in restoration ecology. *Trends in Ecology & Evolution*, **19**, 46–53.

Tansley, A.G. (1935) The use and abuse of vegetational concepts and terms. *Ecology*, **16**, 284–307.

Vermeij, G.J. (2005) Invasion as expectation: a historical fact of life, in *Species Invasions: Insights into Ecology, Evolution, and Biogeography* (eds D.F. Sax, J.J. Stachowicz and S.D. Gaines), Sinauer Associates, Inc. Publishers, Sunderland, Massachusetts, 315–340.

Vitousek, P.M., Mooney, H.A., Lubchenco, J. and Melillo, J.M. (1997) Human domination of Earth's ecosystems. *Science*, **277**, 494–499.

Wardle, D.A., Bargett, R.D., Callaway, R.M. and Van Der Putten, W.H. (2011) Terrestrial ecosystem responses to species gains and losses. *Science*, **332**, 1273–1277.

Weiss, S.B. (1999) Cars, cows, and checkerspot butterflies: nitrogen deposition and managment of nutrient-poor grasslands for a threatened species. *Conservation Biology*, **13**, 1476–1486.

Wilkinson, D.M. (2004) The parable of Green Mountain: Ascension Island, ecosystem construction, and ecological fitting. *Journal of Biogeography*, **31**, 1–4.

Williams, J.W. and Jackson, S.T. (2007) Novel climates, no-analog communities, and ecological surprises. *Frontiers in Ecology and the Environment*, **5**, 475–482.

Woodcock, D. (2003) To restore the watersheds: Early twentieth-century tree planting in Hawai'i. *Annals of the Association of American Geographers*, **93**, 624–635.

Ziegler, A.C. (2002) *Hawaiian Natural History and Evolution*. University of Hawai'i Press, Honolulu, HI, USA.

Chapter 6

DEFINING NOVEL ECOSYSTEMS

Richard J. Hobbs[1], Eric S. Higgs[2] and Carol M. Hall[2]
(on behalf of, and with input from, participants of the Pender Island Workshop)

[1]Ecosystem Restoration and Intervention Ecology (ERIE) Research Group, School of Plant Biology, University of Western Australia, Australia

[2]School of Environmental Studies, University of Victoria, Canada

---

A WORKING DEFINITION THAT HAS ENDURED MANY CONVERSATIONS AND REFLECTIONS IS AS FOLLOWS

A novel ecosystem is a system of abiotic, biotic and social components (and their interactions) that, by virtue of human influence, differ from those that prevailed historically, having a tendency to self-organize and manifest novel qualities without intensive human management. Novel ecosystems are distinguished from hybrid ecosystems by practical limitations (a combination of ecological, environmental and social thresholds) on the recovery of historical qualities.

What characterizes a novel ecosystem? Can a novel ecosystem be defined in a way that is both useful and general? These questions created heated debate both at the workshop and during the book-writing process. As workshop leaders, we deliberately steered discussion away from trying to come up with an agreed definition during the workshop in May 2011. Definitions inevitably provoke strong opinions and attachments to particular wordings and emphases. Yet, how it is possible to discuss a topic effectively if you do not first define what you are talking about? Without clear definitions,

terms and concepts may become relatively useless either because there is no clarity around their use or because they can become 'panchrestons' or 'plastic words', or terms may be used so broadly and in so many contexts that they become vague and ambiguous (Poerksen 1995; Lindenmayer and Fischer 2007). Confused sets of ideas, multiple and normative meanings and academic discussions couched in highly abstract language can result. On the other hand, Reiners and Lockwood (2009) argue that ambiguity can spawn 'creative pluralism' and attempts to enforce

---

rigid definitions compromise the utility of a concept. Sometimes ecologists get hung up on definitions when they should be asking if an idea, however used, has proven or may prove useful in practice or alternatively limits understanding.

As with many emerging concepts, it is perhaps easier to indicate what novel ecosystems are not: (1) a system that would have occupied that space in the past (i.e. part of a historical range of variability); (2) managed intensively for specific production or built over; or (3) managed with the purpose of reproducing the historical ecosystem (i.e. classic restoration).

The earlier definitions of what is and what is not a novel ecosystem were viewed and agreed on by many, but not all, of the participants. This definition captures the essence of what we consider to be important factors characterizing novel ecosystems. We recognize also, however, that the term is in the public domain and being used in multiple ways already; any attempt to corral this process is likely to have at best limited success. However, the definition is ultimately our attempt at a working characterization.

Along the way towards this definition, various aspects of novel ecosystems were hotly debated. For instance, is novelty a continuum or are there clear breakpoints where it is clear that one system is novel and another is not? Several participants strongly suggested that novelty occurred along a continuum, while others posited that being able to classify systems as novel or not novel had clear management utility that was lacking in a continuum approach. Central to this discussion was the question of whether clear thresholds exist in the state space illustrated by Figure 3.2, as put forward by Hobbs et al. (2009). The idea of threshold dynamics has proven very useful in considering ecosystem management choices and techniques (e.g. Hobbs & Harris 2001; Suding & Hobbs 2009); including thresholds in considerations of novel ecosystems therefore seemed appropriate. This line of thought was carried through the book into chapters dealing with management interventions (e.g. Chapters 3, 4, 5, 18 and 24). Nevertheless, valid concerns remain about the actual ability to observe, measure and diagnose ecological thresholds and hence to make decisions effectively using the schemes outlined in Chapters 3, 18 and 24. This issue is not restricted to discussions on novel ecosystems, but is a live topic of urgent debate and research in applied ecology in general (Andersen et al. 2008; Samhouri et al. 2010; Bestelmeyer et al. 2011). Perhaps novel ecosystems can display a dual

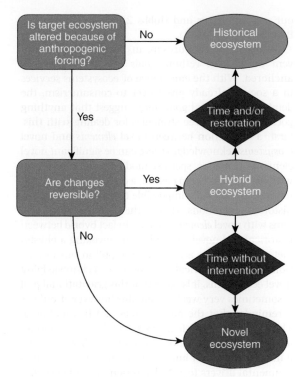

Figure 6.1 A simplified view of the definition that illustrates the relationship between historical, hybrid and novel ecosystems. Many people contributed to the development of this figure and the ideas supporting it, especially Jim Harris, Joe Mascaro, Steve Murphy and Cara Nelson.

character in the same way as physics lives with the dual notion of light as both a wave and a particle.

Figure 6.1 presents a simplified view of the definition by illustrating the relationship between historical, hybrid and novel ecosystems, based on the degree of change from historical conditions and reversibility of that change. These ideas are expanded upon and pursued in more detail in subsequent chapters (see Chapter 18).

Why is it important to sort this out? Identifying an ecosystem as novel signals that management is restricted to goals associated with novelty, and that recourse to hybrid or historical ecosystems is no longer practical. In doing so, we leave behind traditional goals that underpin conservation and restoration, notably connection to historically rooted ideals. These goals embed deeply held values that inform and explain our actions (Higgs and Hobbs 2010; Thompson and Bendik-Keymer 2012).

Accepting new natures means that we need to reinvent or at least rethink goals and how these are anchored. With the emergence of ecosystems services in a society already given over to consumerism, the demise of historical goals may suggest that anything goes. There are three strategies for dealing with this. First, a distinction between novel *elements* and novel *ecosystems* acknowledges there can be significant novel elements (invasive species, modified soil conditions) without the ecosystem passing a critical threshold that renders it practically impossible to return to hybrid or historical conditions. Hence, there are many ecosystems with novel *elements* that are in fact hybrid between historical and novel. Second, again invoking a physics metaphor, there is a gravitational pull in our discussions toward historical conditions. In acknowledging novel ecosystems, it is plain that this gravitational pull is sometimes very weak; it remains however, if only as a reminder that the past matters and has mattered. Third, the definition features social drivers of novelty alongside ecological and environmental drivers. While it may be practically impossible to reverse certain environmental drivers (e.g. N deposition) in some cases, it is more likely that social drivers can be addressed. Not that these are easy in the case of deeply embedded economic constraints, political intransigence or cultural values. Thresholds matter, but they should not be trotted out as an excuse to avoid resolute action.

REFERENCES

Andersen, T., Carstensen, J., Hernandez-Garcia, E. and Duarte, C.M. (2008) Ecological thresholds and regime shifts: approaches to identification. *Trends in Ecology and Evolution*, **24**, 49–57.

Bestelmeyer, B.T., Goolsby, D.P. and Archer, S.R. (2011) Spatial perspectives in state-and-transition models: a missing link to land management? *Journal of Applied Ecology*, **48**, 746–757.

Higgs, E.S. and Hobbs, R.J. (2010) Wild design: Principles to guide interventions in protected areas, in *Beyond Naturalness: Rethinking Park and Wilderness Stewardship in an Era of Rapid Change* (eds D. Cole and L. Yung), Island Press, Washington, DC, 234–251.

Hobbs, R.J. and Harris, J.A. (2001) Restoration ecology: Repairing the Earth's ecosystems in the new millennium. *Restoration Ecology* **9**, 239–246.

Hobbs, R.J., Higgs, E. and Harris, J.A. (2009) Novel ecosystems: implications for conservation and restoration. *Trends in Ecology and Evolution*, **24**, 599–605.

Lindenmayer, D.B. and Fischer, J. (2007) Tackling the habitat fragmentation panchreston. *Trends in Ecology & Evolution*, **22**, 127–132.

Poerksen, U. (1995) *Plastic Words: The Tyranny of a Modular Language*. The Pennsylvania University Press, University Park, PA.

Reiners, W.A. and Lockwood, J.A. (2009) *Philosophical Foundations for the Practices of Ecology*. Cambridge University Press, Cambridge.

Samhouri, J.F., Levin, P.S. and Ainsworth, C.H. (2010) Identifying thresholds for ecosystem-based management. *PLoS ONE*, **5**, e8907.

Suding, K.N. and Hobbs, R.J. (2009) Threshold models in restoration and conservation: A developing framework. *Trends in Ecology and Evolution*, **24**, 271–279.

Thompson, A. and Bendik-Keymer, J. (2012) *Ethical Adaptation to Climate Change: Human Virtues of the Future*. MIT Press, Cambridge, MA.

Part III

What We Know (and Don't Know) about Novel Ecosystems

Chapter 7

PERSPECTIVE: ECOLOGICAL NOVELTY IS NOT NEW

Stephen T. Jackson

Department of Botany and Program in Ecology, University of Wyoming, USA and Southwest Climate Science Center, US Geological Survey, Tucson, Arizona, USA

A group of ecologists touring the world 14 or 15 thousand years ago, while the ice sheets were in rapid retreat, would have seen many communities and ecosystems alien to their modern eyes. In the central Urals, they'd see pied lemmings living together with steppe lemmings and gray hamsters. This wouldn't seem right; today's pied lemmings live in tundra along the Arctic coast of Siberia, while the hamsters and steppe lemmings live in temperate steppe 1000 km to the south (Stafford et al. 1999). In the American Midwest, they'd find open forests of spruce together with oak, ash and other temperate hardwoods. For our ecologists, these forests would defy classification; there are no comparable forests anywhere in North America today (Williams et al. 2001). Our ecologists would be flummoxed repeatedly as they encountered peculiar groupings of plants, insects, mollusks, birds, mammals and other groups from the tropics to the high latitudes (Jackson and Williams 2004).

Quaternary paleoecologists have been puzzling over these enigmatic assemblages since the 1960s. Variously called 'disharmonious' or 'intermingled' fauna and flora (Graham 1986) and 'no-analog' assemblages (Overpeck et al. 1992), they all represent ancient communities and ecosystems that lack precise modern counterparts (Jackson and Williams 2004). They were particularly prevalent during the global transition from glacial to interglacial conditions 20,000 to 10,000 years ago. Underlying causes have been extensively debated, and the search for comprehensive explanation continues. It is probably no coincidence that the spatial and temporal occurrence of no-analog assemblages corresponds to the occurrence of climates that lack modern analogs (Williams et al. 2001; Williams and Jackson 2007). Such 'no-analog climates' represent combinations of climate variables with no modern counterpart; a prominent example is the hyperseasonal climate (warm summers, cold winters) of mid-continental North America during the glacial–interglacial transition (Williams and Jackson 2007). Megafaunal decline and herbivory release may also have played a role in development of no-analog vegetation, at least in some regions (Gill et al. 2009, 2012).

Novel Ecosystems: Intervening in the New Ecological World Order, First Edition. Edited by Richard J. Hobbs, Eric S. Higgs, and Carol M. Hall.
© 2013 John Wiley & Sons, Ltd. Published 2013 by John Wiley & Sons, Ltd.

Reversing the perspective, from looking at the past through a modern lens to considering the future through a lens calibrated to some past period, our ecologists are likely to have a similar alien-world experience. Looking forward from virtually any point in time, future ecological states are likely to be novel in at least some regard. Ecological history is an unfolding narrative of changing environments and changing ecological states. Most terrestrial ecosystems on Earth are relatively young; none seem to have escaped major changes in vegetation composition and structure (and accordingly, function) during the last 12,000 years (Jackson 2006, 2012). These include ecosystems least affected by human activities; climatic changes of the past centuries and millennia have driven changes in vegetation. Although most species have shifted their geographic distributions in response to these climatic changes, ecological communities have not been conserved but instead have gone through successive reconfigurations and recombinations (Jackson and Overpeck 2000). Historians recognize that "the past is a foreign country" (Lowenthal 1985); similarly, the future will always be a foreign country. In the light of the past, there is nothing new about novel ecosystems.

Novel ecosystems are arising today from a variety of causes, as amply discussed in this volume. None of those causes – climatic change, invasions by transformative species, declines of keystone species, alteration of biogeochemical cycles, severe disturbances of land-cover – are new to the planet.

Change, including rapid and disruptive change, is a natural feature of the world we find ourselves in. Figure 7.1 is a conceptual diagram showing how ecological novelty may behave as a function of time. The black arrow indicates the present, which could correspond to any reference point in past or future time. Looking backward from the reference point in a 'smooth world', ecological novelty might increase as a linear or curvilinear function of temporal distance from the observer. This would describe a gradually changing world with steadily accumulating and irreversible change (e.g. monotonically changing environment; organic evolution). Looking forward to the future, this pattern might be mirrored by similarly smooth changes. Novelty would be far in the future, and develop slowly. Under some scenarios, novelty might even cycle to some extent as a function of time; for example, global environments and vegetation during the last glacial maximum were more similar to those

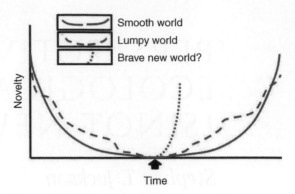

Figure 7.1 Ecological novelty as a function of time: past, present and future. The arrow denotes the location of the observer ("the present").

of previous glacial maxima 100,000 and more years ago than to the succeeding interglacial period, just a few thousand years away (see Jackson and Overpeck 2000, fig. 7).

A 'lumpy world' (Fig. 7.1) comprises alternating periods of gradual and rapid environmental and ecological change. Looking to the past in such a world, there may be 'jumps' in novelty corresponding to periods of rapid change, with slower change occurring during other time intervals. Similar patterns will emerge in the future, but they will not predictably mirror the past changes; all depends on when and at what rates future changes unfold. The near-future will have high novelty if a rapid environmental change is imminent. The 'lumpy world' corresponds most closely to the nature of environmental and ecological change of the past 20,000 to 40,000 years, in which the Earth emerged from glacial–interglacial conditions and underwent other gradual and rapid climate changes.

A new kind of novelty seems to be arising in which human activities are accelerating all of the processes that lead to novel ecosystems. Unique features of this 'new novelty' include widespread extent and rapid increases in magnitude of multiple drivers and, consequentially, new interactions among drivers and between drivers and biota. Looking to the near-future in the 'brave new world' (Fig. 7.1), ecological novelty may arise very rapidly due to unprecedentedly rapid changes in climate, introductions of exotic species, human land use and resource exploitation, alteration of global biogeochemical cycles and other factors.

Ecological novelty (natural or not) may be monstrous, particularly when changes are large, widespread and rapid. Our recognition of novel ecosystems is coincident with our recognition that we inhabit a dynamic world that has been in continual environmental and ecological flux throughout its history. This coincidence poses challenges that cut to the very core of our most deeply held values – and even values systems – as ecologists, conservationists and managers. We must accept the reality of novel ecosystems, but that doesn't necessarily mean we must embrace all novel ecosystems. Instead, we need to clarify our values, order our priorities and understand novel ecosystems and their consequences. Ecological novelty is widely perceived as a threat to conservation and restoration, and indeed it can be. However, it also comprises a reality and, more importantly, an opportunity. If we can understand them better, we can leverage novel ecosystems and other aspects of ecological novelty to advantage in pursuing broad goals in conservation and climate-change adaptation.

REFERENCES

Gill, J.L., Williams, J.W., Jackson, S.T., Lininger, K.B. and Robinson, G.S. (2009) Pleistocene megafaunal collapse, novel plant communities, and enhanced fire regimes in North America. *Science*, **326**, 1100–1103.

Gill, J.L., Williams, J.W., Donnelly, J.P., Jackson, S.T. and Schellinger, G.C. (2012) Climatic and megaherbivore controls on late-glacial vegetation dynamics: a new, high-resolution, multi—proxy record from Silver Lake, Ohio. *Quaternary Science Reviews*, **34**, 66–80.

Graham, R.W. (1986) Response of mammalian communities to environmental changes during the late Quaternary, in *Community Ecology* (eds J. Diamond and T.J. Case), Harper and Row, New York, 300–313.

Jackson, S.T. (2006) Vegetation, environment, and time: the origination and termination of ecosystems. *Journal of Vegetation Science*, **17**, 549–557.

Jackson, S.T. (2012) Conservation and resource management in a changing world: Extending historical range-of-variability beyond the baseline, in *Historical Environmental Variation in Conservation and Natural Resource Management* (eds J.A. Wiens, G.D. Hayward, H.D. Safford and C. Giffen), Wiley-Blackwell, 92–109.

Jackson, S.T. and Overpeck, J.T. (2000) Responses of plant populations and communities to environmental changes of the late Quaternary. *Paleobiology*, **26** (Supplement), 194–220.

Jackson, S.T. and Williams, J.W. (2004) Modern analogs in Quaternary paleoecology: here today, gone yesterday, gone tomorrow? *Annual Review of Earth and Planetary Sciences*, **32**, 495–537.

Lowenthal, D. (1985) *The Past is a Foreign Country*. Cambridge University Press, Cambridge, UK.

Overpeck, J.T., Webb, R.S. and Webb, T. III. (1992) Mapping eastern North American vegetation change of the past 18 ka: no-analogs and the future. *Geology*, **20**, 1071–1074.

Stafford, T.W. Jr, Semken, H.A. Jr, Graham, R.W., Klippel, W.F., Markova, A., Smirnov, N.G. and Southon, J. (1999) First accelerator mass spectrometry ^{14}C dates documenting contemporaneity of nonanalog species in late Pleistocene mammal communities. *Geology*, **27**, 903–906.

Williams, J.W. and Jackson, S.T. (2007) Novel climates, no-analog communities, and ecological surprises. *Frontiers in Ecology and the Environment*, **5**, 475–482.

Williams, J.W., Shuman, B.N., Webb, T. III. (2001) Dissimilarity analyses of late-Quaternary vegetation and climate in eastern North America. *Ecology*, **82**, 3346–3362.

Chapter 8

THE EXTENT OF NOVEL ECOSYSTEMS: LONG IN TIME AND BROAD IN SPACE

Michael P. Perring[1] and Erle C. Ellis[2]

[1]Ecosystem Restoration and Intervention Ecology (ERIE) Research Group, School of Plant Biology, University of Western Australia, Australia

[2]Geography & Environmental Systems, University of Maryland, USA

8.1 INTRODUCTION

Novel ecosystems are not new. Recognition of their extent and significance, particularly in conservation and restoration circles, has been slow however and their place in conservation management engenders debate (Hobbs et al. 2009). In this chapter, we use spatially explicit historical models to investigate how novel ecosystems have spread across the terrestrial biosphere in the last 8000 years, and we also show their likely current distribution in both the terrestrial and marine realms. Our aim here is to demonstrate that novel ecosystems are now widespread, can be ancient and deserve greater consideration in our efforts to manage and conserve the biosphere for humanity and other species over the long term. We discuss how archeological and paleoecological approaches might confirm the patterns of spread we have modeled, as well as detailing future areas of research. In particular, we highlight the difficulties associated with inferring novel ecosystem extent and in distinguishing between hybrid and novel ecosystems from currently available data.

We consider the application of alternative land classification systems, and also discuss the tension that arises when applying the novel ecosystem concept (see Chapters 5 and 6) to spatially explicit models of the biosphere. Our message remains clear however: human-caused environmental change is here to stay and will continue to affect the world's ecosystems by rearranging biota and altering abiotic conditions. If humanity intends to prosper over the long term while retaining the evolutionary heritage of our planet, ecologists need to understand where novel ecosystems have come from and how to manage these fundamentally altered and dynamic ecosystems by applying coupled socio-ecological theory to real-world practice.

8.2 MAPPING THE SPATIAL EXTENT AND TEMPORAL TRAJECTORY OF NOVEL ECOSYSTEMS

Chapter 5 outlined how novel ecosystems are formed when intensively and productively used land or sea is abandoned, or 'wild' land or sea is altered by human activities (Hobbs et al. 2006). Human actions can be intentional or accidental, and their influence on ecosystems can be observed onsite or offsite. To delimit novel ecosystems the ecological changes caused by human action must represent the formation of novel persistent states, representing the crossing of critical thresholds away from their state as undisturbed by human activity (see also Chapters 5 and 6 and Hobbs et al. 2009). The recognition of these critical thresholds can aid managers in their decisions as to how to manage these systems (see Chapters 3 and 18).

However, mapping such thresholds and therefore the extent of novel ecosystems at a global scale is more than difficult given the data that are currently available. Mapping is difficult even at local and regional scales in many regions, particularly as 'irreversible thresholds' refers to economic and social as well as ecological barriers (see Chapters 3 and 18).

Here we instead investigate the use of proxy variables in an attempt to delimit the current extent of novel ecosystems both on land and at sea. From these proxies, such as the spatial extent of human use of land, novel ecosystem extent may be inferred and estimated; the quality of these proxy estimates which (on land at least) do not incorporate data on biotic change demands careful consideration however, as we discuss here. We utilize a map of current human-induced ecological effects on the marine sphere to suggest the global extent of novel marine ecosystem development. We then use spatially explicit historical models of global population and land use to infer novel ecosystem extent on land, both now and over the past 8000 years. Finally, we integrate these estimates to outline how to create a synthetic overview of global novel ecosystem extent and suggest other lines of evidence that could corroborate or revise the patterns we have elucidated here.

8.3 MARINE NOVEL ECOSYSTEMS

Characterizing the spatial extent and temporal trajectory of novel ecosystems in the marine realm is not easily tractable. There is no permanent human population density within marine areas, nor are there specific types of land use except for some areas designated as marine protected zones or sites of fish farms; utilizing population and land-use proxies to delimit the extent of marine novel ecosystems is therefore not a viable option. However, human use of the marine environment through fishing or through the transfer of goods has undoubtedly created novel marine ecosystems directly. Species transfer, particularly of microscopic organisms including viruses and cholera bacteria (Ruiz et al. 2000), around the world has been aided by the release of ballast water (Carlton and Geller 1993). Invasions by these and macroscopic organisms have potentially large ecological and evolutionary consequences (Grosholz 2002). Demand for particular fish species has led to declining 'mean trophic levels' (i.e. fewer top-level predators and increased catches of fish species lower down the food web which, by definition, have a lower trophic position than the predators) in the world's oceans (Pauly et al. 1998; Jennings et al. 2001, 2002) and consequent community changes (Myers and Worm 2003). These changes threaten the delivery of ocean ecosystem services (Worm et al. 2006), although the magnitude of the problem is sometimes debated (Holker et al. 2007; Wilberg and Miller 2007).

In addition to direct onsite influences on marine systems, offsite influences have also created novel marine ecosystems. Coral reef systems appear to be particularly susceptible to human-induced disruption with community change observed due to ocean temperature rise, pollution of surface water from agricultural and industrial run-off (Fabricius 2005) and disease (Chapter 12). *In extremis*, high-temperature events potentially lead to massive coral bleaching depending on the susceptibility of the reef due to other factors such as species composition and the light environment (Fitt et al. 2001; Smith et al. 2005). One of the most noticeable effects of terrestrial human land-use intensification has been the creation of algal blooms through non-point source pollution with the subsequent formation of hypoxia in susceptible ocean basins. Nitrogen and phosphorus run-off have increased dramatically since the 1950s; humanity has quadrupled the environmental flow of phosphorus (Elser and Bennett 2011) and nitrogen output from rivers had increased 20-fold as much as 15 years ago (Howarth et al. 1996). Hypoxia formed from this nutrient input leads to community change and the creation of novel interactions between tolerant species, typified by the annual formation of the Gulf of Mexico 'dead zone' due to fertilizer and pesticide discharge from the Mississippi and Atchafalaya Rivers (Rabalais et al. 2002).

One globally pervasive human-caused change has been ocean acidification precipitated by rising atmospheric CO_2 concentration. This alters ocean ecosystems (Beaufort et al. 2011) but its effect together with temperature rise on biodiverse coral reefs has been particularly highlighted (Hoegh-Guldberg et al. 2007; Pandolfi et al. 2011). Irreversible reef decline has been mooted as a logical endpoint of the changes humanity is witnessing, due to an inability to form carbonate with decreasing aragonite saturation state (Hoegh-Guldberg et al. 2007). Instead, increasingly novel ecosystems will form in the reef's place. Recent work suggests that more nuanced changes may occur however, with some reef species showing the potential to adapt, and carbonate formation actually increasing despite the change in saturation state. The current rapidly changing biosphere does not however appear to have a precursor in the fossil record that otherwise shows coral reefs persisting in a high CO_2 concentration atmosphere (Pandolfi et al. 2011).

Whatever the exact trajectories of change, novel marine ecosystems around the globe have likely been created as shown by the extent of ecological change and geographic spread of the examples. However, can we map the temporal trajectory and spatial extent of these systems across the world's marine realm? We do not have access to data that can describe the temporal spread of marine novel ecosystems. Recent work by Halpern et al. (2008b) has however revealed the current extent of ecological change in the marine realm, as wrought by human influence. Halpern et al. (2008b) combined 17 anthropogenic drivers of ecological change (including commercial fishing, oil rigs, shipping lanes, invasive species, land-based impact from non-point source pollution, sea-surface temperature change, UV change and ocean acidification) that they could map globally from the last decade, with the known distribution of 14 ecosystem types and the modeled distribution of a further 6 ecosystem types. They then arrived at an intensity of ecological impact (Halpern et al. 2008b; Fig. 8.1), which may potentially be used as a proxy for the potential extent of novel ecosystems in marine systems: the greater the ecological impact, the more likely there has been change that has moved systems beyond currently irreversible thresholds. This allows us to make a first estimate of novel ecosystem extent in ocean ecosystems.

Figure 8.1 shows that no ocean is free of human influence and consequent ecological change. Large areas of little ecological impact remain however, particularly near the poles (although note Blight and Ainley 2008). Furthermore, there are particular areas that are heavily influenced by human activity which may delimit novel ecosystem extent if we assume that the greater ecological change that this human activity has wrought corresponds to the presence of novel ecosystems. These novel systems are concentrated around coastlines, which is unsurprising given the conjunction of high population densities leading to direct use and the high input of inorganic and organic pollutants from draining watersheds and geographic gradients over short distances, leading to multiple ecosystem types in a single square. Based on degree of ecological change, the areas of greatest novel ecosystem extent include the North and Norwegian (Baltic) Seas, the South and East China Seas, the Eastern Caribbean, the Mediterranean, the North American eastern seaboard, the Persian Gulf, the Bering Sea and the waters around Sri Lanka. Interestingly, it is the combination of drivers of ecological change by ecosystem types that reveals the pattern of novel ecosystem extent; examining the multiple drivers in isolation, and without taking into account the degree of ecological change they cause, leads to a different intensity pattern that may not be so reflective of ecological change (Halpern et al. 2008b). Taking an average ecological impact score, rather than summing together all the ecosystem types (as shown in Fig. 8.1) does not change the pattern markedly (Halpern et al. 2008b).

Although some have criticized aspects of the approach (Blight and Ainley 2008; Heath 2008), such criticisms are unlikely to change the patterns shown substantially and, if anything, suggest that the analysis downplays the extent of ecological change (Halpern et al. 2008a; Selkoe et al. 2008). This is particularly the case since inclusion of other drivers, currently not available at the global scale, will likely exacerbate the patterns shown. Omitted drivers include recreational, illegal, unregulated, unreported and historical fishing, aquaculture, disease, coastal engineering, point source pollution, atmospheric pollution and tourism (Halpern et al. 2008b). Although we cannot at this time show a trajectory of ecological change and novel ecosystem formation in the marine sphere, we can observe that currently there has been (sometimes large) ecological change wrought upon essentially the whole ocean ecosystem by multiple drivers. The potential of these

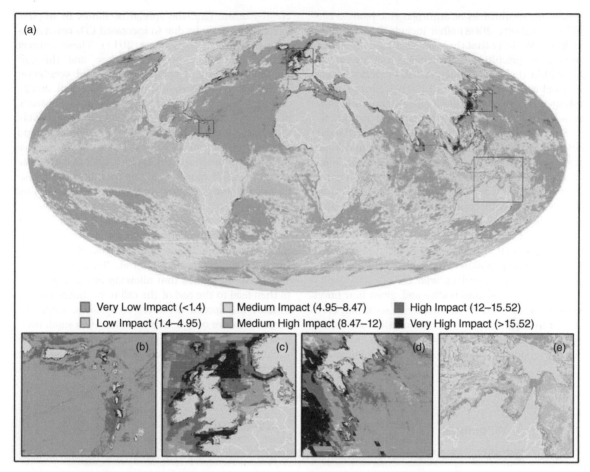

Figure 8.1 (a) Ecological impact of anthropogenic drivers on marine ecosystems (from Halpern et al. 2008b). Assuming greater ecological impact from on- and offsite anthropogenic drivers corresponds to a greater likelihood of novel ecosystem presence, this figure details the potential current spatial extent of marine novel ecosystems. More detailed views of highly affected areas: (b) Caribbean; (c) North Sea; and (d) China Sea and of a relatively unaffected area (e) north of Australia. See text and Halpern et al. (2008b) for further details. From Halpern et al. (2008b). Reprinted with permission from American Association for Advancement of Science.

changed systems to maintain themselves is unknown, but the impracticality of 'going back' (see Chapter 5) suggests that they are potentially novel ecosystems.

8.4 TERRESTRIAL NOVEL ECOSYSTEMS

8.4.1 Methods

We use available spatially explicit population and land-use models for the last 8000 years to characterize the spread of novel ecosystems across the terrestrial biosphere (Ellis et al. 2010; Ellis 2011). To accomplish this, Earth's ice-free land surface is stratified into grid cells of equal latitude and longitude (5 arc minutes) within which human populations and areas of land used for agriculture (crops, pastures) and urban settlements are estimated using two spatially explicit historical models (HYDE 3.1: Klein Goldewijk et al. 2010, 2011 and KK10: Kaplan et al. 2011) and compared with a base map of potential vegetation (Ramankutty and Foley 1999). Grid cells that include human populations and/or use of land for agriculture or settlements (anthromes or anthropogenic biomes; Ellis and Ramankutty 2008) often include areas without such use. We infer that these areas of land remaining unused for agriculture or settlements that are embedded within these anthrome grid cells represent areas of novel ecosystem (Ellis et al. 2010). Unused areas within anthromes typically include remnant ecosystems left unused due to their unsuitability for agriculture or infrastructure (e.g. steep slopes, wetlands), infrequently used areas such as woodlots and lands abandoned after use. In these cases, unused lands usually retain elements of the native biota. They are less disturbed by human activity than lands used directly, and may therefore be considered to constitute novel ecosystems (Ellis et al. 2010; Fig. 8.2). In each grid cell we therefore delimit novel ecosystem extent as 1 minus proportion used; where proportion used is 0 (i.e. there is no population and/or no direct use of the land), we classify the cell as 'wild'.

Our approach is conservative and, given how novel ecosystems are defined (see Chapter 6), some may even say controversial since we do not consider an important additional aspect of novel ecosystem formation: the conversion of 'wild' land by human influences

created offsite (e.g. the spread of lianas in tropical forest may partly be due to increased CO_2 concentration; Schnitzer and Bongers 2011). These indirect changes are likely pervasive before and through the Holocene (e.g. trophic cascades and vegetation change caused by Paleolithic hunters; Rule et al. 2012), although such effects could also occur in the absence of human influence given analogues through glacial–interglacial cycles (Ellis 2011). These indirect and offsite drivers of novel ecosystem formation deserve further consideration in mapping and estimating the extent of novel ecosystems. Consideration could also be given to the condition that, when cells containing 'used' land are completely abandoned, these may be reclassified as 'wild' in later time periods of our current scheme, thereby underestimating the persistence of indirect human influence (e.g. retention of non-native species; Williams 2008; Ellis et al. 2012). However, it could also be argued that allowing any use in a cell to then lead to the rest of the cell to be 'novel' overestimates the spread of novel ecosystems; since the largest cells are only $85\,km^2$, this is only a small overestimation. A further source of estimation error arises because the techniques cannot focus on the biotic

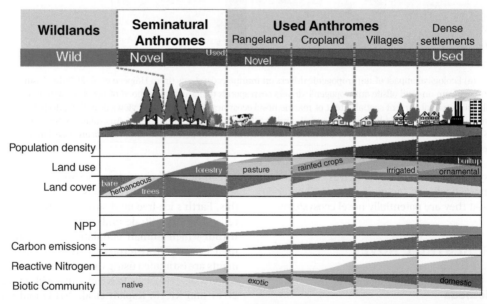

Figure 8.2 Degrees of novelty associated with variations in population density, land use and land cover across anthromes. Hypothetical response of selected ecosystem variables (NPP: Net Primary Production) to these variations is illustrated in the lower portion of the figure. Reproduced from Ellis (2011).

changes (both past and ongoing) within cells and hence cannot provide information on the biotic component of ecosystem change, particularly the presence or extent of non-native species. This also limits the ability to distinguish between novel and hybrid systems, and we do not aim to do so in this chapter.

The two models we use to estimate land use (HYDE 3.1 and KK10) differ significantly in how they model per capita use of land over time. The HYDE land-use model allocates land to mapped historical populations by assuming fairly stable land use per capita over time based on contemporary levels of land use. This produces conservative estimates of early land use because land use per capita is generally much higher under earlier agricultural conditions, declining by an order of magnitude or more as population densities increase and land use intensifies (Williams 2008; Ruddiman and Ellis 2009; Ellis 2011). KK10 predicts early land use from an empirical model of prehistoric European population and land-clearing relationships, which include a nearly 10-fold decrease in per capita land use as the modern era is approached. Full details on these models are documented in the source literature (Klein Goldewijk et al. 2010, 2011; Kaplan et al. 2011).

We show how long cells have had intensive use (which potentially indicates when novel ecosystems first appeared) and the current (AD 2000) extent of novel ecosystems within used cells. We also present and discuss the different patterns of spread by global regions (based on Ellis and Ramankutty 2008) and as shown by the different models. Briefly, these regions are based on similarity in historical patterns of population increase and spread and land use. North America, Australia and New Zealand are therefore considered as one region due to their similar colonial history preceded by millennia of indigenous occupation, in comparison to the developed area of Europe which has some similarity to the Near East in terms of a long period of agricultural settlement and intensive use.

8.4.2 Results

Land has been intensively used for millennia in many regions (Fig. 8.3a, b), with KK10 suggesting that this older intensive use was more extensive than HYDE 3.1. Intensive use appears to have spread from different centers with present-day eastern China, the Indian sub-continent, Iraq, southern Quebec and the northeastern United States, Central America and land associated with the Andean mountain chain having the most historical intensive use. Interestingly, Africa has had limited intensive use despite being regarded as the cradle of humanity (Manica et al. 2007), although present-day Ethiopia has had a long period of intensive use (over 5000 years). This history of intensive use suggests that novel ecosystems have been present on the terrestrial biosphere, at least in certain areas, for millennia. Contrary to these areas of ancient use, some areas have remained 'wild' throughout recorded history, particularly the Amazon, central areas of Australia (although not according to the HYDE model) and northern boreal regions. In between these extremes, the vast majority of the terrestrial biosphere has been converted to intensive use sometime in the last 250 years.

A long history of intensive use does not necessarily lead to large areas of novel ecosystems. Rather, some of the greatest extents of novel ecosystems are found in areas that have most recently been converted from 'wild' lands (Fig. 8.4), particularly in African and South American tropical forest areas and around the islands and peninsulas of southeast Asia. Western North America also has large extents of novel ecosystems, despite their more recent history of intensive use. Some areas are so intensively used that no part of a cell has any unused, and thereby 'novel', area. This is particularly the case for urban centers, a large swathe of central North America and parts of the Arabian Peninsula. It is likely that these urban centers, for example, contain unused areas that could be considered novel ecosystems (see also Chapter 38), but the coarse scale of the land-use models precludes capture of such detail.

An interesting avenue of future research would be quantifying the relationship between how long an area has been used and novel ecosystem extent, as well as how novel the ecosystems are. This would likely show that the intensity of direct human impact does not necessarily equate to a greater novel ecosystem area; however, areas with more intensive use may have ecosystems with greater novelty embedded within them due to increased anthropogenically forced ecological changes.

Both models show similar novel ecosystem amount at 8000 years before present, although KK10 shows slightly more than 20% of the ice-free land being 'novel' as compared to under 20% in HYDE 3.1 (Fig. 8.5a, b). Increasing human population and consequent changes in land use lead to an increase in novel

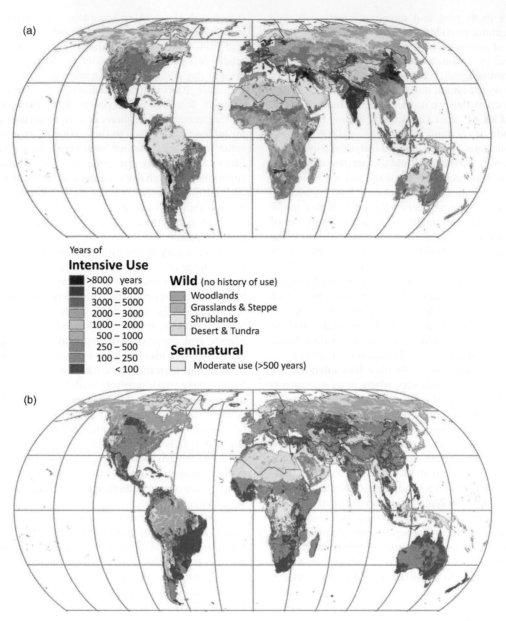

(a)

Years of
Intensive Use

- ■ >8000 years
- 5000 – 8000
- 3000 – 5000
- 2000 – 3000
- 1000 – 2000
- 500 – 1000
- 250 – 500
- 100 – 250
- ■ < 100

Wild (no history of use)
- Woodlands
- Grasslands & Steppe
- Shrublands
- Desert & Tundra

Seminatural
- Moderate use (>500 years)

(b)

Figure 8.3 Years of intensive use across the terrestrial biosphere based on the (a) KK10 and (b) HYDE 3.1 historical models of population density, land cover and land use. Reproduced from Ellis (2011).

ecosystem extent in both models as we move towards the present. Both models agree that about 50% of the ice-free surface of the terrestrial globe is the maximum extent of novel ecosystem extent (48% and 53% 100 years ago in HYDE 3.1 and KK10, respectively).

Interestingly, both models show a drop in novel ecosystem extent in the very recent past (to 36% in HYDE 3.1 and 28% in KK10), probably as a consequence of more intensive use of the land across the globe. Despite this drop, there is still more 'novel' ice-free land than 'wild'

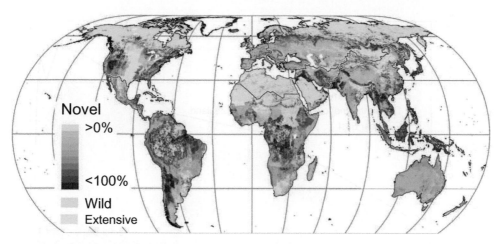

Figure 8.4 Current (AD 2000) extent of novel terrestrial ecosystems, estimated as unused lands within used grid cells based on the HYDE 3.1 dataset (not significantly distinct from KK10). Areas estimated as 100% used ("Extensive" use) or without evidence of human populations or land use ("Wild") are highlighted separately; these landscapes likely also include novel ecosystems as a result of underestimation of human presence, species invasions and environmental change ("wild"), and failure to account for smaller unused areas embedded within used grid cells. Eckert IV projection.

lands in the HYDE dataset (25.3% 'wild'). 'Wild' land still outweighs 'novel' ecosystem extent in KK10 (39% 'wild'), mainly due to the amount simulated in central Australia. The global increase in novel ecosystem area masks some interesting regional variations, as well as the reduction in 'wild' areas and the exponential increase in 'used' lands.

Differences between the models are apparent (compare Fig. 8.6a with Fig. 8.6b) essentially as a result of the different treatment of per capita impact of human population on land use; there is a much more rapid erosion of 'wild' land area in KK10 compared to HYDE 3.1. The more conservative HYDE 3.1 misses some potentially important events that characterize the spread of anthropogenic impact on the terrestrial globe. Particularly noticeable is the spike in 'wild' land that occurs in Latin America and, to a lesser extent, in North America 300–500 years ago. This spike corresponds to the depopulation of these areas due to disease and warfare brought by European settlers, and likely affected the global atmospheric concentration of carbon dioxide (Kaplan et al. 2011). Although KK10 can resolve this depopulation, it is likely that the increase in 'wild' was in fact an increase in novel ecosystem extent. However, depending on how 'wild', 'novel' and 'used' is defined for the purposes of mapping, such total

land abandonment is misclassified in terms of its likely actual appearance.

An interesting line of further research would be how regional differences in unused land area during different time periods relate to the extent of novel ecosystems. In other words, are all unused lands in particular regions 'novel' due to the region being used intensively, while other regions have less novel ecosystem extent due to substantial areas of 'wild' unused lands? Again, research aimed at removing some of the ambiguity surrounding our understanding of 'wild' lands, and particularly the effect of offsite influences, may aid in furthering our understanding.

8.5 CORROBORATING THE SPREAD OF NOVEL ECOSYSTEMS

The results we show here are based on population estimates and likely land conversion based on estimated per capita requirements for pasture and crops through history. The different treatment of per capita land use of the two models leads to different patterns of novel ecosystem spread; more land gets used earlier in time in KK10 than HYDE 3.1. The different trajectories highlight the uncertainty surrounding our knowledge

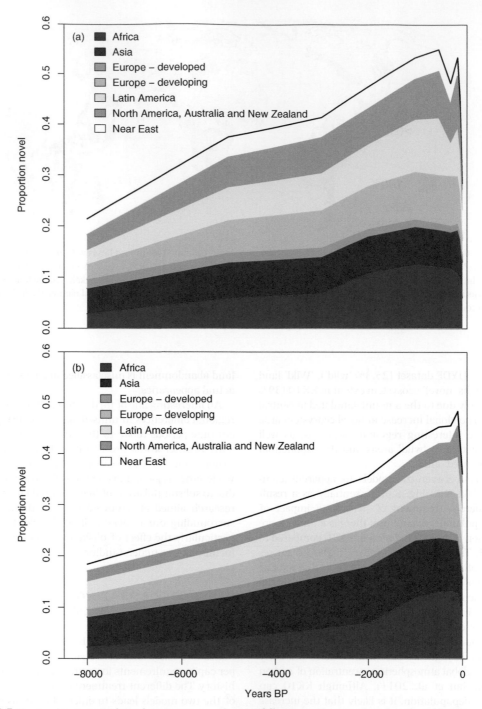

Figure 8.5 Temporal trajectory of novel ecosystem spread across different world regions based on the (a) KK10 and (b) HYDE 3.1 historical datasets.

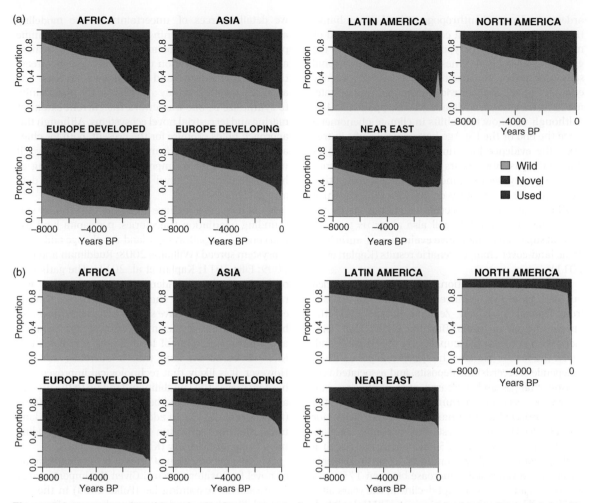

Figure 8.6 Proportion of Earth's ice-free land area categorized as 'wild' (unpopulated, unused lands), 'used' (populated and/or used for agriculture and settlements) and 'novel' (unused but embedded within used lands) in terrestrial regions based on the (a) KK10 and (b) HYDE historical models.

of prehistoric and historical land-use change and population growth (Williams 2008). Various lines of evidence could corroborate the patterns of spread that we outline (see Ellis 2011 for a review), in particular novel anthropogenic ecological forms and geological and archeological indicators that have been left through the action of natural and anthropogenically novel processes (Ellis 2011). For instance, tillage can lead to the formation of anthrosols or combustion can lead to concentrated deposits of charcoal and ash. These formations can be dated and hence confirm dates of intensive land use. Pollen records can also be dated and show vegetation changes that may include declines in fire-sensitive trees and increases in species that benefit from disturbance (McWethy et al. 2009). Other proxies include total nitrogen and phosphorus, as well as organic carbon accumulation (Ellis 2011).

Overcoming biases in the distribution of ecological research (Martin et al. 2012) may also aid in furthering our understanding of novel ecosystem formation and spread. In addition, natural scientists need to engage with anthropologists and archeologists in order to understand anthropogenic land-cover change (Kaplan et al. 2011) and consequently the spread of novel ecosystems. This type of collaboration and the gathering of multiple lines of evidence at regional scales for paleoreconstructions (e.g. AUS2k; Turney et al. 2008) will likely provide further backing to statements such as from Williams (2008; pp. 347–348): "Although we know that shifts in climate phenomena since the end of the Ice Age have caused shifts in tree taxa, the evidence for human initiation of changes during the past 6000 years is also clear. The evidence may be patchy, yet the conclusion is inescapable that humans were the primary engine of a change which was far greater than suspected and greater than some would care to admit." It will also allow us to move beyond superficial, qualitative evaluation of anthropogenic land-cover change scenario results (Kaplan et al. 2011).

Assigning causal relationships needs to be carried out with care. For instance, macrocharcoal deposits from five lakes in New Zealand clearly reveal the appearance of an 'Initial Burning Period' that coincided with Polynesian occupation of the South Island. Given the lack of climate change in this period (from independent records) the deposits, and associated vegetation changes and increased rates of erosion, can clearly be assigned to human influence (McWethy et al. 2009). On the other hand, a global fire history for the last 2000 years suggests that declining biomass burning from AD 1 to 1750 was associated with a cooling climate rather than due to anthropogenic effects. However, a sharp increase between 1750 and 1870, and an equally abrupt decline in the years after 1870, indicated anthropogenic control (Marlon et al. 2008).

8.6 TACKLING UNCERTAINTY AND FUTURE APPROACHES

There are a number of lines of uncertainty surrounding the trajectory of novel ecosystem spread that we have proposed here. We have already discussed some of these issues, such as the classification of 'wild' lands, and the likelihood that at least some parts of these lands are also 'novel' due to offsite human influences. In addition, we are not able to assess the degree of novelty in the novel ecosystems that are embedded within anthromes using the approach here; instead, we detail likely novel ecosystem area. In this section,

we detail sources of uncertainty in the modeling approach before broadening our discussion to examine potential future approaches that may allow a more accurate characterization of novel ecosystem spread, the degree of novelty and their drivers. At least one approach we suggest may also provide the opportunity to synthesize in one framework the current extent of marine and terrestrial novel ecosystems. Although this approach holds promise for the present, both for characterization and from a management perspective, it will be far harder to improve our understanding of historical spread of novel ecosystems given the data available however.

Numerous authors have noted the uncertainty surrounding population trajectories, per capita land use and consequent pathways of land-use change and novel ecosystem spread (Williams 2008; Ruddiman and Ellis 2009; Ellis 2011; Kaplan et al. 2011). The gathering of further lines of evidence mentioned earlier will reduce uncertainty in the reconstructions of novel ecosystem spread we have shown here. There may also be ways to improve the modeling itself; for instance, agricultural suitability of land in both HYDE 3.1 and KK10 is based on land that is currently suitable. However, it is likely that technological improvements have increased the suitability of land up to the present and such changes could be added to modeling efforts (Kaplan et al. 2011). In addition, the turnover of land under shifting cultivation could be varied depending upon likely productivity (Kaplan et al. 2011).

One obvious omission at present is the identification of novel freshwater systems. Given the importance of freshwater in sustaining life (Folke 2003) in the terrestrial biosphere, understanding human effects on these water bodies is important. It is likely that where small water bodies exist in intensively used landscapes, then they will be novel ecosystems. Larger water bodies that comprise many grid squares may not be novel ecosystems, depending upon surrounding terrestrial land use and population density. Novel ecosystem extent in freshwater areas may therefore be inferred from the surrounding terrestrial map, but further work would be required to corroborate these contentions. The inference of greater novel ecosystem extent in freshwater bodies with greater population density and land use may also have implications for the likelihood of novel ecosystem extent formed offsite. 'Wild' ecosystems downwind of large population centers may exhibit greater ecological change compared to those without such influences, although change will depend at least in part upon the susceptibility of the species in the ecosystem and landscape context. Such inferences need to be made with care and with appropriate caveats.

These broader concerns about inferring novel ecosystem extent from proxy data such as population density and land use naturally lead to the question of whether alternative classification schemes exist. As has already been highlighted the ambiguity about wild areas in particular, and the likelihood of novel ecosystems existing in such areas but being overlooked with the use of our current classification scheme, argues for consideration of alternative approaches. We suggest that one way to more accurately portray the current extent of novel ecosystems on land may be to consider the ecological effects of multiple drivers simultaneously, in a similar manner to that pursued for the marine sphere (Halpern et al. 2008b). In addition to describing the extent of ecological impact with a greater ecological impact more suggestive of the presence of novel ecosystems, such an approach would allow the synthesizing of marine and terrestrial spheres in one framework. However, our understanding of multiple interacting drivers on given ecosystems is somewhat limited, although frameworks are being proposed (e.g. Suding et al. 2008). In addition, coming up with global estimates of just one driver of ecological change (such as plant invasions by non-natives) and its consequent ecological effects is a problem with no easy solution, although attempts have been initiated (Ellis et al. 2012).

The mapping effort outlined earlier potentially indicates ways to approach management; for example, in coastal areas multiple anthropogenic drivers are assailing ecosystems and coordinated management actions will be required (Halpern et al. 2008b). Elsewhere, anthropogenic drivers are spatially uncorrelated and so independent regulation and conservation management tools may be more effective (Halpern et al. 2008b). Further discussion on the management of novel ecosystems can be found in Chapters 3 and 18.

It is possible that anthropogenic domesticates and species invasions (Ellis et al. 2012) and their relation to the rise of transportation systems and commodity markets (Verburg et al. 2011) are strong candidates for confirming the emergence and trajectory of novel ecosystems terrestrially at local, regional and global scales. Such tying of ecological and social systems together has been alluded to before. As Alessa and Chapin (2008) noted in their review of Ellis and Ramankutty's

(2008) initial approach to mapping the extent of human influence on the terrestrial biosphere, is population density the most powerful variable in explaining variation in socio-ecological properties? In his review of forest clearance through human occupation of the terrestrial biosphere, Williams (2008) raised the same issue and noted that considering population density alone does not explain clearing (and by implication, the formation of novel ecosystems). He went on to invoke a number of other factors, some of which have more contemporary relevance (such as misdirected past policies of aid agencies and governments) while others are of more general relevance such as inequality in asset distribution (Williams 2008).

These considerations also raise the question of whether our approach to categorizing novel ecosystems is valid. The suggested definition of a novel ecosystem (Chapter 6) incorporates three facets: difference, currently irreversible thresholds and the tendency of the system to self-organize and persist (and evolve) without intensive human management (see also Chapter 24). Ideally, and as Chapter 24 discusses, these attributes should all be measured to assess the presence (or otherwise) of a novel ecosystem. We do not explicitly measure these attributes and, indeed, cannot at a global scale. Instead, in the terrestrial realm we adopted a pragmatic approach that the direct use of land by humans creates novel ecosystems in remaining pockets of unused lands (see also Ellis et al. 2010).

As discussed earlier, other means of classifying could potentially exist and hold promise for more accurately delineating novel ecosystem spread at the global scale; with the data currently available, we used the approach presented. We would argue that the pockets of unused land in used areas certainly possess facets of novelty, as their fragmented nature prevents a return to some previous state and they therefore exhibit both difference and currently irreversible thresholds. These thresholds may be more or less likely to be reversed depending upon the degree of fragmentation and the extent of ecological change, as well as the ecological context. Persistence is far harder to measure, and certainly cannot be assessed from our approach. Chapter 9 presents a case study where novel ecosystems are mapped on a regional scale on the basis of biotic difference.

In the marine sphere, direct use of land is not a viable approach and instead we reviewed the latest research that shows the large ecological and evolutionary impact that multiple human-caused environmental changes are having. We contend that it is reasonable to suggest that the greater the ecological effect in a given system, the more likely the system is novel. However, only on-the-ground (or in-the-water) measurements could confirm these assertions. Again, we could pose the question to what extent difference, irreversibility and persistence are shown by these marine systems, and do they therefore constitute novel ecosystems? Again, we would answer that difference and the current irreversibility of the ecosystem state are both present. Whether the marine systems will maintain their properties and retain their ability to evolve in the face of future environmental change is unknown at present.

8.7 CONCLUSIONS

We have shown that novel ecosystems are spread throughout the globe in terrestrial areas, currently occupying between 28% and 36% of the ice-free land surface depending upon the model used. Novel ecosystems are also likely found in marine spheres, given the number of simultaneous drivers and degree of ecological change found around the world's oceans. For terrestrial systems we have evidence for the long-term presence of novel ecosystems in many areas, and the associated land clearance has likely had impacts on the atmosphere and potentially the climate for millennia (Kaplan et al. 2011). Although improvements will almost certainly be made to our current estimates as more archeological and anthropological evidence is gathered, and alternative frameworks may also be developed, our conclusions will likely not change: novel ecosystems are not new and they have been broadly spread, particularly on land, for millennia.

Such awareness of the broad spatial distribution and long temporal presence of these human-created and -altered systems questions current conservation and restoration norms of bemoaning the loss of wilderness and returning to some historical baseline of a state 'unaltered by humans'. Over large parts of the globe, the 'wilderness' that people refer back to never existed; setting a historical baseline is therefore problematic. Instead, ecologists, conservationists, policy makers and the broader public need to start engaging with not-so-novel systems in a far more proactive manner and perhaps philosophically accept that nature has become embedded within human systems more broadly (Ellis 2011). As the desired results of nature conservation and management will depend on the

goals and aspirations of the managers and the context within which they operate, accepting a permanent role for humans as stewards of the biosphere opens up possibilities for adaptive management of novel ecosystems based on more flexible interpretations of historical reference.

Theory and practice of novel ecosystem management occupies the bulk of the remainder of this book. This chapter has shown the widespread importance of novel ecosystems in the biosphere, while hinting at ways to improve our understanding of their extent and ecological significance at global and regional scales.

REFERENCES

Alessa, L. and Chapin, F.S. (2008) Anthropogenic biomes: a key contribution to earth-system science. *Trends in Ecology and Evolution*, **23**, 529–531.

Beaufort, L., Probert, I., de Garidel-Thoron, T., Bendif, E.M., Ruiz-Pino, D., Metzl, N., Goyet, C., Buchet, N., Coupel, P., Grelaud, M., Rost, B., Rickaby, R.E.M. and de Vargas, C. (2011) Sensitivity of coccolithophores to carbonate chemistry and ocean acidification. *Nature*, **476**, 80–83.

Blight, L.K. and Ainley, D.G. (2008) Southern Ocean not so pristine. *Science*, **321**, 1443.

Carlton, J.T. and Geller, J.B. (1993) Ecological roulette: The global transport of non-indigenous marine organisms. *Science*, **261**, 78–82.

Ellis, E.C. (2011) Anthropogenic transformation of the terrestrial biosphere. *Philosophical Transactions of the Royal Society A*, **369**, 1010–1035.

Ellis, E.C. and Ramankutty, N. (2008) Putting people in the map: anthropogenic biomes of the world. *Frontiers in Ecology and Environment*, **6**, 439–447.

Ellis, E.C., Klein Goldewijk, K., Siebert, S., Lightman, D. and Ramankutty, N. (2010) Anthropogenic transformation of the biomes, 1700 to 2000. *Global Ecology and Biogeography*, **19**, 589–606.

Ellis, E.C., Antill, E.C. and Kreft, H. (2012) All is not loss: Plant biodiversity in the Anthropocene. *PLoS ONE*, **7**, e30535.

Elser, J. and Bennett, E. (2011) A broken biogeochemical cycle. *Nature*, **478**, 29–31.

Fabricius, K.E. (2005) Effects of terrestrial runoff on the ecology of corals and coral reefs: review and synthesis. *Marine Pollution Bulletin*, **50**, 125–146.

Fitt, W.K., Brown, B.E., Warner, M.E. and Dunne, R.P. (2001) Coral bleaching: interpretation of thermal tolerance limits and thermal thresholds in tropical corals. *Coral Reefs*, **20**, 51–65.

Folke, C. (2003) Freshwater for resilience: a shift in thinking. *Philosophical Transactions of the Royal Society B: Biological Sciences*, **358**, 2027–2036.

Grosholz, E. (2002) Ecological and evolutionary consequences of coastal invasions. *Trends in Ecology and Evolution*, **17**, 22–27.

Halpern, B.S., Kappel, C.V., Micheli, F., Selkoe, K.A., D'Agrosa, C., Bruno, J., Casey, K.S., Ebert, C.M., Fox, H.E., Fujita, R., Heinemann, D., Lenihan, H.S., Madin, E.M.P., Perry, M., Selig, E.R., Spalding, M., Steneck, R., Walbridge, S. and Watson, R. (2008a). Response. *Science*, **321**, 1444–1445.

Halpern, B.S., Walbridge, S., Selkoe, K.A., Kappel, C.V., Micheli, F., D'Agrosa, C., Bruno, J.F., Casey, K.S., Ebert, C., Fox, H.E., Fujita, R., Heinemann, D., Lenihan, H.S., Madin, E.M.P., Perry, M.T., Selig, E.R., Spalding, M., Steneck, R. and Watson, R. (2008b) A global map of human impact on marine ecosystems. *Science*, **319**, 948–952.

Heath, M.R. (2008). Comment on 'A global map of human impact on marine ecosystems'. *Science*, **321**, 1446a–1446b.

Hobbs, R.J., Arico, S., Aronson, J., Baron, J.S., Bridgewater, P., Cramer, V.A., Epstein, P.R., Ewel, J.J., Klink, C.A., Lugo, A.E., Norton, D., Ojima, D., Richardson, D.M., Sanderson, E.W., Valladares, F., Vila, M., Zamora, R. and Zobel, M. (2006) Novel ecosystems: theoretical and management aspects of the new ecological world order. *Global Ecology and Biogeography*, **15**, 1–7.

Hobbs, R.J., Higgs, E. and Harris, J.A. (2009) Novel ecosystems: implications for conservation and restoration. *Trends in Ecology and Evolution*, **24**, 599–605.

Hoegh-Guldberg, O., Mumby, P.J., Hooten, A.J., Steneck, R.S., Greenfield, P., Gomez, E., Harvell, C.D., Sale, P.F., Edwards, A.J., Caldeira, K., Knowlton, N., Eakin, C.M., Iglesias-Prieto, R., Muthiga, N., Bradbury, R.H., Dubi, A. and Hatziolos, M.E. (2007) Coral reefs under rapid climate change and ocean acidification. *Science*, **318**, 1737–1742.

Holker, F., Beare, D., Dorner, H., di Natale, A., Ratz, H.-J., Temming, A. and Casey, J. (2007) Comment on 'Impacts of biodiversity loss on ocean ecosystem services'. *Science*, **316**, 1285c.

Howarth, R.W., Billen, G., Swaney, D., Townsend, A., Jaworski, N., Lajtha, K., Downing, J.A., Elmgren, R., Caraco, N., Jordan, T., Berendse, F., Freney, J., Kudeyarov, V., Murdoch, P. and Zhao-Liang, Z. (1996) Regional nitrogen budgets and riverine N & P fluxes for the drainages to the North Atlantic Ocean: Natural and human influences. *Biogeochemistry*, **35**, 75–139.

Jennings, S., Pinnegar, J.K., Polunin, N.V.C. and Warr, K.J. (2001) Impacts of trawling disturbance on the trophic structure of benthic invertebrate communities. *Marine Ecology Progress Series*, **213**, 127–142.

Jennings, S., Greenstreet, S.P.R., Hill, L., Piet, G.J., Pinnegar, J.K. and Warr, K.J. (2002) Long-term trends in the trophic structure of the North Sea fish community: evidence from stable-isotope analysis, size-spectra and community metrics. *Marine Biology*, **141**, 1085–1097.

Kaplan, J.O., Krumhardt, K.M., Ellis, E.C., Ruddiman, W.F., Lemmen, C. and Klein Goldewijk, K. (2011) Holocene carbon emissions as a result of anthropogenic land cover change. *The Holocene*, **21**, 775–791.

Klein Goldewijk, K., Beusen, A. and Janssen, P. (2010) Long-term dynamic modeling of global population and built-up area in a spatially explicit way: HYDE 3.1. *The Holocene*, **20**, 565–573.

Klein Goldewijk, K., Beusen, A., van Drecht, G. and de Vos, M. (2011) The HYDE 3.1 spatially explicit database of human-induced global land-use change over the past 12,000 years. *Global Ecology and Biogeography*, **20**, 73–86.

Manica, A., Amos, W., Balloux, F. and Hanihara, T. (2007) The effect of ancient population bottlenecks on human phenotypic variation. *Nature*, **448**, 346–348.

Marlon, J.R., Bartlein, P.J., Carcaillet, C., Gavin, D.G., Harrison, S.P., Higuera, P.E., Joos, F., Power, M.J. and Prentice, I.C. (2008) Climate and human influences on global biomass burning over the past two millennia. *Nature Geoscience*, **1**, 697–702.

Martin, L.J., Blossey, B. and Ellis, E. (2012) Mapping where ecologists work: Biases in the global distribution of terrestrial ecological observations. *Frontiers in Ecology and Environment*, **10**, 195–201.

McWethy, D.B., Whitlock, C., Wilmshurst, J.M., McGlone, M.S. and Li, X. (2009) Rapid deforestation of South Island, New Zealand, by early Polynesian fires. *The Holocene*, **19**, 883–897.

Myers, R.A. and Worm, B. (2003) Rapid worldwide depletion of predatory fish communities. *Nature*, **423**, 280–283.

Pandolfi, J.M., Connolly, S.R., Marshall, D.J. and Cohen, A.L. (2011) Projecting coral reef futures under global warming and ocean acidification. *Science*, **333**, 418–422.

Pauly, D., Christensen, V., Dalsgaard, J., Froese, R. and Torres, R. Jr. (1998) Fishing down marine food webs. *Science*, **279**, 860–863.

Rabalais, N.N., Turner, R.E. and Wiseman, W.J. Jr. (2002) Gulf of Mexico hypoxia; a.k.a. 'The Dead Zone'. *Annual Review of Ecology and Systematics*, **33**, 235–263.

Ramankutty, N. and Foley, J.A. (1999) Estimating historical changes in global land cover: Croplands from 1700 to 1992. *Global Biogeochemical Cycles*, **13**, 997–1027.

Ruddiman, W.F. and Ellis, E.C. (2009) Effect of per-capita land use changes on Holocene forest clearance and CO_2 emissions. *Quaternary Science Reviews*, **28**, 3011–3015.

Ruiz, G.M., Rawlings, T.K., Dobbs, F.C., Drake, L.A., Mullady, T., Huq, A. and Colwell, R.R. (2000) Global spread of microorganisms by ships. *Nature*, **408**, 49–50.

Rule, S., Brook, B.W., Haberle, S.G., Turney, C.S.M., Kershaw, A.P. and Johnson, C.N. (2012) The aftermath of megafaunal extinction: ecosystem transformation in Pleistocene Australia. *Science*, **335**, 1483–1486.

Schnitzer, S.A. and Bongers, F. (2011) Increasing liana abundance and biomass in tropical forests: emerging patterns and putative mechanisms. *Ecology Letters*, **14**, 397–406.

Selkoe, K.A., Kappel, C.V., Halpern, B.S., Micheli, F., D'Agrosa, C., Bruno, J., Casey, K.S., Ebert, C., Fox, H.E., Fujita, R.,

Heinemann, D., Lenihan, H.S., Madin, E.M.P., Perry, M.T., Selig, E.R., Spalding, M., Steneck, R., Walbridge, S. and Watson, R. (2008) Response to Comment on 'A global map of human impact on marine ecosystems'. *Science*, **321**, 1446c.

Smith, D. J., Suggett, D.J. and Baker, N.R. (2005) Is photoinhibition of zooxanthellae photosynthesis the primary cause of thermal bleaching in corals? *Global Change Biology*, **11**, 1–11.

Suding, K.N., Lavorel, S., Chapin F.S. III, Cornelissen, J.H.C., Diaz, S., Garnier, E., Goldberg, D., Hooper, D.U., Jackson, S.T. and Navass, M.L. (2008) Scaling environmental change through the community-level: a trait-based response-and-effect framework for plants. *Global Change Biology*, **14**, 1125–1140.

Turney, C.S.M., Duncan, R., Nicholls, N., Moberg, A. and Pollack, H. (2008) Towards an Australasian climate reconstruction for the past two millennia. *PAGES News*, **16**, 34.

Verburg, P.H., Ellis, E.C. and Letourneau, A. (2011) A global assessment of market accessibility and market influence for global environmental change studies. *Environmental Research Letters*, **6**, 034019.

Wilberg, M.J. and Miller, T.J. (2007) Comment on 'Impacts of biodiversity loss on ocean ecosystem services'. *Science*, **316**, 1285b.

Williams, M. (2008) A new look at global forest histories of land clearing. *Annual Review of Environment and Resources*, **33**, 345–367.

Worm, B., Barbier, E.B., Beaumont, N., Duffy, J.E., Folke, C., Halpern, B.S., Jackson, J.B.C., Lotze, H.K., Micheli, F., Palumbi, S.R., Sala, E., Selkoe, K.A., Stachowicz, J.J. and Watson, R. (2006) Impacts of biodiversity loss on ocean ecosystem services. *Science*, **314**, 787–790.

Chapter 9

CASE STUDY: GEOGRAPHIC DISTRIBUTION AND LEVEL OF NOVELTY OF PUERTO RICAN FORESTS

Sebastián Martinuzzi[1], Ariel E. Lugo[2], Thomas J. Brandeis[3] and Eileen H. Helmer[2]

[1]Department of Forest and Wildlife Ecology, University of Wisconsin–Madison, USA
[2]International Institute of Tropical Forestry, USDA Forest Service, Río Piedras, Puerto Rico
[3]Southern Research Station, USDA Forest Service, Knoxville, Tennessee, USA

A set of questions that emerge time and again during discussions of novel forests deal with the spatial cover of these systems. Are they dominant on landscapes? How broad is their geographic extent? Are they limited in ecological space to particular geoclimatic conditions? Responses to these questions require large-scale and georeferenced inventories of forests over a variety of geoclimatic conditions, regardless of level of human activity. The taxonomy of component trees also needs to be known so that questions about the nature of the species mix of forests, i.e. if the species are native or introduced as well as their relative importance, can be evaluated with confidence. Knowing the history of human activity over the same spatial scales as those of the forests being inventoried also helps to answer these questions.

Puerto Rico is an ideal place to address questions about the spatial distribution of novel forests. The island has been flown for aerial photography since the 1930s. This allows spatial as well as temporal analyses of its forest cover, which has experienced a dramatic change over the past centuries (Lugo 2004, fig. 1; see also Martinuzzi et al. 2009 for a 200-year analysis of mangrove forest cover change). The different land covers between the time when the island was mostly agricultural to today, with its 11% urban cover, is also well known (Helmer et al. 2002; Kennaway and Helmer 2007; Martinuzzi et al. 2007). In addition, forest cover of the island has been assessed several times in relation to climate and land cover (e.g. Lugo 2002; Helmer 2004; Lugo and Helmer 2004; Lugo and Brandeis 2005; Brandeis et al. 2009). Moreover, tree taxonomy in Puerto Rico is well known (Little et al. 1974; Francis and Liogier 1991) and the whole island has been inventoried three times between the 1980s and 2004 (Birdsey and Weaver 1982, 1987; Brandeis et al. 2007).

Given the wealth of information about forest cover and species composition available to us, we set out to develop a map of forest cover in Puerto Rico and to convert that map into novel forest cover maps. We used the presence of introduced tree species and their Importance Value (IV, a composite measure of abundance) as the criteria for novelty. We also included those locations where introduced species as a group accounted for 100% of the IV of vegetation in our mapping. As part of the mapping process we also estimated the extent of native, mixed native and introduced and purely introduced forest species assemblages present in the island. The mapping of novel forests allowed us to compare their distribution with that of native forests and to make observations about the relationship between human activity and forest cover in this tropical island.

9.1 METHODS

Our study combined recent land cover and forest inventory data. We used the results from the last forest inventory of Puerto Rico (Brandeis et al. 2009) to evaluate the abundance of introduced species across the different forest types of the island, which served as a surrogate for novelty. We also used a recent land cover map to spatially represent the information on introduced species. Our criteria for novelty included three different indexes, which we summarized by forest type: (1) the proportion of field plots belonging to tree species assemblages with any presence of introduced species; (2) the contribution of introduced species to the IV, for which we used the median across the plots (IV of a species included its relative basal area and stem density and its frequency, normalized to 100% for the sum of all species); and (3) the proportion of field plots belonging to assemblages composed entirely of introduced species. The higher these values, the more novel are the forests. These criteria were used to create maps of novel forests and to estimate the total area of native, mixed and purely introduced forest assemblages present in the island.

The forest inventory data were extracted from Brandeis et al. (2009) who identified the different tree species assemblages of the island and summarized their frequency by geoclimatic units, which are equivalent to forest types. Our plot data were at the species assemblage level. According to Brandeis et al. (2009) there are 14 major tree species assemblages whose presence and abundance change by geoclimatic unit or forest type. For each assemblage, the authors described the species composition including the distinction of native versus introduced species and the IV of each species. From knowledge of the assemblage species composition and the number of plots per geoclimatic unit, we were able to quantify our indexes for novelty for most forest types in the island. Forest types with low area coverage i.e. urban forests, forested wetlands, upper montane forests and abandoned palm plantations, were not well represented in the inventory. We therefore reviewed local studies to gain information about their species composition and hence their level of novelty (Figueroa et al. 1984; Lugo 1998; Lugo and Brandeis 2005; Gould et al. 2006; Table 9.1).

The forest cover map that we used to overlay species data (Fig. 9.1) was a simplification of a detailed land cover map developed by Gould et al. (2008). We added the classes 'urban forest' and 'upper montane forest' to distinguish two extreme conditions of anthropogenic presence, i.e. maximum for the urban, minimum for the upper montane. Urban forests were mapped by identifying the portion of the forest cover that was embedded in the island's urban areas (i.e. cities and towns, as mapped by Martinuzzi et al. 2007). Urban areas are hotspots for introduced species, and most of coastal Puerto Rico is urban. The upper montane forest, on the other hand, encompasses high-elevation wet and rain forests that have escaped from human disturbances due to their low suitability for agriculture, extreme rainfall and strong legal protection (see Helmer 2004; Kennaway and Helmer 2007). These areas include palo colorado, sierra palm and elfin forests, and other mature forest types located above the 600 m elevation (e.g. tabonuco forests; all of them mapped based on Gould et al. 2008). Finally, we were able to relate land cover and forest inventory data because both efforts used the same baseline information (i.e. geoclimatic units) to classify (with GIS) or sample (in the field) vegetation. In other words, the different forest types of the island were named, mapped and inventoried in a consistent fashion, making it possible to combine available data.

In summary, we first reclassified the forest cover map based on the three different criteria for novelty, using

Table 9.1 Abundance of introduced tree species in different forest types of Puerto Rico, as measured by three different plot-level inventory attributes. N: sample size; N/A: data not available; LM: lower montane; AVS: alluvial/volcanic/sedimentary substrate.

Forest type	Field plots belonging to assemblages with some introduced tree species (%)	Contribution of introduced species to IV (%)	Proportion of field plots belonging to assemblages composed entirely of introduced tree species (%)	N	Data source
Urban	N/A	66	N/A	6	Lugo and Brandeis (2005)
Dry/AVS	87	100	60	15	Brandeis et al. (2009)
Dry/limestone	86	13	43	14	Brandeis et al. (2009)
Dry, moist/ultramafic	63	10	38	8	Brandeis et al. (2009)
Moist/AVS	83	24	32	81	Brandeis et al. (2009)
Moist, wet/limestone	78	15	19	72	Brandeis et al. (2009)
Wet/AVS	82	24	18	68	Brandeis et al. (2009)
Wet, LM/ultramafic	17	0	0	6	Brandeis et al. (2009)
Upper montane	0	0	0	16	Gould et al. (2006); Brandeis et al. (2009)
Abandoned palm	100	53	N/A	2	Figueroa et al. (1984)
Forested wetlands	0	0	0	N/A	Lugo (1998)

the information for each species assemblage that we summarized by forest zone as described earlier. This resulted in three maps of novel forests where the original forest types are now represented by a value depicting the importance of introduced species. We used these maps to evaluate general patterns of forest novelty across the island and to compare forest types. We then quantified the extent of native, mixed and introduced forests by multiplying the proportion of plots in native, mixed or introduced species assemblages by the total area of the corresponding forest cover type. We considered species assemblages as native, mixed or introduced based on the contribution of introduced species to the IV: 0% for native; >0% and <100% for mixed; and 100% for introduced.

9.2   RESULTS AND DISCUSSION

9.2.1   Spatial patterns of novel forests

The novel forest maps (Fig. 9.2a–c) revealed two species composition attributes of Puerto Rican forests. First, introduced species exhibit a widespread distribution as they are present almost everywhere. Second, the maps depict the spatial patterns or spatial distribution of novelty. Only a few parts of the island contain forest types without introduced species. These locations with native forests include mountain peaks (upper montane forest type) and areas of land–ocean interface with forested wetlands. The rest of the island's forests all have some level of introduced species presence.

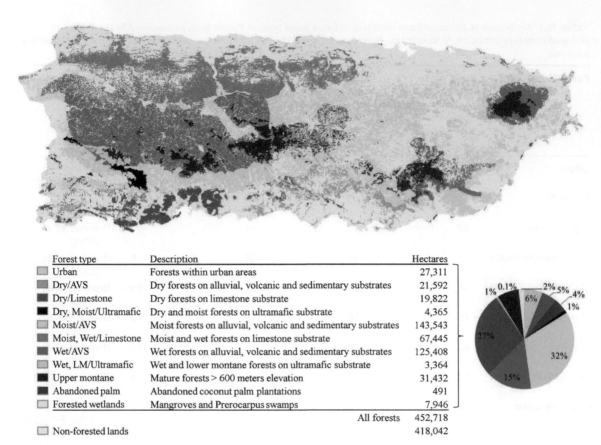

Forest type	Description	Hectares
Urban	Forests within urban areas	27,311
Dry/AVS	Dry forests on alluvial, volcanic and sedimentary substrates	21,592
Dry/Limestone	Dry forests on limestone substrate	19,822
Dry, Moist/Ultramafic	Dry and moist forests on ultramafic substrate	4,365
Moist/AVS	Moist forests on alluvial, volcanic and sedimentary substrates	143,543
Moist, Wet/Limestone	Moist and wet forests on limestone substrate	67,445
Wet/AVS	Wet forests on alluvial, volcanic and sedimentary substrates	125,408
Wet, LM/Ultramafic	Wet and lower montane forests on ultramafic substrate	3,364
Upper montane	Mature forests > 600 meters elevation	31,432
Abandoned palm	Abandoned coconut palm plantations	491
Forested wetlands	Mangroves and Prerocarpus swamps	7,946
	All forests	452,718
Non-forested lands		418,042

Figure 9.1 Forest land-cover types of Puerto Rico, including description and extent. From Gould et al. (2008).

The high presence of introduced species is particularly visible in Fig. 9.2a, which displays forest types based on the number of field plots belonging to species assemblages with introduced species. Most of the island's forests have many plots with introduced species, i.e. 80% or more. Upper elevation wet areas (i.e. regions for the upper montane and wet ultramafic forests) as well as forested wetlands were the only regions with very low or no presence of introduced species. The northern karst region (with the moist/wet limestone forest) had intermediate values.

In terms of the contribution of introduced species to IV, the highest values were observed in some of the lowlands including dry forests in the south (i.e. on alluvial/volcanic/sedimentary substrate or AVS), urban areas and abandoned palm plantations (Fig. 9.2b). In these forests the contribution of introduced species represented >50% of the IV. The areas with the lowest contributions of introduced species to the IV were located on limestone and ultramafic substrates, upland forests and forested wetlands.

The southern forests were the areas supporting the highest proportions of plots composed entirely of introduced species (i.e. 40–60%), followed by some of the eastern moist forests (i.e. on AVS substrate; Fig. 9.2c). The rest of the island contains typically lower values (<20% or zero).

9.2.2 Extension of native, mixed and introduced forests

Our study revealed that in Puerto Rico about 50% of the forest cover is mixed (i.e. corresponding to mixed

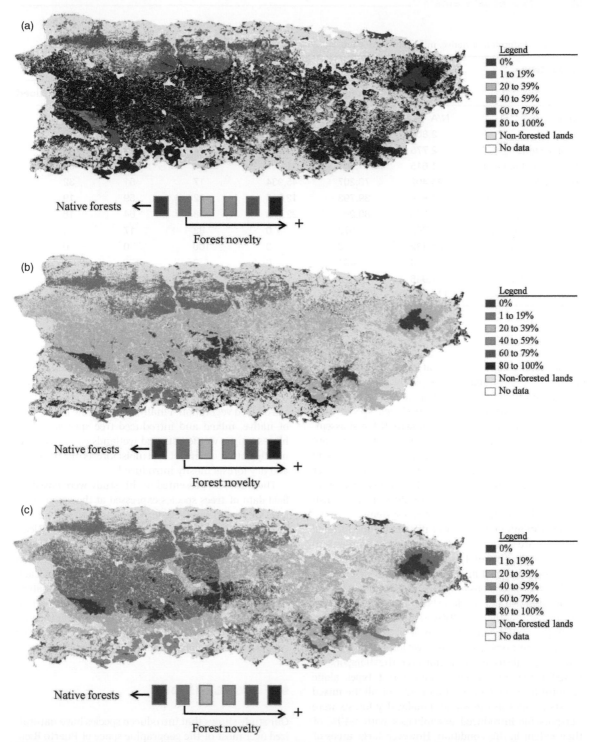

Figure 9.2 (a) Geographic patterns of forest novelty based on the proportion of field plots with some presence of introduced tree species. (b) Geographic patterns of forest novelty based on the contribution of introduced species to the IV estimator. (c) Geographic patterns of forest novelty based on the proportion of field plots composed entirely of introduced species. The values are calculated and displayed by forest type.

Table 9.2 Extension of native, mixed and purely introduced forests. Totals do not include urban forests.

Forest type	Hectares			In Percent		
	Native	**Mixed**	**Introduced**	**Native**	**Mixed**	**Introduced**
Urban	N/A	N/A	N/A	N/A	N/A	N/A
Dry/AVS	2,807	5,830	12,955	13	27	60
Dry/limestone	2,775	8,523	8,523	14	43	43
Dry, moist/ultramafic	1,615	1,091	1,659	37	25	38
Moist/AVS	24,402	73,207	45,934	17	51	32
Moist, wet/limestone	14,838	39,793	12,815	22	59	19
Wet/AVS	22,573	80,261	22,573	18	64	18
Wet, LM/ultramafic	2,792	572	0	83	17	0
Upper montane	31,432	0	0	100	0	0
Abandoned palm	0	491	0	0	100	0
Forested wetlands	7,946	0	0	100	0	0
Total	111,180	209,768	104,459	26	49	25

species assemblages), 25% is native (i.e. composed only of native species) and 25% is purely introduced (i.e. composed only of introduced species). For the mixed forests, the contribution of introduced species was about 25% of the IV (data not shown). If forest assemblages with any presence of introduced species are considered novel, then three-quarters of the island forests are novel (Table 9.2). Our study did not consider whether changes in forest species composition were reversible or not, and hence we did not distinguish potential hybrid systems from novel systems.

Upper montane forests, wet ultramafic and forested wetlands are the hotspots for native vegetation assemblages, totaling about 42,000 ha. These areas support very low presence of introduced species or no presence at all. Large extensions of native vegetation can also be found in moist and wet forests on limestone and AVS substrates (another 62,000 ha) however, but representing a small fraction (22% or less) of the total area of these forest types. Moist and wet forests on AVS and limestone substrates were however the hotspots for mixed forest assemblages. These forest types alone account for near 193,000 ha or 92% of all the mixed forests present in the island. Finally, dry forests were hotspots for introduced assemblages with >40% of their extent in this condition. However, large areas of introduced vegetation are also present in moist and wet forests. Information for urban forests was limited but,

based on IV values in Table 9.1, we expect these forests to have a high presence of introduced species; they may therefore be composed mostly of mixed and/or introduced vegetation. Finally, based on the proportion of native, mixed and introduced tree species assemblages, upper wet forests and wetlands can be seen as mostly native, other wet and moist forests mostly mixed and dry forests mostly introduced.

The numbers presented in this study were based on field data of trees species expressed at the assemblage level and were valuable for assessing the geographic distribution and level of novelty of the Puerto Rican forests. However, the species assemblage level data excluded rare species and masked natural variations in species composition that may occur within each assemblage. Estimating the ultimate area of native, mixed and introduced forest vegetation will require analysis of the complete information for each individual plot. For these reasons, our numbers should be considered as an approximation.

9.3 CONCLUSIONS

Our study shows that introduced species have naturalized over most of the geographic space of Puerto Rico. However, forests in extreme environments, i.e. high rainfall, nutrient-poor soils (i.e. ultramafic substrate) or flooded, are still dominated by native vegetation; examples of areas dominated by native vegetation still occur on all forest types of the island. Novel forests do not appear to be restricted by geoclimatic conditions. Instead, they appear to be associated with the residual effects of past and present anthropogenic activity, which have in turn extended to most parts of the island. Two-thirds of the Puerto Rican forests may contain novel forest assemblages. We also found that the area of native forest species assemblages is now equaled by the area of forests that are classified as completely introduced. There is a critical need to study the functioning of novel forests in Puerto Rico and elsewhere in the tropics where they occur.

ACKNOWLEDGMENTS

This work was carried out in cooperation with the University of Puerto Rico.

REFERENCES

Birdsey, R.A. and Weaver, P.L. (1982) The forest resources of Puerto Rico. Resource Bulletin SO-85, USDA Forest Service Southern Forest Experiment Station, New Orleans, LA.

Birdsey, R.A. and Weaver, P.L. (1987) Forest area trends in Puerto Rico. United States Department of Agriculture Forest Service Southern Forest Experiment Station Research Note SO-331, New Orleans, LA.

Brandeis, T.J., Helmer, E.H. and Oswalt, S.N. (2007) The status of Puerto Rico's forests, 2003. USDA Forest Service, Southern Research Station Resource Bulletin SRS-119, Asheville, NC.

Brandeis, T.J., Helmer, E., Marcano Vega, H. and Lugo, A.E. (2009) Climate shapes the novel plant communities that form after deforestation in Puerto Rico and the U.S. Virgin Islands. *Forest Ecology and Management*, **258**, 1704–1718.

Figueroa, J.C., Totti, L., Lugo, A.E. and Woodbury, R.O. (1984) Structure and composition of moist coastal forests in Dorado, Puerto Rico. Research Paper SO-202, USDA, Forest Service, Southern Forest Experiment Station, New Orleans, LA.

Francis, J.K. and Liogier, H.A. (1991) Naturalized exotic tree species in Puerto Rico. General Technical Report SO-82, USDA, Forest Service, Southern Forest Experiment Station, New Orleans, LA.

Gould, W.A., González, G. and Carrero Rivera, G. (2006) Structure and composition of vegetation along an elevational gradient in Puerto Rico. *Journal of Vegetation Science*, **17**, 563–574.

Gould, W.A., Alarcón, C., Fevold, B., Jiménez, M.E., Martinuzzi, S., Potts, G., Quiñones, M., Solórzano, M. and Ventosa, E. (2008) The Puerto Rico Gap Analysis Project. Volume 1: Land cover, vertebrate species distributions, and land stewardship. IITF-GTR-39, US Forest Service, International Institute of Tropical Forestry, PR.

Helmer, E.H. (2004) Forest conservation and land development in Puerto Rico. *Landscape Ecology*, **19**, 29–40.

Helmer, E.H., Ramos, O., López, T.D.M., Quiñones, M. and Díaz, W. (2002) Mapping forest type and land cover of Puerto Rico, a component of the Caribbean biodiversity hotspot. *Caribbean Journal of Science*, **38**, 165–183.

Kennaway, T. and Helmer, E.H. (2007) The forest types and ages cleared for land development in Puerto Rico. *GIScience and Remote Sensing*, **44**, 356–382.

Little, E.L., Woodbury, R.O. and Wadsworth, F.H. (1974) Trees of Puerto Rico and the Virgin Islands, volume 2. USDA Forest Service Agriculture Handbook 449, Washington, DC.

Lugo, A.E. (1998) Mangrove forests: a tough system to invade but an easy one to rehabilitate. *Marine Pollution Bulletin*, **37**, 427–430.

Lugo, A.E. (2002) Can we manage tropical landscapes? An answer from the Caribbean perspective. *Landscape Ecology*, **17**, 601–615.

Lugo, A.E. (2004) The outcome of alien tree invasions in Puerto Rico. *Frontiers in Ecology and the Environment*, **2**, 265–273.

Lugo, A.E. and Helmer, E. (2004) Emerging forests on abandoned land: Puerto Rico's new forests. *Forest Ecology and Management*, **190**, 145–161.

Lugo, A.E. and Brandeis, T.J. (2005) A new mix of alien and native species coexist in Puerto Rico's landscapes, in *Biotic Interactions in the Tropics: Their Role in the Maintenance of Species Diversity* (eds D.F.R.P. Burslem, M.A. Pinard and S.E. Hartley), Cambridge University Press, Cambridge, England, 484–509.

Martinuzzi, S., Gould, W.A. and Ramos González, O.M. (2007) Land development, land use, and urban sprawl in Puerto Rico integrating remote sensing and population census data. *Landscape and Urban Planning*, **79**, 288–297.

Martinuzzi, S., Gould, W.A., Lugo, A.E. and Medina, E. (2009) Conversion and recovery of Puerto Rican mangroves: 200 years of change. *Forest Ecology and Management*, **257**, 75–84.

Chapter 10

NOVEL ECOSYSTEMS AND CLIMATE CHANGE

Brian M. Starzomski

School of Environmental Studies, University of Victoria, Canada

10.1 INTRODUCTION

Climate change science has long recognized the possibility for dramatic changes in ecosystems due to shifting climate. The most dramatic changes have been labeled novel, 'no-analog' or 'non-analog' systems, combinations of abiotic conditions with no modern equivalents that may lead to communities of species we have never seen living together before. For example, using multimodel simulations of IPCC emissions scenarios, Williams et al. (2007) show that between 4% and 48% of the world's surface may experience loss of climate conditions or become no-analog/novel ecosystems under predicted rates of climate change. The consequences of this are profound: how do we adapt to and manage ecosystems we have never seen before? How are our concepts of ecological restoration altered in the face of climate change?

Species respond individualistically to climate changes, and therefore accurate predictions of the future states of communities and ecosystems are difficult and probably impossible (Araújo et al. 2005; Araújo and Rahbek 2006; Fitzpatrick and Hargrove 2009). By definition,

communities of species that have their components changing independently due to anthropogenic impacts become hybrid then novel ecosystems (sensu Hobbs et al. 2009). In addition to all the other threats to biodiversity (lumped into the 'Big 5' threats which also include land-use change, invasive species, nitrogen deposition and overexploitation; Sala et al. 2000), climate change has important consequences for biodiversity. Cumulative effects of the 'Big 5' make for significant challenges for management of species. This review will focus on the specifics arising from climate change.

The concept of novel ecosystems may provide us with a useful tool for understanding and managing the impacts of climate change on species, communities and ecosystems. Novel ecosystems are characterized by non-historical species configurations that arise due to anthropogenic environmental change (such as climate change), land-use alteration, species invasions or a combination of all three. They are a consequence of human activity but do not depend on human intervention for their maintenance (Hobbs et al. 2006; Chapter 5; Fig. 3.2). Novel ecosystems also are not practically

Novel Ecosystems: Intervening in the New Ecological World Order, First Edition. Edited by Richard J. Hobbs, Eric S. Higgs, and Carol M. Hall.
© 2013 John Wiley & Sons, Ltd. Published 2013 by John Wiley & Sons, Ltd.

reversible to the original state. Through alteration of abiotic conditions, climate change on its own is one of the generators of novel ecosystems, creating change that likely cannot be reversed. Moving along the *x* axis in Figure 3.2 (abiotic conditions), in many places we have crossed from historical to hybrid ecosystems already; it is currently impossible to say when we will cross into full global novel ecosystems, however (or if we ever will, on such a broad scale). While it is possible to follow trajectories parallel to the *x* axis with a minimum of alteration to biotic components of ecosystems, the biotic and abiotic axes are in fact dependent to a greater or lesser extent: virtually all of our understanding of the factors driving community composition suggests independence is very unlikely. Under a novel ecosystems framework we may shift our focus from preserving individual species, habitats or ecosystems as we have understood them in recent history, to a realization that disturbed and changed ecosystems can have significant biodiversity value in and of themselves (see Chapters 3, 18 and 39; Thompson and Starzomski 2007). We may shift resources from attempting to stop certain changes, or attempting to reverse those changes, and instead learn to embrace novelty through maintaining significant functions. In many locations we are already doing this, in fact. There may be biodiversity value in maintaining some ecosystem changes to support significant biodiversity despite the lack of a completely pristine system (Zavaleta et al. 2001; Graves and Shapiro 2003; Schlaepfer et al. 2011).

10.2 IMPACTS OF CLIMATE CHANGE ON BIODIVERSITY

There is strong evidence that climate change is having and will continue to have a global impact on natural systems and biodiversity (Bradshaw and Holzapfel 2001; McCarty 2001; Walther et al. 2002; Parmesan and Yohe 2003; Thomas et al. 2004; Alleaume-Benharira et al. 2006; Parmesan 2006; IPCC 2007; Loarie et al. 2009; Meier et al. 2012). Northern hemisphere polar sea ice has reached historical lows, permafrost limits and treelines continue to move poleward or upward in elevation, and disturbance events such as fire and pest outbreaks have increased in frequency (McCarty 2001; Walther et al. 2002; Root et al. 2003; Parmesan 2006; IPCC 2007). These changes pose substantial challenges for a society that has adapted to a

relatively stable period in the Earth's climatic history. Despite the fact that there have been more than 20 ice-age events in the last 1.8 million years, humans have been fortunate to avoid major oscillations in global climate in the most recent 2000 years (Pielou 1991). This is changing.

10.2.1 Global patterns

Climate changes influence the distributions of species through altered species-specific physiological, reproductive and phenological responses to temperature and precipitation shifts (McCarty 2001; Walther et al. 2002; Parmesan 2006; Bellard et al. 2012). Indirect effects of climate change include decreased fitness due to changes in habitat, food, disease and parasitism. Much biotic change is occurring in an asynchronous fashion with prey and predator species changing differently, consequently disrupting important interspecific interactions within and between trophic levels (Walther et al. 2002). All of these factors may lead to ecosystem disruption (McCarty 2001; Parmesan 2006) and novel ecosystem development. Climate change probably now impacts biodiversity at the same rate as land-use change (Thomas et al. 2004; Parmesan 2006); their combined impacts are certainly substantial (e.g. Travis 2003; Kerr and Cihlar 2004; Kerr and Deguise 2004; Venter et al. 2006).

10.2.2 Species range shifts

Species are adapting to the effects of climate change by moving in the directions predicted by climate models and climate change scenarios (Rosenzweig et al. 2007). Studies have shown that 59% of studied species have responded with phenological or range shifts in the past 140 years (Parmesan 2006). Generally, these directions are upward (in mountains) and poleward (to exploit emerging poleward habitat).

Long-term datasets demonstrate changes in the distributions of species. Butterflies (Parmesan et al. 1999), birds (Hitch and Leberg 2007) and trees (Webb 1992; Pitelka 1997; Iverson and Prasad 1998, 2002; Davis and Shaw 2001; Iverson et al. 2004a, b; Parmesan 2006; Aitken et al. 2008) are among the taxa that show clear poleward and upward range shifts. On average, species move north at 6.1 km per decade and up in mountains at 6.1 m per decade (Parmesan and Yohe 2003).

Specific studies demonstrate the processes behind these changes. Danby and Hik (2007) show that the treeline in southwest Yukon has increased in density and moved upward in elevation since the early 20th century. Increases in stand density were correlated with higher summer temperature and were strongly correlated 30–50 years after the strongest warming event, demonstrating the important role of tree recruitment in explaining these patterns (Danby and Hik 2007). In other words, current impacts of climate change may take decades to be detected in long-lived slow-growing species. While higher treelines may not be novel ecosystems per se, this example demonstrates how changes can be latent for many years.

Movement for many plants and animals may be constrained by soil processes. In particular, mycorrhizal associations with plant roots may be an important process in establishment of species beyond their current ranges (Lafleur et al. 2010), although we know little about the distribution of local soil characteristics like these. In some cases however, research has shown that soil fungal associations may be appropriate beyond current range boundaries (Kernaghan and Harper 2001). Still, we know little about these processes, and they may well play a significant role in determining whether species can move poleward and upward as quickly as some models suggest (de Vries et al. 2012).

These changes have important consequences for high-latitude countries. For example, many species and habitats in Canada are located on the northern edge of their ranges (e.g. Carolinian forest and associated species in southern Ontario, dry interior species in the Okanagan, Garry Oak habitat in southwestern British Columbia and coastal plain flora and fauna in Nova Scotia; Kerr and Cihlar 2004; Kerr and Deguise 2004; Freemark et al. 2006; Venter et al. 2006). High-latitude countries may have an important role to play in conserving and restoring leading-edge species that provide important dispersing propagules in response to climate change (see Box 10.1; Gibson et al. 2009). These countries may also be subject to the greatest relative changes in communities and ecosystems, and may face the greatest challenges from developing novel ecosystems. When dealing with changes in species range shifts, the first adaptation thought is generally to ensure a high degree of landscape connectivity.

10.2.3 Changes in phenology

Changes in phenology are common responses to climate change (Root and Hughes 2005; Parmesan 2006; Rosenzweig et al. 2007). For example, breeding date for a variety of birds has advanced by up to a month (Crick et al. 1997; Winkel and Hudde 1997; McCleery and Perrins 1998; Brown et al. 1999; Dunn and Winkler 1999; Slater 1999). Other studies show advanced migration dates for bird species (Mason 1995; Bradley et al. 1999; Inouye et al. 2000). Parmesan (2006) and Rosenzweig et al. (2007) provide comprehensive reviews of the results of climate-related phenological changes across invertebrates, vertebrates and plants.

To demonstrate the novel ecosystem components of phenological changes, we can look at changes in abiotic and biotic components of the life history of the pied flycatcher (*Ficedula hypoleuca*) in the Netherlands. Both and colleagues (Both and Visser 2001; Both et al. 2006) have described a dramatic decline of up to 90% in pied flycatchers. This is attributed to a difference in the timing of events for two species that had historically been tightly coupled: the flycatcher and its caterpillar prey. The pied flycatcher times its migration by abiotic cues, using length of day to determine when to leave its wintering grounds in northern Africa. Its migration cue is therefore an abiotic one unrelated to climate change. However, the food source it times breeding around (the tent caterpillar) has a biotic cue to determine the initiation of its spring activity and period of maximum abundance. This cue has advanced with a warming climate. Caterpillars have subsequently become most numerous ahead of the spring arrival of the flycatcher. The mismatch of biotic and abiotic cues across species has lead to substantial declines in the flycatcher. In contrast, populations of pied flycatchers which breed further north in Scandinavia, and subsequently arrive on their breeding grounds in synchrony with abundance of prey, have not declined (Both et al. 2006). These species-level individualistic responses and mismatches make multispecies community and ecosystem-level prediction difficult. They also provide practitioners of ecological restoration with significant non-local challenges to their work.

10.2.4 Changes in food webs

Food webs are altered due to the individualistic responses of species to changing climate (Fig. 10.2;

Box 10.1 Climate change, species at risk and the border: embracing leading-edge species as propagules for novel ecosystems

The 49th parallel that forms the border between the United States and Canada cuts through multiple ecoregions, effectively leaving the majority of these ecoregions south of the border with small northern extents poking into Canada (Fig. 10.1). Associated with these are the majority of the country's at-risk species, endangered and threatened species that are rare partly because their habitat has small spatial extent in Canada. In many cases (but not all), these species are much more common to the south. The Committee on the Status of Endangered Wildlife in Canada (COSEWIC), which decides on the status of species at-risk, has always struggled with whether to list these species that may be rare in the northern part of their global range within Canada but relatively common to the south (Bunnell et al. 2004; Gibson et al. 2009). Climate change is acting on these regions and, as elsewhere in the world, species

are moving poleward. Canada probably has a global responsibility to protect and maintain populations of species at risk along this border, as these individuals may already be pre-adapted to the cooler conditions north of their range. They also provide the leading edge of species moving with climate change. Leading edges of species ranges and ecoregions are locations where the impacts of climate change are likely to occur very strongly and produce novel assemblages of species the most rapidly. These novel ecosystems are likely to make the decisions surrounding species at-risk in these regions very difficult, as new species colonize Canada from the south and formerly at-risk species expand northward. By actively managing these species, Canada plays a global role in ensuring that leading-edge peripheral species are available to colonize poleward habitats as part of biodiversity adaptation to climate change.

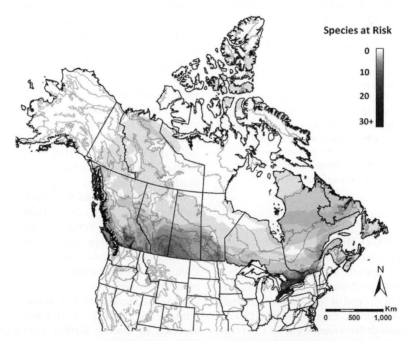

Figure 10.1 Map of northern North America showing ecozone boundaries (light gray lines) and concentrations of at-risk species in Canada (shown in shades of gray; darker indicating more at-risk species). Note the concentrations of at-risk species in Canada that correspond to the southern border (the 49th parallel through much of western North America) and the parts of ecoregions that have only a minority of their area in Canada.

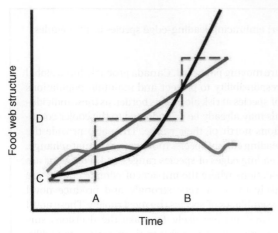

Figure 10.2 Food webs change through time by processes of ecological drift (green line), although these processes have been altered as a consequence of climate change. As climate changes in time (along the *x* axis), limits of species physiology, species range and phenology (which can change together or independently) will be crossed, resulting in novel altered food web structure (blue, red and black lines). Points A and B correspond to the points where, moving along the *x* axis (abiotic conditions) on Fig. 3.2, we cross from a historical to hybrid system (point A) and then from a hybrid to a novel system (point B). Points C and D show the upper and lower bounds in historical food web structure. Movement beyond these bounds results in novel communities characterized by altered community membership or drastically altered dominance–diversity relationships. The blue line shows changes that occur with a regular pace that we may be able to adapt to. The black line shows accelerating change in food web structure with changes in climate. The red line shows discontinuous change that poses the greatest challenges to management, where changes are abrupt, strong and unpredictable (i.e. surprising).

Root and Schneider 2002). This has been observed historically, for example. Reconstructions of past dominant vegetation in eastern North America show different communities developed as trees repopulated from glacial refugia at the end of the Wisconsin glacial period about 15,000 years before present (Pielou 1991; Davis and Shaw 2001; Overpeck et al. 1991) or at similar changes in Europe (Stone et al. 2008). In the recent past, North America was covered with communities of tree species that we do not see together in the modern climate (for examples see Davis 1969 and Pielou 1991).

Individualistic responses to climate change also result in breakdown of trophic structure. For example, with increased atmospheric CO_2 the defensive structure of soybean is reduced and the invasive Japanese beetle (*Popillia japonica*) and a variant of western corn rootworm (*Diabrotica virgifera virgifera*) have increased success on the plant. Increased levels of CO_2 can indirectly facilitate the success of an invasive species, further affecting food web structure (Zavala et al. 2008) and complicating restoration activities. Species that once played a minor role in the food web can gain increased importance (e.g. Berlow 1999), ameliorating the impacts of climate change on system function but also altering the structure of the system in novel ways.

10.2.5 Changes in climate–ecosystem relations

Some studies have shown dramatic changes in the interaction of ecosystems and the climate (e.g. Chapin et al. 2005). Changes in albedo associated with amount of snow cover cause increased warming through a positive feedback mechanism. As the length of the snow-free year increases, more energy is absorbed by the dark-colored ground, further increasing warming. This causes a feedback loop that increases the length of the snow-free year. This has been shown in a synthesis of experiments in northern Alaska (Chapin et al. 2005), and may signal increased permafrost melting and carbon release (Jorgenson et al. 2001; Serreze et al. 2002; Zhang et al. 2005). Changes to the hydrologic cycle have been detected through vegetation composition and ecosystem functioning, as well as carbon dioxide and methane fluxes that occur together with permafrost degradation (McNamara et al. 1998; Jorgenson et al. 2001; Christensen et al. 2004; Smith et al. 2005). These results are worrying, as habitat relations may change over a large part of the near-polar regions; there is a very real possibility of increased C emissions from melting permafrost and the biotic material frozen within it (Davidson and Janssens 2006). The increased C emissions from these habitats (Zhuang et al. 2006; Euskirchen et al. 2007) are expected to lead to further warming in a positive feedback.

Changes in disturbance rates such as pest, fire and drought are also being seen (Parmesan 2006; Kurz et al. 2008; Lemmen et al. 2008). Increased drying and temperature increases in locations around the globe may lead to increases in fire ignition and extent (Krawchuk et al. 2009). In places such as western North America, climate changes can also lead to exacerbation of forest insect pests such as the mountain pine beetle (*Dendroctonus ponderosae*) (Kurz et al. 2008). These indirect effects of climate change pose significant problems for novel ecosystem development. Large-scale disturbances such as fire and insect defoliation do not guarantee that the same ecosystems will re-develop after the disturbance (Holling 1973; Folke et al. 2004; Scheffer 2009). There is a very real threat of wholesale ecosystem change after disturbances like these, where the novel ecosystems that form are very different to the familiar ecosystem that existed pre-disturbance.

10.3 PREDICTIONS FOR FUTURE ECOSYSTEMS

A primary means for understanding the impacts of climate change is to develop predictions, typically through climate-envelope models. An important example is Thomas et al. (2004), which showed that between 15% and 37% of all species (on the 20% of the Earth's surface that they studied) may be committed to extinction by 2050.

Climate-envelope models are parameterized with the results of analyses of the contemporary relationships between where species are located and the climatic conditions at those locations (Iverson and Prasad 1998, 2002; Peterson et al. 2002; Pearson and Dawson 2003; Araújo and Guisan 2006). By constructing a relationship between where species are currently found and the climatic conditions there, a model can be constructed that, when simulated using projected climate data, predicts the future ranges of species (Pearson and Dawson 2003; Araújo and Guisan 2006; Araújo and Luoto 2007; Williams et al. 2007). Examples abound, including changes in biodiversity habitat in Mexico (Peterson et al. 2002), Europe (Berry et al. 2002; Thuiller et al. 2006) and eastern and western North America (Iverson and Prasad 1998, 2002; Currie 2001; Shafer et al. 2001) and models of the predicted impacts on protected areas (Araújo et al. 2004) and in the habitats of various animal species (Kerr and Packer 1998; Levinsky et al. 2007).

Parameterizing climate-envelope models is of course difficult, and their predictions are anything but certain (Davis et al. 1998; Hampe 2004; Araújo et al. 2005; Araújo and Rahbek 2006; Pearson et al. 2006; Botkin et al. 2007; Morin and Thuiller 2009; Buckley et al. 2010, 2011). Because present-day distributions of species tell us nothing about a species' ability to adapt to new climatic conditions, all predictions are wrong in ways we are still struggling to understand (Fitzpatrick and Hargrove 2009). Do these models over- or underpredict the capacity for species to adapt to climate change and the novel ecosystems that may develop? Some of the most recent predictions include examples of novel assemblages of species developing over half of a jurisdiction (see for example Stralberg et al. 2009 for California birds). Significant changes in community composition have certainly occurred through time (Davis 1969; Overpeck et al. 1991; Pielou 1991), but never have these changes occurred with such speed and with so many other threats to species (Travis 2003; Parmesan 2006; Loarie et al. 2009). The cumulative effects of the threats facing biodiversity mean that development of novel ecosystems is a certainty. What these predictive models offer are scenarios for what we might expect in the future and are useful for avoiding overwhelming ecological surprises (Scheffer 2009).

10.4 CLIMATE CHANGE, MANAGED RELOCATION AND NOVEL ECOSYSTEMS

One proposed solution for avoiding ecological surprises is to design the ecosystem we think may be present in the future. Managed relocation (often called 'assisted migration' or even 'assisted invasion') seeks to do this by moving at-risk species into what is proposed to be future habitat. By definition, managed relocation creates novel ecosystems through the anthropogenic introduction of at-risk species to poleward (or upward) habitats (McLachlan et al. 2007; Hewitt et al. 2011). Perhaps the loudest arguments related to climate change and novel ecosystems have played out in the realm of managed relocation/assisted migration in the past few years (McLachlan et al. 2007; Ricciardi and Simberloff 2009; Sax et al. 2009). At the very least, consideration of managed relocation for species at risk of extinction due to climate change requires massive amounts of data on the risks of a relocated species becoming invasive (the primary concern in this debate; see Richardson et al. 2009; Hewitt et al. 2011). Nevertheless, managed relocation is likely to become a component of management without sufficient data being available. As Hewitt et al. (2011) point out, little of the scientific debate has trickled out of the literature to the general public. Assisted migration is underway in several situations (including, most famously, *Torreya taxifolia* and, in another guise, gardens around the world), and the perceived risk trade-off appears to underwhelm all but a few commentators (e.g. Ricciardi and Simberloff 2009). Certainly the risks can be great and should not be glossed over. The best way forward in dealing with managed relocation is probably to collect as many data as possible to decrease the risks of moving species whose conservation has no other option. On continents, at least, the risks associated with these movements may be less than on islands (Sax et al. 2009; Chapter 4). Novel ecosystems are developing primarily through natural movements of species in response to abiotic and biotic changes in the environment; managed relocation is a relatively minor component of this.

10.4.1 How are our concepts of ecological restoration altered in the face of climate change?

Perhaps the fundamental issue facing the field of ecological restoration is that there are no longer historical reference ecosystems that can be used to guide restoration activities. Here, too, acceptance of ecosystem function as the primary guide for restoration may be most important (Harris et al. 2006). The development of novel ecosystems refocuses our efforts from species assemblages to ecosystem function.

Along these lines, invasive species are normally considered negative components of systems to be restored, and are often among the first targets of ecological restoration. Invasive species have been considered the 'drivers' or 'passengers' of change (sensu MacDougall and Turkington 2005), and are most often recognized as being the passengers (that is, many are a consequence of anthropogenic change on a system, more so than the primary driver of change in an ecosystem). Perhaps we are already on the road to accepting that species moving as a consequence of climate change are also passengers. We have yet to be concerned about how the arrival of new 'native' species will impact the ecological health of poleward regions, at least at the level we have been worried about invasives. Perhaps we have already accepted novel assemblages of species as healthy and unavoidable, and need to get on with figuring out how to rearrange our management to deal with these new species.

10.4.2 Contribution of novel ecosystems concept to climate change understanding

Robert MacArthur asked if we would ever be able to predict species identity in community ecology: "Will the explanation . . . degenerate into a tedious set of case histories, or is there some common pattern running through them all?" (MacArthur 1972). His work offered 'MacArthur's Paradox': a neutral theory for community membership (with E.O. Wilson, the theory of island biogeography; MacArthur and Wilson 1967) and a competition-based niche model (with R. Levins, MacArthur and Levins 1967). We have made substantial progress in understanding the structure of communities since then, although we still do not have a tool to predict the identity of species in a community. The tools we have developed, that often either predict the number of species (but not their identities) or variety of traits in a community, suggest that the novel ecosystem concept (where function is valued over specifies identity) may be important for managing climate change impacts on biodiversity. First, community ecology research shows the importance of history in determining the membership of communities (through priority effects; Fukami 2004; Starzomski et al. 2008). The community ecology literature is filled with neutral theories that will predict the number, but not identity, of species in a community (e.g. MacArthur and Wilson 1967; Hubbell 2001). In addition, theories that may be used to determine community membership are numerous (e.g. Vellend 2010), and many of those theories directly contradict others (e.g. Ricklefs 2011). Finally, some work suggests that while communities may develop along similar lines in terms of their trait composition, this may not hold for species identity (Fukami et al. 2005). To the extent that the novel ecosystems concept might allow us to make some useful generalizations about maintaining the trait or functional equivalence of a community, it is an important component of our management toolbox.

As has been pointed out in other chapters, the real benefit of the novel ecosystems concept is its pragmatism. The ability to value species or communities of species that we typically ignore or outright seek to destroy (e.g. Rodrigues Fody in Marris 2011; Chapter 3; Fig. 3.1) may be its greatest strength with regard to climate change. To begin with, the novel ecosystems concept provides us with a theoretical framework that allows us to value novel assemblages of species rather than always 'restoring' a collection of species to some historical condition: some authors have termed this a focus on 'ecosystem assemblages or functioning ecosystems' (Harris et al. 2006). Embracing the dynamic aspects of communities and ecosystems may allow us to value those changes we see more than we have in the past, where conservation and preservation of the status quo seemed much more important (Hobbs et al. 2009, 2010). The novel ecosystems concept also provides a framework for valuing the functional role of a species over its identity, independent of its other anthropogenic attributes (e.g. whether it is an introduced or invasive species; Chapter 3). While these decisions may never be easy (one introduced species may be beneficial while another may not; e.g. Severns and Warren 2008), we do have the capacity to be specific about the species' function under this framework.

10.4.3   Embracing novel assemblages

Perhaps the most important impact of climate change on biodiversity is the creation of novel assemblages of species. This is not new in the history of the Earth, although the speed at which it is happening (Loarie et al. 2009) is probably the most difficult thing for scientists and managers to deal with. At the species level this may have local impacts on management of species at risk, or cause continent-wide changes in how species at-risk are listed or valued (Gibson et al. 2009). While we may envision a hierarchy of importance of species in terms of their management values (all biodiversity may be equal but some is more equal: endangered species should have highest priority), we may use the novel ecosystems framework to value functional roles most highly. This could take the form of maintaining the capacity of a system to provide the functions (i.e. ecosystem services) we may be most interested in and, if the species or assemblage provides that function (and does not simultaneously cause a decline in other valued ecosystem attributes), then we may value those species. The Rodrigues Fody example already mentioned earlier shows how this may work. Other examples abound (e.g. Graves and Shapiro 2003; Rodriguez 2006). This is not to say that we throw out the baby with the proverbial bathwater: much research shows that those regions that are already dominated or damaged by

local human activity (i.e. not by global effects such as climate change) suffer the brunt of the impacts of, for example, invasive species. It therefore remains key to maintain as much as possible of the most resilient and valuable wild areas relatively untouched by the non-climate-change impacts of human activity (Hobbs et al. 2010). In other words, creation of protected areas remains a priority. Further, since ecological restoration can bring back significant biodiversity values (Rey Benayas et al. 2009), we must not be afraid of utilizing functional equivalents in our restoration activities.

Although climate-envelope modeling can make predictions about potential communities of species (based on climate–species relationships), no tool currently exists to tell us exactly what species may be found in a given area in the future. Moving beyond a focus on restoring historical assemblages of species and embracing novelty in species assemblages must become part of our management toolbox. The question for novel ecosystems and climate change can therefore be broken into two parts: do we manage for adaptation to climate change using novel ecosystems, or do we manage to avoid novel ecosystems? This chapter makes the point that we gain much more by focusing on the former. To the extent that we have passed the point where it is possible to avoid their development, we must embrace novel ecosystems and add them to our conservation and management toolbox.

10.4.4   Novel ecosystems in the current adaptation mix

The responses to climate change typically fall under the umbrella of 'adaptation', and most developed strategies for adaptation to climate change involve the following:
1. gathering knowledge about climate change impacts on biodiversity through a combination of observation and projections of likely future scenarios;
2. minimizing the impacts of identified threats;
3. incorporating knowledge and harm-minimization strategies into management activities in an adaptive management framework;
4. implementing emissions mitigation strategies; and
5. developing new monitoring programs with identification of thresholds for management.
While these management techniques do not preclude the development and existence of novel ecosystems, in general they do not provide much guidance for dealing

with novel ecosystems encountered under climate change. Nevertheless, when a central component of novel ecosystems management – embracing change – is included in the mix, managing climate change effects on ecosystems becomes much more flexible. Embracing change is not always simple, something which highlights the importance of the social aspect of conservation. Ensuring that communities of people who enjoy and use the resource are as much part of the process of management as they are part of the landscape is key to successful management (Heller and Zavaleta 2009).

Sutherland et al. (2006) for the United Kingdom and Lucier et al. (2006) and Fleishman et al. (2011) in the United States provide lists of important conservation and ecological restoration questions. While the originals include more than 150 questions, some can be restated and developed with a climate change and novel ecosystems focus to show where we can focus research efforts to understand important roles of future novel ecosystems.

• What are the benefits of protected areas for carbon sequestration when novel ecosystems might be expected? Does the carbon sequestration function continue under novel ecosystems?
• How effective is the current protected areas network for protecting species into the future?
• Are novel ecosystems likely to be net carbon sinks or sources with climate change?
• What are the benefits of restoring damaged areas to historical assemblages? How does this compare to natural regeneration? What reference ecosystems do we use? Preservation of remaining intact habitats is the priority, but how is the role of restoring degraded ecosystems changed in light of climate change and the development of novel ecosystems?
• How does climate change interact with other ecological pressures (e.g. habitat loss and fragmentation) to create synergistic and cumulative effects on novel ecosystems?
• How can we increase the resilience of habitats and species to climate change by using novel ecosystems?
• What are the likely changes in disturbance rates and intensities in novel ecosystems? How does this impact biodiversity?
• What time lags can be expected between climate change and development of novel ecosystems?
• How many species extinctions might we expect in novel ecosystems?
• What are the relative strengths of biotic and abiotic changes in structuring novel ecosystems?

• Which ecosystems or ecosystem properties are most susceptible to slow and fast changes due to climate change?
• What new novel ecosystems might we expect following predicted fire/pest outbreaks and changed climate regimes?
• How do we scale observations for individual species to the habitat and landscape scale so that we might predict the development of novel ecosystems?
• What measures should be adopted as thresholds in relation to climate change and novel ecosystem development? How will these thresholds be monitored and used to design adaptation strategies?
• How will the relationship between the conservation value of a protected area versus outside the protected area change with novel ecosystems?

In the end, the novel ecosystems concept is important in light of climate change because it allows for: (1) recognizing when we have crossed a community composition line where it would be too expensive or impossible to go back; (2) a focus on managing for functional equivalents; and (3) prepping our management regimes for future states where we expect novel components and, using predictions and scenarios, planning for versions of these. We can also recognize places where these changes are likely to happen fastest or be most important (e.g. for peripheral populations or species' range edges).

10.5   CONCLUSIONS AND WHERE TO FROM HERE

Climate change is causing the development of novel ecosystems. Species respond individualistically to the impacts of climate changes and therefore the development of novel assemblages of species in the landscape is a fundamental observation of the impacts of climate change on biodiversity. Embracing novelty does not mean forsaking current biodiversity values. On the contrary, the large-scale dynamics that drive the structure of communities may become more obvious because of the movement of organisms on the landscape, at rates faster than we have been ready to ascribe as 'normal' (we seem to have no trouble accepting that invasive species move rapidly around the landscape).

Accepting arguments from functional equivalence might be key to properly managing our ecosystems. To the extent that we are comfortable with patterns in the landscape (biodiversity) leading to processes that we need/enjoy (biodiversity, ecosystem function and services), then we may accept changed background patterns to maintain those processes and functions (i.e. species change but a functional equivalence remains). Novel ecosystems explicitly allow for this, and this pragmatism may be their greatest strength in providing a framework for management in the face of climate change. At the very least we require continued protection of intact ecosystems (including a high degree of landscape connectivity, as mentioned earlier), and updated management for human-dominated ecosystems. Increasing protection of intact areas, as well as maintenance of connectivity between them, are priorities to maximize ecosystem resilience in the face of climate change. We may need to further develop climate-envelope models and the like to provide realistic simulations and predictions for future biodiversity–climate scenarios. Ecological restoration activities may come to rely more on the outputs of these predictive models to restore to novel states rather than historical references.

REFERENCES

Aitken, S.N., Yeaman, S., Holliday, J.A., Wang, T. and Curtis-McLane, S. (2008) Adaptation, migration or extirpation: Climate change outcomes for tree populations. *Evolutionary Applications*, **1**, 95–111.

Alleaume-Benharira, M., Pen, I.R. and Ronce, O. (2006) Geographical patterns of adaptation within a species' range: interactions between drift and gene flow. *Evolutionary Biology*, **19**, 203–215.

Araújo, M.B. and Guisan, A. (2006) Five (or so) challenges for species distribution modelling. *Journal of Biogeography*, **33**, 1677–1688.

Araújo, M.B. and Rahbek, C. (2006) How does climate change affect biodiversity? *Science*, **313**, 1396–1397.

Araújo, M.B. and Luoto, M. (2007) The importance of biotic interactions for modelling species distributions under climate change. *Global Ecology and Biogeography*, **16**, 743–753.

Araújo, M.B., Cabeza, M., Thuiller, W., Hannah, L. and Williams, P.H. (2004) Would climate change drive species out of reserves? An assessment of existing reserve-selection methods. *Global Change Biology*, **10**, 1618–1626.

Araújo, M.B., Pearson, R.G., Thuiller, W. and Erhard, M. (2005) Validation of species-climate impact models under climate change. *Global Change Biology*, **11**, 1504–1513.

Bellard, C., Bertelsmeier, C., Leadley, P., Thuiller, W. and Courchamp, F. (2012) Impacts of climate change on the future of biodiversity. *Ecology Letters*, **15**, 365–377.

Berlow, E.L. (1999) Strong effects of weak interactions in ecological communities. *Nature*, **398**, 330–334.

Berry, P.M., Dawson, T.P., Harrison, P.A. and Pearson, R.G. (2002) Modelling potential impacts of climate change on the bioclimatic envelope of species in Britain and Ireland. *Global Ecology and Biogeography*, **11**, 453–462.

Both, C. and Visser, M.E. (2001) Adjustment to climate change is constrained by arrival date in a long-distance migrant bird. *Nature*, **411**, 296–298.

Both, C., Bouwhuis, S., Lessells, C.M. and Visser, M.E. (2006) Climate change and population declines in a long-distance migratory bird. *Nature*, **441**, 81–83.

Botkin, D., Saxe, H., Araújo, M.B., Betts, R., Bradshaw, R., Cedhagen, T., Chesson, P., Davis, M.B., Dawson, T., Etterson, J., Faith, D.P., Guisan, A., Ferrier, S., Hansen, A.S., Hilbert, D., Kareiva, P., Margules, C., New, M., Skov, F., Sobel, M.J. and Stockwell, D. (2007) Forecasting effects of global warming on biodiversity. *Bioscience*, **57**, 227–236.

Bradley, N.L., Leopold, A.C., Ross, J. and Huffaker, W. (1999) Phenological changes reflect climate change in Wisconsin. *Proceedings of the National Academy of Sciences*, **96**, 9701–9704.

Bradshaw, W.E. and Holzapfel, C.M. (2001) Genetic shifts in photoperiodic response correlated with global warming. *Proceedings of the National Academy of Sciences*, **98**, 14509–14511.

Brown, J.L., Li, S.H. and Bhagabati, N. (1999) Long-term trend toward earlier breeding in an American bird: a response to global warming? *Proceedings of the National Academy of Sciences*, **96**, 5565–5569.

Buckley, L.B., Urban, M.C., Angilletta, M.J., Crozier, L.G., Rissler, L.J. and Sears, M.W. (2010) Can mechanism inform species' distribution models? *Ecology Letters*, **13**, 1041–1054.

Buckley, L.B., Waaser, S.A., MacLean, H.J. and Fox, R. (2011) Does including physiology improve species distribution model predictions of responses to recent climate change? *Ecology*, **92**, 2214–2221.

Bunnell, F.L., Campbell, R.W. and Squires, K.A. (2004) Conservation priorities for peripheral species: the example of British Columbia. *Canadian Journal of Forest Resources*, **34**, 2240–2247.

Chapin, F.S., Sturm, M., Serreze, M.C., McFadden, J.P., Key, J.R., Lloyd, A.H., McGuire, A.D., Rupp, T.S., Lynch, A.H., Schimel, J.P., Beringer, J., Chapman, W.L., Epstein, H.E., Euskirchen, E.S., Hinzman, L.D., Jia, G., Ping, C.-L., Tape, K.D., Thompson, C.D.C., Walker, D.A. and Welker, J.M. (2005) Role of land-surface changes in arctic summer warming. *Science*, **310**, 657–660.

Christensen, T.R., Johansson, T., Okerman, H.J.A., Mastepanov, M., Malmer, N., Friborg, T., Crill, P. and Svensson, B.H. (2004) Thawing sub-arctic permafrost: Effects on vegetation and methane emissions. *Geophysical Research Letters*, **31**, L04501.

Crick, H.Q.P., Dudley, C., Glue, D.E. and Thomson, D.L. (1997) UK birds are laying eggs earlier. *Nature*, **388**, 526.

Currie, D.J. (2001) Projected effects of climate change on patterns of vertebrate and tree species richness in the conterminous United States. *Ecosystems*, **4**, 216–225.

Danby, R.K. and Hik, D.S. (2007) Variability, contingency and rapid change in recent subarctic alpine tree line dynamics. *Journal of Ecology*, **95**, 352–363.

Davidson, E.A. and Janssens, I.A. (2006) Temperature sensitivity of soil carbon decomposition and feedbacks to climate change. *Nature*, **440**, 165–173.

Davis, A.J., Jenkinson, L.S., Lawton, J.L., Shorrocks, B. and Wood, S. (1998) Making mistakes when predicting shifts in species range in response to global warming. *Nature*, **391**, 783–786.

Davis, M.B. (1969) Climatic changes in southern Connecticut recorded by pollen deposition at Rogers Lake. *Ecology*, **50**, 409–422.

Davis, M.B. and Shaw, R.G. (2001) Range shifts and adaptive responses to quaternary climate change. *Science*, **292**, 673–678.

de Vries, F.T., Liiri, M.E., Bjørnlund, L., Bowker, M.A., Christensen, S., Setälä, H.M. and Bardgett, R.D. (2012) Land use alters the resistance and resilience of soil food webs to drought. *Nature Climate Change*, doi: 10.1038/nclimate1368.

Dunn, P.O. and Winkler, D.W. (1999) Climate change has affected breeding date of tree swallows throughout North America. *Proceedings of the Royal Society B, London*, **266**, 2487–2490.

Euskirchen, E.S., McGuire, A.D. and Chapin, F.S. III (2007) Energy feedbacks of northern high-latitude ecosystems to the climate system due to reduced snow cover during 20th century warming. *Global Change Biology*, **13**, 2425–2438.

Fitzpatrick, M.C. and Hargrove, W.W. (2009) The projection of species distribution models and the problem of non-analog climate. *Biodiversity and Conservation*, **18**, 2255–2261.

Fleishman, E., Blockstein, D.E., Hall, J.A., Mascia, M.B., Rudd, M.A., Scott, J.M., Sutherland, W.J., Bartuska, A.M., Brown, A.G., Christen, C.A., Clement, J.P., Dellasala, D., Duke, C.S., Eaton, M., Fiske, S.J., Gosnell, H., Haney, J.C., Hutchins, M., Klein, M.L., Marqusee, J., Noon, B.R., Nordgren, J.R., Orbuch, P.M., Powell, J., Quarles, S.P., Saterson, K.A., Savitt, C.C., Stein, B.A., Webster, M.S. and Vedder, A. (2011) Top 40 priorities for science to inform US conservation and management policy. *BioScience*, **61**, 290–300.

Folke, C., Carpenter, S., Walker, B., Scheffer, M., Elmqvist, T., Gunderson, L. and Holling, C.S. (2004) Regime shifts, resilience, and biodiversity in ecosystem management. *Annual Review of Ecology, Evolution, and Systematics*, **35**, 557–581.

Freemark, K.E., Meyers, M., White, D., Warman, L.D., Kister, A.R. and Lumban-Tobing, P. (2006) Species richness and biodiversity conservation priorities in British Columbia, Canada. *Canadian Journal of Zoology*, **84**, 20–31.

Fukami, T. (2004) Assembly history interacts with ecosystem size to influence species diversity. *Ecology*, **85**, 3234–3242.

Fukami, T., Martijn Bezemer, T., Mortimer, S.R. and Putten, W.H. (2005) Species divergence and trait convergence in experimental plant community assembly. *Ecology Letters*, **8**, 1283–1290.

Gibson, S.Y., van der Marel, R.C. and Starzomski, B.M. (2009) Climate change and conservation of leading-edge peripheral populations. *Conservation Biology*, **23**, 1369–1373.

Graves, S.D. and Shapiro, A.M. (2003) Exotics as host plants of the California butterfly fauna. *Biological Conservation*, **110**, 413–433.

Hampe, A. (2004) Bioclimate envelope models: What they detect and what they hide. *Global Ecology and Biogeography*, **13**, 469–471.

Harris, J.A., Hobbs, R.J., Higgs, E. and Aronson. J. (2006) Ecological restoration and global climate change. *Restoration Ecology*, **14**, 170–176.

Heller, N.E. and Zavaleta, E.S. (2009) Biodiversity management in the face of climate change: A review of 22 years of recommendations. *Biological Conservation*, **142**, 14–32.

Hewitt, N., Klenk, N., Smith, A.L., Bazely, D.R., Yan, N., Wood, S., MacLellan, J.I., Lipsig-Mumme, C. and Henriques, I. (2011) Taking stock of the assisted migration debate. *Biological Conservation*, **144**, 2560–2572.

Hitch, A.T. and Leberg, P.T. (2007) Breeding distributions of North American bird species moving north as a result of climate change. *Conservation Biology*, **21**, 534–539.

Hobbs, R.J., Arico, S., Aronson, J., Baron, J., Bridgewater, P., Cramer, V.A., Epstein, P.R., Ewel, J.J., Klink, C.A., Lugo, A.E., Norton, D., Ojima, D., Richardson, D.M., Sanderson, E.W., Valladares, F., Vila, M., Zamora, R. and Zobel, M. (2006) Novel ecosystems: theoretical and management aspects of the new ecological world order. *Global Ecology and Biogeography*, **15**, 1–7.

Hobbs, R.J., Higgs, E.S. and Harris, J.A. (2009) Novel ecosystems: implications for conservation and restoration. *Trends in Ecology & Evolution*, **24**, 599–605.

Hobbs, R.J., Cole, D.N., Yung, L., Zavaleta, E.S., Aplet, G.H., Chapin, F.S., Landres, P.B., Parsons, D.J., Stephenson, N.L., White, P.S., Graber, D.M., Higgs, E.S., Millar, C.I., Randall, J.M., Tonnessen, K.A. and Woodley, S. (2010) Guiding concepts for park and wilderness stewardship in an era of global environmental change. *Frontiers in Ecology and the Environment*, **8**, 483–490.

Holling, C.S. 1973. Resilience and stability of ecological systems. *Annual Review of Ecology and Systematics*, **4**, 1–23.

Hubbell, S.P. (2001) *The Unified Neutral Theory of Biodiversity and Biogeography*. Princeton University Press, Princeton, NJ.

Inouye, D.W., Barr, B., Armitage K.B. and Inouye, B.D. (2000) Climate change is affecting altitudinal migrants and hibernating species. *Proceedings of the National Academy of Science of the United States of America*, **97**, 1630–1633.

Intergovernmental Panel on Climate Change (IPCC) (2007) Climate change 2007: Impacts, adaptation and vulnerability. Working Group II contribution to the Intergovernmental Panel on Climate Change fourth assessment report. Geneva: World Meteorological Organization. Available: http://www.ipcc.ch/pdf/assessment-report/ar4/syr/ar4_syr_spm.pdf.

Iverson, L.R. and Prasad, A.M. (1998) Predicting abundance of 80 tree species following climate change in the eastern United States. *Ecological Monographs*, **68**, 465–485.

Iverson, L.R. and Prasad, A.M. (2002) Potential redistribution of tree species habitat under five climate change scenarios in the eastern US. *Forest Ecology and Management*, **155**, 205–222.

Iverson, L.R., Schwartz, M.W. and Prasad, A.M. (2004a) Potential colonization of newly available tree-species habitat under climate change: An analysis for five eastern US species. *Landscape Ecology*, **19**, 787–799.

Iverson, L.R., Schwartz, M.W. and Prasad, A.M. (2004b) How fast and far might tree species migrate under climate change in the eastern United States. *Global Ecology and Biogeography*, **13**, 209–219.

Jorgenson, M.T., Racine, C.H., Walters, J.C. and Osterkamp, T.E. (2001) Permafrost degradation and ecological changes associated with a warming climate in central Alaska. *Climatic Change*, **48**, 551–579.

Kernaghan, G. and Harper, K.A. (2001) Community structure of ectomycorrhizal fungi across an alpine/subalpine ecotone. *Ecography*, **24**, 181–188.

Kerr, J.T. and Cihlar, J. (2004) Patterns and causes of species endangerment in Canada. *Ecological Applications*, **14**,743–753.

Kerr, J.T. and Deguise, I. (2004) Habitat loss and limits to recovery of endangered wildlife. *Ecology Letters*, **7**, 1163–1169.

Kerr, J.R. and Packer, L. (1998) The impact of climate change on mammal diversity in Canada. *Environmental Monitoring and Assessment*, **49**, 263–270.

Krawchuk, M.A., Moritz, M.A., Parisien, M.-A., Van Dorn, J. and Hayhoe, K. (2009) Global pyrogeography: the current and future distribution of wildfire. *PLoS ONE*, **4**, e5102.

Kurz, W.A., Dymond, C.C., Stinson, G., Rampley, G.J., Neilson, E.T., Carroll, A.L., Ebata, T. and Safranyik, L. (2008) Mountain pine beetle and forest carbon feedback to climate change. *Nature*, **452**, 987–990.

Lafleur, B., Paré, D., Munson, A.D. and Bergeron, Y. (2010) Response of northeastern North American forests to climate change: Will soil conditions constrain tree species migration? *Environmental Reviews*, **18**, 279–289.

Lemmen, D.S., Warren, F.J., Lacroix J. and Bush E. (eds) (2008) *From Impacts to Adaptation: Canada in a Changing Climate 2007*. Government of Canada, Ottawa, ON. 448 pp.

Levinsky, I., Skov, F., Svenning, J.-C. and Rahbek, C. (2007) Potential impacts of climate change on the distributions and diversity patterns of European mammals. *Biodiversity Conservation*, **16**, 3803–3816.

Loarie, S.R., Duffy, P.B., Hamilton, H., Asner, G.P., Field, C.B. and Ackerly, D.D. (2009) The velocity of climate change. *Nature*, **462**, 1052–1055.

Lucier, A., Palmer, M., Mooney, H., Nadelhoffer, K., Ojima, D. and Chavez, F. (2006) Ecosystems and Climate Change: Research Priorities for the U.S. Climate Change Science Program. Recommendations from the Scientific Community. Report on an Ecosystems Workshop, prepared for the Ecosystems Interagency Working Group. Special Series No. SS-92-06, University of Maryland Center for Environmental Science, Chesapeake Biological Laboratory, Solomons, MD, USA.

MacArthur, R.H. (1972) *Geographical Ecology: Patterns in the Distribution of Species*. Princeton University Press.

MacArthur, R.H. and Levins, R. (1967) The limiting similarity, convergence, and divergence of coexisting species. *American Naturalist*, **101**, 377–387.

MacArthur, R.H. and Wilson, E.O. (1967) *The Theory of Island Biogeography*. Princeton University Press.

MacDougall, A.S. and Turkington, R. (2005) Are invasive species the drivers or passengers of change in degraded ecosystems? *Ecology*, **86**, 42–55.

Marris, E. (2011) *Rambunctious Garden*. Bloomsbury, US.

Mason, C.F. (1995) Long-term trends in the arrival dates of spring migrants. *Bird Study*, **42**, 182–189.

McCarty, J.P. (2001) Ecological consequences of recent climate change. *Conservation Biology*, **15**, 320–331.

McCleery, R.H. and Perrins, C.M. (1998) Temperature and egg-laying trends. *Nature*, **391**, 30–31.

McLachlan, J.S., Hellmann, J.J. and Schwartz, M.W. (2007) A framework for debate of assisted migration in an era of climate change. *Conservation Biology*, **21**, 297–302.

McNamara, J.P., Kane, D.L. and Hinzman, L.D. (1998) An analysis of streamflow hydrology in the Kuparuk River basin, Arctic Alaska: A nested watershed approach. *Journal of Hydrology*, **206**, 39–57.

Meier, E.S., Lischke, H., Schmatz, D.R. and Zimmermann, N.E. (2012) Climate, competition and connectivity affect future migration and ranges of European trees. *Global Ecology and Biogeography*, **21**, 164–178.

Morin, X. and Thuiller, W. (2009) Comparing niche-and process-based models to reduce prediction uncertainty in species range shifts under climate change. *Ecology*, **90**, 1301–1313.

Overpeck, J.T., Bartlein, P.J. and Webb, T. (1991) Potential magnitude of future vegetation change in eastern North America: Comparisons with the past. *Science*, **254**, 692–695.

Parmesan, C. (2006) Ecological and evolutionary responses to recent climate change. *Annual Review of Ecology, Evolution, and Systematics*, **37**, 637–669.

Parmesan, C. and Yohe, G. (2003) A globally coherent finger-print of climate change impacts across natural systems. *Nature*, **421**, 37–42.

Parmesan, C., Ryrholm, N., Stefanescu, C., Hill, J.K., Thomas, C.D., Descimon, H., Huntley, B., Kaila, L., Kullberg, J., Tammaru, T., Tennent, J., Thomas, J.A. and Warren, M. (1999) Poleward shift of butterfly species' ranges associated with regional warming. *Nature*, **399**, 579–583.

Pearson, R.G. and Dawson, T.P. (2003) Predicting the impacts of climate change on the distribution of species: Are bioclimate envelope models useful? *Global Ecology and Biogeography*, **12**, 361–371.

Pearson, R.G., Thuiller, W., Araújo, M.B., Martinez-Meyer, E., Brotons, L., McClean, C., Miles, L., Segurado, P., Dawson, T.P. and Lees, D.C. (2006) Model-based uncertainty in species range prediction. *Journal of Biogeography*, **33**, 1704–1711.

Peterson, A.T., Ortega-Huerta, M.A., Bartley, J., Sánchez-Cordero, V., Soberón, J., Buddemeier, R.H. and Stockwell, D.R.B. (2002) Future projections for Mexican faunas under global climate change scenarios. *Nature*, **416**, 626–629.

Pielou, E.C. (1991) *After the Ice Age: The Return of Life to Glaciated North America*. University of Chicago Press.

Pitelka, L.F. (1997) Plant migration and climate change. *American Scientist*, **85**, 464–473.

Rey Benayas, J.M., Newton, A.C., Díaz, A. and Bullock, J.M. (2009) Enhancement of biodiversity and ecosystem services by ecological restoration: a meta-analysis. *Science*, **325**, 1121–1124.

Ricciardi, A. and Simberloff, D. (2009) Assisted colonization is not a viable conservation strategy. *Trends in Ecology & Evolution*, **24**, 248–253.

Richardson, D.M., Hellmann, J.J., McLachlan, J.S., Sax, D.F., Schwartz, M.W., Gonzalez, P., Brennan, E.J., Camacho, A., Root, T.L., Sala, O.E., Schneider, S.H., Ashe, D.M., Clark, J.R., Early, R., Etterson, J.R., Fielder, E.D., Gill, J.L., Minteer, B.A., Polasky, S., Safford, H.D., Thompson, A.R. and Vellend, M. (2009) Multidimensional evaluation of managed relocation. *Proceedings of the National Academy of Sciences*, **106**, 9721–9724.

Ricklefs, R.E. (2011) A biogeographical perspective on ecological systems: some personal reflections. *Journal of Biogeography*, **38**, 2045–2056.

Rodriguez, L.F. (2006) Can invasive species facilitate native species? Evidence of how, when, and why these impacts occur. *Biological Invasions*, **8**, 927–939.

Root, T.L. and Schneider, S.H. (2002) Climate change: overview and implications for wildlife, in *Wildlife Responses to Climate Change: North American Case Studies* (eds T.L. Root and S.H. Schneider), Island Press, 1–56.

Root, T.L. and Hughes, L. (2005) Present and future phenological changes in wild plants and animals. Chapter 5 in *Climate Change and Biodiversity* (eds T.E. Lovejoy and L. Hannah), Yale University Press, 61–69.

Root, T.L., Price, J.T., Hall, K.R., Schneider, S.H., Rosenzweig, C. and Pounds, J.A. (2003) Fingerprints of global warming on wild animals and plants. *Nature*, **421**, 57–60.

Rosenzweig, C., Casassa, G., Karoly, D.J., Imeson, A., Liu, C., Menzel, A., Rawlins, S., Root, T.L., Seguin, B. and Tryjanowski, P. (2007) Assessment of observed changes and responses in natural and managed systems. *Climate Change 2007: Impacts, Adaptation and Vulnerability*. Contribution of Working Group II to the Fourth Assessment Report of the Intergovernmental Panel on Climate Change (eds M.L. Parry, O.F. Canziani, J.P. Palutikof, P.J. van der Linden and C.E. Hanson), Cambridge University Press.

Sala, O.E., Chapin, F.S. III, Armesto, J.J., Berlow, E., Bloomfield, J., Dirzo, R., Huber-Sanwald, E., Huenneke, L.F., Jackson, R.B. and Kinzig, A. (2000) Global biodiversity scenarios for the year 2100. *Science*, **287**, 1770–1774.

Sax, D.F., Smith, K.F. and Thompson, A.R. (2009) Managed relocation: a nuanced evaluation is needed. *Trends in Ecology and Evolution*, **24**, 472–473.

Scheffer, M. (2009) *Critical Transitions in Nature and Society*. Princeton University Press.

Schlaepfer, M.A., Sax, D.F. and Olden, J.D. (2011) The potential conservation value of non-native species. *Conservation Biology*, **25**, 428–437.

Serreze, M.C., Bromwich, D.H., Clark, M.P., Etringer, A.J., Zhang, T. and Lammers, R. (2002) Large-scale hydroclimatology of the terrestrial Arctic drainage system. *Journal of Geophysical Research*, **108**(D2), 8160.

Severns, P.M. and Warren, A.D. (2008) Selectively eliminating and conserving exotic plants to save an endangered butterfly from local extinction. *Animal Conservation*, **11**, 476–483.

Shafer, S.L., Bartlein, P.J. and Thompson, R.S. (2001) Potential changes in the distributions of western North America tree and shrub taxa under future climate scenarios. *Ecosystems*, **4**, 200–215.

Slater, F.M. (1999) First-egg date fluctuations for the Pied Flycatcher *Ficedula hypoleuca* in the woodlands of mid-Wales in the twentieth century. *Ibis*, **141**, 489–506.

Smith, L.C., Sheng, Y., MacDonald, G.M. and Hinzman L.D. (2005) Disappearing Arctic lakes. *Science*, **308**, 1429.

Starzomski, B.M., Parker, R.L. and Srivastava, D.S. (2008) On the relationship between regional and local species richness: a test of saturation theory. *Ecology*, **89**, 1921–1930.

Stone, G.N., Van Der Ham, R.W.J. and Brewer, J.G. (2008) Fossil oak galls preserve ancient multitrophic interactions. *Proceedings of the Royal Society B: Biological Sciences*, **275**, 2213–2219.

Stralberg, D., Jongsomjit, D., Howell, C.A., Snyder, M.A., Alexander, J.D., Wiens, J.A. and Root, T.L. (2009) Reshuffling of species with climate disruption: a no-analog future for California birds? *PLoS ONE*, **4**, e6825.

Sutherland, W.J., Armstrong-Brown, S., Armsworth, P.R. et al. (2006) The identification of 100 ecological questions of high policy relevance in the UK. *Journal of Applied Ecology*, **43**, 617–627.

Thomas, C.D., Cameron, A., Green, R.E., Bakkenes, M., Beaumont, L.J., Collingham, Y.C., Erasmus, B.F.N., Ferreira de Siqueira, M., Grainger, A., Hannah, L., Hughes, L., Huntley, B., van jarrsveld, A.S., Midgley, G.F., Miles, L., Ortega–Huerta, M.A., Peterson, A.T., Phillips, O.L. and Williams, S.E. (2004) Extinction risk from climate change. *Nature*, **427**, 145–148.

Thompson, R.M. and Starzomski, B.M. (2007) What does biodiversity do? A review for policy-makers and managers. *Biodiversity and Conservation*, **16**, 1359–1378.

Thuiller, W., Lavorel, S., Sykes M.T. and Araújo, M.B. (2006) Using niche-based modelling to assess the impact of climate change on tree functional diversity in Europe. *Diversity and Distributions*, **12**, 49–60.

Travis, J.M.J. (2003) Climate change and habitat destruction: a deadly anthropogenic cocktail. *Proceedings of the Royal Society B, London*, **270**, 467–473.

Vellend, M. (2010) Conceptual synthesis in community ecology. *The Quarterly Review of Biology*, **85**, 183–206.

Venter, O., Brodeur, N.H., Nemiroff, L., Belland, B., Dolinsek, I.J. and Grant, J.W.A. (2006) Threats to endangered species in Canada. *Bioscience*, **56**, 1–8.

Walther, G–R., Post, E., Convey, P., Menzel, A., Parmesan, C., Beebee, T., Fromentin, J.–M., Hoegh–Guldberg, O. and Bairlein, F. (2002) Ecological responses to recent climate change. *Nature*, **416**, 389–395.

Webb, T. III. (1992) Past changes in vegetation and climate: Lessons for the future, in *Global Warming and Biological Diversity* (eds R.L. Peters and T.E. Lovejoy), Yale University Press, 59–75.

Williams, J.W., Jackson, S.T. and Kutzbach, J.E. (2007) Projected distributions of novel and disappearing climates by 2100 A.D. *Proceedings of the National Academy of Sciences*, **104**, 5738–5742.

Winkel, W. and Hudde, H. (1997) Long-term trends in reproductive traits of tits (*Parus major, P. caeruleus*) and Pied Flycatchers *Ficedula hypoleuca*. *Journal of Avian Biology*, **28**, 187–190.

Zavala, J.A., Casteel, C.L., DeLucia, E.H. and Berenbaum, M.R. (2008) Anthropogenic increase in carbon dioxide compromises plant defense against invasive insects. *Proceedings of the National Academy of Sciences*, **105**, 5129–5133.

Zavaleta, E., Hobbs, R.J. and Mooney, H.A. (2001) Viewing invasive species removal in a whole-ecosystem context. *Trends in Ecology & Evolution*, **16**, 454–459.

Zhang, T., Frauenfeld, O.W., Serreze, M.C. et al. (2005) Spatial and temporal variability in active layer thickness over the Russian Arctic drainage basin. *Journal of Geophysical Research*, **110**, D16101.

Zhuang Q., Melillo, J.M., Sarofim, M.C. et al. (2006) CO_2 and CH_4 exchanges between land ecosystems and the atmosphere in northern high latitudes over the 21st Century. *Geophysical Research Letters*, **33**, 1–5.

Chapter 11

PLANT INVASIONS AS BUILDERS AND SHAPERS OF NOVEL ECOSYSTEMS

David M. Richardson and Mirijam Gaertner

Centre for Invasion Biology, Department of Botany and Zoology, Stellenbosch University, South Africa

11.1 INTRODUCTION

As the other chapters in this book clearly show, novel ecosystems mean different things to different people (see Chapter 6). The initial outline of the concept by Hobbs et al. (2006) has proved an attractive stand on which to hang a diverse collection of hats. Alien plant invasions are sufficiently widespread and their influences sufficiently pervasive to ensure that they are influential builders and shapers of novel ecosystems in many parts of the world. This chapter describes some issues and concepts of alien plant invasions relating to novel ecosystems.

Very few ecosystems are currently free of introduced species (also termed 'alien' or 'non-native' species, i.e. those that arrived as a result of human-mediated dispersal). The contribution of introduced species to biotas and their relative roles in human-induced changes to ecosystems differ markedly between regions. Biotas of many regions are increasingly dominated by alien species (e.g. many islands have more alien than native species of plants and some groups of animals). A proportion of introduced species spread and may be

termed 'invasive' (Richardson et al. 2000; Blackburn et al. 2011); such species are an increasing problem for biodiversity conservation worldwide. Complex interactions between elements of global change often make it difficult to rank the importance of individual causes/drivers of biodiversity decline. Nevertheless, invasive species are widely acknowledged to be a fundamental driver of rapid ecosystem change (termed 'degradation' in some contexts).

The effects of invasive species in different biomes, ecosystems and habitats have been reviewed and debated in many recent publications; most studies agree that invasive species cause significant impacts at all levels of biological organization (e.g. Vilà et al. 2011). Invasion of many types of alien plants leads to decreased plant species diversity at local scales (Vila et al. 2006; Gaertner et al. 2009; Hejda et al. 2009) and altered ecosystem productivity. Invasions also change nutrient cycling (Ehrenfeld 2003; Liao et al. 2008) and fire regimes (Brooks et al. 2004). A recent topic of debate is whether invasive species are more often 'drivers' or 'passengers' of degradation. Some invaders clearly *drive* ecosystem change, but in some cases inva-

Novel Ecosystems: Intervening in the New Ecological World Order, First Edition. Edited by Richard J. Hobbs, Eric S. Higgs, and Carol M. Hall.
© 2013 John Wiley & Sons, Ltd. Published 2013 by John Wiley & Sons, Ltd.

sive species may be more appropriately viewed as *symptoms* or *passengers* of altered disturbance regimes or other ecosystem changes (Bauer 2012). This distinction has fundamental implications for understanding invasion dynamics and the trajectories of effects and for devising management strategies.

Invasive species are listed as important drivers and shapers of novel ecosystems in much of the literature on this concept (e.g. Lindenmayer et al. 2008). Indeed, many widely discussed examples of novel ecosystems are those where invasive species are: important or dominant agents of ecological novelty; have exerted obvious influence on the structure and functioning of the ecosystem; and are the main focus of management actions. Some examples are tree invasions in South African fynbos, invasions of African grasses in tropical Australia and invasive plants transforming the structure of Hawaiian rain forests.

The ubiquity of introduced and invasive species globally raises several questions. Do any and all ecosystems affected (i.e. invaded) by alien species (even one) qualify as novel ecosystems? If so, then virtually every ecosystem is novel (see Chapter 41 for discussion). Should some level or degree of novelty in structure or function be evident and demonstrable before ecosystems with non-native species should be termed 'novel'? Is the novel ecosystem concept useful, and/or how could it be adapted for describing or categorizing invaded ecosystems and elucidating invasion processes or in guiding management?

This chapter examines how the novel ecosystem concept has been applied with reference to sites affected by alien plant invasions in different parts of the world. We summarize what is known about the mechanisms and processes whereby plant invasions are known to generate impacts in invaded ecosystems and how such drivers and impacts link with key facets of current discussions and debates around novel ecosystems. Finally, we discuss the usefulness of the novel ecosystems concept for understanding and managing plant invasions in the face of rapid global change.

11.2 REVIEW OF APPLICATION OF NOVEL ECOSYSTEMS TO ECOSYSTEMS AFFECTED BY ALIEN PLANT INVASIONS

Perusal of the *c.* 200 papers that have cited Hobbs et al. (2006) shows that a range of meanings or con-

notations are attached to the term and concept 'novel ecosystems' when applied to ecosystems affected by invasive plants.

Many studies invoke the novel ecosystem concept and associated terminology simply to denote sites, habitats or ecosystems invaded by alien species, sometimes making observations and drawing broad generalizations about the (potential) implications of the presence or abundance of alien species (e.g. Pennington et al. 2010; Kowarik 2011). Some emphasize the unknown but potentially profound implications for ecological and evolutionary processes resulting from the new combination of species. Some stress the importance of introduced species as components of novel ecosystems in general terms because of the "new ecologies that can have both cultural and environmental values" (Kull et al. 2012). The novel ecosystem label is also used to denote some particular functional connotation associated with plant invasions, usually describing combinations of native and alien species that confer some functionality or influence the trajectory of succession. For example, Colon et al. (2011) argue that alien-dominated forest sites in Puerto Rico "diversify with native species thus evolving towards new forest types with novel species combinations". Lugo (2009) writes that "natural processes have remixed or reassembled native and introduced plant and animal species into novel communities adapted to anthropogenic environmental conditions". Other examples include discussion of sites where alien species: "hinder recovery of native communities on former agricultural land" (Tognetti et al. 2009). Kueffer et al. (2007) describe how alien species modify conditions of the regeneration niche in mid-altitude forests of Mahe in the Seychelles due to the loss of key resources provided by native species such as decaying wood and organic matter accumulation on rocks.

An increasing number of management-focused studies invoke the novel ecosystem concept in arguing that the removal of alien species from certain ecosystems is impractical or impossible and that a mind-shift in management is needed. These studies generally call for more pragmatic approaches to management whereby goals are defined in terms of ecosystem functions and services, and not on the basis of native/alien species composition. Examples of ecosystems where this line of reasoning have been advanced include: the heavily invaded fern-sedge zone in the highlands of the Galapagos (Jäger et al. 2009) and heavily invaded riparian habitats in strongly human-modified landscapes

(Richardson et al. 2007; Meek et al. 2010). Some papers emphasize the synergistic relations between invasive species and other facets of global change and the complexity of untangling individual drivers of change. For example, Schweiger et al. (2010) explain that alien species can both partly compensate for other negative effects of climate change but may also amplify them in some cases.

In summary, ideas related to novel ecosystems are generally not applied very rigorously for sites invaded by alien plants. There is little evidence that the advent of the novel ecosystem concept has driven any major conceptual or practical advances for elucidating the drivers or trajectories of ecosystem change caused by invasive species or for managing plant invasions. There are, however, signs of re-thinking approaches for managing invasive species, for example, to manage for function rather than composition. In the next section we discuss some current views on the drivers of ecosystem change through the agency of plant invasions and how aspects considered in such studies relate to concepts and ideas about novel ecosystems.

11.3 PLANT INVASIONS: DRIVERS OF IMPACTS

Adding *any* 'new' species (i.e. one that evolved elsewhere and did not or could not arrive there without human assistance) to an ecosystem will have some effect or impact; it will bring about some change to that ecosystem. This is because every species interacts with every other species in a community in some way by feeding on them, providing food or other resources or influencing habitat. Some additions may cause subtle changes, at least in the short term, whereas some generate dramatic changes that lead to rapid and radical transformation of the structure and functioning of invaded ecosystems.

Much recent work has described the dynamics of particular invasive plant species in certain ecosystems and on categorizing the various mechanisms whereby the invasive species cause impacts (see Levine et al. 2003 for a review). Most research has focused on describing the impacts caused by species that have had obvious, and in some cases dramatic, influences in invaded ecosystems. We review the biotic and abiotic processes underlying impacts of alien plant invasions and discuss the native ecosystem response.

The outcomes of invasions on the native biota can be observed at population, community and ecosystem levels. At the population level, invasions can have immediate effects on the performance and phenotypic traits of individual plant species. Such impacts may manifest in reduced growth or reproduction or changes in native species morphology (Parker et al. 1999). Invasions can also have direct genetic impacts on native species through hybridization or introgression, and indirect genetic impacts by altering patterns of natural selection within native populations. Community-level effects include initial impacts (e.g. increased biomass, nutrient addition, species displacement and altered abundance), secondary impacts (e.g. decomposition processes and changed seed bank dynamics) and tertiary impacts resulting from diverse interactions among species (e.g. Le Maitre et al. 2011). The most frequently reported initial impact of biological invasions at the community level is a reduction in plant species richness or diversity. The nature and magnitude of such impacts are related to features of the invading species and the invaded ecosystem. A secondary impact of invasion on plant community composition is the depletion and homogenization of the native soil seed bank. Plant invasions can significantly alter richness, diversity and composition of native seed banks, resulting in long-term changes for the plant community. Tertiary impacts involve infiltration and diverse modifications of native ecosystem networks by invasive species (Traveset and Richardson 2006; Milton et al. 2007). Invasive alien plants can also impact on soil microbes, indirectly changing soil processes such as nitrification.

The impacts of alien plant invasions at the ecosystem level can be categorized following the approach by Richardson et al. (2000) in summarizing effects of 'transformer species' of plant invaders (those that change the character, condition, form or nature of ecosystems over a substantial area relative to the extent of that ecosystem). Richardson et al. (2000) listed the following categories of transformers that may be distinguished among the most damaging plant invaders worldwide: (a) excessive users of resources (water; water and light; light and oxygen); (b) donors of limiting resources (nitrogen); (c) fire promoters/suppressors; (d) sand stabilizers; (e) erosion promoters; (f) colonizers of intertidal mudflats/sediment stabilizers; (g) litter accumulators; and (h) salt accumulators/redistributors. Each type of transformer species exerts influence in the invaded ecosystem in a different way.

Figure 11.1 illustrates idealized patterns for three important categories of transformer species. Three examples are selected to show trends and trajectories of invader abundance and changes in impact over time since the arrival of the alien species. In the next section we describe salient aspects of these patterns as a means of elucidating the dynamics of impact.

Invasive alien species can utilize substantial amounts of water and other limited resources and thus affect plant communities. Decreased availability of resources such as water, light or nutrients can lead to replacement of already established native species in a community or prevent the establishment of new individuals, thereby potentially compromising ecosystem resilience

(Davis et al. 2000; Yurkonis et al. 2005). An often-cited example of an excessive user of resources is invasive salt cedar trees (*Tamarix* spp.) in riparian ecosystems in the south-western USA (Fig. 11.1a). Salt cedars consume large quantities of water through evapotranspiration (Dahm et al. 2002). Reduction in groundwater results in reduced water availability for native species and can lead to soil salinization, giving salt-tolerant salt cedar trees a competitive advantage (Shafroth et al. 1995; Nagler et al. 2008). Once a threshold is reached, dry, saline soils in concert with increasing invader abundance may trigger decreases in native species abundance. This can trigger positive feedback loops with increasing invader abundance

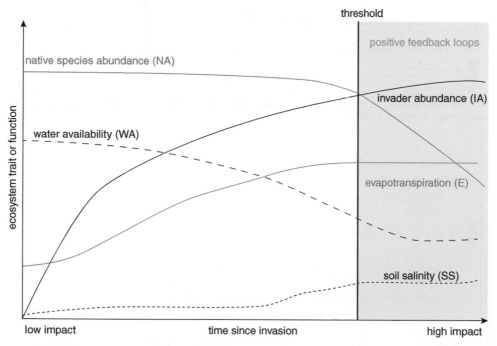

(a): Transformer species (user of resources): e.g. *Tamarix* spp. invasion in the Western United States

Figure 11.1 Changes to ecosystems due to alien plant invasions. (a) Patterns associated with invasion of salt cedar trees (*Tamarix* spp.) in riparian ecosystems in the south-western USA as an example for excessive water use through evapotranspiration (E) (Dahm et al. 2002) resulting in reduced water availability (WA) and soil salinization (SS) (Shafroth et al. 1995), followed by increases in invader abundance (IA) and decreases in native species abundance (NA). (b) Key changes associated with *Morella faya* invasion in Hawaii where the invader adds substantial amounts of total and available nitrogen (TAN) to naturally nitrogen-free young volcanic soils, thereby promoting its own growth and potentially the proliferation of other weedy species (OWS) (Vitousek et al. 1987) and reducing germination of native tree species (Walker and Vitousek 1991). (c) Effects of Gamba grass (*Andropogon gayanus*) invasion on fire regimes in northern Australian savannas. High invader biomass (IB) leads to increases in fire intensity (FI) and frequency (Rossiter et al. 2003), resulting in soil nutrient losses (SN) (Rossiter-Rachor et al. 2008) and a decline of native tree recruitment (NR) (Setterfield 2002).

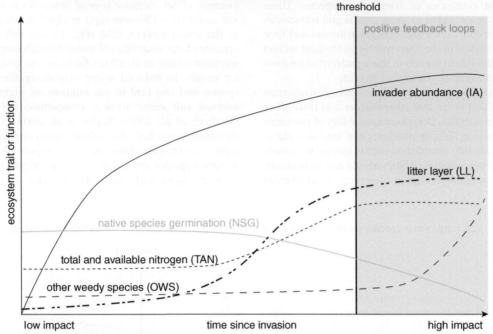

(b): Transformer species (donor species): e.g. Morella faya invasion in Hawaii

(c): Transformer species (changes in disturbance regimes): e.g. Gamba grass *(Andropogon gayanus)* invasions in northern Australia

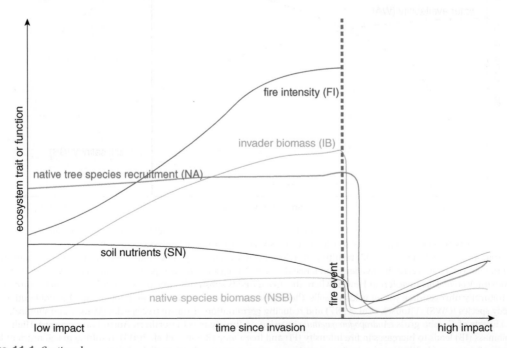

Figure 11.1 *Continued*

resulting in increased soil salinity and further decreases in native species abundance.

Many invasive plants accumulate resources such as nitrogen either through N-fixation or through increased biomass and elevated net primary production (i.e. production of litter with higher decomposition rates than native species; Ehrenfeld 2003). A prominent example is invasion of the fire tree (*Morella faya*) in young volcanic soils in Hawaii (Fig. 11.1b). *Morella faya* is a N-fixing species which can add substantial amounts of total and available nitrogen to the soil. In Hawaii *Morella* fertilizes soils that are naturally nitrogen-free, radically altering ecosystem-level processes (Vitousek et al. 1987). Its ability to fix nitrogen gives *Morella* an enormous advantage over native tree species under nitrogen-limited field conditions (Walker and Vitousek 1991). Elevated nitrogen levels can create positive feedback loops that promote further proliferation of the invader and other weedy species. Once a certain threshold is reached, changed ecosystem conditions (e.g. thick litter layer) can reduce native species germination. Due to the ability of *Morella* to form dense canopies, native species eventually become displaced and *Morella* forms mono-specific stands with no understorey (Walker and Vitousek 1991).

Invasive alien species may also suppress or promote fire, thereby driving rapid ecosystem transformation (Brooks et al. 2004). For example, Gamba grass (*Andropogon gayanus*) invasion in northern Australian savanna ecosystems alters fuel characteristics of the savanna with invaders having four times greater biomass than native species (Fig. 11.1c). High fuel loads support fires on average eight times more intense than those fueled by native grasses (Rossiter et al. 2003). Changed fire regimes decrease or eliminate native species recruitment (Setterfield 2002) and cause significant fire-mediated nutrient losses (Rossiter-Rachor et al. 2008). By altering soil nutrient cycles according to its preferences, Gamba grass establishes positive feedback loops enhancing its competitive superiority and persistence (Rossiter-Rachor et al. 2009).

11.4 MOVING FORWARD: INCORPORATING RESILIENCE AND THRESHOLDS

The previous section showed that effects of plant invasions on communities can be very complex, with out-comes determined by the type of ecosystem invaded and the alien species involved. The three examples in Figure 11.1 discussed earlier emphasize the importance of incorporating aspects of resilience and thresholds in describing the progression of impacts in invaded ecosystems. We now expand on these aspects in discussing a potentially profitable way of applying concepts associated with novel ecosystems to elucidate invasion dynamics and management strategies.

The concept of ecosystem resilience and associated thresholds has been suggested as a helpful framework for identifying the degree of ecosystem degradation. In the context of ecosystem restoration, resilience is defined as the ability of an ecosystem to recover from disturbance without human intervention (Westman 1978). As discussed earlier, loss of ecosystem resilience generated by biological invasions can be triggered by biotic interactions leading to changes in community composition and structure. Depending on the type of ecosystem invaded and the identity of the invading species, structural changes can be followed and/or accompanied by changes in ecosystem functions (e.g. changes in nutrient cycling).

Most plant invasions change structural and functional conditions in concert, creating a switch from negative to positive feedback loops. If positive feedback loops are established and the system has shifted to an alternative state, the system can be referred to as a novel ecosystem (sensu Hobbs et al. 2006). The point at which the dominance of the negative (regulating) feedbacks that maintain ecosystem resilience is replaced by the dominance of positive (supportive) feedbacks that lead to losses in resilience can be considered a threshold (Briske et al. 2006). Threshold crossings are the result of several interacting ecosystem components, and identifying when such points are reached is fundamentally difficult (see also Chapters 3, 18 and 24). Some invaded ecosystems may never reach a threshold and therefore never switch to an alternative ecosystem state. However, many invasions do cross one or more thresholds, producing an alternative ecosystem state. In practice, observing changes in ecosystem function over time or over a range of sites with different levels of invasion allows us to identify levels of ecosystem functioning associated with thresholds (Le Maitre et al. 2011). To identify whether an ecosystem has crossed one or more thresholds it is necessary to evaluate ecosystem response which can also be interpreted by determining the degree of invasion (King and Hobbs 2006; Gaertner et al. 2012).

Figure 11.2 (a) Marked alteration in vegetation structure associated with invasion of natural fynbos shrublands by Australia *Acacia* species in South Africa's Cape Floristic Region. Invasive species are: (b) *Acacia cyclops*; (c) *A. saligna*; (d) *A. longifolia* and (e) *A. pycnantha* (E). Photographs: D.M. Richardson.

The example of Australian *Acacia* species as invasive species in South African fynbos ecosystems (Fig. 11.2) is instructive for exploring these ideas further. Widespread invasions of Australian acacias and other species have affected fynbos ecosystems over much of the Cape Floristic Region and are one of the main threats to biodiversity across the region (Rouget et al. 2003). The impacts of these invaders have been reasonably well studied (reviews in Richardson and van Wilgen 2004; Le Maitre et al. 2011). Australian acacias have huge invasive potential and strong persistence due to massive long-lived seed banks (Richardson and Kluge 2008). Many species resprout after fire and/or possess other mechanisms, including fire-stimulated germination, that ensure persistence and proliferation under any possible fire regime in these ecosystems. Invaded ecosystems have radically increased biomass and

changed fuel properties (van Wilgen and Richardson 1985). Nutrient cycling is drastically changed due to nitrogen fixation together with increased litter production and mineralization (Yelenik et al. 2004). Structural changes associated with these invasions have profound implications for the biota of these ecosystems (e.g. Veldtman et al. 2011) and also change their potential to deliver a range of goods and services, thereby altering human perceptions relating to value (Kull et al. 2012).

Acacia saligna is one of the most invasive and widespread of the suite of Australian acacias in South Africa; the invasion dynamics and impacts of this species have been studied most thoroughly. Early in the invasion of fynbos by *A. saligna*, changes occur to the structure of the community. With increasing time and densification of the invasion, such altered biotic structure generates changes to biotic functions through the suppression of native species. Further increasing dominance of invasive acacias and decreasing native species diversity and abundances affects ecosystem resilience. Once the invasive acacias form dense thickets, a biotic threshold is passed and the system occupies an alternative state; native species composition and structure are changed and alien *Acacia* propagules dominate the soil seed bank. However, although biotic and some abiotic components have been altered, the system is still amenable to autogenic recovery with human intervention (mechanical clearing of *Acacia*) as remnant fynbos elements retain regeneration capability and their seedlings can cope with the elevated soil nitrogen levels (Musil 1993; Fig. 11.3a).

Dense *A. saligna* stands alter ecosystem functioning via elevated biomass, litter production, nitrification and changed nutrient-cycling rates. This in turn promotes growth of alien species, enhancing their competitive advantage over native species. An abiotic threshold is reached after a certain level of invasion (typically after one and three fire cycles in lowland and mountain fynbos, respectively). This condition can be defined as a *hybrid state*. At this stage, restoration to a target ecosystem state focusing on biodiversity is still possible but would possibly require reintroducing native species to return lost functional groups. In some cases, reduction of soil nutrient may be necessary (Fig. 11.3b).

Structural and functional changes to biotic and abiotic components cause positive feedback loops: increased soil nutrient levels lead to enhanced growth of *A. saligna* and high total biomass, which results in high-intensity fires that further promote recruitment of *A. saligna*. A third threshold is reached when the invader forms a virtual monoculture that suppresses growth of all but the most weedy native and other nitrophilous alien species. The latter may form an understorey that can carry annual fires. In this state, the system has shifted to a *novel ecosystem*. Restoration of the ecosystem in this state is very difficult and will require major management input and careful planning (Fig. 11.3c).

11.5 MANAGEMENT IMPLICATIONS

We contend that considering invaded ecosystems as novel ecosystems, drawing in insights from refinements to the original framework of Hobbs et al. (2006)

as set out in this book (see Chapters 3, 5 and 6), has potential for guiding management of these ecosystems. Distinguishing different degrees of ecosystem degradation (i.e. identifying certain patterns in the invasion process) and defining different ecosystem states (e.g. hybrid ecosystem and novel ecosystems; see Chapter 18) can guide decisions on whether restoration is feasible and affordable and, if so, how the restoration should be approached. Prior to decision-making it is important to establish restoration goals. Depending on the degree of degradation, restoration goals will range from re-establishing a natural ecosystem state (focusing on biodiversity components) to restoring ecosystem processes and functions. The intensity of management intervention required is proportional to the duration of invasion (Holmes et al. 2000); identifying the degree of ecosystem degradation can therefore also help to adapt management interventions to the specific needs of an invaded ecosystem. Ultimately, it would be desirable to identify certain parameters that characterize an invaded ecosystem as a novel ecosystem. Restoration of heavily degraded ecosystems will need major management input and hence substantial resources. The decision-making process for restoration of such systems therefore requires careful consideration (see Chapter 22 for case study). Motives for managing invaded ecosystems are manifold and include ecological arguments such as halting or slowing declines in native species richness (Gaertner et al. 2009; Hejda et al. 2009) and largely economic arguments relating to mitigating reduced river flows (Le Maitre et al. 1996, 2002).

Increasingly, multiple (and sometimes conflicting) criteria need to be considered: those relating to conservation objectives on the one hand and those relating to the provision of goods and services on the other. In many cases however, the first question that needs to be asked is whether management required to achieve the stated goal is practically feasible. Many invaded ecosystems are altered to such an extent that recovery options are fundamentally compromised by major obstacles such as soil degradation or secondary invasion (Zavaleta et al. 2001). In such cases restoration is only possible with major management inputs, and even then may have unwanted or unexpected outcomes. Secondly, in some cases invaders are integrated into native ecosystem networks where they fulfill important ecological functions (Traveset and Richardson 2006). Removing such species will largely or entirely eliminate a function which has become necessary for

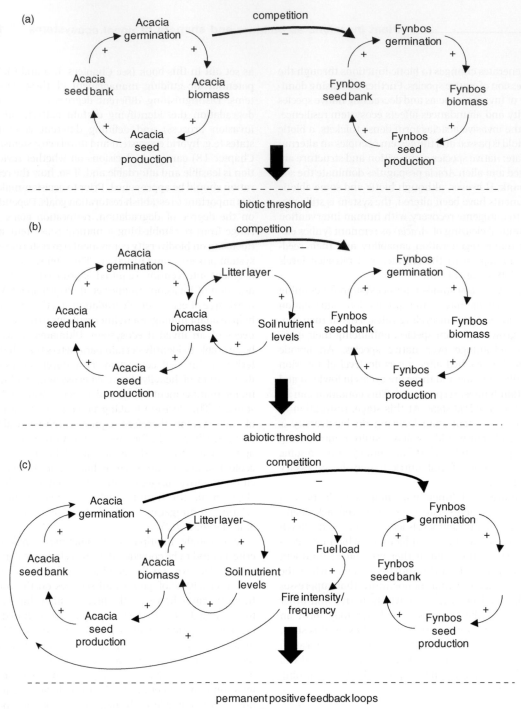

Figure 11.3 *Acacia saligna* invasion of fynbos, illustrating the different ecosystem states and thresholds as invasion intensifies. (a) Disturbed fynbos under recent *Acacia* invasion with biotic changes: *A. saligna* rapidly accumulates biomass because of its rapid growth rate and its capacity to produce large, persistent seed banks and hence out-compete native plants. At this stage no special intervention is required for restoration as the system is still able to recover autogenically. (b) Hybrid ecosystem: after further *Acacia* invasion, biotic changes (both structural and functional) are accompanied by abiotic changes: *Acacias* accumulate nitrogen through nitrogen-fixing, increased biomass and production of litter with higher decomposition rates than native species. Elevated nitrogen levels create reinforcing feedback loops that promote proliferation of the invader, further preventing the re-establishment of native plants. Management will require reintroducing native species to return lost functional groups, and in some cases, soil nutrient conditions will need to be manipulated. (c) Novel ecosystem: permanent positive feedback loops are established. Besides the nutrient-cycling feedback loop, a fire feedback loop is established and the system has shifted to an alternative stable state. Effective restoration at this stage will require major intervention.

other biota from that ecosystem. Thirdly, resources for management are scarce and managers must prioritize actions. In some cases, if the available management actions are no longer feasible, resources may be better spent in less degraded systems. Alternatively, different interventions may be developed that optimize functions and services from the novel ecosystem.

11.6 CONCLUSIONS

The novel ecosystems concept has been widely applied to ecosystems invaded by alien plants, but the label is mostly applied very loosely with reference to either ecosystems with high numbers of alien species or where impacts associated with alien species are pronounced and create many types of management challenges. New approaches need to be developed to map important processes, mechanisms and trajectories associated with alien plant invasions onto contemporary models for conceptualizing novel ecosystems. Using the example of widespread invasions of Australian acacias into South African fynbos, we have shown that the application of concepts of resilience and thresholds can accommodate the objective and robust delineation of historical, hybrid and novel ecosystems as defined by Hobbs et al. (2009). The challenge will be to integrate such systems with existing frameworks for biological invasions (e.g. Blackburn et al. 2011) and practical ways of quantifying levels of invasion (e.g. Catford et al. 2012).

REFERENCES

Bauer, J. (2012) Invasive species: "back-seat drivers" of ecosystem change? *Biological Invasions*, **14**,1295–1304.

Blackburn, T.M., Pyšek, P., Bacher, S., Carlton, J.T., Duncan, R.P., Jarošík, V., Wilson, J.R.U. and Richardson, D.M. (2011) A proposed unified framework for biological invasions. *Trends in Ecology & Evolution*, **26**, 333–339.

Briske, D.D., Fuhlendorf, S.D. and Smeins, F.E. (2006) A unified framework for assessment and application of ecological thresholds. *Rangeland Ecology and Management*, **59**, 225–236.

Brooks, M.L., D'Antonio, C.M., Richardson, D.M., Grace, J.B., Keeley, J.E., DiTomaso, J.M., Hobbs, R.J., Pellant, M. and Pyke, D. (2004) Effects of invasive alien plants on fire regimes. *Bioscience*, **54**, 677–688.

Catford, J.A., Vesk, P., Richardson, D.M. and Pyšek, P. (2012) Quantifying levels of biological invasion: towards the objec-tive classification of invaded and invasible ecosystems. *Global Change Biology*, **18**, 44–62.

Colon S.M., Lugo, A.E. and Gonzalez, O.M.R. (2011) Novel dry forests in southwestern Puerto Rico. *Forest Ecology and Management*, **262**, 170–177.

Dahm, C.N., Cleverly, J.R., Allred Coonrod, J.E., Thibault, J.R., McDonnell D.E. and Gilroy, D.J. (2002) Evapotranspiration at the land/water interface in a semi-arid drainage basin. *Freshwater Biology*, **47**, 831–843.

Davis, M.A., Grime, J.P. and Thompson, K. (2000) Fluctuating resources in plant communities: a general theory of invasibility. *Journal of Ecology*, **88**, 528–534.

Ehrenfeld, J.G. (2003) Effects of exotic plant invasions on soil nutrient cycling processes. *Ecosystems*, **6**, 503–523.

Gaertner M., den Breeÿen, A., Hui, C. and Richardson, D.M. (2009) Impacts of alien plant invasions on species richness in Mediterranean-type ecosystems: a meta-analysis. *Progress in Physical Geography*, **33**, 319–338.

Gaertner, M., Holmes P.M. and Richardson, D.M. (2012) Biological invasions, resilience and restoration, in *Restoration Ecology – The New Frontier* (eds J. van Andel and J. Aronson), Wiley-Blackwell, Oxford, UK, 265–280.

Hejda, M., Pyšek, P. and Jarošík, V. (2009) Impact of invasive plants on the species richness, diversity and composition of invaded communities. *Journal of Ecology*, **97**, 393–403.

Hobbs, R.J., Arico, S., Aronson, J., Baron, J.S., Bridgewater, P., Cramer, V.A., Epstein, P.R., Ewel, J.J., Klink, C.A., Lugo, A.E., Norton, D., Ojima, D., Richardson, D.M., Sanderson, E.W., Valladares, F., Vilà, M., Zamora, R. and Zobel, M. (2006) Novel ecosystems: theoretical and management aspects of the new ecological world order. *Global Ecology and Biogeography*, **15**, 1–7.

Hobbs, R.J., Higgs, E. and Harris, J.A. (2009) Novel ecosystems: Implications for conservation and restoration. *Trends in Ecology & Evolution*, **24**, 599–605.

Holmes, P.M., Richardson, D.M., Van Wilgen, B.W. and Gelderblom, C. (2000) Recovery of South African fynbos vegetation following alien woody plant clearing and fire: Implications for restoration. *Australian Ecology*, **25**, 631–639.

Jäger H., Kowarik, I. and Tye, A. (2009) Destruction without extinction: Long-term impacts of an invasive tree species on Galapagos highland vegetation. *Journal of Ecology*, **97**, 1252–1263.

King, E.G. and Hobbs, R.J. (2006) Identifying linkages among conceptual models of ecosystem degradation and restoration: Towards an integrative framework. *Restoration Ecology*, **14**, 369–378.

Kowarik, I. (2011) Novel urban ecosystems, biodiversity, and conservation. *Environmental Pollution*, **159**, 1974–1983.

Kueffer, C., Schumacher, E., Fleischmann, K., Edwards, P.J. and Dietz, H. (2007) Strong below-ground competition shapes tree regeneration in invasive *Cinnamomum verum* forests. *Journal of Applied Ecology*, **95**, 273–282.

Kull, C.A., Tassin, J., Moreau, S., Ramiarantsoa, H.R., Blanc-Pamard, C. and Carriere, S.M. (2012) The introduced flora of Madagascar. *Biological Invasions*, **14**, 875–888.

Le Maitre, D.C., van Wilgen, B.W., Chapman, R.A. and McKelly, D.H. (1996) Invasive plants and water resources in the Western Cape Province, South Africa: Modelling the consequences of a lack of management. *Journal of Applied Ecology*, **33**, 161–172.

Le Maitre, D.C., van Wilgen, B.W., Gelderblom, C.M., Bailey, C., Chapman, R.A. and Nel, J.A. (2002) Invasive alien trees and water resources in South Africa: Case studies of the costs and benefits of management. *Forest Ecology and Management*, **160**, 143–159.

Le Maitre, D.C., Gaertner, M., Marchante, E., Ens, E.J., Holmes, P.M., Pauchard, A., O'Farrell, P.J., Rogers, A.M., Blanchard, R., Blignaut, J. and Richardson, D.M. (2011) Impacts of invasive Australian acacias: implications for management and restoration. *Diversity and Distributions*, **17**, 1015–1029.

Levine, J.M., Vilà, M., D'Antonio, C.M., Dukes, J.S., Grigulis, K. and Lavorel, S. (2003) Mechanisms underlying the impacts of exotic plant invasions. *Proceedings of the Royal Society of London Series B-Biological Sciences*, **270**, 775–781.

Liao, C.Z., Peng, R.H., Luo, Y.Q., Zhou, X.H., Wu, X.W., Fang, C.M., Chen, J.K. and Li, B. (2008) Altered ecosystem carbon and nitrogen cycles by plant invasion: A meta-analysis. *New Phytologist*, **177**, 706–714.

Lindenmayer, D.B., Fischer, J., Felton, A., Crane, M., Michael, D., Macgregor, C., Montague-Drake, R., Manning, A. and Hobbs, R.J. (2008) Novel ecosystems resulting from landscape transformation create dilemmas for modern conservation practice. *Conservation Letters*, **1**, 29–135.

Lugo, A.E. (2009) The emerging era of novel tropical forests. *Biotropica*, **41**, 589–591.

Meek, C., Richardson, D.M. and Mucina L. (2010) A river runs through it: Land use and the composition of vegetation along a riparian corridor in the Cape Floristic Region, South Africa. *Biological Conservation*, **143**, 156–164.

Milton, S.J., Wilson, J.R.U., Richardson, D.M., Seymour, C.L., Dean, W.R.J., Iponga, D.M. and Procheş, S. (2007) Invasive alien plants infiltrate bird-mediated shrub nucleation processes in arid savanna. *Journal of Ecology*, **95**, 648–661.

Musil, C.F. (1993) Effect of invasive Australian acacias on the regeneration, growth and nutrient chemistry of South-African lowland fynbos. *Journal of Applied Ecology*, **30**, 361–372.

Nagler, P.L., Glenn, E.P., Didan, K., Osterberg, J., Jordan, F. and Cunningham, J. (2008) Wide-area estimates of stand structure and water use of *Tamarix* spp. on the Lower Colorado River: Implications for restoration and water management projects. *Restoration Ecology*, **16**, 136–145.

Parker, I.M., Simberloff, D., Lonsdale, W.M., Goodell, K., Wonham, M., Kareiva, P.M., Williamson, M.H., Von Holle, B., Moyle, P.B., Byers, J.E. and Goldwasser, L. (1999) Impact: toward a framework for understanding the ecological effect of invaders. *Biological Invasions*, **1**, 3–19.

Pennington, D.N., Hansel, J.R. and Gorchov, D.L. (2010) Urbanization and riparian forest woody communities: Diversity, composition, and structure within a metropolitan landscape. *Biological Conservation*, **143**, 182–194.

Richardson, D.M. and van Wilgen, B.W. (2004) Invasive alien plants in South Africa: how well do we understand the ecological impacts? *South African Journal of Science*, **100**, 45–52.

Richardson, D.M. and Kluge, R.L. (2008) Seed banks of invasive Australian *Acacia* species in South Africa: Role in invasiveness and options for management. *Perspectives in Plant Ecology, Evolution and Systematics*, **10**, 161–177.

Richardson, D.M., Pyšek, P., Rejmánek, M., Barbour, M.G., Panetta, D.F. and West, C.J. (2000) Naturalization and invasion of alien plants – concepts and definitions. *Diversity and Distributions*, **6**, 93–107.

Richardson, D.M., Holmes, P.M., Esler, K.J., Galatowitsch, S.M., Stromberg, J.C., Kirkman, S.P., Pyšek, P. and Hobbs, R.J. (2007) Riparian zones – degradation, alien plant invasions and restoration prospects. *Diversity and Distributions*, **13**, 126–139.

Rossiter N.A., Setterfield, S.A., Douglas, M.M. and Hutley, L.B. (2003) Testing the grass-fire cycle: alien grass invasion in the tropical savannas of northern Australia. *Diversity and Distributions*, **9**, 169–176.

Rossiter-Rachor, N.A., Setterfield, S.A., Douglas, M.M., Hutley, L.B. and Cook, G.D. (2008) *Andropogon gayanus* (Gamba grass) invasion increases fire-mediated nitrogen losses in the tropical savannas of northern Australia. *Ecosystems*, **11**, 77–88.

Rossiter-Rachor, N.A., Setterfield, S.A., Douglas, M.M., Hutley, L.B., Cook, G.D. and Schmidt, S. (2009) Invasive *Andropogon gayanus* (gamba grass) is an ecosystem transformer of nitrogen relations in Australian savanna. *Ecological Applications*, **19**, 1546–1560.

Rouget, M., Richardson, D.M., Cowling, R.M., Lloyd, J.W. and Lombard, A.T. (2003) Current patterns of habitat transformation and future threats to biodiversity in terrestrial ecosystems of the Cape Floristic Region, South Africa. *Biological Conservation*, **112**, 63–85.

Setterfield, S.A. (2002) Seedling establishment in an Australian tropical savanna: Effects of seed supply, soil disturbance and fire. *Journal of Applied Ecology*, **39**, 949–959.

Schweiger, O., Biesmeijer, J.C., Bommarco, R., Hickler, T., Hulme, P.E., Klotz, S., Kuhn, I., Moora, M., Nielsen, A., Ohlemuller, R., Petanidou, T., Potts, S.G., Pyšek, P., Stout, J.C., Sykes, M.T., Tscheulin, T., Vila, M., Walther, G.R., Westphal, C., Winter, M., Zobel, M. and Settele, J. (2010) Multiple stressors on biotic interactions: how climate change and alien species interact to affect pollination. *Biological Reviews*, **85**, 777–795.

Shafroth, P.B., Friedman, J.M. and Ischinger, L.S. (1995) Effects of salinity on establishment of *Populus fremontii*

(Cottonwood) and *Tamarix ramosissima* (Saltcedar) in Southwestern United-States. *Great Basin Naturalist*, **55**, 58–65.

Tognetti, P.M., Chaneton, E.J., Omacini, M., Trebino, H.J. and León, R.J.C. (2009) Exotic vs. native plant dominance over 20 years of old-field succession on set-aside farmland in Argentina. *Biological Conservation*, **143**, 2494–2593.

Traveset, A. and Richardson, D.M. (2006) Biological invasions as disruptors of plant reproductive mutualisms. *Trends in Ecology & Evolution*, **21**, 208–216.

van Wilgen, B.W. and Richardson, D.M. (1985) The effects of alien shrub invasions on vegetation structure and fire behaviour in South African fynbos shrublands: A simulation study. *Journal of Applied Ecology*, **22**, 955–966.

Veldtman, R., Lado, T.F., Botes, A., Proches, S., Timm, A.E., Geertsema, H. and Chown, S.L. (2011) Creating novel food webs on introduced Australian acacias: indirect effects of galling biological control agents. *Diversity and Distributions*, **17**, 958–967.

Vila, M., Tessier, M., Suehs, C.M., Brundu, G., Carta, L., Galanidis, A., Lambdon, P., Manca, M., Medail, F., Moragues, E., Traveset, A., Troumbis, A.Y. and Hulme, P.E. (2006) Local and regional assessments of the impacts of plant invaders on vegetation structure and soil properties of Mediterranean islands. *Journal of Biogeography*, **33**, 853–861.

Vilà, M., Espinar, J.L., Hejda, M., Hulme, P.E., Jarošík, V., Maron, J.L., Pergl, J., Schaffner, U., Sun, Y. and Pyšek, P. (2011) Ecological impacts of invasive alien plants: a meta-analysis of their effects on species, communities and ecosystems. *Ecology Letters*, **14**, 702–708.

Vitousek, P.M., Walker, L., Whiteaker, L., Mueller-Dombois, D. and Matson, P. (1987) Biological invasion of *Myrica faya* alters ecosystem development in Hawaii. *Science*, **238**, 802–804.

Walker, L.R. and Vitousek, P.M. (1991) An invader alters germination and growth of a native dominant tree in Hawaii. *Ecology*, **72**, 1449–1455.

Westman, W.E. (1978) Measuring the inertia and resilience of ecosystems. *BioScience*, **28**, 705–710.

Yelenik, S.G., Stock, W.D. and Richardson, D.M. (2004) Ecosystem level impacts of invasive *Acacia saligna* in the South African fynbos. *Restoration Ecology*, **12**, 44–51.

Yurkonis, K.A., Meiners, S.J. and Wachholder, B.E. (2005) Invasion impacts diversity through altered community dynamics. *Journal of Ecology*, **93**, 1053–1061.

Zavaleta, E.S., Hobbs, R.J. and Mooney, H.A. (2001) Viewing invasive species removal in a whole-ecosystem context. *Trends in Ecology & Evolution*, **16**, 454–459.

Chapter 12

INFECTIOUS DISEASE AND NOVEL ECOSYSTEMS

Laith Yakob

School of Population Health, University of Queensland, Australia

12.1 INTRODUCTION

Through direct and indirect anthropogenic effects, novel ecosystems are evolving and bringing with them emerging infectious diseases. An emerging pathogen can be defined as the causative agent of an infectious disease following its appearance in a new host population or whose incidence is increasing in an existing host population as a result of long-term changes in its underlying epidemiology (Woolhouse and Dye 2001). A recent survey showed that 177 out of 1407 recognized species of human pathogens are regarded as emerging or re-emerging (Woolhouse and Gowtage-Segueria 2005). Novel host-pathogen pairings are also of increasing concern in livestock (Cleaveland et al. 2001), wildlife (Daszak et al. 2000) and plants (Anderson et al. 2004). A cursory list demonstrating the breadth of impact of anthropogenic drivers on infectious disease is provided at Table 12.1. The table shows viral, bacterial, fungal and other parasitic pathogens of plants and animals that have already emerged within a diverse range of ecosystems. In some cases, these novel elements arising from anthropogenic

changes to the nature, quality and connectivity of host species have contributed to the emergence of hybrid or novel ecosystems (see Chapters 4, 5 and 6).

The contributions of habitat destruction, exotic species invasion and climate change on infectious disease ecology are difficult to disentangle. In this chapter, two case studies are considered. The first, African Highland malaria, offers a case study for the multiplicity of anthropogenic changes to natural ecosystems and their effects on infectious disease epidemiology. Persistence of vector-borne diseases such as malaria depends not only on host-pathogen interactions, but also on the climatic and topographical requirements of the vector species; this results in a particularly close association between transmission and the environment. Malaria is also one of the most-researched infectious diseases, facilitating a fuller mechanistic understanding of the key environmental processes that shape its global distribution.

There is a natural bias in the scientific literature towards human pathogens in linking infectious diseases with novel ecosystems because of the amenability of these subjects to observation. Inaccessibility

Novel Ecosystems: Intervening in the New Ecological World Order, First Edition. Edited by Richard J. Hobbs, Eric S. Higgs, and Carol M. Hall.
© 2013 John Wiley & Sons, Ltd. Published 2013 by John Wiley & Sons, Ltd.

Table 12.1 Wide range of emerging pathogens arising from anthropogenic causes.

Pathogen	Disease, Host	Anthropogenic cause
Prion proteins: PrP^{Sc}	Several diseases in myriad hosts e.g. bovine spongiform encephalopathy in cattle ('mad cow disease').	Hypothesized to have originated from UK farmers feeding cattle infected sheep offal (Wilesmith et al. 1988; Smith and Bradley 2003).
Viruses: dengue	Epidemic Dengue haemorrhagic fever in humans and other primates.	Vector (*Aedes* spp.) introductions and co-circulating virus strains through trade and human travel (Gubler 1998).
Bacteria: *Clostridium difficile*	Potentially fatal infection of humans with several animal reservoir species.	Hyper virulent strains are associated with widespread use of antibiotics disrupting normal gut micro-biome (McDonald et al. 2005).
Water moulds: *Aphanomyces astaci*	Crayfish plague in freshwater river systems all across Europe.	Transportation of US spp. of crayfish into European farms with spill-over into wider network of waterways (Alderman and Polglase 1988).
Fungi: *Batrachochytridium dendrobatiditis*	Chytridiomycosis is a fatal disease causing widespread declines in numerous amphibian species.	Spread through international trade in exotic frog spp. with regional distributions associated with deforestation (Mazzoni et al. 2003; Becker and Zamudio 2011).
Protozoans: *Plasmodium relictum*	Avian malaria has caused mass mortalities in several endemic species of the Hawaiian islands.	Accidental introduction of the mosquito vector to Hawaii (*Culex quinquefasciatus*) in 1826 (Warner 1968).
Helminths: *Globodera rostochiensis* and *G. pallida*	Potato cysts causing huge economic losses.	Now globally distributed, these pests are hypothesized to have originated with the potato from S. America (Hodda and Cook 2009).
Arthropods: *Lepeophtheirus salmonis*	This louse is an ectoparasite that feeds on salmon surface tissues, stressing and potentially killing the host.	Widespread infestations that are threatening wild salmon populations are associated with salmon farms (Krkošek et al. 2007).

continues to hamper the study of marine systems. Consequently, marine infectious disease understanding is less developed than for terrestrial systems and particularly nascent when compared with human epidemiology. This brings us to the second case study in this chapter: novel Caribbean coral reef ecosystems. Temporal dynamics of this drastically altered marine system demonstrate not only the effect that infectious disease can have in reshaping ecosystems, but also the recursive nature of the relationship between infectious disease and novel ecosystems.

12.2 AFRICAN HIGHLAND MALARIA

Malaria parasites are diverse, i.e. from several species of *Plasmodium*. Birds and mammals are infected through anopheline mosquito (Diptera: Culicidae) bites and reptiles by sandfly (Diptera: Psychodidae) bites. Human malaria is caused by four main species: *Plasmodium falciparum*, *P. vivax*, *P. malariae* and *P. ovale*. Together they give rise to approximately 250 million infections each year, killing nearly 1 million people (WHO 2010). Infections are generally constrained geographically by topographical and/or climatic factors which limit the persistence of either the vector or parasite. The *P. falciparum* parasite, which is primarily responsible for human morbidity and mortality, requires ambient temperatures of 20°C in order to develop within the vector (Mollneaux 1988) and the mosquito vector requires warm, standing pools of water in which pre-adult stages can develop (Gillies and De Meillon 1968). The term malaria is derived from 'bad air' which emanates from swamps and

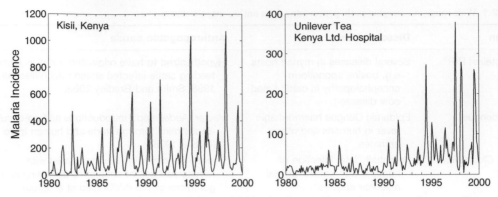

Figure 12.1 Increased frequency and severity of African highland malaria epidemics in two sites of western Kenya. From Zhou et al. (2004). © 2004 National Academy of Sciences, USA.

marshlands. It is this requirement of the mosquitoes that historically led to the association of the disease with stagnant water; the connection between humans, the environment and disease was made thousands of years ago in 'Airs, Water and Places' (Hippocrates, c. 460–370 BC). However, it was not until 1897 that Ross first observed malaria parasites within the mosquito (Ross 1923).

Shaped by numerous anthropogenic factors, the contours of malaria-associated death and disease (Dobson 1997) have shifted over the last century. The latitudinal extent of infection has shrunk through swamp drainage and mass application of dichlorodiphenyltrichloroethane (DDT) in North America (www.cdc.gov/malaria/about/history) and through drainage of swamp lands and improvements to public health services in Europe (Carter and Mendis 2002). The same century, however, has witnessed range expansion in many tropical regions, with 75% of malaria mortalities now occurring in African children. It is increasingly recognized that climate change and anthropogenic changes to land use and land cover have played important roles in eliciting re-emergent and newly emergent malaria transmission in these regions.

The range of malaria has altitudinal limits (Garnham 1945) as incurred by the same climatic constraints that impose its latitudinal limits. Historically, temperatures have been too low in East African highlands (>1500 m above sea level) to support the sustained development of the parasite in its vector. In recent decades, however, malaria epidemics of increasing

severity and frequency have occurred in the highland communities of Uganda (Mouchet et al. 1998), Ethiopia, Tanzania (Matola et al. 1987), Rwanda (Loevinsohn 1994), Madagascar (Lepers et al. 1988) and western Kenya (Fig. 12.1). The recent and future role of climate change in providing more permissive conditions in these highland environs has led to heated debate within the epidemiological community (Hay et al. 2002; Pascual et al. 2006; Lafferty 2009). The main points of contention are the appropriate level of scale at which climate variables are evaluated and the relative role of climate as part of a complex system with many moving parts. What remains uncontended, however, is the evidence for direct anthropogenic changes to the environment drastically altering malaria transmission.

Concordant with the marginal (c. 1°C) but biologically relevant (Paaijmans et al. 2009) regional warming in African highlands over the last few decades has been a rapid increase in the human population. This mass immigration and growth has resulted in drastic and widespread changes in the ecosystem. Between 1980 and 2010, 4.5 million ha of rainforest was cleared from East Africa (FAO 2011). Deforestation to clear space for high-density human settlement and agriculture affects the epidemiology of malaria via several avenues (Fig. 12.2). First, agriculture and human settlement bring with them novel mosquito breeding sites such as human-made stream edges, streambed pools and drainage channels at communal water supply points (Munga et al. 2009). Also, water accumulating in indentations and tracks left by people

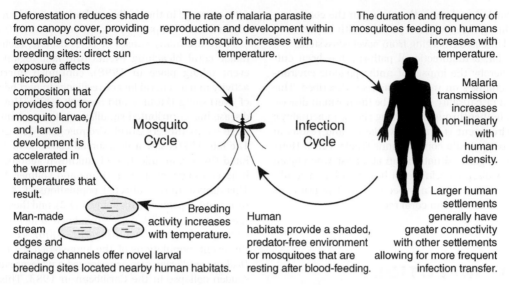

Figure 12.2 The malaria transmission cycle and mosquito vector lifecycle, how they are linked with the environment and how they have been affected by anthropogenic changes to the African highlands.

and livestock acts as an ideal habitat for *Anopheles gambiae* mosquito larvae, giving literal meaning to the human footprint of infectious disease in novel (or hybrid) ecosystems. Second, in reducing canopy cover, deforestation further increases ambient temperature of the microclimate, accelerating the vector and parasite development (Patz and Olson 2006). Third, like most other infectious diseases, the persistence of malaria is dependent on the density of available hosts (Anderson and May 1991). This is perhaps best understood in ecological terms whereby hosts might be considered as habitats and pathogens as the species of interest. Intuitively, the persistence and extinction of species are highly dependent on the connectivity of habitats within the metapopulation (Hanski and Gilpin 1991).

In fact, the degree of specialization on humans of *Anopheles gambiae* (the most important malaria vector species complex) at all stages of their lifecycle suggests a very recent and human-driven origin. Human population expansion in sub-Saharan Africa during the 5th millennium BC, the development of sedentary agriculture and the associated deforestation are hypothesized to have facilitated this specialization (Ayala and Coluzzi 2005). Recent genomic data provides evidence for ongoing diversification of this vector species, with new

ecological speciation driven by anthropogenic environmental change (White et al. 2011). Concurrent with mosquito adaptation to humans has been the speciation of human malaria parasites. The phylogenetic history of human malaria parasite species suggests they descended from malaria species of monkeys and apes (Carter and Mendis 2002; Liu et al. 2010). As humans increasingly encroached upon the habitat and supplanted predecessor host species, it is likely that there would have been strong selective pressure for both the vector and parasite to switch to humans as the preferred mammalian sources of blood and hosts of infection.

Genetic and evolutionary invasion analyses demonstrate that a newly emergent fifth human malaria species (*Plasmodium knowlesi*) appears to be in the process of switching from the natural hosts of long-tailed macaques to humans in SE Asia through human settlement and associated changes in land use (Cox-Singh et al. 2008; Yakob et al. 2010). Importantly, exposure to novel pathogens is classically understood to risk particularly virulent infections within hosts (Anderson and May 1982). Indeed, a recent review demonstrated that 10% of human infections with *P. knowlesi* develop life-threatening complications (Daneshvar et al. 2009). Of concern is the possibility that knowlesi malaria is the canary in the coal mine, highlighting the significant public health implications of novel diseases emerging from novel ecosystems.

Thus far, a unidirectional pathway has been outlined whereby the impact of anthropogenic environmental changes on malaria has been described. This direction of causality seems to be the norm in disease ecology (either through reporting bias or in actuality): the environment is drastically altered and results in novel host-parasite combinations (Table 12.1). However, the opposite situation can also arise whereby an infectious disease decimates a host population resulting in the creation of a novel ecosystem. This brings us to the novel Caribbean coral reef.

12.3 INFECTIOUS DISEASE AND THE NOVEL CARIBBEAN CORAL REEF

Laith Yakob and Peter J. Mumby

Scleractinian corals are colonial marine animals that secrete a calcareous skeleton. Over time, the process of calcification creates reefs which form the largest natural structures on Earth and provide habitat for thousands of organisms. In this respect, corals are 'ecosystem engineers', the marine equivalent of the canopy-forming trees of rainforests (Davidson 1998). They constitute an important keystone species of the marine environment yet they are currently experiencing an unprecedented rate of decline (Pandolfi et al. 2003; see also Chapter 8). Evidence suggests that infectious disease has had a big hand in drastically altering these ecosystems, most notably within the Caribbean (although coral bleaching, hurricanes and algal blooms have also been important).

For thousands of years, Caribbean coral reefs were built by large, often long-lived corals whose life history strategy included annual mass spawning events that allowed large-scale dispersal of larvae in the plankton for a month or more. Typical framework-building species included the massive (mound-shaped) coral *Montastraea* spp and branching species *Acropora cervicornis* and *A. palmata* (Edmunds and Elahi 2007). Unfortunately, the abundance of these corals has declined dramatically in the last 40 years from about

50% coverage in the Caribbean basin to less than 10% (Bythell and Sheppard 1993; Gardner et al. 2003).

There are many causes of coral decline, including major coral bleaching events (the first region-wide event taking place in 1998), continued hurricane activity and a general loss of resilience in part because of overfishing (Mumby and Steneck 2008). However, disease has contributed significantly to the demise of large corals (Rogers 2008; Cròquer and Weil 2009). The late 1970s saw a widespread epizootic of White Band Disease, an infection of unknown aetiology but believed to be a bacterium (Kline and Vollmer 2011). This disease mostly affected populations of *A. cervicornis* and *A. palmata* (Gladfelter 1982), and these coral species are now listed as threatened under the US endangered species act (Hogarth 2006). Following this event, populations of the long-spined sea urchin *Diadema antillarum* also experienced a widespread and sudden collapse in the Caribbean in 1983. This was described as "the most extensive epidemic documented for a marine invertebrate" (Lessios et al. 1984). These urchins are functionally important herbivores that maintain algae in a cropped state, thereby reducing the competitive exclusion of corals. The sudden loss of urchins led to a massive algal bloom, particularly in areas where fishing had already depleted the natural stock of herbivorous fishes (Hughes 1994; Steneck 1994). Throughout the region, *Diadema* populations were reduced by as much as 98% (Knowlton 2001), with little sign of recovery in the subsequent 20 years (Edmunds and Carpenter 2001). The identity of the aetiological agent for this disease remains unknown. What was immediately realized, however, was the impact this event was likely to have on the region-wide ecology as cover by the dominant algal genera of *Lobophora* and *Dictyota* rapidly increased in some areas by as much as 88% (Hughes 1994).

Retarded recovery of the reef-builders and the urchin populations are linked. The urchins had been integral to the facilitation of growth and recruitment of corals by reducing macroalgal competition (Hughes et al. 1999; Nugues and Bak 2006; Foster et al. 2008). Concurrently, the exclusion of macroalgae by the presence of live coral cover provides a more desirable substrate for juvenile urchin growth (Bak 1985). Urchin populations were likely prevented from rebounding by maintained low-levels of infection in pockets throughout the Caribbean (Forcucci 1994), strong density dependence in their recruitment (Karlson and Levitan 1990) or elevated post-settlement mortality because of high densities of predators in the new algal canopy (Mumby and Steneck 2008). At the same time, recovery of reef-building corals has been hampered by their slow population turnovers relative to macroalgal competitors (Mumby 2009). When healthy, the great longevity of these species might only have necessitated occasional recruitment events for the population to persist (Edmunds and Elahi 2007). For this reason, however, ecological theory suggests that species with these 'slow' life histories tend to be much more vulnerable to extinction through a delayed reaction to perturbation. Using a classic epidemiological approach, a recent study describes how these species are also likely to be highly susceptible to infectious disease by simple virtue of their demographics (Yakob and Mumby 2011). For a pathogen to spread, a host must survive long enough to become infected and then to pass on the infection. Long-lived coral species are thusly predisposed to infectious disease outbreaks.

The overall result in the Caribbean coral reef ecosystem has not been simply less of the same, but the emergence of a novel assemblage of corals. As the large framework-builders declined, largely because of a combined predisposition to disease and failure to recover, the dominance of what were minor components of the coral assemblage has increased (Green et al. 2008). Typically, the minor components of reef assemblages were diminutive corals that brood their larvae year-round and achieve a high success in recruitment, at approximately a tenfold higher density than the spawning species that built the reef. Examples include the corals *Agaricia agaricites* and *Porites astreoides* which are increasingly dominating reefs of the region (Fig. 12.3). These small-bodied fast-growing species have taken advantage of the window of opportunity provided by the benthic clearance of the massive species (Green et al. 2008). This shift in the composition of coral assemblages is deleterious from the viewpoint of both biodiversity and ecosystem services. For example, a reef dominated by small-bodied corals will have less structural complexity, which means that it offers less surface area and fewer refugia for other taxa and therefore has lower diversity and fewer fish (Gratwicke and Speight 2005). Fossil records indicate that these 'novel assemblages' are indeed novel as they have not been seen since at least the Late Pleistocene (Pandolfi and Jackson 2006; see also Chapter 7).

Disease outbreaks are still occurring in the Caribbean region (Porter et al. 2001; Weil 2004; Cròquer and Weil 2009) and future projections indicate that a third

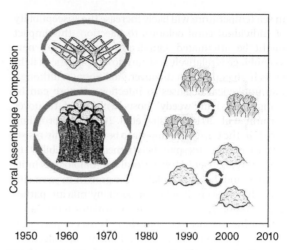

Figure 12.3 The recent switch in the dominant coral species within Caribbean reefs. Species with low demographic rates, *Acropora cervicornis* (top left) and *Montastraea annularis* (bottom left), have been replaced by species with high demographic rates, *Agaricia* spp. (top right) and *Porites astreoides* (bottom right). From Yakob and Mumby (2011). © 2011 National Academy of Sciences, USA.

of the global reef corals are at risk of extinction through a combination of climate change and disease (Carpenter et al. 2008). Estimating the future role of infectious disease within these novel ecosystems has several hurdles. The causes of the recent spate of epizootics are unknown; determining causality in a large and complex ecosystem is not trivial (Plowright et al. 2008). The collapse of many Caribbean coral reefs was preceded by dwindling fish stocks and increased nutrient and sediment runoff from land (Hughes 1994; Jackson et al. 2001). In other words, anthropogenic harm reduced functional redundancy within key ecological processes and elicited environmental conditions that compromised coral health prior to the epizootics.

Epidemiological theory suggests that the rapid turnovers of the current dominant coral species might impose demographic resistance to infection (Diekmann and Heesterbeck 2002; Yakob and Mumby 2011). This is important because the novel assemblage will function quite differently to its predecessor. Coral susceptibility to infection is linked to thermal stress, resulting in a pessimistic outlook for future reefs under climate change (Bruno et al. 2007). While continued increases in sea temperature will likely increase the susceptibility of individual coral colonies to infection, this impact might be attenuated significantly because the new assemblage is relatively resistant to an epizootic in light of its high population turnover. Additionally, although immunological defenses to infectious disease can be compromised in 'weedy' hosts with rapid turnover (Cronin et al. 2010; Keesing et al. 2010), the brooding corals in the Caribbean appear to be intrinsically resistant to disease (despite heightened susceptibility to bleaching; McField 1999). Coral microbiology is in its infancy and, because isolation for growth in pure cultures has proven difficult for so many marine pathogens, demonstration of Koch's postulates (criteria for establishing causal relationships between pathogen and disease) only exist for a handful of infections (Rosenberg and Kushmaro 2011). This means that, despite a rapidly expanding interest in coral infectious disease (Sokolow 2009), the identities of most coral pathogens remain unknown. From basic biology to complex ecology, numerous key epizoological processes still urgently require elucidation. This will not only allow for better estimation of the current and future role of disease in these threatened, marine ecosystems, but will enable guidelines for appropriate action to surface.

12.4 FINAL REMARKS

Ecosystem change is both the cause and effect of emerging infectious disease (Fig. 12.4). While understanding causal relationships is the ultimate goal for ecologists and epidemiologists alike this merely facilitates, and is not a prerequisite of, intervention. Associations are sometimes so apparent and reproduc-ible that action can be taken to attenuate disease without complete understanding of the underlying processes. 'Miasma', 'ague' and 'remittent fever' were all associated with proximity to stagnant water, long before the malaria parasite was discovered and the mosquito implicated in its spread. The logical interventions were to drain the swamps or keep away from them. Environmental management has historically been important in eliminating local disease transmission, and interest has been recently rekindled in this method as part of an integrated 21st century approach to controlling malaria (Yakob and Yan 2009, 2010).

This bodes well for future efforts in ameliorating the feedbacks between infectious disease and hybrid or novel ecosystems; where patterns are evident in descriptive longitudinal data, hypotheses can be formulated and tested. Directionality between disease and ecosystem novelty might remain obscured by the muddied waters of Caribbean coral reef epizoology, but immediate action can still be taken. Models have shown the multiplicative effect that macroalgal growth and infectious disease can have in compromising coral health. While the aetiology of infection might still require elucidation for many coral infections, appropriate intervention is clear: reduce algal competition by protecting the key herbivores.

There is an ongoing paradigm shift in ecology, with greater emphasis on ecosystem disequilibrium (White and Pickett 1985; Levin 1999). Since they represent phases in a continuing feedback process, the distinction between where 'cause' finishes and 'effect' begins will blur. Our environment is rapidly changing and new pathogenic threats will continue to emerge. Improved interfacing between scientific disciplines will allow for a more holistic understanding of the dynamic nature of humans, their environment and infectious disease.

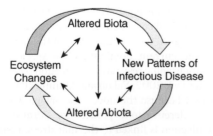

Figure 12.4 The feedbacks between novel ecosystems and infectious disease.

REFERENCES

Alderman, D.J. and Polglase, J.L. (1988) Pathogens, parasites and commensals, in *Freshwater Crayfish: Biology, Management and Exploitation* (D.M. Holdrich and R.S. Lowery, eds), Timber Press, OR, p. 176–183.

Anderson, P.K., Cunningham, A.A., Patel, N.G., Morales, F.J., Epstein, P.R. and Daszak, P. (2004) Emerging infectious diseases of plants: pathogen pollution, climate change and agrotechnology drivers. *Trends in Ecology Evolution*, **19**, 535–544.

Anderson, R.M. and May, R.M. (1982) Co-evolution of hosts and parasites. *Parasitology*, **85**, 411–426.

Anderson, R.M. and May, R.M. (1991) *Infectious Diseases of Humans: Dynamics and Control*. Oxford University Press, Oxford and New York.

Ayala, F.J. and Coluzzi, M. (2005) Chromosome speciation: Humans, Drosophila, and mosquitoes. *Proceedings of the National Academy of Sciences of the United States of America*, **102**, 6535–6542.

Bak, R.P.M. (1985) Recruitment patterns and mass mortalities in the sea urchin Diadema antillarum. *Proceedings of Fifth International Coral Reef Congress*, Tahiti, 267–272.

Becker, C.G. and Zamudio, K.R. (2011) Tropical amphibian populations experience higher disease risk in natural habitats. *Proceedings of the National Academy of Science USA*, **108**, 9893–9898.

Bruno, J.F., Selig, E.R., Casey, K.S., Page, C.A., Willis, B.L., Harvell, C.D., Sweatman, H. and Melendy, A.M. (2007) Thermal stress and coral cover as drivers of coral disease outbreaks. *PLoS Biol*, **5**, e124.

Bythell, J.C. and Sheppard, C.R.C. (1993) Mass mortality of caribbean shallow corals. *Marine Pollution Bulletin*, **26**, 296–297.

Carpenter, K.E., Abrar, M., Aeby, G. et al. (2008) One-third of reef-building corals face elevated extinction risk from climate change and local impacts. *Science*, **321**, 560–563.

Carter, R. and Mendis, K.N. (2002) Evolutionary and historical aspects of the burden of malaria. *Clinical Microbiology Reviews*, **15**, 564–594.

Cleaveland, S., Laurenson, M.K. and Taylor, L.H. (2001) Diseases of humans and their domestic mammals: pathogen characteristics, host range and the risk of emergence. *Philosophical Transactions of the Royal Society of London. Series B: Biological Sciences*, **356**, 991–999.

Cox-Singh, J., Davis, T.M.E., Lee, K.-S., Shamsul, S.S.G., Matusop, A., Ratnam, S., Rahman, H.A., Conway, D.J. and Singh, B. (2008) Plasmodium knowlesi malaria in humans is widely distributed and potentially life threatening. *Clinical Infectious Diseases*, **46**, 165–171.

Cronin, J.P., Welsh, M.E., Dekkers, M.G., Abercrombie, S.T. and Mitchell, C.E. (2010) Host physiological phenotype explains pathogen reservoir potential. *Ecology Letters*, **13**, 1221–1232.

Cròquer, A. and Weil, E. (2009) Changes in Caribbean coral disease prevalence after the 2005 bleaching event. *Diseases of Aquatic Organisms*, **87**, 33–43.

Daneshvar, C., Davis, T.M.E., Cox-Singh, J., Rafa'ee, M.Z., Zakaria, S.K., Divis, P.C.S. and Singh, B. (2009) Clinical and laboratory features of human Plasmodium knowlesi infection. *Clinical Infectious Diseases*, **49**, 852–860.

Daszak, P., Cunningham, A.A. and Hyatt, A.D. (2000) Emerging infectious diseases of wildlife: threats to biodiversity and human health. *Science*, **287**, 443–449.

Davidson, O.G. (1998) *The Enchanted Braid*. Wiley, New York.

Diekmann, O., and Heesterbeck, J.a.P. (2002) *Mathematical Epidemiology of Infectious Diseases: Model Building, Analysis and Interpretation*. Wiley, New York.

Dobson, M.J. (1997) *Contours of Death and Disease in Early Modern England*. Cambridge University Press.

Edmunds, P.J. and Carpenter, R.C. (2001) Recovery of Diadema antillarum reduces macroalgal cover and increases abundance of juvenile corals on a Caribbean reef. *Proceedings of the National Academy of Sciences*, **98**, 5067–5071.

Edmunds, P.J. and Elahi, R. (2007) The demographics of a 15-year decline in cover of the Caribbean reef coral *Montastraea annularis*. *Ecological Monographs*, **77**, 3–18.

FAO (2011) *State of the World's Forests*. Food and Agriculture Organization of the United Nations, Rome.

Forcucci, D. (1994) Population density, recruitment and 1991 mortality event of Diadema antillarum in the Florida keys. *Bulletin in Marine Science*, **54**, 917–928.

Foster, N.L., Box, S.J. and Mumby, P.J. (2008) Competitive effects of macroalgae on the fecundity of the reef-building coral Montastraea annularis. *Marine Ecology Progress Series*, **367**, 143–152.

Gardner, T.A., Côté, I.M., Gill, J.A., Grant, A. and Watkinson, A.R. (2003) Long-term region-wide declines in Caribbean corals. *Science*, **301**, 958–960.

Garnham, P.C.C. (1945) Malaria epidemics at exceptionally high altitudes in Kenya. *British Medical Journal*, **11**, 45–47.

Gillies, M.T. and De Meillon, B. (1968) *The Anophelinae of Africa South of the Sahara*. South African Institute of Medical Research, Johannesburg.

Gladfelter, W.B. (1982) White band disease in Acropora palmata: implications for the structure and growth of shallow reefs. *Bulletin in Marine Science*, **32**, 639–643.

Gratwicke, B. and Speight, M.R. (2005) Effects of habitat complexity on Caribbean marine fish assemblages. *Marine Ecology-Progress Series*, **292**, 301–310.

Green, D.H., Edmunds, P.J. and Carpenter, R.C. (2008) Increasing relative abundance of Porites astreoides on Caribbean reefs mediated by an overall decline in coral cover. *Marine Ecology Progress Series*, **359**, 1–10.

Gubler, D.J. (1998) Dengue and dengue hemorrhagic fever. *Clinical Microbiology Reviews*, **11**(3), 480–496.

Hanski, I. and Gilpin, M. (1991) Metapopulation dynamics: brief history and conceptual domain. *Biological Journal of the Linnean Society*, **42**, 3–16.

Hay, S.I., Cox, J., Rogers, D.J., Randolph, S.E., Stern, D.I., Shanks, G.D., Myers, M.F. and Snow, R.W. (2002) Climate change and the resurgence of malaria in the East African highlands. *Nature*, **415**, 905–909.

Hodda, M. and Cook, D.C. (2009) Economic impact from unrestricted spread of potato cyst nematodes in Australia. *Phytopathology*, **99**, 1387–1393.

Hogarth, W.T. (2006) Endangered and threatened species: final listing determinations for the Elkhorn coral and Staghorn coral. *Federal Register*: 26852–26861.

Hughes, T.P. (1994) Catastrophes, phase shifts, and large-scale degradation of a caribbean coral reef. *Science*, **265**, 1547–1551.

Hughes, T., Szmant, A.M., Steneck, R., Carpenter, R. and Miller, S. (1999) Algal blooms on coral reefs: what are the causes? *Limnology and Oceanography*, **44**, 1583–1586.

Jackson, J.B.C., Kirby, M.X., Berger, W.H. et al. (2001) Historical overfishing and the recent collapse of coastal ecosystems. *Science*, **293**, 629–637.

Karlson, R.H. and Levitan, D.R. (1990) Recruitment-limitation in open populations of Diadema antillarum: an evaluation. *Oecologia*, **82**, 40–44.

Keesing, F., Belden, L.K., Daszak, P., Dobson, A., Harvell, C.D., Holt, R.D., Hudson, P., Jolles, A., Jones, K.E., Mitchell, C.E., Myers, S.S., Bogich, T. and Ostfeld, R.S. (2010) Impacts of biodiversity on the emergence and transmission of infectious diseases. *Nature*, **468**, 647–652.

Kline, D.I. and Vollmer, S.V. (2011) White Band Disease (type I) of Endangered Caribbean Acroporid Corals is Caused by Pathogenic Bacteria. *Scientific Reports*, **1**, article no. 7, doi: 10.1038/srep00007.

Knowlton, N. (2001) Sea urchin recovery from mass mortality: New hope for Caribbean coral reefs? *Proceedings of the National Academy of Sciences*, **98**, 4822–4824.

Krkošek, M., Ford, J.S., Morton, A., Lele, S., Myers, R.A. and Lewis, M.A. (2007) Declining wild salmon populations in relation to parasites from farm salmon. *Science*, **318**, 1772–1775.

Lafferty, K.D. (2009) The ecology of climate change and infectious diseases. *Ecology*, **90**, 888–900.

Lepers, J.P., Deloron, P., Fontenille, D. and Coulanges, P. (1988) Reappearance of falciparum malaria in central highland plateaus of Madagascar. *Lancet*, **1**, 586–587.

Lessios, H.A., Robertson, D.R. and Cubit, J.D. (1984) Spread of diadema mass mortality through the Caribbean. *Science*, **226**, 335–337.

Levin, S.A. (1999) Towards a science of ecological management. *Conservation Ecology*, **3**, 6.

Liu, W., Li, Y., Learn, G.H. et al. (2010) Origin of the human malaria parasite Plasmodium falciparum in gorillas. *Nature*, **467**, 420–425.

Loevinsohn, M.E. (1994) Climatic warming and increased malaria incidence in Rwanda. *Lancet*, **343**, 714–718.

Matola, Y.G., White, G.B. and Magayuka, S.A. (1987) The changed pattern of malaria endemicity and transmission at Amani in the eastern Usambara mountains, north-eastern Tanzania. *Journal of Tropical Medicine and Hygiene*, **90**, 127–134.

Mazzoni, R., Cunningham, A.A., Daszak, P., Apolo, A., Perdomo, E. and Speranza, G. (2003) Emerging pathogen of wild amphibians in frogs (*Rana catesbeiana*) farmed for international trade. *Emerging Infectious Diseases*, **9**, 995–998.

McDonald, L.C., Killgore, G.E., Thompson, A., Owens, R.C., Kazakova, S.V., Sambol, S.P., Johnson, S. and Gerding, D.N. (2005) An epidemic toxin gene-variant strain of *Clostridium difficile*. *New England Journal of Medicine*, **353**, 2433–2441.

McField, M.D. (1999) Coral response during and after mass bleaching in Belize. *Bulletin of Marine Science*, **64**, 155–172.

Mollneaux, L. (1988) The epidemiology of human malaria as an explanation of its distribution, including some implications for its control, in *Malaria: Principles and Practice of Malariology* (eds W.H. Wernsdorfer and I. McGregor), Churchill Livingstone, New York.

Mouchet, J., Manguin, S., Sircoulon, J., Laventure, S., Faye, O., Onapa, A.W., Carnevale, P., Julvez, J. and Fontenille, D. (1998) Evolution of malaria in Africa for the past 40 years: impact of climatic and human factors. *Journal of the American Mosquito Control Association*, **14**, 121–130.

Mumby, P.J. (2009) Phase shifts and the stability of macroalgal communities on Caribbean coral reefs. *Coral Reefs*, **28**, 761–773.

Mumby, P.J. and Steneck, R.S. (2008) Coral reef management and conservation in light of rapidly-evolving ecological paradigms. *Trends in Ecology & Evolution*, **23**, 555–563.

Munga, S., Yakob, L., Mushinzimana, E., Zhou, G., Ouna, T., Minakawa, N., Githeko, A. and Yan, G. (2009) Land use and land cover changes and spatiotemporal dynamics of anopheline larval habitats during a four-year period in a highland community of Africa. *The American Journal of Tropical Medicine and Hygiene*, **81**, 1079–1084.

Nugues, M.M. and Bak, R.P.M. (2006) Differential competitive abilities between Caribbean coral species and a brown alga: a year of experiments and a long-term perspective. *Marine Ecology Progress Series*, **315**, 75–86.

Paaijmans, K.P., Read, A.F. and Thomas, M.B. (2009) Understanding the link between malaria risk and climate. *Proceedings of the National Academy of Sciences*, **106**, 13844–13849.

Pandolfi, J.M. and Jackson, J.B.C. (2006) Ecological persistence interrupted in Caribbean coral reefs. *Ecology Letters*, **9**, 818–826.

Pandolfi, J.M., Bradbury, R.H., Sala, E. et al. (2003) Global trajectories of the long-term decline of coral reef ecosystems. *Science*, **301**, 955–958.

Pascual, M., Ahumada, J., Chaves, L.F., Rodó, X. and Bouma, M.J. (2006) Malaria resurgence in East African Highlands: temperature trends revisited. *Proceedings of the National Academy of Sciences*, **103**, 5829–5834.

Patz, J.A., and Olson, S.H. (2006) Malaria risk and temperature: Influences from global climate change and local land use practices. *Proceedings of the National Academy of Sciences*, **103**, 5635–5636.

Plowright, R.K., Sokolow, S.H., Gorman, M.E., Daszak, P. and Foley, J.E. (2008) Causal inference in disease ecology: investigating ecological drivers of disease emergence. *Frontiers in Ecology and Evolution*, **6**, 420–429.

Porter, J.W., Dustan, P., Jaap, W.C., Patterson, K.L., Kosmynin, V., Meier, O.W., Patterson, M.E. and Parsons, M. (2001)

Patterns of spread of coral disease in the Florida Keys. *Hydrobiologia*, **460**, 1–24.

Rogers, C. (2008) Coral bleaching and disease should not be underestimated as causes of Caribbean coral reef decline. *Proceedings of the Royal Society of London Series B Biological Sciences*, **276**, 197–198.

Rosenberg, E. and Kushmaro, A. (2011) Microbial diseases of corals: pathology and ecology, in *Coral Reefs: An Ecosystem in Transition* (eds Z. Dubinsky and N. Stambler), Springer, Dordrecht, Heidelberg, London, New York.

Ross, R. (1923) *Memoirs*. John Murray, London.

Sokolow, S. (2009) Effects of a changing climate on the dynamics of coral infectious disease: a review of the evidence. *Diseases of Aquatic Organisms*, **87**, 5–18.

Smith, P.G. and Bradley, R. (2003) Bovine spongiform encephalopathy (BSE) and its epidemiology. *British Medical Bulletin*, **66**(1), 185–198.

Steneck, R.S. (1994) Is herbivore loss more damaging to reefs than hurricanes? Case studies from two Caribbean reef systems (1978–1988), in *Global Aspects of Coral Reefs: Health, Hazards, and History* (ed. R.N. Ginsburg), University of Miami, Florida, C32–C37.

Warner, R.E. (1968) The role of introduced diseases in the extinction of the endemic Hawaiian avifauna. *Condor*, **70**, 101–120.

Weil, E. (2004) Coral reef diseases in the wider Caribbean, in *Coral Health and Disease* (eds E. Rosenberg and Y. Loya), Springer Verlag, New York.

White, B.J., Collins, F.H. and Besansky, N.J. (2011) Evolution of Anopheles gambiae in relation to humans and malaria. *Annual Review of Ecology, Evolution and Systematics*, **42**, 111–132.

White, P.S. and Pickett, S.T.A. (1985) Natural disturbance and patch dynamics: an introduction, in *The Ecology of Natural Disturbance and Patch Dynamics* (eds P.S. White and S.T.A. Pickett), Academic Press, New York.

Wilesmith, J., Wells, G., Cranwell, M. and Ryan, J. (1988) Bovine spongiform encephalopathy: epidemiological studies. *Veterinary Record*, **123**(25), 638–644.

WHO (2010) *World Malaria Report*. World Health Organisation, Geneva.

Woolhouse, M.E.J. and Dye, C. (2001) Preface. *Philosophical Transactions of the Royal Society of London. Series B: Biological Sciences*, **356**, 981–982.

Woolhouse, M.E. and Gowtage-Seguería, S. (2005) Host range and emerging and reemerging pathogens. *Emerging Infectious Diseases*, **11**, 1842–1847.

Yakob, L. and Yan, G. (2009) Modeling the effects of integrating larval habitat source reduction and insecticide treated nets for malaria control. *PLoS ONE*, **4**, e6921.

Yakob, L. and Yan, G. (2010) A network population model of the dynamics and control of African malaria vectors. *Transactions of the Royal Society of Tropical Medicine and Hygiene*, **104**, 669–675.

Yakob, L. and Mumby, P.J. (2011) Climate change induces demographic resistance to disease in novel coral assemblages. *Proceedings of the National Academy of Sciences*, **108**, 1967–1969.

Yakob, L., Bonsall, M. and Yan, G. (2010) Modelling knowlesi malaria transmission in humans: vector preference and host competence. *Malaria Journal*, **9**, 329.

Zhou, G., Minakawa, N., Githeko, A.K. and Yan, G. (2004) Association between climate variability and malaria epidemics in the East African highlands. *Proceedings of the National Academy of Sciences of the United States of America*, **101**, 2375–2380.

Chapter 13

CASE STUDY: DO FEEDBACKS FROM THE SOIL BIOTA SECURE NOVELTY IN ECOSYSTEMS?

James A. Harris

Environmental Science and Technology Department, Cranfield University, UK

When considering how novel ecosystems come to be established and maintained a recurring theme is that of irreversible thresholds, where return to a previous state is difficult or impossible and persistence in the novel ecosystem can be observed. Identifying such thresholds may prove difficult, but there is one class of interactions that provides a possible mechanism for establishing a threshold and provides a stabilizing force: the interactions between the plant community and the soil biota.

It is well established that the composition of plant communities is intimately linked with, and dependent upon, the soil micro-flora and meso-fauna associated with such communities (Wardle et al. 2004). Many studies in pastures have shown that coupling between plant and microbial communities occurs at a variety of scales, and is certainly modulated by other factors such as pH and grassland management (e.g. Ritz et al. 2004).

Such insight has implications not only for understanding the fundamental relationships between belowground and aboveground communities, but for conservation and restoration programmes which have previously focused almost exclusively on the plant community and relationships to abiotic soil characteristics such as nutrient supply and water relationships. Plant communities certainly develop in the context of, and are therefore likely co-evolve with, soil communities (Schweitzer et al. 2011). The role and importance of the soil biota in providing critical feedback to the plant community and its composition (both positive and negative), profoundly affecting the composition of the plant community, has been widely demonstrated. These feedbacks provide a mechanism for securing novelty in ecosystems by positively encouraging rare genotypes including, in particular, non-native species (e.g. van der Putten et al. 2007).

Reynolds et al. (2003) suggested that positive feedback between plants and soil microbes plays a central role in early successional communities, while negative feedback contributes both to species replacements and to diversification in later successional communities. By

Novel Ecosystems: Intervening in the New Ecological World Order, First Edition. Edited by Richard J. Hobbs, Eric S. Higgs, and Carol M. Hall.
© 2013 John Wiley & Sons, Ltd. Published 2013 by John Wiley & Sons, Ltd.

extrapolation, both types of feedback are an important organizing force for large-scale spatial gradients in species richness. They suggest that the balance of positive feedback as a homogenizing force and negative feedback as a diversifying force may contribute to observed latitudinal (and altitudinal) diversity patterns. The 'Red Queen' hypothesis (e.g. Clay et al. 2008) suggests that plant pathogens become specialized on common host genotypes thus favoring rare genotypes, promoting diversity and ultimately favoring introduced and potentially invasive species.

Soil biota affect plant composition throughout succession; these effects can be a consequence of root-feeding nematodes, insect larvae, pathogenic soil microbes or mutualistic symbionts (De Deyn et al. 2003). Within succession stages, these soil biota can influence species composition and community structure (Van der Heijden et al. 1998). These investigations point to the critical role soil biota have in stabilizing late successional ecosystem composition and function.

Inderjit and van der Putten (2010) have recently described a number of mechanisms that can favor introduced/invasive plant species, classifying them as direct or indirect. Direct effects include:
- native soil communities (including soil pathogens) which resist invasion (Nijjer et al. 2007);
- native soil biota that can create positive feedback (e.g. Callaway et al. 2004); and
- complete or partial release from enemies such as fungi or viruses (e.g. Knevel et al. 2004).

Indirect effects include:
- accumulation of native soil pathogens (e.g. Eppinga et al. 2006);
- disruption of mutualistic associations between symbionts and seedlings (e.g. Rudgers et al. 2007);
- degradation of innocuous novel chemicals released by exotic plants into toxic byproducts by native soil communities (the 'novel biochemical weapons' effect; Bains et al. 2009); and
- nutrient release from exotic litter by decomposers (e.g. Marchante et al. 2008).

As an illustration, the 'novel biochemical weapons' mechanism has been demonstrated to operate within the same species. Bains et al. (2009) showed that an exotic *Phragmites australis* produced tannins that were transformed by the soil microbial community into gallic acid, securing an advantage over the native *P. australis*. This may also hold true for banked versus modern seeds, which may become problematical when sourcing seeds for restoration from long-term seed banks.

Experimentally, the simple manipulation of the biota (e.g. elimination by irradiation) is sufficient in most cases to remove these soil biota plant community feedback effects normally offering advantage to rare genotypes.

These negative soil biota community feedback mechanisms could stabilize novel assemblages of plants, in effect providing an irreversible threshold preventing reversion to a previous configuration and securing novelty. Thinking of soil biota in this way holds significant implications for restoration and intervention programs (Kardol and Wardle 2010). The composition and community history of the soil biota and their interactions with 'exotic' plant genotypes are therefore likely to play a defining role in the emergence of novel ecosystems.

REFERENCES

Bains, G., Kumar, A.S., Rudrappa, T., Alff, E., Hanson, T.E. and Bais, H.P. (2009) Native plant and microbial contributions to a negative plant-plant interaction. *Plant Physiology*, **151**, 2145–2151.

Callaway, R.M., Thelen, G.C., Rodriguez, A. and Holbe, W.E. (2004) Soil biota and exotic plant invasions. *Nature*, **427**, 731–733.

Clay, K., Reinhart, K., Rudgers, J., Tintjer, T., Koslow, J. and Florey, S.L. (2008) Red Queen communities, in *Infectious Disease Ecology: Effects of Disease on Ecosystems and Ecosystems on Disease* (R. Ostfield, F. Keesing and V. Evinor, eds), Princeton University Press, 145–178.

De Deyn, G.B., Raaijmakers, C.E., Zoomer, H.R., Berg, M.P., de Ruiter, P.C., Verhoef, H.A., Bezemer, T.M. and van der Putten, W.H. (2003) Soil invertebrate fauna enhances grassland succession and diversity. *Nature*, **422**, 711–713.

Eppinga, M.B., Rietkerk, M., Dekker, S.C., De Ruiter, P.C. and Van der Putten, W.H. (2006) Accumulation of local pathogens: a new hypothesis to explain exotic plant invasions. *Oikos*, **114**, 168–176.

Inderjit and van der Putten, W.H. (2010) Impacts of soil microbial communities on exotic plant invasions. *Trends in Ecology and Evolution*, **25**, 512–519.

Kardol, P. and Wardle, D.A. (2010) How understanding aboveground-belowground linkages can assist restoration ecology. *Trends in Ecology and Evolution*, **25**, 670–679.

Knevel, I.C., Lans, T., Menting, F.B.J., Hertling, U.M. and van der Putten, W.H. (2004) Release from native root herbivores and biotic resistance by soil pathogens in a new habitat both affect the alien *Ammophila arenaria* in South Africa. *Oecologia*, **141**, 502–510.

Marchante, E., Annelise Kjøllerb, A., Struweb, S., Freitasa, H. (2008) Invasive *Acacia longifolia* Induces changes in the microbial catabolic diversity of sand dunes. *Soil Biology and Biochemistry*, **40**, 2563–2568.

Nijjer, S., Rogers, W.E. and Siemann, E. (2007) Negative plant-soil feedbacks may limit the persistence of an invasive tree due to rapid accumulation of pathogens. *Proceedings of the Royal Society B*, **274**, 2621–2627.

Reynolds, H.L., Packer, A., Bever, J.D. and Clay, K. (2003) Grassroots ecology: Plant-microbe-soil interactions as drivers of plant community structure and dynamics. *Ecology*, **84**, 2281–2291.

Ritz, K., McNicol, J.W., Nunan, N., Grayston, S.J., Millard, P., Atkinson, D., Gollotte, A., Habeshaw, D., Boag, B., Clegg, C.D., Griffiths, B.S., Wheatley, R.E., Glover, L.A., McCaig, A.E. and Prosser, J.I. (2004) Spatial structure in soil chemical and microbiological properties in an upland grassland. *FEMS Microbiology Ecology*, **49**, 191–205.

Rudgers, J.A., Holah, J., Orr, S.P. and Clay, K. (2007) Forest succession suppressed by an introduced plant-fungal symbiosis. *Ecology*, **88**, 18–25.

Schweitzer, J.A., Fischer, D.G., Rehill, B.J., Wooley, S.C., Woolbright, S.A., Lindroth, R.L., Whitham, T.G., Zak, D.R. and Hart, S.C. (2011) Forest gene diversity is correlated with the composition and function of soil microbial communities. *Population Ecology*, **53**, 35–46.

Van der Heijden, M.G.A., Klironomos, J.N., Ursic, M., Moutoglis, P., Streitwolf-Engel, R., Boller, T., Wiemken, A. and Sanders, I. (1998) Mycorrhizal fungal diversity determines plant biodiversity, ecosystem variability and productivity. *Nature*, **396**, 69–72.

van der Putten, W.H., Kowalchuk, G.A., Brinkman, E.P., Doodeman, G.T.A., van der Kaaij, R.M., Kamp, A.F.D., Menting, F.B.J. and Veenendaal, E.M. (2007) Soil feedback of exotic savanna grass relates to pathogen absence and mycorrhizal selectivity. *Ecology*, **88**, 978–988.

Wardle, D.A., Bardgett, R.D., Klironomos, J.N., Setälä, H., van den Putten, W. and Wall, D.H. (2004) Ecological linkages between aboveground and belowground biota *Science*, **304**, 1629–1633.

Chapter 14

FAUNA AND NOVEL ECOSYSTEMS

Patricia L. Kennedy[1], Lori Lach[2], Ariel E. Lugo[3] and Richard J. Hobbs[4]

[1]Department of Fisheries and Wildlife & Eastern Oregon Agriculture & Natural Resource Program, Oregon State University, USA

[2]Ecosystem Restoration and Intervention Ecology (ERIE) Research Group, School of Plant Biology and the Centre for Integrative Bee Research, University of Western Australia, Australia

[3]International Institute of Tropical Forestry, USDA Forest Service, Rio Piedras, Puerto Rico

[4]Ecosystem Restoration and Intervention Ecology (ERIE) Research Group, School of Plant Biology, University of Western Australia, Australia

The concept of novel ecosystems has largely arisen from the field of restoration ecology, which historically has been dominated by plant ecologists (Majer 2009). Most restoration studies focus on flora with relatively little work conducted on fauna (Lugo et al. 2012); fauna may therefore be overlooked when novel ecosystems are discussed and managed. Conversations about novel ecosystems should include these higher trophic levels and not assume their dynamics are represented by the changes occurring in the abiotic and vegetative realms. This chapter was written to ensure a faunal perspective is integrated into discussions of novel ecosystems. In this chapter, we discuss causes and rates of novel ecosystem formation as they pertain to fauna, faunal use of novel ecosystems and how novel ecosystems can be used to achieve faunal conservation goals.

14.1 WHAT IS A NOVEL ECOSYSTEM FROM A FAUNAL PERSPECTIVE?

From a faunal perspective, current novel ecosystems: (1) contain non-native vegetation and/or animals; (2) contain native species mixes not historically present due to rapid recent range expansions; (3) have altered abundances or interactions partially or wholly mediated by anthropogenic activities; (4) may no longer contain historical ecological engineers or keystone species (*sensu* Daily et al. 1993; Brown 1995); and/or (5) have missing or added trophic levels or other interactions.

There are numerous examples in this book as well as the restoration and invasion literature of novelty as a result of changes in plant communities. These novel plant communities can provide habitat features for

Novel Ecosystems: Intervening in the New Ecological World Order, First Edition. Edited by Richard J. Hobbs, Eric S. Higgs, and Carol M. Hall.
© 2013 John Wiley & Sons, Ltd. Published 2013 by John Wiley & Sons, Ltd.

native fauna and this will be discussed in more detail in Section 14.5.3. Novelty created by introduced animals is also well documented in the literature, particularly when the animal(s) are ecological engineers. Famous examples include the brown tree snake (*Boiga irregularis*) invasion of Micronesia (Savidge 1987; Fritts and Rodda 1998), the restoration of beaver (*Castor* spp) throughout the Holarctic (Rosell et al. 2005) and their subsequent introduction in South America (Anderson et al. 2009), feral populations of agricultural animals that have naturalized throughout the globe (Mack et al. 2000) and the red fox, feral cat and European rabbit introductions in Australia (Dickman 1996). Introductions of these non-native animals have resulted in direct and indirect ecosystem effects that have altered the abiotic characteristics of ecosystems as well as abundances or interactions of native fauna. In many instances these changes are irreversible and result in novel ecosystems (Chapter 3).

Less well studied are novel ecosystems that are being created due to human-facilitated rapid range expansions of native species. The reasons for these expansions are not clear but presumably the taxa are responding positively to rapid and widespread anthropogenic creation of new habitat. The presence of these species can result in new biotic interactions that cause changes in abundance of native species of conservation concern. One of the best examples is the recent expansion of the larger and more aggressive North American barred owl (*Strix varia*), which now co-occurs and out-competes the US federally listed threatened spotted owl (*S. occidentalis*) in restored old-growth forests in the Pacific Northwest (Dugger et al. 2011). Historically, the barred owl was limited to eastern North America. In the early 1900s, its range gradually began to extend westward across wooded regions of central Canada and British Columbia then north into southeast Alaska and south into western Montana and Idaho (Livezey 2009). It was first documented in Washington in 1965 and then in Oregon in 1974 (Kelly et al. 2003; Livezey 2009). Between 1974 and 1998, Kelly et al. (2003) estimated that 706 different barred owl territories were established in Oregon.

The barred owl occupies a wide range of forest conditions, and it has been hypothesized that its range expansion was facilitated by forest management practices that reduced availability of older forests and increased forest fragmentation. Using multi-season occupancy models with a large dataset, Dugger et al. (2011) have clearly documented strong interference competition between the two species. Spotted owl territory occupancy increased when the proportion of old forest increased and/or the degree of fragmentation was decreased. In addition, occupancy rates decreased when barred owls were detected, regardless of the territory habitat configuration. Owl territories where barred owls are detected each year continue to increase, so it is clear that the two species have not yet reached equilibrium and the spotted owl may go extinct before this occurs. It is not clear how this competitive interaction will affect trophic dynamics in these forest ecosystems. This may result only in a species replacement or have cascading effects throughout the food web, resulting in novel assemblages.

Changes in species abundances due to rapid environmental change can be far more dramatic than changes in species distributions. Simpson et al. (2011) make a convincing case that fundamental changes in fish abundances have occurred in the northeast Atlantic during the last 30 years of warming. This investigation was based on detailed analyses of surveys conducted in this well-studied area where sea temperatures have risen by 0.04°C per year in the last three decades. Despite relatively minor changes in the ranges of fishes, the authors found 72% of common fish species have exhibited a significant change in abundance. While some species increased in abundance, others declined. Such changes are likely to have much greater effects on the ecology and exploitation of these fish assemblages than any range extensions or contractions. As noted by Harborne and Mumby (2011), the novel assemblages occurring in the northeast Atlantic are not a simple result of climate change but are likely due to species-specific responses to the warming, changed inter-specific interactions from these changed abundances and the level of exploitation which appears to shift based on changing abundances of harvested species (see fig. 1 in Harborne and Mumby 2011 for an illustration of these processes).

Novelty created by the loss of ecological engineers and/or keystone species is well illustrated by recent studies on the ecological effects of the common global practice of eradicating top native vertebrate predators. There is strong evidence eradication of grey wolves (*Canis lupus lupus*) and dingos (*C. l. dingo*) has influenced distribution and densities of a wide range of species that occupy different trophic levels including mesocarnivores, marsupials, rodents, riparian songbirds and palatable plants (Johnson et al. 2007; Ripple and Beschta 2007; Letnic et al. 2009). Similarly, ant and spider manipulation experiments reveal a range of effects these groups have on herbivores, decomposers, parasitoids, pollinators, other predators and plants (Lach 2008; Schmidt-Entling and Siegenthaler 2009; Schmitz 2009; Brechbühl et al. 2010; Mellbrand et al. 2010; Piñol et al. 2010; Sanders and van Veen 2011).

Changes in trophic complexity can also be coincident with the creation of a novel ecosystem. For example, the invasive yellow crazy ant, which acts as both predator and mutualist, has greatly altered food web dynamics on Christmas Island, Australia. The ants consume honeydew from sap-sucking insects in the forest canopy, and thereby achieve high abundance and contribute to rainforest dieback (O'Dowd et al. 2003; Abbott and Green 2007). When they are abundant, the ants prey on the endemic red land crab, which previously lacked significant predation pressure. The loss of herbivorous red land crabs results in development of a dense understory of seedlings (O'Dowd et al. 2003) and also facilitates invasion by the giant African land snail (Green et al. 2011). Through its direct and indirect effects, the ant has dramatically changed the functioning of the island's forest ecosystem.

Another example from Canadian lakes shows how the invasion of mid-trophic-level zooplankton *Bythotrephes* spp. greatly changed trophic complexity. *Bythotrephes* caused a decline in herbivorous zooplankton and an increase in predaceous zooplankton, resulting in an overall increase in trophic position of zooplankton and lake herring. Although the shifts are small, because of biomagnification up the food chain they result in a 30% increase in mercury concentration in lake herring which are consumed by lake trout, which are in turn consumed by anglers (Rennie et al. 2011).

There is some evidence that non-native species can compensate for changes in biotic interactions resulting from eradication of native species. The contribution of novel species to the maintenance of ecosystem function is discussed in several chapters in this book (see Chapters 3 and 18). In terms of replacement or displacement of native species, there are numerous studies although the vast majority are plant-focused (e.g. Westman 1990; Ewel and Putz 2004; Kiers et al. 2010). Pattemore and Wilcove (2011) investigated whether compensation by non-native vertebrate pollinators could maintain pollination in the face of extinctions of native vertebrate pollinators. Using a natural experiment (a comparison of pollination at islands with and without the endemic vertebrate flower visitors), they demonstrated that two recently arrived species in New Zealand, the ship rat (*Rattus rattus*) and the silvereye (*Zosterops lateralis*; a passerine bird), partly maintain pollination for three native forest plant species in northern New Zealand. They also demonstrated that without this compensation, these plants would be significantly more pollen-limited. This study provides empirical evidence that two widespread nonnative species including one known to have severe negative effects on biodiversity (ship rat) can partially play an important role in maintaining at least one ecosystem function. It should be noted, however, that the three plant species studied have very accessible flowers with copious nectar ideal for attracting vertebrate pollinators. This is in contrast to the tubular flower of the endemic (*Rhabdothamnus solandri*), which has diminished following loss of its native pollinators in this system (Anderson et al. 2011; Pattemore and Wilcove 2011; discussed in more detail in Sections 14.4.1 and 14.4.5).

14.2 NOVEL ANIMAL ASSEMBLAGES ARE COMMON HISTORICALLY

While the concept of novel ecosystems is comparatively new, the existence of these ecosystems is not. As noted by Williams and Jackson (2007), no-analog fossil invertebrate and mammal assemblages are pervasive in Quaternary paloecological records and occur in a range of terrestrial and marine ecosystems (see also Chapter 7). The best-known cases are from the most recent glacial–interglacial transition in North America (Edwards et al. 2005). No-analog communities in North America were most extensive between 12,000 and 17,000 years ago and were most prevalent in Alaska and the interior of eastern North America. The development and disappearance of these communities has been linked to climatic changes in these areas (Williams and Jackson 2007).

In addition, many landscapes characterized by a prolonged human presence such as those in Europe support groups of species that may not have lived together without human presence (Lindenmeyer et al. 2008; Jackson 2012). These systems also do not depend on continued human activity for their maintenance (Hobbs et al. 2006). The role of indigenous cultural activities, including agriculture, land clearance, fire setting and hunting in structuring ecosystems before Euro American occupation is under active debate (see Ellis 2011 and Jackson 2012 for comprehensive reviews). In spite of the debate, it is clear that human interactions with ecosystems are exceedingly complex and dynamic and we have altered terrestrial and aquatic ecosystems both intentionally and unintentionally for millennia.

The point of this historical perspective is that goals for faunal conservation are primarily described in the context of native or historical ecosystems. Thus, when attempting to improve conditions for animals via restoration or preservation the target or desired conditions are generally defined as 'natural' or within the 'historical range of variability' for that site. The reference for 'natural' conditions is typically one without effects from the current dominant culture. Others extend this definition and include environmental effects caused by the region's indigenous cultures. Thus, setting restoration or management goals to create or maintain animal assemblages that have occurred within the historical range of variability requires a clear definition of the timeframe used for defining 'historical' as well as acknowledging rapidly changing global conditions (see next section for more details).

14.3 RATE OF NOVEL ECOSYSTEM CREATION

Novel ecosystems are being created at a rapid rate due to climate change, land-use practices, land-cover changes and species eradications and invasions. Ellis (2011) estimates nearly one-third of the terrestrial biosphere has been transformed into novel ecosystems and the majority of these transformations have occurred in the past century (see also Chapter 8).

The rate of formation of future novel ecosystems is difficult to predict but Williams et al. (2007) used recent climate change predictions to identify regions projected to develop novel 21st century climates. Depending on the emissions scenario, they predicted 4–39% of the Earth's terrestrial surface would have novel climates by 2100 AD. Stralberg et al. (2009) examined the potential change in California terrestrial breeding bird communities as a result of regional and local climate change. Their model predictions (which were based on region-specific climate and species distribution models) indicate that, by 2070, over half of California could be occupied by novel assemblages of bird species, implying the potential for dramatic community reshuffling and altered patterns of species interactions.

14.4 CAUSES OF NOVELTY

Human agency is the key driver in the formation of novel ecosystems. As described in Chapter 5, human influence can be *onsite* (e.g. plantings, species eradications, land-cover change) or *offsite* (e.g. CO_2 or N emissions). They may be *deliberate* or *inadvertent*, or they can be *deliberate with inadvertent consequences*. They can be *acts of commission*, a purposeful undertaking to achieve a goal, or *acts of omission*, an intentional lack of activity. We illustrate these mechanisms of novel ecosystem creation with brief discussions of species introductions, species eradications, climate change, land-use practices and a combination of drivers. These particular types of human influence were chosen because of their noted effects on fauna.

14.4.1 Species introductions

Species introductions are a frequent means of human agency. A hybrid or novel ecosystem ensues when species introduced to an ecosystem function and interact differently from the resident biota. The literature on biological invasions is rich with examples of invertebrates and vertebrates that have been brought either intentionally or accidentally to a new landscape and have greatly altered its ecology via competition, predation, herbivory, parasitism or mutualism. The European honeybee (*Apis mellifera*) has been introduced to almost every country in the world and, although it is one of our most useful crop pollinators, it also likely displaces native pollinators and in some cases preferentially visits alien plant species, creating 'invader complexes' (Goulson 2003; Morales and Aizen 2006; Schweiger et al. 2010). In Puerto Rico however, this bee is as effective as the most abundant native avian pollinator *Coereba flaveola* in pollinating *Goetzea elegans*, an endangered and self-incompatible endemic tree (Caraballo Ortiz and Santiago Valentín 2011).

Introducing an additional species to biologically control an invader is another kind of intentional introduction that can increase novelty, whether or not it is successful in controlling the target species. A classic example is the cane toad (*Bufo marinus*) which has been introduced to many regions of the world (particularly the Pacific) for the biological control of agricultural pests. Following the apparent success of the cane toad in eating the beetles that were threatening the sugarcane plantations of Puerto Rico, and the fruitful introductions into Hawaii and the Philippines, there was a strong push for the cane toad to be released in Australia to negate the pests that were ravaging the Queensland cane fields. As a result, toads were collected from Hawaii and approximately 60,000 were released in Australia in the 1930s. The toads became firmly established in Queensland, increasing exponentially in number and extending their range across northern Australia (Shine 2010).

It has since become one of the most intensely studied invasive species in the world. Research on the effect of cane toads on native species (reviewed by Shine 2010 and more recently documented e.g. by Brown et al. 2011; Nelson et al. 2011a, b) indicates the major mechanism of effect is lethal toxic ingestion of toads by frog-eating predators, but the magnitude of effect varies dramatically both spatially and temporally and among predator taxa. No native species have yet gone extinct as a result of toad invasion and many native taxa initially thought to be at risk (e.g. native snakes) are not affected, either because of a physiological ability to tolerate toad toxins or an innate or learned reluctance to consume toads. Some of the species initially severely affected by toad invasion recover within a few decades via aversion learning and longer-term adaptive changes (Phillips and Shine 2004; Phillips et al. 2010). Other indirect trophic effects of the toads have apparently resulted in declines of some native species but increases in others (Shine 2010; Brown et al. 2011).

Accidental introductions can have just as dramatic effects. The varroa mite (*Varroa jacobsoni*) is a major pest of honeybees, and its spread from southeast Asia to almost every country in the world was facilitated by the increase in international trade and travel following World War II. In addition to its direct effects as an ectoparasite, it is associated with many bee viruses (Sammataro et al. 2000), some of which have been implicated in the large-scale deaths of honeybees in North America (Bromenshenk et al. 2010). The four rodent species most commonly inadvertently introduced around the world in the shipping trade are implicated in the extinction of at least 11 insular small mammal species (Harris 2009), and their negative effects on invertebrates are also common, widespread and severe (St Clair 2011).

Conservation translocations are intentional species introductions that are often used to enhance faunal populations. Translocations usually consist of reintroducing native fauna from either wild or captive stock back to their historical range (Armstrong and Seddon 2007). Although this is the ideal, the original habitat may be extensively modified and restoration through reintroduction may not be feasible. In these situations, it might be desirable (and necessary to avoid extinction) to move species outside the natural distribution (Seddon and Soorae 1999), creating novelty in the translocation site.

Alternatively, as noted by Parker et al. (2010), an extension of this approach is to replace extinct species with an appropriate analog species. This approach has merit when the conservation goal is ecological restoration rather than species conservation, but the result is the creation or augmentation of a novel ecosystem. The use of ecological analogs by wildlife biologists is not new. Reintroductions that have successfully used subspecific analogs include the peregrine falcon (*Falco peregrinus*), New Zealand takahe (*Porphyrio mantelli*) and Puerto Rican parrot (*Amazona vittata*). Analogs for congeneric extinct species have also been reintroduced successfully including: (1) yellow-crowned night herons (*Nycticorax violacea*) introduced to Bermuda islands; (2) tundra musk oxen (*Ovibos moschatus*) introduced to the Siberian steppe; and (3) Australian brown quail (*Coturnix ypsilophora*) in New Zealand (Parker et al. 2010).

More controversial recent analog proposals include: (1) Atkinson's (1988) suggestion to introduce a variety of primarily Australian species as ecological replacements for extinct New Zealand wildlife and (2) Donlan et al.'s (2005, 2006) proposal to 'rewild' North America with megafauna replacements from the Pleistocene. These ideas have gained little traction given the uncertainty around the release of any extant introduced species (but see Hansen 2010).

14.4.2 Species eradications

Species eradications are another frequent means of human agency that result in novel ecosystems. These eradications have been deliberate (e.g. removal of predators) or inadvertent (e.g. extinction of Guam avifauna caused by the inadvertent introduction of the brown tree snake) or they can be deliberate with inadvertent consequences (e.g. trophic cascades from removal of wolves). Clearly there are ecological consequences from species eradications, particularly if the species are ecological engineers or keystone species. There is ecological debate about whether every species is required for stable ecosystem function (e.g. Kareiva and Levin 2003). Our objective is not to participate in this debate but to point out that eradication of novel components with the intent of restoring historical conditions may result in unintended negative consequences. The recent experience in the World Heritage island of Macquarie is a sobering example of the need for ecological understanding of the role of invasive species before expensive management actions are undertaken (Bergstrom et al. 2009; Cadotte 2009). A successful eradication program for introduced feral cats resulted in the ecological release of populations of their primary prey, introduced rabbits. The extensive herbivory and burrowing caused by the overabundant rabbit population threatens to exterminate native Macquarie vegetation, which was not a goal of the feral cat eradication.

Another example of potential negative consequences of the removal of novel components is the recent introduction of the non-native biocontrol agent – the tamarisk beetle (*Diorhabda* spp.) – to eradicate tamarisk (*Tamarix* spp.), a genus of non-native trees that has become a dominant component of riparian woodlands in the southwestern US (Friedman et al. 2005). These riparian woodlands support many species of riparian obligates including the southwestern willow flycatcher (*Empidonax traillii extimus*), listed as an endangered subspecies under the US Endangered Species Act. Recently Paxton et al. (2011) used a combination of literature review and demographic modeling to investigate the potential for positive and negative effects of the loss of riparian vegetation from this control program on riparian birds, with a particular focus on the endangered flycatcher. They concluded that species with restricted distributions that include areas dominated by tamarisk (such as the flycatcher) may be negatively affected both in the short and long term. This is because the birds use the tamarisk: in some areas up to 75% of flycatchers nest in tamarisk and reproductive success associated with these nests is indistinguishable from nests in native trees (Sogge et al. 2008). Also, rate of regeneration and/or restoration of native trees may not be fast enough relative to the rate of tamarisk loss due to the permanently altered hydrological characteristics of these riparian areas.

These examples illustrate the complexities associated with the control or eradication of one non-native species: it will not necessarily result in the desired outcome because species interactions may be altered (van Riper et al. 2008; Schlaepfer et al. 2011). Non-native species can become integral to inter-species relationships to such an extent that their eradication can alter well-functioning communities (Ewel and Putz 2004; Prévot-Julliard et al. 2011).

14.4.3   Climate change

The emission of greenhouse gases causing climate change is an offsite human influence that is inescapable even in the most remote landscapes on the planet. Alterations in temperature and rainfall patterns and increased CO_2 are all part of human-induced global climate change that affect the functioning of and interactions in ecosystems (see Chapter 10). Changes in the phenology of species and their latitudinal and altitudinal distribution with increases in temperature have been documented for many species (Visser and Both 2005; Primack et al. 2009). Not all species respond to climate change in a similar way; species interactions are therefore also changing (Primack et al. 2009) and likely further facilitating the development of novelty in ecosystem functioning.

In Section 14.1 we illustrated the changes in species interactions as a result of climate change with the fish assemblages in the northeast Atlantic. Climate change has also led to a mismatch in the temporal and/or spatial occurrence of some terrestrial herbivores and food plants (Schweiger et al. 2010) and predators and prey (Visser and Both 2005; Both et al. 2009). Temporal and spatial incongruities are also expected for parasites and their hosts (Thomson et al. 2010). Mismatches are perhaps especially detrimental for mutualistic interactions, including pollination. In addition to a lack of overlap in spatial and temporal space between plants and pollinators due to climate change (Hegland et al. 2009), mismatches in pollinator body size distribution, nectar quantity and quality, pollinator energy demand, plant community structure and pollinator community structure have been observed or are hypothesized as likely to occur due to climate change (reviewed in Schweiger et al. 2010). Other kinds of interactions likely also have subtle but significant mismatches that may contribute to ecosystem novelty (see Chapter 10 for more discussion on this topic; also see Chapters 15 and 40 for case studies).

14.4.4 Land-use practices

Human populations and their use of land is the largest onsite influence responsible for the creation of novel ecosystems in terrestrial systems. As noted earlier (Section 14.2), predicting the effects of human land use on the biosphere is difficult because human interactions with terrestrial ecosystems are exceedingly complex and dynamic. Humans alter terrestrial ecosystems both intentionally and unintentionally and these alterations depend on a variety of interactions including human population density, technical capacity, mode of resource use and the use of opportunities afforded by native and transformed ecosystems. All of these factors interact and are influenced by the considerable natural variation in terrestrial ecosystem form and process (Ellis 2011). Ellis (2011) has analyzed this effect in detail and concluded human populations and their use of land have transformed most of the terrestrial biosphere into anthropogenic biomes (anthromes), causing a variety of novel ecological patterns and processes.

A good example is the transformation that has occurred on islands in the Caribbean such as Puerto Rico, which is detailed in Lugo et al. (2012; see also Chapter 9). The 8959 km² island was completely forested when Europeans arrived, despite earlier occupation by indigenous people. Puerto Rico's forest cover fell from almost 100% in 1493 to 6% by the late 1940s before recovering to 57% by 2003 due to agricultural abandonment (Brandeis et al. 2007).

The vegetation that emerged from Puerto Rico's abandoned agricultural lands, which in 1966 covered 68.5% of the island, was different from native plant communities present before deforestation (Lugo and Helmer 2004). These plant communities differed in the diversity of introduced species, which have mixed with natives to form novel plant communities. Despite this new mix of species, after several decades these novel forests developed a physiognomy and structure similar to that of mature native forests. Did the island fauna exhibit similar patterns to the flora? According to Lugo et al. (2012) large numbers of introduced faunal species occur throughout the Caribbean, many of which are naturalized and even locally adapted to the novel plant communities that now dominate the islands. Because relatively few of the native and endemic species have become extinct, they suggest that integration of native animals into novel plant and animal communities is possible.

14.4.5 Combination of drivers

With the complexity of human influence, it is not uncommon that novelty can arise from a combination of different human interventions. For example, the range expansion and population eruption of the native mountain pine beetle (*Dendroctonus ponderosae*) that has resulted in the death of millions of hectares of lodgepole pine (*Pinus contorta latifolia*) in central British Columbia is blamed on two anthropogenic factors: warmer winter temperatures that do not kill overwintering larvae and fire suppression (Cudmore et al. 2010; Klingenberg et al. 2010; Safranyik et al. 2010). In turn, the extirpation of these forests is leading to changes in carbon (Pfeifer et al. 2011) and nitrogen cycling (Griffin et al. 2011), lowered probability of crown fire (Simard et al. 2011) and increased mortality in replanted stands due to migration of the Warren root collar weevil (*Hylobius warreni*; Klingenberg et al. 2010).

In Puerto Rico, the expansion of pastures dominated by introduced African grasses and urban grassy areas have not only introduced a new fire disturbance regime to the island, but have also influenced the success and naturalization of introduced granivorous birds. In a forested island with only 2 native granivorous bird species, today the island has 18 granivorous bird species all associated with novel grass communities (Raffaele 1989).

The introduction of novel herbivores, predators and mutualists can all result in the creation of novel ecosystems by changing species composition and interactions, including among plants. As described earlier, invasion of Christmas Island by yellow crazy ants and the range expansion by the mountain pine beetle have both created novel plant assemblages. A change in the pollinator community can likewise result in the creation of novel plant assemblages. The introduction of mammalian predators to New Zealand resulted in the depletion of the endemic avifauna and significantly reduced recruitment of an endemic forest understory shrub (*Rhabdothamnus solandri*) that relied on the native birds for pollination (Anderson et al. 2011). Changes such as these are likely to be slow and difficult to detect, but are nonetheless important drivers in the formation of novel ecosystems.

14.5   WHAT CAN NOVEL ECOSYSTEMS DO?

As the amount and rate of ecosystem alteration increases globally, an increasing proportion of the world's ecosystems move into novel configurations and compositions. The amount of primary habitat available for native species is therefore declining, especially in rapidly changing situations such as urban areas. An increasing number of native fauna, including rare, threatened and endangered species, rely on these novel ecosystems to meet their resource requirements and/or provide landscape connectivity for dispersal (Jones and Bock 2005; Vander Zanden et al. 2006; Morrison 2009). The use of novel ecosystems by native fauna primarily occurs when: (1) these systems provide additional resources to fauna that ameliorate effects of change; (2) the novel system provides structure and resources similar to the native habitat; and/or (3) the native habitat has biotic interactions limiting its suitability.

14.5.1   Help mitigate effects of climate change

As noted previously, there is mounting evidence that climatic changes are having direct effects on animal populations through weather-related effects on vital rates, abundances and distributions. In birds, climate-induced phenological shifts in migration timing (Butler 2003; Murphy-Klassen et al. 2005; MacMynowski et al. 2007) and initiation of breeding (Crick and Sparks 1999; Dunn and Winkler 1999) which influence survival and fecundity, respectively, have been documented. As noted by Seavey et al. (2008), effects of climate change on bird populations may be mediated indirectly through changes in food availability or effects on competitors and predators. These climatic changes may reduce the quality of existing habitat to pre-disturbance species. In these situations, novel ecosystems with a mixture of native and non-native species may provide more resources than what was available to these birds under earlier climate regimes.

An excellent example is the recent changes in blue and great tit (*Cyanistes caeruleus* and *Parus major*) breeding habitat in the UK (summarized in Hobbs et al.

2009). The two titmice species are common nesting species in oak (*Quercus* spp) woodlands throughout the British Isles. Recent evidence demonstrates both species are laying eggs earlier in the season in response to climate cues, but there is no evidence that food resources used by titmice during the breeding season (larval and adult insects) are also emerging earlier. A recent inhabitant of these woodlands, the Turkey oak (*Quercus cerris*), provides the food resources available during the early part of the breeding season. This species was native to the British Isles prior to the last glaciations (~130,000–115,000 years BP), and was reintroduced by humans in the past 300 years. There is no evidence the species is invasive, although there have been concerns raised in conservation circles. Native insects preferred by the titmice use the Turkey oak as a host. However, more importantly, the Turkey oak is one of the few hosts for insects migrating northward in the early spring before the normal food sources of the titmice emerge. In this novel ecosystem, the 'non-native' Turkey oak potentially provides additional resources for the titmice under these changed abiotic conditions (assuming the non-native insects do not negatively affect the availability of the native species later in the season).

14.5.2   Mitigate effects of land-use changes

Most human land-use transformation results in land that is managed intensively for specific production or is built over, and these habitats are not considered novel ecosystems (see definition in Chapter 6). However, small areas of novel ecosystems may occur in pockets at the margins or interstices of productive or built environments; novel species interactions may therefore occur in intensively managed systems. Using agricultural lands and managed plantations, we illustrate the importance of these novel ecosystems to faunal conservation. The importance of novel ecosystems in human settlements is discussed in more detail in Chapter 38.

Approximately 37% of the globally available land area is currently in agricultural production and world food demand is predicted to double by 2050 (Tilman et al. 2001; Green et al. 2005). The need for increased food production has resulted in agricultural intensification and the resulting land-use change is currently the greatest extinction threat to fauna (Donald et al. 2001; Foley et al. 2005; Green et al. 2005; Millennium

Ecosystem Assessment 2005). Cleared landscapes support few species and these species are often habitat generalists and tolerant of humans (Green and Catterall 1998; Hobbs et al. 2003). In these agricultural areas, little remains of pre-settlement conditions and the majority of faunal habitat is provided by novel ecosystems that resulted from these extensive landscape transformations. In most situations, complete restoration of pre-agricultural habitat is likely to be prohibitively expensive on a large spatial scale.

The ecological and economic need to restore faunal habitat has resulted in governments channeling resources into mitigating the effect of intensive agriculture by creating novel ecosystems through voluntary agri-environment programs that reimburse farmers to undertake practices aimed at benefiting fauna, e.g. set-asides that leave field margins uncropped (United States Department of Agriculture 2003; Benton 2007). Agri-environment schemes designed to promote these novel ecosystems have been a major policy instrument for the past two decades, and more than 100 programs that aim specifically at developing more conservation value of agro-ecosystems exist in the member countries of the Organization for Economic Co-operation and Development (Aviron et al. 2009). In Europe and North America, nearly $5.25 billion has been spent annually on agri-environment programs (Donald and Evans 2006; Benton 2007).

The majority of these agri-environment schemes focus on programs that can be implemented on individual farms such as the USDA's Conservation Reserve Program (CRP) and affiliated USDA–state-agency partnerships. These programs provide financial and technical assistance to landowners and managers to remove highly erodible lands from agricultural production and establish permanent cover to achieve natural resource conservation objectives, including improved faunal habitat (Brennan and Kuvlesky 2005; Haufler 2005; Gill et al. 2006). Several dozen conservation practices have been identified under the CRP and related programs that result in diverse flowering plants and sequential bloom throughout the growing season and/or improved nest sites for native bees (USDA 2008). Incorporating patches of semi-natural vegetation such as hedgerows into the agricultural landscape has been demonstrated to benefit native pollinators, including those that are otherwise uncommon in the landscape (Hannon and Sisk 2009).

Although the CRP has been in existence for several decades, few studies experimentally document the success of these programs in restoring wildlife habitat. One exception is a study by Gill et al. (2006) who revegetated 92.4 ha of agricultural lands in northeastern Maryland which were historically coastal, open savannah-grasslands that supported the extinct heath hen (*Tympanuchus cupido cupido*). In 1999 they planted 10 native grass species augmented with "small bags of assorted prairie flowers". Five years after planting they documented a cumulative plant species list of 261 species, 105 of which were introduced. The authors predict cumulative plant species richness will stabilize at about 300 species; the ratio of introduced to native is not clear but will fluctuate depending on management and climatic conditions. The revegetation response suggests a seemingly inexhaustible bank of dormant but viable seeds in the soil, remarkable given the more than 200-year history of intensive pastoral and agricultural use of this area (the most recent of which consisted of row-crop monocultures). However, this system is a novel ecosystem that will always contain a mixture of introduced and native species like most CRP sites.

In terms of fauna response, Gill et al. (2006) monitored grassland birds, taxa of international conservation concern. The bird response to the revegetation was equally impressive. Although these sites did not support the full complement of extant grassland species, this novel ecosystem is providing nesting habitat for several grassland-obligate species of conservation concern. For example, the grasshopper sparrow (*Ammodramus savannarum*) and dickcissel (*Spiza americana*) colonized the restored grasslands in the first year and established sustainable breeding populations within 5 years. Whether or not these novel ecosystems are high-quality habitat is yet to be determined, but the return rates and reproductive success data are suggestive of suitability.

There is an extensive literature on the biodiversity value of plantations (e.g. Lindenmayer and Hobbs 2004; Brockerhoff et al. 2008). For example, since WWII extensive acreages of Eucalyptus woodlands were cleared in Australia for croplands and livestock range. Concurrently, the increasing shortfall projected in timber production emanating from the decline in available domestic native forest resources has provided the commercial stimulus for trebling the rate of farm forestry plantations (to 80,000 ha annually) by 2020. Although forest plantations generally do not support the biodiversity of remnant woodlands, globally they do provide habitat for some species, including some of conservation concern (Newton 1996; Hobbs et al. 2003; Moser and Hilpp 2003).

Lindenmeyer et al. (2008) document the development of a novel assemblage of bird species in southeastern Australia woodlands over a 7-year period in response to recently established introduced pine plantations. Their data suggest these novel ecosystems, which are dominated by an introduced species (radiata pine, *Pinus radiata*), are providing habitat for regionally declining species such as the rufous whistler (*Pachycephala rufiventris*) which likely declined when the original open-woodland *Eucalyptus* were cleared for livestock grazing. The presence of this novel ecosystem also appears to have facilitated the westward expansion of native bird species such as the superb lyrebird (*Menura novaehollandiae*) and the olive whistler (*Pachycephala olivacea*) into nationally endangered open-woodland vegetation types where they have not previously occurred. The effects of these range expansions on resident woodland birds are unknown.

14.5.3   Provide habitat and food resources required by native fauna

Novel ecosystems can provide structural and or nutritional resources for native species. The potential role of novel ecosystems in providing resources for rare native species is likely to be particularly important in situations when restoration of the native species that formerly provided shelter or an energy source is impractical due to limited economic resources or changes in the physical environment (Schlaepfer et al. 2011). For example, Botteri's sparrow (*Aimophila botteri*) is a bird of tall grasslands that disappeared from Arizona following a period of extreme drought and heavy livestock grazing in the 1890s. Botteri's sparrow returned to Arizona in the 1970s but occurs primarily in grasslands dominated by non-native lovegrasses (*Eragrostis* spp); native grasslands with tall structure are restricted to <5% of their former range (Jones and Bock 2005). Jones and Bock (2005) studied the abundance and reproduction of this species across the state and concluded that the novel grasslands were not an ecological trap but provide the tall grass structure required by this species for nesting and survival. Novel ecosystems providing structural resources have been observed in other vertebrates where structure may be a more useful predictor of habitat than floristics. Examples include the: (1) southern emu-wren

(*Stipiturus malachurus*; Wilson and Patton 2004; Maguire 2006); (2) many North American breeding grassland and riparian breeding birds (Gill et al. 2006; Bajema et al. 2009; Kennedy et al. 2009; Fisher and Davis 2010; Paxton et al. 2011); (3) Sarus crane (*Grus antigone*; Sundar 2009); and (4) several marsupials including the red-cheeked dunnart (*Sminthopsis virginiae*; Braithwaite and Lonsdale 1987) and long-nosed bandicoot (*Perameles nasuta*; Chambers and Dickman 2002).

In addition to structural resources, non-native fruit quantity has recently been correlated with abundance of native avian frugivores (Davis 2011; Gleditsch and Carlo 2011) and non-native prey enhances the abundance of native predators (Schlaepfer et al. 2011). In some cases, non-native plants increase the overall diversity and availability of resources for native bees (Winfree et al. 2007; but see Williams et al. 2011).

14.5.4   Mitigate effects of limiting biotic interactions in native habitat

Although to our knowledge this has not been documented, it is possible that novel ecosystems could be refugia from native habitat that has biotic interactions that limits the native species persistence. These biotic interactions could be predation and/or competition and the novel ecosystem could provide shelter from competition and/or predation. In the previous section we discussed the preference of some species, e.g. long-nosed bandicoot, Botteri's sparrow, for the structural characteristics of novel ecosystems. It is not clear if those preferences are: (1) a result of choice from a limited range of alternatives and the novel ecosystem provides the structure that most closely approximates the native habitat; or (2) the novel ecosystem provides a structural environment that reduces the risk of predation and/or competition as compared to the native habitat. Future investigations on faunal use of novel ecosystems should be designed to separate these alternative scenarios.

14.6   UTILIZING NOVEL ECOSYSTEMS TO ACHIEVE FAUNAL CONSERVATION GOALS

Today's managers must recognize that many natural systems of the past are changing forever towards a new natural state thanks to drivers such as climate change, increased urbanization, introduced species and other transformative land-use/land-cover changes that are not reversible with restoration. We agree with Bridgewater et al. (2011) that such ecosystems are a blind spot in global conservation and restoration priorities as well as in discussions of environmental management and sustainable development. We argue that novel ecosystems are now critical for maintenance of faunal diversity at the genetic, species and ecosystem levels, and restoration goals to eliminate novelty might not always benefit faunal conservation. In some cases, novel ecosystems may provide the only habitat available for species of conservation concern. In other cases, novel ecosystems may provide habitat value not provided by native habitats that have evolved under different climatic and landscape conditions. The challenge to managers is to recognize when to intervene and when to roll with the punches in relation to the biotic composition of novel systems. The large cost of reversing the effects of human activities is an incentive to minimize them to the extent possible and, where not possible, to use scientific understanding to recognize when a novel assemblage is persistent and desirable. Chapters 3 and 18 provide some insights and analytical frameworks for assisting with these difficult decisions.

REFERENCES

Abbott, K.L. and Green, P.T. (2007) Collapse of an ant-scale mutualism in a rainforest on Christmas Island. *Oikos*, **116**, 1238–1246.

Anderson, C.B., Pastur, G.M., Lencinas, M.V., Wallem, P., Moorman, M. and Rosemond, A.D. (2009) Do introduced North American beavers *Castor canadensis* engineer differently in southern South America? An overview with implications for restoration. *Mammal Review*, **39**, 33–52.

Anderson, S.H., Kelly, D., Ladley, J.J., Molloy, S. and Terry, J. (2011) Cascading effects of bird functional extinction reduce pollination and plant density. *Science*, **331**, 1068–1071.

Armstrong, D.P. and Seddon, P.J. (2007) Directions in reintroduction biology. *Trends in Ecology and Evolution*, **23**, 20–25.

Atkinson, I.A.E. (1988) Presidential address: opportunities for ecological restoration. *New Zealand Journal of Ecology*, **11**, 1–12.

Aviron, S., Nitsch, H., Jeanneret, P., Buholzer, S., Luka, H., Pfiffner, L., Pozzi, S., Schüpbach, B., Walter, T. and Herzog, F. (2009) Ecological cross compliance promotes farmland

biodiversity in Switzerland. *Frontiers in Ecology and the Environment*, **7**, 247–252.

Bajema, R.A., DeVault, T.L., Scott, P.E. and Lima, S.L. (2009) Reclaimed coal mine grasslands and their significance for Henslow's Sparrows in the American Midwest. *Auk*, **118**, 422–431.

Benton, T.G. (2007) Managing farming's footprint on biodiversity. *Science*, **315**, 341–342.

Bergstrom, D., Kiefer, K., Lucieer, A., Wasley, J., Belbin, L., Pedersen, T. and Chown, S. (2009) Indirect effects of invasive species removal devastate World Heritage island. *Journal of Applied Ecology*, **46**, 73–81.

Both, C., Van Asch, M., Bijlsma, R.G., Van Den Burg, A.B. and Visser, M.E. (2009) Climate change and unequal phenological changes across four trophic levels: constraints or adaptations? *Journal of Animal Ecology*, **78**, 73–83.

Braithwaite, R.W. and Lonsdale, W.M. (1987) The rarity of *Sminthopsis virginiae* (Marsupialia: Dasyuridae) in relation to natural and unnatural habitats. *Conservation Biology*, **1**, 341–343.

Brandeis, T.J., Helmer, E.H. and Oswalt, S.N. (2007) The status of Puerto Rico's forests, 2003. USDA Forest Service, Southern Research Station Resource Bulletin SRS-119, Asheville.

Brechbühl, R., Kropfand, C. and Bacher, C. (2010) Impact of flower-dwelling crab spiders on plant-pollinator mutualisms. *Basic and Applied Ecology*, **11**, 76–82.

Brennan, L.A. and Kuvlesky, W.P. (2005) North American grassland birds: an unfolding conservation crisis? *Journal of Wildlife Management*, **69**, 1–13.

Bridgewater, P., Higgs, E.S., Hobbs, R.J. and Jackson, S.T. (2011) Engaging with novel ecosystems. *Frontiers in Ecology and Evolution*, **9**, 423.

Brockerhoff, E.G., Jactel, H., Parrotta, J.A., Quine, C.P. and Sayer, J. (2008) Plantations and biodiversity: oxymoron or opportunity? *Biodiversity and Conservation*, **17**, 925–951.

Bromenshenk, J.J., Henderson, C.B., Wick, C.H. et al. (2010) Iridovirus and microsporidian linked to honey bee colony decline. *PLoS ONE*, **5**, e13181, doi: 10.1371/journal.pone.0013181.

Brown, G.P., Phillip, B. and Shine, R. (2011) The ecological impact of invasive cane toads on tropical snakes: field data do not support laboratory-based predictions. *Ecology*, **92**, 422–431.

Brown, J.H. (1995) Organisms as engineers: a useful framework for studying effects on ecosystems? *Trends in Ecology and Evolution*, **10**, 51–52.

Butler, C.J. (2003) The disproportionate effect of global warming on the arrival dates of short-distance migratory birds in North America. *Ibis*, **145**, 484–495.

Cadotte, M.W. (2009) Editor's choice: unintended trophic cascades from feral cat eradication. *Journal of Applied Ecology*, **46**, 259–259.

Caraballo Ortiz, M.A. and Santiago Valentín, E. (2011) The breeding system and effectiveness of introduced and native pollinators of the endangered tropical tree *Goetzea elegans* (Solanaceae). *Journal of Pollination Ecology*, **4**, 23–26.

Chambers, L.K. and Dickman, C.R. (2002) Habitat selection of the long-nosed bandicoot, *Perameles nasuta* (Mammalia, Peramelidae), in a patchy urban environment. *Austral Ecology*, **27**, 334–342.

Crick, H.Q.P. and Sparks, T.H. (1999) Climate change related to egg-laying trends. *Nature*, **399**, 423–424.

Cudmore, T.J., Björklund, N., Carroll, A.L. and Lindgren, B.S. (2010) Climate change and range expansion of an aggressive bark beetle: evidence of higher beetle reproduction in naïve host tree populations. *Journal of Applied Ecology*, **47**, 1036–1043.

Daily, G.C., Ehrlich, P.R. and Haddad, N.M. (1993) Double keystone bird in a keystone species complex. *Proceedings of the National Academy of Sciences of the United States of America*, **90**, 592–594.

Davis, M. (2011) Do native birds care whether their berries are native or exotic? No. *BioScience*, **61**, 501–502.

Dickman, C.R. (1996) Impact of exotic generalist predators on the native fauna of Australia. *Wildlife Biology*, **2**, 185–195.

Donald, P.F. and Evans, A.D. (2006) Habitat connectivity and matrix restoration: the wider implications of agri-environment schemes. *Journal of Applied Ecology*, **43**, 209–218.

Donald, P.F., Green, R.E. and Heath, M.F. (2001) Agricultural intensification and the collapse of Europe's farmland bird populations. *Proceedings of the Royal Society of London B*, **268**, 25–29.

Donlan, C.J., Greene, H.W., Berger, J. et al. (2005) Rewilding North America. *Nature*, **436**, 913–914.

Donlan, C.J., Berger, J., Bock, C.E. et al. (2006) Pleistocene rewilding: An optimistic agenda for twenty-first century conservation. *American Naturalist*, **168**, 660–681.

Dugger, K., Anthony, R.G. and Andrews, L.S. (2011) Transient dynamics of invasive competition: barred owls, spotted owls, habitat, and the demons of competition past. *Ecological Applications*, **21**, 2459–2468.

Dunn, P.O. and Winkler, D.W. (1999) Climate change has affected the breeding date of tree swallows throughout North America. *Proceedings of the Royal Society of London B*, **266**, 2487–2490.

Edwards, M.E., Brubaker, L.B., Lozhkin, A.V. and Anderson, P.M. (2005) Structurally novel biomes: a response to past warming in Beringia. *Ecology*, **86**, 1696–1703.

Ellis, E. (2011) Anthropogenic transformation of the terrestrial biosphere. *Philosophical Transactions of the Royal Society A*, **369**, 1010–1035.

Ewel, J.J. and Putz, F.E. (2004) A place for alien species in ecosystem restoration. *Frontiers in Ecology and the Environment*, **2**, 354–360.

Fisher, R.J. and Davis, S.K. (2010) From Wiens to Robel: A review of grassland-bird habitat selection. *Journal of Wildlife Management*, **74**, 265–273.

Foley, J.A., Defries, R., Asner, G.P. et al. (2005) Global consequences of land use. *Science*, **309**, 507–574.

Friedman, J.M., Auble, G.T. and Shafroth, P.B. (2005) Dominance of non-native riaprian trees in western USA. *Biological Invasions*, **7**, 747–751.

Fritts, T.H. and Rodda, G.H. (1998) The role of introduced species in the degradation of island ecosystems: a case history of Guam. *Annual Review of Ecology and Systematics*, **29**, 113–140.

Gill, D.E., Blank, P., Parks, J. et al. (2006) Plants and breeding bird response on a managed Conservation Researve Program grassland in Maryland. *Wildlife Society Bulletin*, **34**, 944–956.

Gleditsch, J.M. and Carlo, T.A. (2011) Fruit quantity of invasive shrubs predicts the abundance of common native avian frugivores in central Pennsylvania. *Diversity and Distributions*, **17**, 244–253.

Goulson, D. (2003) Effects of introduced bees on native ecosystems. *Annual Review of Ecology, Evolution and Systematics*, **34**, 1–26.

Green, P.T., O'Dowd, D.J., Abbott, K.L., Jeffery, M., Retallick, K. and MacNally, R. (2011) Invasional meltdown: invader-invader mutualism facilitates a secondary invasion. *Ecology*, **92**, 1758–1768.

Green, R.E., Cornell, S.J., Scharlemann, J.P.W. and Balmford, A. (2005) Farming and the fate of wild nature. *Science*, **307**, 550–555.

Green, R.J. and Catterall, C.P. (1998) The effects of forest clearing and regeneration on the fauna of Wivenhoe Park, south-east Queensland. *Wildlife Research*, **25**, 677–690.

Griffin, J.M., Turner, M.G. and Martin, S. (2011) Nitrogen cycling following mountain pine beetle disturbance in lodgepole pine forests of Greater Yellowstone. *Forest Ecology and Management*, **261**, 1077–1089.

Hannon, L.E. and Sisk, T.D. (2009) Hedgerows in an agri-natural landscape: potential habitat value for native bees. *Biological Conservation*, **142**, 2140–2154.

Hansen, D.M. (2010) On the use of taxon substitutes in rewilding projects on islands, in *Islands and Evolution* (eds V. Pérez-Mellado and C. Ramon), Institut Menorquí d'Estudis, Recerca, 111–146.

Harborne, A.R. and Mumby, P.J. (2011) Novel ecosystems: altering fish assemblages in warming waters. *Current Biology*, **21**, R822–R824.

Harris, D.B. (2009) Review of negative effects of introduced rodents on small mammals on islands. *Biological Invasions*, **11**, 1611–1630.

Haufler, J.B. (ed.) (2005) Fish and wildlife benefits of Farm Bill conservation programs: 2000–2005 update. *The Wildlife Society Technical Review*, 05–2.

Hegland, S.J., Nielsen, A., Lázaro, A., Bjerknes, A.-L. and Totland, Ø. (2009) How does climate warming affect plant-pollinator interactions? *Ecology Letters*, **12**, 184–195.

Hobbs, R., Catling, P.C., Wombey, J.C., Clayton, M., Atkins, L. and Reid, A. (2003) Faunal use of bluegum (*Eucalyptus globulus*) plantations in southwestern Australia. *Agroforestry Systems*, **58**, 195–212.

Hobbs, R.J., Arico, S., Aronson, J. et al. (2006) Novel ecosystems: theoretical and management aspects of the new ecological world order. *Global Ecology and Biogeography*, **15**, 1–7.

Hobbs, R.J., Higgs, E. and Harris, J.A. (2009) Novel ecosystems: implications for conservation and restoration. *Trends in Ecology and Evolution*, **24**, 599–605.

Jackson, S.T. (2012) Conservation and resource management in a changing world: extending historical range-of-variability beyond the baseline, in *Historical Environmental Variation in Conservation and Natural Resource Management* (eds. J.A. Wiens, G.D. Hayward, H.D. Safford and C. Giffen), Wiley-Blackwell.

Johnson, C.N., Isaac, J.L. and Fisher, D.O. (2007) Rarity of a top predator triggers continent-wide collapse of mammal prey: dingoes and marsupials in Australia. *Proceedings of the Royal Society B*, **274**, 341–346.

Jones, Z.F. and Bock, C.E. (2005) The Botteri's Sparrow and exotic Arizona grasslands: an ecological trap or habitat regained? *Condor*, **107**, 731–741.

Kareiva, P. and Levin, S. (eds) (2003) *The Importance of Species. Perspectives on Expendability and Triage*. Princeton University Press, Princeton.

Kelly, E.G., Forsman, E.D. and Anthony, R.G. (2003) Are barred owls displacing spotted owls? *Condor*, **105**, 45–53.

Kennedy, P.L., DeBano, S.J., Bartuszevige, A.M. and Lueders, A.S. (2009) Effects of native and nonnative grassland plant communities on breeding passerines: implications for restoration of northwest bunchgrass prairie. *Restoration Ecology*, **17**, 515–525.

Kiers, T.E., Palmer, T.M., Ives, A.R., Bruno, J.F. and Bronstein, J.L. (2010) Mutualisms in a changing world: an evolutionary perspective. *Ecology Letters*, **13**, 1459–1474.

Klingenberg, M.D., Lindgren, B.S., Gillingham, M.P. and Aukema, B.H. (2010) Management response to one insect pest may increase vulnerability to another. *Journal of Applied Ecology*, **47**, 566–574.

Lach, L. (2008) Argentine ants displace floral arthropods in a biodiversity hotspot. *Diversity and Distributions*, **14**, 281–290.

Letnic, M., Koch, F., Gordon, C., Crowther, M.S. and Dickman, C.R. (2009) Keystone effects of an alien top-predator stem extinctions of native mammals. *Proceedings of the Royal Society B*, **276**, 3249–3256.

Lindenmayer, D.B. and Hobbs, R.J. (2004) Fauna conservation in Australian plantation forests – a review. *Biological Conservation*, **119**, 151–168.

Lindenmeyer, D.B., Fischer, J., Felton, A., Crane, M., Michael, D., Macgregor, C., Montague-Drake, R., Manning, A. and Hobbs, R.J. (2008) Novel ecosystems resulting from landscape transformation create dilemnas for modern conservation practice. *Conservation Letters*, **1**, 129–135.

Livezey, K.B. (2009) Range expansion of barred owls. Part I: chronology and distribution. *American Midland Naturalist*, **161**, 49–56.

Lugo, A.E. and Helmer, E. (2004) Emerging forests on abandoned land: Puerto Rico's new forests. *Forest Ecology and Management*, **190**, 145–161.

Lugo, A.E., Carlo, T.A. and Wunderle, J.M. (2012) Natural mixing of species: novel plant- animal communities on Caribbean Islands. *Animal Conservation*, **15**, 233–241.

Mack, R.N., Simberloff, D., Lonsdale, W.N., Evans, H., Clout, M. and Bazzaz, F.A. (2000) Biotic invasions: causes, epidemiology, global consequences, and control. *Ecology*, **10**, 689–710.

MacMynowski, D.P., Root, T.L., Ballard, G. and Guepel, G.R. (2007) Changes in spring arrival of Nearctic–Neotropical migrants attributed to multiscalar climate. *Global Change Biology*, **13**, 2239–2251.

Maguire, G.S. (2006) Fine-scale habitat use by the southern emu-wren (*Stipiturus malachurus*). *Wildlife Research*, **33**, 137–148.

Majer J.D. (2009) Animals in the restoration process-progressing the trends. *Restoration Ecology*, **17**, 315–319.

Mellbrand K., Östman, Ö. and Hambäck, P.A. (2010) Effects of subsidized spiders on coastal food webs in the Baltic Sea area. *Basic and Applied Ecology*, **11**, 450–458.

Millennium Ecosystem Assessment. (2005) *Ecosystems and Human Well-Being: Synthesis*. Island Press, Washington, DC.

Morales, C.L. and Aizen, M.A. (2006) Invasive mutualisms and the structure of plant-pollinator interactions in the temperate forests of north-west Patagonia, Argentina. *Journal of Ecology*, **94**, 171–180.

Morrison, M. (2009) *Restoring Wildlife. Ecological Concepts and Practical Applications*. Island Press, Washington DC.

Moser, B.W. and Hilpp, C.K. (2003) Wintering use of hybrid poplar plantations in northeastern Oregon. *Journal of Raptor Research*, **37**, 286–291.

Murphy-Klassen, H.M., Underwood, T.J., Sealy, S.G. and Czyrnyj, A.A. (2005) Long-term trends in spring arrival dates of migrant birds at Delta Marsh, Manitoba, in relation to climate change. *Auk*, **122**, 1130–1148.

Nelson, D.W.M., Crossland, M.R. and Shine, R. (2011a) Behavioural responses of native predators to an invasive toxic prey species. *Austral Ecology*, **36**, 605–611.

Nelson, D.W.M., Crossland, M.R. and Shine, R. (2011b) Foraging responses of predators to novel toxic prey: effects of predator learning and relative prey abundance. *Oikos*, **120**, 152–158.

Newton, I. (1996) Sparrowhawks in conifer plantations, in *Raptors in Human Landscapes. Adaptations to Built and Cultivated Environments* (eds D. Bird, D. Varland and J. Negro), Academic Press, London, 191–199.

O'Dowd, D.J., Green, P.T. and Lake, P.S. (2003) Invasional 'meltdown' on an oceanic island. *Ecology Letters*, **6**, 812–817.

Parker, K.A., Seabrook-Davison, M. and Ewen, J.G. (2010) Opportunities for nonnative ecological replacements in ecosystem restoration. *Restoration Ecology*, **18**, 269–273.

Pattemore, D.E. and Wilcove, D.S. (2011) Invasive rats and recent colonist birds partially compensate for the loss of endemic New Zealand pollinators. *Proceedings of the Royal Society B*, doi: 10.1098/rspb.2011.2036.

Paxton, E.H., Theimer, T.C. and Sooge, M.K. (2011) Tamarisk biocontrol using tamarisk beetles: potential consequences for riparian birds in the souhwestern United States. *Condor*, **113**, 255–265.

Pfeifer, E.M., Hicke, J.A. and Meddens, A.J.H. (2011) Observations and modeling of aboveground tree carbon stocks and fluxes following a bark beetle outbreak in the western United States. *Global Change Biology*, **17**, 339–350.

Phillips, B.L. and Shine, R. (2004) Adapting to an invasive species: toxic cane toads induce morphological change in Australian snakes. *Proceeding of the National Academy of Sciences of the USA*, **101**, 17150–17155.

Phillips, B.L., Greenlees, M.J., Brown, G.P. and Shine, R. (2010) Predator behavior and morphology mediates the impact of an invasive species: cane toads and death adders in Australia. *Animal Conservation*, **13**, 53–59.

Piñol, J., Espadaler, X., Cañellas, N., Martínez-Vilalta, J., Barrientos, J.A. and Sol, D. (2010) Ant versus bird exclusion effects on the arthropod assemblage of an organic citrus grove. *Ecological Entomology*, **35**, 367–376.

Prévot-Julliard, A.-C., Clavel, J., Teillac-Deschamps, P. and Julliard, R. (2011) The need for flexibility in conservation practices: exotic species as an example. *Environmental Management*, **47**, 315–321.

Primack, R.B., Ibáñez, I., Higuchi, H., Lee, S.D., Miller-Rushing, A.J., Wilson, A.M. and Silander, J.A. Jr. (2009) Spatial and interspecific variability in phenological responses to warming temperatures. *Biological Conservation*, **142**, 2569–2577.

Raffaele, A.H. (1989) The ecology of native and introduced granivorous birds in Puerto Rico, in *Biogeography in the West Indies: Past, Present, and Future* (ed. C.A. Woods), Sandhill Crane Press, Gainesville, 541–566.

Rennie, M., Strecker, A. and Palmer, M.E. (2011) *Bythotrephes* invasion elevates trophic position of zooplankton and fish: implications for contaminant biomagnification. *Biological Invasions*, **13**, 2621–2634.

Ripple, W.J. and Beschta, R.L. (2007) Restoring Yellowstone's aspen with wolves. *Biological Conservation*, **138**, 514–519.

Rosell, F., Bozsér, O., Collen, P. and Parker, H. (2005) Ecological impact of beavers *Castor fiber* and *Castor canadensis* and their ability to modify ecosystems. *Mammal Review*, **35**, 248–276.

Safranyik, L., Carroll, A.L., Régnière, J., Langor, D.W., Riel, W.G., Shore, T.L., Peter, B., Cooke, B.J., Nealis, V.G. and Taylor, S.W. (2010) Potential for range expansion of mountain pine beetle into the boreal forest of North America. *Canadian Entomologist*, **142**, 415–442.

St Clair, J.J.H. (2011) The impacts of invasive rodents on island invertebrates. *Biological Conservation*, **144**, 68–81.

Sammataro, D., Gerson, U. and Needham, G. (2000) Parasitic mites of honey bees: life history, implications, and impact. *Annual Review of Entomology*, **45**, 519–548.

Sanders, D. and van Veen, F.J.F. (2011) Ecosystem engineering and predation: the multi-trophic impact of two ant species. *Journal of Animal Ecology*, **80**, 569–576.

Savidge, J.A. (1987) Extinction of an island forest avifauna by an introduced snake. *Ecology*, **68**, 660–668.

Schlaepfer, M.A., Sax, D.F. and Olden, J.D. (2011) The potential conservation value of non-native species. *Conservation Biology*, **25**, 428–437.

Schmidt-Entling, M.H. and Siegenthaler, E. (2009) Herbivore release through cascading risk effects. *Biology Letters*, **5**, 773–776.

Schmitz, O.J. (2009) Effects of predator functional diversity on grassland ecosystem function. *Ecology*, **90**, 2339–2345.

Schweiger, O., Biesmeijer, J.C., Bommarco, R. et al. (2010) Multiple stressors on biotic interactions: how climate change and alien species interact to affect pollination. *Biological Reviews*, **85**, 777–795.

Seavey, N.E., Dybala, K.E. and Snyder, M.A. (2008) Climate models and ornithology. *Auk*, **125**, 1–10.

Seddon, P.J. and Soorae, P.S. (1999) Guidelines for subspecific substitutions in wildlife restoration projects. *Conservation Biology*, **13**, 177–184.

Shine, R. (2010) The ecological impact of invasive cane toads (*Bufo marinus*) in Australia. *Quarterly Review of Biology*, **85**, 253–291.

Simard, M., Romme, W.H., Griffin, J.M. and Turner, M.G. (2011) Do mountain pine beetle outbreaks change the probability of active crown fire in lodgepole pine forests? *Ecological Monographs*, **81**, 3–24.

Simpson, S.D., Jennings, S., Johnson, M.P., Blanchard, J.L., Schön, P.-J., Sims, D.W. and Genner, M.J. (2011) Continental shelf-wide response of a fish assemblage to rapid warming of the sea. *Current Biology*, **21**, 1565–1570.

Sogge, M.K., Sferra, S.J. and Paxton, E.H. (2008) Tamarix as habitat for birds: implications for riparian restoration in the southwestern United States. *Restoration Ecology*, **16**, 146–154.

Stralberg, D., Jongsomjit, D., Howell, C.A., Snyder, M.A., Alexander, J.D., Wiens, J.A. and Root, T.L. (2009) Reshuffling of species with climate disruption: a no-analog future for California birds? *PLoS ONE*, **4**, e6825. doi: 10.1371/journal.pone.0006825.

Sundar, K.S.G. (2009) Are rice paddies suboptimal breeding habitat for Sarus Cranes in Uttar Pradesh, India? *Condor*, **111**, 611–623.

Thomson, L.J., Macfadyen, S. and Hoffman, A.A. (2010) Predicting the effects of climate change on natural enemies of agricultural pests. *Biological Control*, **52**, 296–306.

Tilman, D., Fargione, J., Wolff, B., D'Antonio, C., Dobson, A., Howarth, R., Schindler, D., Schlesinger, W.H., Simberloff, D.

and Swackhamer, D. (2001) Forecasting agriculturally driven global environmental change. *Science*, **292**, 281–284.

United States Department of Agriculture (USDA) (2003) Highlights of expected Conservation Reserve Program (CRP) changes for 2003 and beyond. National Bulletin 300-3-4. USDA Natural Resources Conservation Service, Washington DC.

United States Department of Agriculture (USDA) (2008) Using farm bill programs for pollinator conservation. Technical Note No. 78. http://plants.usda.gov/pollinators/Using_Farm_Bill_Programs_for_Pollinator_Conservation.pdf.

van Riper, C. III, Paxton, K.L., O'Brien, C., Shafroth, P.B. and McGrath, L.J. (2008) Rethinking avian response to Tamarix on the lower Colorado River: a threshold hypothesis. *Restoration Ecology*, **16**, 155–167.

Vander Zanden, M.J., Olden, J.D. and Gratton, C. (2006) Food-web approaches in restoration ecology, in *Foundations of Restoration Ecology* (eds D.A. Falk, M.A. Palmer and J.B. Zedler), Island Press, Washington DC, 165–189.

Visser, M.E. and Both, C. (2005) Shifts in phenology due to global climate change: The need for a yardstick. *Proceedings of the Royal Society B*, **272**, 2561–2569.

Westman, W.E. (1990) Park management of exotic plant species: problems and issues. *Conservation Biology*, **4**, 251–260.

Williams, J.W. and Jackson, S.T. (2007) Novel climates, no-analog communities, and ecological surprises. *Frontiers in Ecology and Environment*, **5**, 475–482.

Williams, J.W., Jackson, S.T. and Kutzbach, J.E. (2007) Projected distributions of novel and disappearing climates by 2100 AD. *Proceedings of the National Academy of Sciences of the United States of America*, **104**, 5738–5742.

Williams, N.M., Cariveau, D., Winfree, R. and Kremen, C. (2011) Bees in disturbed habitats use, but do not prefer, alien plants. *Basic and Applied Ecology*, **12**, 332–341.

Wilson, D. and Patton, D.C. (2004) Habitat use by the southern emu-wren, *Stipiturus malachurus* (Aves: Maluridae), in South Australia and evaluation of vegetation as a potential translocation site for *S. m. intermedius*. *Emu*, **104**, 37–43.

Winfree, R., Griswold, T. and Kremen, C. (2007) Effect of human disturbance on bee communities in a forested ecosystem. *Conservation Biology*, **21**, 213–223.

Chapter 15

CASE STUDY: ECOSYSTEM TRANSFORMATIONS ALONG THE COLORADO FRONT RANGE: PRAIRIE DOG INTERACTIONS WITH MULTIPLE COMPONENTS OF GLOBAL ENVIRONMENTAL CHANGE

Timothy R. Seastedt[1], Laurel M. Hartley[2] and Jesse B. Nippert[3]

[1]Department of Ecology and Evolutionary Biology, University of Colorado, USA
[2]Department of Integrative Biology, University of Colorado, USA
[3]Division of Biology, Kansas State University, USA

". . . in order to reliably predict the effects of global environmental change (GEC) on community and ecosystem processes, the greatest single challenge will be to determine how biotic and abiotic context alters the direction and magnitude of GEC effects on biotic interactions."

Tylianakis et al. 2008

Novel Ecosystems: Intervening in the New Ecological World Order, First Edition. Edited by Richard J. Hobbs, Eric S. Higgs, and Carol M. Hall.
© 2013 John Wiley & Sons, Ltd. Published 2013 by John Wiley & Sons, Ltd.

15.1   INTRODUCTION

Biotic change in the 21st century is underway and is occurring via mechanisms that, because of their complexity, have been difficult for ecologists to encapsulate into general theory. At one extreme, biotic change is insidious, that is, slow change resulting from chronic low-intensity directional climate drivers interacting with relatively intact and resilient communities. The other endpoint involves the popularized 'tipping points' where the community rapidly transforms into a recognizably different state. As a terrestrial example, 'desertification' of the southwest is perhaps the best historical example of a rapid transition to a new community type (Schlesinger et al. 1990). These transformations can be the result of high-intensity short-duration events (drought, fire, overgrazing, etc.) acting alone, or the result of events operating on systems already undergoing responses to a suite of global environmental change factors.

In this chapter we describe a case study where a transformation is occurring as a result of events operating on a system already undergoing responses to a suite of global environmental change factors. In the Front Range of Colorado, intensive grazing by native black-tailed prairie dogs (*Cynomys ludovicianus*) in urban and suburban landscapes is interacting with climate change, nutrient deposition and non-native plant invasion to result in what we believe will be novel communities.

Responsible ecosystem stewardship demands that we understand – and hopefully predict – causal factors explaining the rate and trajectories of change (Chapin et al. 2010). Unfortunately, ecosystem complexity can result in 'ecological surprises' (Paine et al. 1998; Hastings and Wysham 2010) such as rapid die-back of forests in much of the Rocky Mountain region because of multiple sources of plant stress (Breshears et al. 2009). Paine et al. (1998) noted that multiple perturbations are sources of rapid changes, and that "understanding these . . . synergisms will be basic to environmental management decisions." Ecological surprises – and sometimes tipping points – can be generated directly by directional drivers but may, in particular, be facilitated when the drivers cause secondary (trophic level) impacts that amplify the direct effects. The opposite response is also possible (e.g. Post and Pedersen 2008) but, with the potential for invasive species to amplify trophic interactions, surprises should be expected.

Here we present an intriguing example of the ecological complexity initiated by global change factors.

Within the framework of the environmental changes documented later, we identify mechanisms responsible for plant community change and how these mechanisms are influenced by the presence or absence of an important consumer, the black-tailed prairie dog.

Directional changes in climate and atmospheric chemistry are altering the environment of the Colorado Front Range, a region of increasing urbanization located at the junction of mountain foothills and mixed-grass to short-grass prairie. Among these directional changes are elevated average temperature (c. 2.5°F; Ray et al. 2008), higher rates of nitrogen deposition (Baron et al. 2000; Fenn et al. 2003), increased carbon dioxide concentrations (Morgan et al. 2007) and expansion of the length of the growing season (Archer and Predick 2008) are hypothesized and in some cases known to affect the distribution and abundance of grassland communities. These communities are perhaps more sensitive to changes than other systems because of the historical mix of species with C_3 or C_4 photosynthetic pathways that exhibit different CO_2, water and nitrogen (N) use efficiencies that in turn affect phenology and competitive interactions (e.g. Sage and Kubien 2003). In addition, these species evolved with a frequent fire return interval that has been suppressed by human activities. Since reduced fire intervals can enhance plant-available N on grassland sites and these sites are now exposed to perhaps an order of magnitude higher rate of inorganic N deposition than historical levels, the relative availability of this resource has been enhanced. Rapid-growing N-loving species clearly benefit from this change (e.g. Clark and Tilman 2008; Cherwin et al. 2009).

Growing season, as evidenced by remote sensing and a variety of phenological metrics, has expanded over the last decades (Myneni et al. 1997; Parmesan 2006; Kreyling 2010) and long-term records of flowering date in the western US show that spring blooms occur 2–3 days earlier per decade (Cayan et al. 2001). More recently, analysis of seasonal changes in northern hemisphere carbon dioxide concentrations indicates an expansion of the growing season of almost a month over the last 50 years (P. Tans, personal communication, 2005). These directional changes have occurred within the context of a temperate zone, continental climate that is characterized by very high interannual variability in precipitation (Knapp et al. 2001). Detection of directional changes can be further confounded by multiple decadal-scale oscillations in climate (Kitzberger et al. 2007).

A preliminary analysis of precipitation records indicates that no absolute changes in annual precipitation for the Front Range can be documented; however, winters are significantly wetter (Lawton 2010; J. Prevey, personal communication, 2012). The 110 year Boulder Colorado precipitation record indicates that the October–March 6 month interval produces an average of only 163 mm of 482 mm (34%) of annual precipitation. However, Lawton (2010) reported that the most recent 30 years of records had winter precipitation averaging 199 mm, 122% of the previous 70-year average. While climate change model predictions for this region contain high uncertainty regarding precipitation, the warming alone will alter the seasonality of plant water availability (Smith and Wagner 2006). Further, as monthly temperatures in Boulder now average above freezing year-round, the percentage of precipitation as snow is less and snow that does fall will likely melt rather than be retained for significant periods of time. Exactly how this affects plant water availability remains unstudied, but the exposed surface may allow for winter growth of cold-tolerant species.

The impacts of these cumulative environmental changes are intrinsically interesting and merit careful study. For example, the longer growing season and wetter winters appear to favor an increase in the abundance of winter annual plants (Lawton 2010). Among these, *Bromus tectorum* (cheatgrass) is best known. While introduced cheatgrass has long been a problem in the winter-wet Great Basin ecosystems of North America (Leopold 1949), its emergence onto the western edge of the Great Plains as a dominant species of this more summer-wet biome was unexpected and appears a recent phenomenon (Bush et al. 2007; Buckner and Downey 2009). The increase in winter annuals is believed to be a result of adequate or increased winter precipitation and increased nitrogen deposition, in addition to the expanded growing season.

These species have not been particularly successful at invading low-nutrient soils (Cherwin et al. 2009), but have done well in more fertile soil areas (Lawton 2010). Among the other common winter annuals are *Alyssum parviflorum* (annual peppergrass) and *Erodium cicutarium* (storksbill). Both of these are non-native annual forbs that appear to compete with the annual grasses for this new niche. Only a single native annual grass, Sixweeks fescue (*Vulpia octoflora*), is occasionally found in this group (Lawton 2010). Recent modeling

efforts and a synthesis of existing literature suggest that soil-moisture-mediated competition and competitive exclusion are likely to occur between the winter annuals and native perennials (Everard et al. 2010). The presence of a new suite of cool-season-adapted species may therefore not only exploit a new resource opportunity, but also compete with the native species that would otherwise have benefited from the environmental changes.

The Front Range has also been the location for invasions by a number of C_3 (cool season) perennial species. Of these, *Convolvulus arvensis* (field bindweed) appears to become a dominant species within the community only when competition from other common plant species is severely reduced. Until recently this species was regarded as an agricultural weed, one that exploited disturbance and high nutrient availability, but was not a concern in less-disturbed areas. One mechanism that facilitates dominance by bindweed and other introduced plants is grazing and burrowing by *Cynomys ludovicianus*, the black-tailed prairie dog (Magle and Crooks 2008). Prairie dogs are identified as keystone species and have been discussed for potential listing as threatened and endangered, yet they have facilitated winter dust storms at numerous sites along the Colorado Front Range (Figs 15.1 and 15.2). High densities of prairie dogs in the past, even under drought conditions, did not produce documented winter dust storms. This phenomenon was not known to occur prior to the increased dominance of invasive plant species in the Colorado Front Range. One interpretation is that we are observing the first stages of a catastrophic regime shift (e.g. Scheffer and Carpenter 2003) similar to the desertification phenomenon observed in the Southwest (Schlesinger et al. 1990).

Prairie dogs are common within the grassland remnants found along the Colorado Front Range. These animals have been intensively studied elsewhere due to their ability to perform as keystone species (Wiens 2009), maintain grasslands (Weltzin et al. 1997) and function as ecosystem engineers (Kotliar et al. 1999; VanNimwegen et al. 2008). The black-tailed prairie dog in particular has been well studied because of its widespread dominance or former dominance in the mixed- and short-grass prairie region of the Great Plains, and because the species has been perceived as a competitor with cattle (Whicker and Detling 1988; Derner et al. 2006; Wiens 2009). Its impacts on vegetation in both natural areas (e.g. Fahnestock and

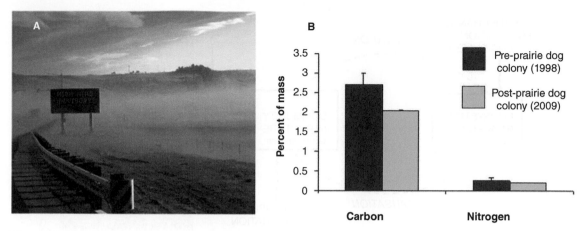

Figure 15.1 A. (a) Dust emissions from prairie dog colony south of Boulder, CO, 7 January 2009 causes a highway hazard (courtesy of US Geological Survey). This photo was published on the cover of the October 2010 issue of *Frontiers in Ecology and the Environment*. (b) Total carbon and nitrogen content of top 10 cm of soil measured before and after the site experienced substantial soil erosion.

Figure 15.2 Left: a native prairie occupied by prairie dogs for a decade as seen in September 2008. Center: by spring 2009 most plant cover was gone and severe wind erosion was noted. Right: plague removed the colony by late spring 2009, resulting in a meadow of native fringed sage (*Artemesia frigida*), annual sunflowers (*Helianthus* sp.) and non-native bindweed (*Convovulus arvensis*).

Detling 2002) and at the urban interface (e.g. Magle and Crooks 2008) are well known.

The wealth of studies that have been produced by Detling and colleagues since the 1980s confirm a consistent pattern of impacts produced by this species (e.g. Whicker and Detling 1988; Fahnestock and Detling 2002; Hartley et al. 2009). First, aboveground production appears largely unchanged despite very large changes in community composition, with a decrease in the absolute and relative abundance of the grasses and an increase in herbaceous dicots (forbs). In both short- and mixed-grass prairies, the effects of prairie dog grazing on plant community composition increases with the length of time the area is occupied. Individual prairie dog colonies shift spatially on the landscape over time, with the result being that long-term grazed

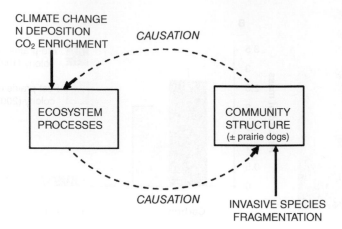

Figure 15.3 New plant species and a new environment form communities that interact with and modify grazing intensity by prairie dogs (courtesy of Mark Bradford); cheatgrass is the dominant plant cover.

areas are periodically abandoned (Augustine et al. 2008). Now, however, with habitat fragmentation in the Front Range and lack of acceptable adjacent habitats, the animals may remain on areas for longer time intervals. As prairie dog colonies become more static, these animals engineer habitats dominated by mostly non-native forb and shrub species (Fig. 15.3). Until very recently however, these changes had not caused large-scale erosion events. Further, we observed colonies to remain on heavily impacted areas even though sites were not bounded by any human or natural barriers to dispersal. The animals may prefer remaining on what appear to be degraded habitats rather than invade adjacent grasslands.

Prairie dogs have been discussed for federal listing as an endangered species (Kotliar et al. 1999) yet their presence is a concern at the urban–wildland interface, particularly in the Colorado Front Range. These animals form unusually high densities adjacent to urban areas (Johnson and Collinge 2004) and in particular appear to overgraze areas that have been restored from agricultural lands. In 2006, a restored grassland site in Fort Collins, Colorado experienced extensive wind erosion on areas colonized by prairie dogs such that the animals were removed and the site proactively restored. In the winter of 2008/2009, four similar areas in Fort Collins were affected (Pankratz 2009). Similar dust storms began on prairie dog colonies in the prairies surrounding the Boulder area in

2008/2009 and included sites that had been restored as well as native prairie areas.

Prairie dogs do not hibernate, but they greatly restrict their foraging activities during cold weather. This behavior presumably results from the fact that energy loss is reduced by staying in burrows rather than attempting to forage senescent vegetation under cold temperatures. Exactly how the duration and intensity of foraging has been influenced by the longer growing season and by a change in food quality (available energy and protein content) of the senescent vegetation in this region also remains unknown, but we believe that prairie dogs now use invasive species for grazing during the non-growing season in ways that contribute to the denudation of the landscape. While quantitative studies are lacking, the observations of either only non-native vegetation or essentially no vegetation on colonies attests to a transformation from historical patterns.

The lack of vegetation cover on prairie dog colonies in and adjacent to urban areas in mid-winter now appears to cause wholesale wind and water erosion of mixed-grass prairie in some years. This change appears well beyond the reduction in cover and increased forb abundance reported for prairie dog colonies of native prairie in earlier studies (Fahnestock and Detling 2002), or the erosion caused by isolated prairie dog mounds. This is a new phenomenon with wholesale implications for biodiversity and ecosystem services of these areas, i.e. the generation of novel ecosystems (Hobbs et al. 2006, 2009; Seastedt et al. 2008). The 'new' dynamics caused by the interactions of fragmentation, climate change and invasive species are hypothesized to convert a keystone species into an ecosystem transformer, and shift perennial grasslands to shrublands and landscapes dominated by non-native forbs (Fig. 15.4).

In the fall of 2008, several prairie dog colonies in Boulder County were observed that consisted largely of recently germinated annual plants accompanied with senescent fields of bindweed. By spring, these areas were devoid of vegetation. Soils were resampled at one site that had been part of a vegetation study in 1999, and the top 10 cm of soils had lost 25% of their organic C and N over the time period (Fig. 15.1b). The dust storms did not reoccur in 2009/2010 due to less wind, the presence of snow cover during much of the winter and the fact that the plague decimated many prairie dog colonies. Ironically, plague (*Yersinia pestisi*) is a disease accidentally introduced from Eurasia in the last century (Antolin et al. 2002), but its recurrence in the Front Range now appears to be the mechanism providing some persistence in vegetation diversity and dynamics because colony areas are periodically relieved of grazing pressure when plague epizootics occur (Hartley et al. 2009). Studies underway at the University of Colorado (L. Sackett, personal communication, 2012) suggest that resistance to the plague may be developing in the prairie dog species. In any event, the majority of colonies in our area already appear to be recovering from a current cycle of plague that began in Boulder County in 2006. By the winters of 2010/2011 and 2011/2012, minor dust storms were again being generated from colonies in the area.

The dialog about protecting prairie dogs as endangered species is ongoing, but the dialog about protecting ecosystem services provided by the prairie dogs within the realities imposed by global environmental change has yet to be initiated. While we confirm the scientific consensus that prairie dogs are critically important species in grasslands (Wiens 2009), we also believe that failing to recognize how this species interacts with new climate regimes and new plant species causes additional conservation problems. This problem is viewed as a model system of how trophic interactions can interact with global environmental change drivers to produce novel outcomes not predicted by climate drivers alone (Paine et al. 1998; Suttle et al. 2007). The concept of 'keystone species' in an era of rapid

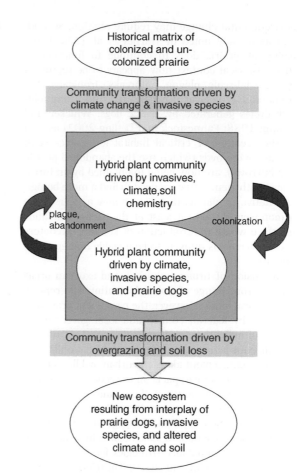

Figure 15.4 The short- and mixed-grass prairies of the Colorado Front Range are diverging from their historical configuration, with prairie dogs altering soils in ways that favor species that historically were only minor components of the vegetation. Colonization (a relatively slow process) and colony removal by plague (a rapid transformation) and abandonment connect the two new systems, but return to the historical grasslands is not viewed as a viable option. Once prairie dogs have created novel communities, their legacy effects on sites that are 'plagued out' may differ significantly from sites only impacted by invasive species. The yellow arrows are viewed as 'points of no return' (without large amounts of human intervention) but the second conversion to an eroded landscape more common to the southwestern US remains uncertain.

environmental change may require revision. Second, and as already noted by Hartley et al. (2009), the ability of prairie dogs to continue to function in their historical context now appears to be regulated by decadal-scale plague outbreaks. Ironically, the traditional ecological functions (i.e. the graminoid-forb cycles generated by prairie dogs; Whicker and Detling 1988; Fahnestock and Detling 2002) as well as the creation of critical habitat for species such as burrowing owls (which use the abandoned prairie dog burrows), are now being maintained by an introduced pathogen. While we need to find a more benign alternative to plague, this disease now appears to be a management aid (at least at the urban–wildlands interface where reintroduction of keystone predators such as ferrets is unlikely to be a viable management option).

Expansion of urban, suburban and exurban areas on the Front Range of Colorado continues to replace agricultural lands. However, the permanent grassland holdings by federal, county and local governments consist of many thousands of hectares and will remain a significant percentage of this landscape. Conflicts of use and management along the urban–wildland interface that have large impacts on the ecological services and conservation values of these lands are common and, as indicated here, are increasing. We see this example as a model system to engage educators, managers, policy makers and stakeholders in a dialog for developing sustainability strategies for the Colorado Front Range (Chapin et al. 2010). The science provided by this study should inform the dialog required for decisions to 'manage for resilience' versus 'manage for change' and for considering the values and consequences of new ecosystem states (West et al. 2009). As noted by Chapin et al. (2010), there exists a compelling need to have "researchers and managers collaborate through adaptive management to create continuous learning loops". The current interactions among prairie dogs, new plant species and other global environmental change factors provide a compelling and relevant focus for such a learning activity.

ACKNOWLEDGMENTS

The study of interactions between climate-invasive plant species and prairie dogs was supported by the National Science Foundation grant, DEB-1120390.

REFERENCES

Antolin, M.F., Gober, P., Luce, B., Biggins, D.E., Van Pelt, W.E., Seery, D.B., Lockhart, M. and Ball, M. (2002) The influence of sylvatic plague on North American wildlife at the landscape level, with special emphasis on black-footed ferret and prairie dog conservation. *Transactions of the North American Wildlife and Natural Resources Conference*, **67**, 104–128.

Archer, S.R. and Predick, K.I. (2008) Climate Change and Ecosystems of the Southwestern United States. *Rangelands*, **30**, 23–28.

Augustine, D.J., Dinsmore, S.J., Wunder, M.B., Dreitz, V.J. and Knopf, F.L. (2008) Response of mountain plovers to plague-driven dynamics of black-tailed prairie dog colonies. *Landscape Ecology*, **23**, 689–697.

Baron, J.S., Rueth, H.M., Wolfe A.M. et al. (2000) Ecosystem responses to nitrogen deposition in the Colorado Front Range. *Ecosystems*, **3**, 352–368.

Breshears, D.D., Myers, O.B., Meyer, C.W., Barnes, F.J., Zou, C.B., Allen, C.D., McDowell, N.G. and Pockman, W.T. (2009) Tree die-off in response to global change-type drought: mortality insights from a decade of plant water potential measurements. *Frontiers in Ecology and the Environment*, **7**, 885–189.

Buckner, D. and Downey, S. (2009) Patterns of annual brome abundance in reclaimed and native rangelands in the northern Great Plains: a case study from the Big Sky Mine, southeastern MT, in *Revitalizing the Environment: Proven Solutions and Innovative Approaches May 30–June 5, 2009* (ed. R.I. Barnhisel), National Meeting of the American Society of Mining and Reclamation, Billings, MT: Published by ASMR, 3134 Montevista Rd. Lexington KY.

Bush, R.T., Seastedt, T.R. and Buckner, D. (2007) Plant community response to the decline of diffuse knapweed in a Colorado grassland. *Ecological Restoration*, **25**, 169–174.

Cayan, D.R., Kammerdiener, S.A., Dettinger, M.D., Caprio, J.M. and Peterson, D.H. (2001) Changes in the onset of spring in the western United States. *Bulletin of the American Meteorological Society*, **82**, 399–415.

Chapin, F.S., Carpenter, S.R., Kofinas, G.P. et al. (2010) Ecosystem stewardship: sustainability strategies for a rapidly changing planet. *Trends in Ecology and Evolutionary Biology*, **25**, 241–49.

Cherwin, K.L., Seastedt, T.R. and Suding, K.N. (2009) Effects of nutrient manipulations and grass removal on cover, species composition, and invasibility of a novel grassland in Colorado. *Restoration Ecology*, **17**, 818–826.

Clark, C.M. and Tilman, D. (2008) Loss of plant species after chronic low-level nitrogen deposition to prairie grasslands. *Nature*, **451**, 712–715.

Derner, J.D., Detling, J.K. and Antolin, M.F. (2006) Are livestock weight gains affected by black-tailed prairie dogs? *Frontiers in Ecology and the Environment*, **4**, 459–464.

Everard, K., Seabloom, E.W., Harpole, W.S. and de Mazancourt, C. (2010) Plant water use affects competition for nitrogen:

why drought favors invasive species in California. *American Naturalist*, **175**, 85–97.

Fahnestock, J.T. and Detling, J.K. (2002) Bison-prairie dog-plant interactions in a North American mixed-grass prairie. *Oecologia*, **132**, 86–95.

Fenn, M.E., Baron, J.S., Allen, E.B. et al. (2003) Ecological effects of nitrogen deposition in the western United States. *BioScience*, **53**, 404–420.

Hartley, L.M., Detling, J.K. and Savage, L.T. (2009) Introduced plague lessens the effects of an herbivorous rodent on grassland vegetation. *Journal of Applied Ecology*, **46**, 861–869.

Hastings, A. and Wysham, D.B. (2010) Regime shifts in ecological systems can occur with no warning. *Ecology Letters*, **13**, 464–472.

Hobbs, R.J., Arico, S., Aronson, J. et al. (2006) Novel ecosystems: theoretical and management aspects of the new ecological world order. *Global Ecology and Biogeography*, **15**, 1–7.

Hobbs, R.J., Higgs, E. and Harris, J.A. (2009) Novel ecosystems and their implications for conservation and restoration. *Trends in Ecology and Evolutionary Biology*, **24**, 599–605.

Johnson, W.C. and Collinge, S.K. (2004) Landscape effects on black-tailed prairie dog colonies. *Biological Conservation*, **115**, 487–497.

Kitzberger, T., Brown, P.M., Heyerdahl, E.K., Swetnam, T.W. and Veblen, T.T. (2007) Contingent Pacific-Atlantic Ocean influence on multi-century wildfire synchrony over western North America. *Proceedings of the National Academy of Sciences*, **104**, 543–548.

Knapp, A.K., Briggs, J.M. and Koelliker, J.T. (2001) Frequency and extent of water limitation to primary production in a mesic temperate grassland. *Ecosystems*, **4**, 19–28.

Kotliar, N.B., Baker, B.W., Whicker, A.D. and Plumb, G. (1999) A critical review of the assumptions about the prairie dog as a keystone species. *Environmental Management*, **24**, 177–192.

Kreyling, J. (2010) Winter climate change: a critical factor for temperate vegetation performance. *Ecology*, **91**, 1939–1948.

Lawton, W.C. (2010) Global environmental change factors and noxious weed management shape a novel grassland ecosystem on the Colorado Front Range. MSc thesis, University of Colorado, Boulder.

Leopold, A. (1949) *A Sand County Almanac*. Oxford University Press, Oxford.

Magle, S.B. and Crooks, K.R. (2008) Interactions between black-tailed prairie dogs (Cynomys ludovicianus) and vegetation in habitats fragmented by urbanization. *Journal of Arid Environments*, **72**, 238–246.

Morgan, J.A., Milchunas, D.G., LeCain, D.R., West, M. and Mosier, A.R. (2007) Carbon dioxide enrichment alters plant community structure and accelerates shrub growth in shortgrass steppe. *Proceedings of the National Academy of Science*, **104**, 14724–14729.

Myneni, R.B., Keeling, C.D., Tucker, C.J., Asrar, G. and Nemani, R.R. (1997) Increased plant growth in the northern high latitudes from 1981 to 1991. *Nature*, **386**, 698–702.

Paine, R.T., Tegner, M.J. and Johnson, E.A. (1998) Compounded perturbations yield ecological surprises. *Ecosystems*, **1**, 535–545.

Pankratz, H. (2009) Fort Collins plans prairie dog cull. *Denver Post*. http://www.denverpost.com/news/ci_11689469, Accessed August 2012.

Parmesan, C. (2006) Ecological and evolutionary responses to recent climate change. *Annual Review of Ecology, Evolution, and Systematics*, **37**, 637–669.

Post, E. and Pedersen, C. (2008) Opposing plant community responses to warming with and without herbivores. *Proceedings of National Academy of Sciences*, **105**, 12353–12358.

Ray, A.J., Barsugli, J.J., Averyt, K.B., Wolter, K., Hoerling, M., Doesken, N., Udall, B. and Webb, R.S. (2008) Climate change in Colorado. A report to the Colorado Water Conservation Board. URL: http://wwa.colorado.edu/CO_Climate_Report/index.html (accessed August 2012).

Sage, R.F. and Kubien, D.S. (2003) Quo vadis C_4? An ecophysiological perspective on global change and the future of C_4 plants. *Photosynthesis Research*, **77**, 209–225.

Scheffer, M. and Carpenter, S.R. (2003) Catastrophic regime shifts in ecosystems: linking theory to observation. *Trends in Ecology and Evolution*, **18**, 648–656.

Schlesinger, W.H., Reynolds, J.F., Cunningham, G.L., Huennke, L.F., Jarrell, W.M., Virginia, R.A. and Whitford, W.G. (1990) Biological feedbacks in global desertification. *Science*, **8**, 1043–1048.

Seastedt, T.R., Hobbs, R.J. and Suding, K.N. (2008) Management of novel ecosystems: are novel approaches required? *Frontiers in Ecology and the Environment*, **6**, 547–553.

Smith, J.B. and Wagner, C. (2006) Climate change and its implications for the Rocky Mountain region. *Journal of the American Water Works Association*, **98**, 80–92.

Suttle, K.B., Thomsen, M.A. and Power, M.E. (2007) Species interactions reverse grassland responses to climate change. *Science*, **315**, 640–642.

Tylianakis, J.M., Didham, R.K., Bascompte, J. and Wardle, D.A. (2008) Global change and species interactions in terrestrial ecosystems. *Ecology Letters*, **11**, 1351–1363.

VanNimwegen, R.E., Kretzer, J. and Cully, J.F. (2008) Ecosystem engineering by a colonial mammal: how prairie dogs structure rodent communities. *Ecology*, **89**, 3298–3305.

Weltzin, J.F., Archer, S. and Heitschmidt, R.K. (1997) Small mammal regulation of vegetation structure in a temperate savanna. *Ecology*, **78**, 751–763.

West, J.M., Julius, S.H., Kareiva, P., Enquist, C., Lawler, J.J., Petersen, B., Johnson, A.E. and Shaw, M.R. (2009) US natural resources and climate change: concepts and approaches for management adaptation. *Environmental Management*, **44**, 1001–1021.

Whicker, A.D. and Detling, J.K. (1988) Ecological consequences of prairie dog disturbances. *Bioscience*, **38**, 778–785.

Wiens, J.A. (2009) Landscape ecology as a foundation for sustainable conservation. *Landscape Ecology*, **24**, 1053–1065.

Chapter 16

PERSPECTIVE: PLUS ÇA CHANGE, PLUS C'EST LA MÊME CHOSE

Stephen D. Murphy

Department of Environment and Resource Studies, University of Waterloo, Canada

As a boy I tried to sort out what my place was: entertainer, technologist, naturalist or all three? The latter it seems. An important inspiration in my life was my maternal grandfather, Harold Arthur, who was a forester and naturalist and even amateur Vaudevillian. I learned from him and his tales and from meeting some of his friends, many of whom were members of First Nations. From them, I gleaned a multilayered relationship with the pre-European landscape. There were spiritual angles that clashed or coexisted with pragmatic trade needs. There was a conservation ethos and a business ethos. The idealistic view of First Nations in harmony with nature was something the First Nations members I met would ruefully discuss, and an outsider like me had to tread lightly with this topic. Still, I learned enough oral history and traditional knowledge – and was permitted to acknowledge such – to understand a rich and complex history of First Nations.

Fast forward ten years later. I am listening to people (experts all) discuss disjuncts: species that are in places far away from their known distribution or 'where they should be'. Many just so stories are promulgated: weird glaciation effects, mass changes in water flows, sudden windborne gusts, die-off of the surrounding populations so it only looks like they were disjuncts. I raise my hand and ask: "Isn't it more likely that the disjuncts are from trade via First Nations? If you look at the species, most are ones useful for food or medicine and the rest tend to be 'weeds' or 'pests' brought along with transporting what we would now call commercially viable plants."

Several told me that I was out of my mind. How could First Nations do that without cars or big ships or St Lawrence seaway or planes? Besides, they always fought; they never traded. They could not possibly have done that. Note the use of the word "they". The term is accurate in the sense that most of us are not First Nations members, but there seemed to be derogatory subtext. By listening to First Nations, testing out distributions and tracing patterns of migration, we could find evidence that disjunct migration was assisted. Outside of ecological circles at the time, historian William Cronon, ethnobotanist Nancy Turner and many other scholars were examining intricate relationships of First Nations with ecosystems.

Novel Ecosystems: Intervening in the New Ecological World Order, First Edition. Edited by Richard J. Hobbs, Eric S. Higgs, and Carol M. Hall.
© 2013 John Wiley & Sons, Ltd. Published 2013 by John Wiley & Sons, Ltd.

Those who provided evidence against such stories had however already influenced me. I was also influenced by Monty Python's Flying Circus and couldn't help thinking that the discussion of disjuncts sounded like the Holy Grail skit where there was debate over whether swallows were able to carry large coconuts: it was that surreal. Not all disjuncts may originate from trade by First Nations, but the inability to consider cultural actions as a viable force before our contemporary era was rather astounding, especially given the extreme ideas needed to derive an ecological explanation (as extreme as swallows carrying those coconuts).

What lesson was learnt? People have been adding novel elements to ecosystems since we first rumbled out of Africa. Other species did, and do, the same. Novel ecosystems are not unique to the post-European colonization era in North America or any colonization era anywhere. They are simply more noticeable over the last 300 years because our technologies and values allowed influence on mass scales and at an increasing pace. Just as we now influence global climate, so too do we influence the biomes. Have we known that all along? Perhaps, but an occasional reminder is helpful.

Another lesson, perhaps more important, is that ecologists must not divorce themselves from social context just as they argue that sociologists, for example, cannot ignore ecological constraints. We would do well not only to listen to 'others' such as First Nations members, but also to acknowledge that others think differently and operate from contrasting paradigms. Novel ecosystems as a concept offer a great chance for interdisciplinary work and understanding what should or should not be done. Although I sort-of recognized this as a boy while trying to decide between arts, technology and science, I truly understood it the day I listened to the tales of the disjuncts.

Part IV

When and How to Intervene

Chapter 17

PERSPECTIVE: FROM RIVETS TO RIVERS

Joseph Mascaro

Department of Global Ecology, Carnegie Institution for Science, Stanford, California, USA

"As you walk from the terminal toward your airliner, you notice a man on a ladder busily prying rivets out of its wing. Somewhat concerned, you saunter over to the rivet popper and ask him just what the hell he's doing."

Paul and Anne Ehrlich, 1981

I met Shahid Naeem in 2003. I had just graduated from The University of Michigan and was biding my time in a weed-plucking ecology chain-gang at Cedar Creek, Minnesota. The same day, Shahid wowed a room full of interns with the best demonstration of a scientific theory I've ever seen.

Cedar Creek, like most field stations, is a place old computers go to die. Shahid pulled some strings to acquire a 486 that was on life support for a public-gutting session. He jammed a projector cable into it, broadcast the sorry-looking Windows 3.1 background in front of the confused audience, pulled the dust cover off the motherboard and said: "Let's imagine this computer is an ecosystem. A lake or a forest. Just like an ecosystem, this computer is made up of thousands of interconnected parts, each of which has its own properties but each of which also affects the functioning of the whole machine." He then explained that he was going to begin to pull pieces out of it, one-by-one, to see what happened. And that's just what he did. Each time he yanked a capacitor, he'd wiggle the mouse so that we could all see that ecosystem function was maintained. And sure enough, after the third yank the mouse froze and within a few seconds the screen went dark. Presto! I understood the biodiversity–ecosystem function paradigm.

Shahid had of course demonstrated that while extinction may not initially destabilize ecosystem function, continued extinction will do so *inevitably*. The analogy to a computer was apt for the decade. In the 1980s, the analogy chosen by Paul and Anne Ehrlich was an airplane. By drawing us in with the terror of flying onboard a partially disassembled machine, they succeeded masterfully in raising awareness of the extinction crisis. This 'rivet-popper hypothesis' was the genesis of the biodiversity–ecosystem

Novel Ecosystems: Intervening in the New Ecological World Order, First Edition. Edited by Richard J. Hobbs, Eric S. Higgs, and Carol M. Hall.
© 2013 John Wiley & Sons, Ltd. Published 2013 by John Wiley & Sons, Ltd.

function paradigm, and Shahid was among the first to demonstrate the pattern experimentally.

Three months later, I accepted a USGS internship in Hawaii. After a white-knuckled flight, during which I studied the structural integrity of the starboard wing a little too intently, I arrived at the extinction capital of the United States. The first days were depressing; botanical talks were a who's-who of toxic, noxious, invasive plants that were swallowing the island. Tours of preserves took us to hibiscus patches that were among the last dozen of a critically endangered species, or fields of alien grasses that were carrying fires so frequently that they killed all the stately Ohia trees. There is a delicate quality to the Hawaiian flora and fauna, and watching it disappear before your eyes is painful and demoralizing. This is what you see when you visit Hawaii with a gaggle of ecologists: ecosystems crashing all around you, their wreckage strewn across the landscape. The rivet-popper analogy seems all too apt.

Except there's one problem: Hawaii has more – not fewer – rivets than it used to. A *lot* more. The best data suggest that Hawaii has lost less than 100 angiosperm species to extinction but has gained at least 1000 naturalized angiosperms, effectively doubling the size of its flora.

Ecosystem function hasn't stopped. It hasn't collapsed or crashed into a mountain. Consider productivity, the simplest of ecosystem functions, which is booming in lowland rainforests on Hawaii Island as a clan of pantropical alien trees crank sunlight into matter and pour it onto the forest floor. Cycles of carbon, nitrogen and phosphorous are churning. In Pahoa and along the Hamakua coast, production and nutrient cycling have been restored to valleys and old agricultural fields once scarred by fire and tilling. The roster of species in these novel forests is the same who's-who of invasive plants that are swallowing the island. Introduced mangroves are straining sediment and building habitat that native fishes utilize. Alien birds are dispersing native shrubs in exotic tree plantations. The list goes on.

The emergence of novel ecosystems *in no way* sets aside concerns over extinction, but it does highlight a key limitation to the Ehrlich and Naeem analogies: unlike airplanes and computers, ecosystems absorb new components. Ecosystem function does not solely reflect species loss, as implied by the popping of rivets and capacitors: it also reflects species additions.

When we step outside the notional machine with its exhaustive list of perfectly complete parts, we can see the novel ecosystem for what it is: not a disaster or meltdown, but the next reach of a river in time.

REFERENCE

Ehrlich, P.R. and Ehrlich, A.H. (1981) *Extinction: The Causes and Consequences of the Disappearance of Species*. Random House, New York.

Chapter 18

INCORPORATING NOVEL ECOSYSTEMS INTO MANAGEMENT FRAMEWORKS

Kristin B. Hulvey[1], Rachel J. Standish[1],
Lauren M. Hallett[2], Brian M. Starzomski[3],
Stephen D. Murphy[4], Cara R. Nelson[5],
Mark R. Gardener[6], Patricia L. Kennedy[7],
Timothy R. Seastedt[8] and Katharine N. Suding[2]

[1]Ecosystem Restoration and Intervention Ecology (ERIE) Research Group, School of Plant Biology, University of Western Australia, Australia

[2]Department of Environmental Science, Policy & Management, University of California, Berkeley, USA

[3]School of Environmental Studies, University of Victoria, Canada

[4]Department of Environment and Resource Studies, University of Waterloo, Canada

[5]Department of Ecosystem and Conservation Sciences, College of Forestry and Conservation, University of Montana, USA

[6]Charles Darwin Foundation, Galapagos Islands, Ecuador, and School of Plant Biology, University of Western Australia, Australia

[7]Department of Fisheries and Wildlife & Eastern Oregon Agriculture & Natural Resource Program, Oregon State University, USA

[8]Department of Ecology and Evolutionary Biology, University of Colorado, USA

18.1 INTRODUCTION

How might the existence of hybrid and novel ecosystems alter ecosystem management? Non-native species invasions, climate change, pollution and land development all are creating ecosystems that consist of new combinations of species, and often have altered functioning and structure. A common goal in ecosystem management is to maintain native populations and traditional functions by removing the species, disturbances and conditions that lead to degradation (Grumbine 1997), and thus return ecosystems to their

Novel Ecosystems: Intervening in the New Ecological World Order, First Edition. Edited by Richard J. Hobbs, Eric S. Higgs, and Carol M. Hall.
© 2013 John Wiley & Sons, Ltd. Published 2013 by John Wiley & Sons, Ltd.

pre-disturbance trajectories or states. The emergence of novel ecosystems forces managers to reconsider this paradigm because, at times, no amount of management action will reverse ecological changes. New management goals may continue to recognize the value of protecting species and ecosystem processes, although they might not include continuity with the historical system. In these cases, managers might choose to utilize non-traditional or alternative management strategies derived through a broad suite of planning tools to reach management goals.

This chapter aims to provide a framework that helps managers, whether scientists or stewards, navigate the decisions that lead to new management approaches in hybrid and novel ecosystems. We first present a decision-making flowchart (Fig. 18.1) that can be used

as a roadmap to navigate possible management actions. We also explore the role of both ecological and social barriers in the creation and maintenance of hybrid and novel ecosystems (Fig. 18.2). Finally, five case studies (Chapters 19–23) highlight examples of challenging decision points (Box 18.1) which managers will likely face as they work to incorporate hybrid and novel ecosystems into strategies for restoration, conservation and management.

18.2   THE NOVEL ECOSYSTEM DECISION FRAMEWORK

An overview of the decision framework is given in Figure 18.1, where various options for intervention are

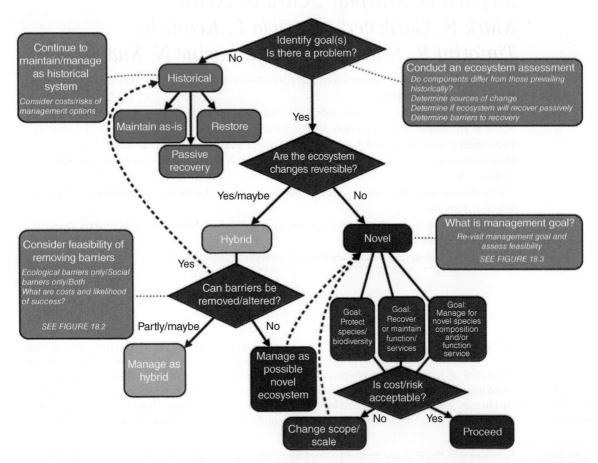

Figure 18.1 Flowchart showing major decisions to be made on interventions in historical, hybrid and novel ecosystems.

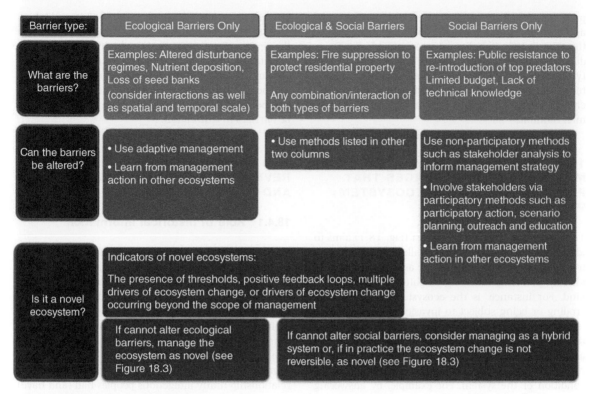

Figure 18.2 Identifying novel ecosystems.

Box 18.1 Difficult questions raised when managing novel ecosystems

Ecological
• Do sufficient data exist to define a reference state/historical baseline?
• How do we determine if ecosystem changes are reversible?
• How do we make management decisions when we have incomplete knowledge about the target ecosystem and the drivers causing ecosystem change?
• How do we manage novel ecosystems to maintain past or obtain future ecosystem values when we cannot alter the drivers of change (e.g. climate change)?
• How do we maintain desirable characteristics of a hybrid ecosystem that we might lose if we implement restoration to a historical state?

Social
• When interventions to restore ecosystems are technically possible, when is it acceptable for economic and social factors to prevent action?
• How do we balance competing desires for ecosystems to be restored to a historical state or for intrinsic values versus managed to provide currently desired goods and services?
• How do constraints on managers' time and budgets affect their ability to identify ecosystem novelty?
• When managing novel ecosystems with limited budgets, how do we decide which species and functions to protect and maintain?

presented based on ecosystem conditions and management goals. Broadly, the flowchart, along with the text and case studies, aims to provide the foundations for a dialogue among practitioners of all types, focused on the management of changing ecosystems. More specifically, this flowchart highlights critical decision points for determining management interventions.

18.3   IDENTIFYING NEW MANAGEMENT CHALLENGES THAT PREVENT HISTORICAL ECOSYSTEM PERSISTENCE

The first question in the flowchart (Fig. 18.1) aims to identify if new management challenges prevent historical ecosystem persistence, by asking managers to identify problems that require intervention of some kind. For instance, is the ecosystem losing species, eroding or being subject to invasion by one or more non-native species? An assessment that includes identification of ecosystem changes, the drivers of change and the barriers (ecological and social) to ecosystem recovery is an important first step in determining the condition of the system. For example, by answering questions such as whether ecosystem components differ from those prevailing historically and what is responsible for the change(s), managers will be able to determine whether slight alterations to current actions might be sufficient to improve management or if a new course of action is needed. Such an information-gathering process is included as a necessary step in many ecosystem management projects and management plans (e.g. US Fish & Wildlife Service 2002a, b; TNC 2007). Ecosystems that maintain a high degree of historical continuity (or have the capacity for restoring historical continuity) may be managed well with current methods, and management can proceed using traditional restoration and management approaches.

If new and significant changes are identified, managers can use information gathered in the ecosystem assessment to help decide if the ecosystem is: (1) a hybrid ecosystem that can be restored; (2) a hybrid ecosystem where alterations in current management methods can reverse some but not all change so that the ecosystem exhibits both historical and novel characteristics; or (3) a novel ecosystem where changes are so severe that there is little continuity with historical systems and changes cannot be reversed with any

amount of active intervention. Determining which of these three ecosystem types exists can greatly influence the options for management, which we will expand on later in the chapter. Since deciding among these options is not always easy, the following section focuses on the identification of ecological and social factors that both influence ongoing ecosystem change and can be barriers to restoration.

18.4   ARE ECOSYSTEM CHANGES REVERSIBLE? IDENTIFYING HYBRID AND NOVEL ECOSYSTEMS

18.4.1   Role of historical information

The question of whether ecosystem changes are reversible (Fig. 18.1) implies that ecosystems have changed from some former state. To inform management goals (including restoration) it is helpful to define this former state, although this process can be remarkably difficult. Managers outside of Europe often choose ecosystems that existed before the arrival of European settlers, aiming for ecosystems that were 'untouched' by humans. Historically this failed to recognize the close relationship many indigenous peoples had (and have) with the land, although this is improving (e.g. Berkes et al 1998). Furthermore, the effects of climate change are clear (see Chapter 10); many managers realize restoration references need to consider not just what was historical in the ecosystem, but also the extent to which a system's biophysical envelope may have changed because of factors other than anthropogenic drivers.

Despite these challenges, we argue that working to understand the historical structure and conditions in an ecosystem is useful for planning current management strategy. This is because information on historical states can both inform restoration of hybrid ecosystems where changes are determined to be reversible as well as highlight where very few, if any, changes can be reversed, therefore indicating the presence of a novel ecosystem. In short, information about ecosystem changes can help determine the next steps for ecosystem management and intervention.

Importantly, historical ecosystem references always provide more of a guide than a strict template for determining current management action. Due to the dynamic nature of ecosystems and imperfect knowledge of past conditions, using information from multiple reference sites or using multiple types of historical information will often yield the most accurate model from which to assess novelty (Swetnam et al. 1999; Chapter 24). The case study on meadows in Atlantic Canada (Chapter 19) provides an example of how incomplete historical information about ecosystems can affect restoration outcomes. How managers addressed these informational shortcomings is discussed and an after-the-fact summary of the pros and cons of working to restore ecosystems to historical states is provided.

18.4.2 Identification of barriers to ecosystem recovery

Both ecological and social factors can influence ongoing ecosystem change and can be barriers to restoration (Fig. 18.2). Examples of ecological barriers include reduced seed banks that lead to population decline of native species, altered disturbance regimes that favor non-native species spread or the presence of non-native species that prevent recruitment of desired native species. Examples of social barriers that can affect management decisions include limited budgets for restoration, social norms and human welfare needs, and gaps in knowledge about the efficacy of management actions. In the following sections we describe a number of ecological and social factors that can signal the presence of a novel ecosystem.

18.4.2.1 Ecological barriers

18.4.2.1.1 The presence of thresholds
Chapter 3 focused mainly on ecological barriers in the form of thresholds: tipping points where an ecosystem moves from one ecological state to another. An ecological state is defined by the abiotic and biotic attributes of an ecosystem along with the feedbacks and dynamics that contribute to these attributes. When a tipping point is crossed, changing feedbacks and dynamics result in a new array of attributes. Examples of possible tipping points include species extinctions or distribution shifts, habitat fragmentation, nutrient deposition, changing disturbance regimes, increases in numbers and abundances of invasive species or the development of colonization barriers such as shrinking or absent seed banks, (e.g. Scheffer 2009).

The threshold concept is broadly useful for identifying novel ecosystems because the recognition of state shifts indicates the possibility that traditional management actions may no longer produce desired

management outcomes. Recognition of an altered ecosystem state starts the process of determining what the underlying drivers of change might be. When these drivers cannot be addressed through management interventions, managers may decide to treat the ecosystem as if it were novel and consider the management actions presented in Figure 18.3. Chapters 3 and 24 offer a more detailed discussion of thresholds, as well as Suding and Hobbs (2009) and the Thresholds Database at: http://www.resalliance.org/index.php/thresholds_database.

18.4.2.1.2 Positive feedback loops
The presence of positive feedback loops (also called amplifying feedbacks) may also be good indicators of novel ecosystems. Positive feedback loops occur when an initial ecosystem change results in continued changes in the same direction. These feedbacks continually drive an ecosystem away from its original state, making it hard for managers to return it to the original state (Suding and Hobbs 2009). Such positive feedback loops are often associated with threshold dynamics in ecosystems (Suding and Hobbs 2009).

A classic example of a positive feedback resulting in the shift of an ecosystem state is the invasion of non-native species that alter disturbance regimes such that the new regime favors the ongoing spread of the invader (D'Antonio and Vitousek 1992). An example is the invasion of *Bromus tectorum* (cheatgrass) which was introduced into the US in the late 1800s and had invaded grass and shrubland ecosystems throughout the west by the 1930s (Mack 1981; Menakis et al. 2002). Areas invaded by cheatgrass have uncharacteristically large accumulations of litter; this build-up of fuel increases the fire-return interval in summer-dry areas, which favors the quick-growing grass and leads to a conversion of native ecosystems to cheatgrass-dominated grasslands. Other examples of positive feedback loops include the simultaneous build-up of a non-native seed bank and depletion of a native seed bank, such as that which occurs on old fields in southwestern Australia (Standish et al. 2007).

18.4.2.1.3 Multiple interacting drivers of ecosystem change (cumulative effects)
Changes to ecosystem structure and function stem from a variety of internal and external drivers. Examples of internal drivers include management systems, land conversion, species harvesting and some forms of pollution. Examples of external drivers include

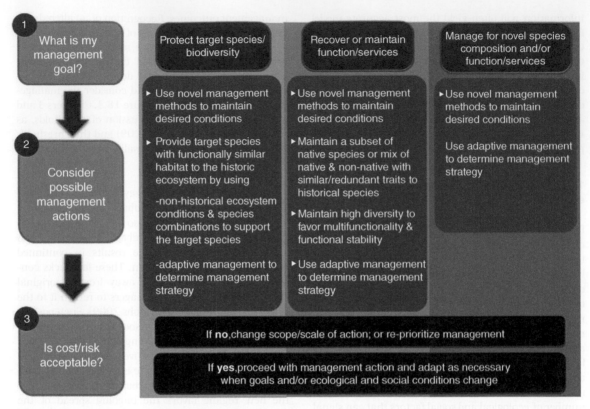

Figure 18.3 Novel ecosystem management options.

climate change, nitrogen deposition and non-native species invasion. Often multiple drivers cumulatively impact an ecosystem, resulting in a cascade of changes in the abiotic and biotic characteristics of the ecosystem, as well as possible changes to ecosystem dynamics (Suding et al. 2004). In these cases, novel ecosystems may result from the inability of managers to simultaneously address all drivers in a way that will mitigate ongoing ecosystem changes.

18.4.2.1.4 Drivers of ecosystem change occurring at scales beyond the scope of management
Novel ecosystems may occur when managers are unable to address ecosystem change drivers because of a disjunct between the scale at which the driver is operating and the scale at which management can be targeted. Managers may have little control over the activities and developments that occur outside their reserve boundaries but impact ecosystems within their jurisdiction.

Because management actions can be limited to the individual reserve, park or other area within their jurisdiction, finding management solutions to address change drivers that are not under their control may be difficult. If a mismatch exists in scale of driver and scale of management, restoring an ecosystem to a reference state is highly unlikely. In these cases, managers may need to rely on tools that can span such property boundaries, such as environmental regulation and policy or conservation easements and other incentive-based programs. Chapter 20 describes an example of how mismatches in the scale of driver and scale of management action can limit the ability of managers to restore ecosystems and can lead to novel ecosystems.

18.4.2.2 Social barriers

At times, management strategies solely consider ecological barriers to recovery. This may be due to an assumption by managers, and more particularly natural scientists, that social factors are not as important as ecological barriers for ecosystem recovery. The reality is that social factors can be both as important to ecosystem management and as difficult to address as ecological barriers.

Deciding when social barriers to ecosystem recovery are insurmountable is beyond the scope of this chapter. However, understanding the impact of social factors on the ability of managers to reach management goals is key in creating achievable management strategies (Keough and Blahna 2006). For example, in the case study of ecosystem management on St George Island, Alaska (see Chapter 21), the management of introduced reindeer depends on the collaboration of a diverse group of stakeholders with, at times, contrasting management goals. As a result, the current management strategy perpetuates the presence of the reindeer, and thus a hybrid ecosystem, rather than restoring the ecosystem to its historical trajectory (i.e. complete removal of reindeer on St George Island). Undoubtedly, social factors will be an ongoing consideration in management decisions, and successful management strategies in changing and novel ecosystems will likely require increased communication and collaboration among ecologists, land owners, policy makers, managers and other ecosystem stewards.

A number of social factors that affect management decisions are described below.

18.4.2.2.1   *Limited management budget*
Limited budgets, and thus insufficient funds for restoration, are one of the most pervasive management constraints and can lead to the maintenance of hybrid ecosystems. While resource shortfalls may sometimes be alleviated by strategic planning that allows for readjustment of funding assignments or the development of collaborations, budgets may be controlled by government allotments or organizational processes that prevent the dedication of sufficient resources to restore hybrid systems to reference states. Additionally, realistic levels of resourcing are not always available to meet the objectives of a project.

18.4.2.2.2   *Social norms and property systems*
When ecosystems span a mosaic of land uses and types of owners, the values of the people living on the land and/or managing the land will affect ecosystem management. In particular, it may be difficult to coordinate actions across property boundaries when adjacent landowners use properties for different purposes or value different aspects of the environment (Wiens 2009). For example, stopping the spread of a nonnative plant may require the coordinated effort of adjoining property owners (Aslan et al. 2009); if some owners value the plant while others consider it a threat to native vegetation, it may be hard to coordinate management action (Gardener et al. 2010).

In other cases, norms based on strongly held community beliefs may prevent managers from undertaking action that leads to ecosystem restoration. For example, in some cases managers have the knowledge to reestablish self-sustaining populations of top predators to areas that were historically part of their distribution. A number of social factors can however influence management possibilities including: the belief that top predators can be a threat to human and/or livestock safety; beliefs that an increase in predators results in declines in valued prey populations (see Wasser et al. 2011 for a test of this assumption); or perceptions that the reintroduction of predators with legal protection is an infringement of private property rights (e.g. Bangs and Fritts 1997; Smith and Ferguson 2006; Levy 2009).

When social norms differ across landscapes or are contrary to recovery, restoration may at times be limited to public lands or require increased communication between managers and landowners to determine acceptable and appropriate management goals. There are a number of tools managers can use to incorporate social considerations into management decisions including stakeholder analysis, participatory action, scenario planning (Peterson et al. 2003), outreach and education. These tools help stakeholders understand differing social values and build relationships and collaborations. Adaptive management can also be used to fill knowledge gaps. At a broader scale, environmental regulations and policy may help to address some of the social barriers to ecosystem recovery.

18.4.2.2.3   *Gaps in knowledge*
A variety of types of knowledge gaps can hinder people's ability to restore ecosystems. For example, insufficient knowledge of ecosystem change drivers can prevent the recognition of reversible change. Such knowledge gaps might be addressed through ecosystem assessments as described earlier in the chapter.

Incomplete understanding of the effects of widely used but unsuccessful management strategies is a second knowledge gap that may prevent ecosystem recovery. One example of this is where management activities can have a narrow focus on the restoration of plants and abiotic conditions, leading to the belief that successful revegetation is synonymous with successful faunal restoration. This is an extension of the Field of Dreams hypothesis (Palmer et al. 1997), originally based on wetland restoration where "getting the hydrology and soil right" seemed to be the most important ingredient for restoration success. Some studies point out that passive fauna recolonization of revegetated sites does not occur in all cases however, due to factors such as the specialized habitat needs of individual faunal species or the isolation of restored sites from source populations (Kanowski et al. 2006). In these cases active intervention is in fact required. Careful consideration of final recovery goals is needed to ensure actions can lead to desired outcomes.

Finally, knowledge gaps may stem from insufficient technology or methodologies to reverse ecosystem changes. In such cases, the ability to restore ecosystems might change as methodologies are developed through targeted research or as a consequence of new technologies. Some ecosystems might initially be managed as novel because of this, but managed in a way that preserves opportunities for later restoration.

In cases where management action is irreversibly constrained by social barriers, managers have a couple of options. The first option is to continue with the status quo. This may mean mitigating degradation when possible and, at times, allowing the system to passively move to a novel ecological state. A second option is to maintain the ecosystem in a hybrid state, regardless of the ecological possibility of returning it to its historical trajectory. This option might entail reassessment of management goals and reprioritization of management actions (see Chapter 21).

18.4.3   When uncertainty persists

Despite best efforts to understand if an ecosystem change is reversible, this may remain unclear. Perhaps the drivers of ecosystem change are not yet clearly understood, or maybe social and economic constraints are such that pursuing management to restore ecosystems is out of the question. In such situations, managers may be unwilling to label the target ecosystem *novel*. Instead, they might bet that future innovations in management techniques, results from an ongoing

ecosystem assessment or coalitions with stakeholders will eventually allow for ecosystem recovery. Because of this, Figure 18.1 includes a pathway at the decision point focusing on 'reversibility', accounting for this uncertainty (i.e. an answer of 'yes/maybe' to the question of whether ecosystem changes are reversible is possible). A number of possible management options for when such uncertainty persists are provided in the section on managing hybrid ecosystems (Section 18.5.1).

18.5   NAVIGATING THE MANAGEMENT OF HYBRID AND NOVEL ECOSYSTEMS

After barriers to ecosystem recovery are identified and an ecosystem is determined to be either hybrid or novel, managers will have a number of choices of how to proceed with management actions (Fig. 18.1).

18.5.1   Hybrid management choices

When an ecosystem is determined to be hybrid, managers have a number of possible options (Fig. 18.1). Although changes in active management techniques are an option, it is also possible that the ecosystem might recover passively. Recent work highlights how scarce restoration resources can be conserved through careful decisions about where to focus restoration activities in those ecosystems that require no or minimal active intervention to recover, versus those ecosystems that need active management (Prach and Hobbs 2008; Sawtschuk et al. 2010; Holl and Aide 2011). Whether an ecosystem will recover without intervention depends on a number of factors including ecosystem type, level of degradation and connectivity with the surrounding landscape (Prach and Hobbs 2008). For example, passive restoration is not always effective for restoring higher trophic levels. If management goals include restoration of fauna, then some form of active restoration may be necessary (Majer 2009). An ecosystem assessment that includes identification of ecosystem changes, the drivers of change and the barriers (ecological and social) to ecosystem recovery will provide the best available information to determine the possibility of passive recovery.

When passive recovery is not an option, management options for the hybrid ecosystem may include: (1) fully restoring the ecosystem through active intervention; (2) managing the ecosystem in a hybrid state either by partially restoring it or deciding to allow ecosystem changes to persist; or (3) deciding to manage the ecosystem as novel despite possibilities of restoring it to a historical state (Fig. 18.1). Which of these options is chosen will likely depend on the ability of the manager to remove/alter the ecological and social barriers present in the ecosystem (see Fig. 18.2).

As mentioned in Section 18.4.3, there is also the possibility that, after identifying barriers to recovery, managers may continue to be uncertain if ecosystem changes are reversible. What are possible management options in this case? One choice is to mitigate ongoing degradation, while simultaneously developing the ecological and social capacity that will allow for desired management action. Such a tactic is common for species extinct in the wild but existing in captivity/seed banks. In addition to the educational, ethical and research benefits of these facilities, many of these also devote resources to continual propagation and genetic management of the captive population for the time when the extinction driver is eliminated and reintroduction is possible.

Another example of this strategy would be for managers to control weeds through targeted spraying while gathering information both on drivers of weed spread and on methods that can be used to overcome ecological and social barriers to ecosystem recovery. A drawback of this type of ongoing stop-gap action is that it can become the final project objective, rather than a temporary goal pursued while methods are found to stop and reverse ongoing degradation. If this occurs, such management will require continued resource input with a slim hope of ecosystem recovery. Importantly, this tactic could prevent managers from identifying an ecosystem as novel, thereby preventing the implementation of an alternative management strategy that could be more likely to achieve management goals.

Another management choice in cases where managers are uncertain about the reversibility of ecosystem change is to preserve the foundational building blocks of ecosystems while simultaneously developing the ecological and social capacity that will allow for required management action. This may involve identifying key ecosystem characteristics – abiotic, biotic or dynamic – that influence ecosystem structure and function, focusing for example on actions such as preservation of soil, vegetation cover, keystone and engineer species, genetic diversity, habitat connectivity and maintenance of disturbance regimes. Although a focus on such ecosystem foundations does not guarantee that recovery goals will be met at some point in the future, conserving integral characteristics of ecosystems may offer a greater potential for future restoration than simple mitigation of ongoing degradation by ensuring that some of the self-sustaining aspects of the ecosystem remain intact.

Lastly, acknowledging uncertainty allows managers to move forward with management action. Managers might choose to manage the ecosystem as novel, but continue to collect information about barriers to ecosystem recovery. When developing management strategies based on chosen goals, managers might opt for conservative management based on actions that can be easily reversed rather than those that may be harder to control. For example, a manager deciding whether or not to plant a non-native species to maintain an existing ecosystem function may decide that the consequences for neighboring ecosystems are too great to warrant its introduction.

18.5.2 Novel management choices

Managers also have a number of management choices when ecosystems are recognized as being novel (Fig. 18.1 & Fig. 18.3). While the restoration to a historical state in these cases is very unlikely, an ecosystem identified as *novel* is not devoid of conservation value. It is possible that a subset of native species and historical functions may be present, and some traditional goals of ecosystem management can still be pursued.

As described in Chapter 3, the management of novel ecosystems may require the development of new techniques and methods to address unique and multiple drivers of change. For example, such approaches might target new species interactions to mitigate the effects of irreversible change drivers such as by using grazing to reduce impacts of non-native plant invasion. While such methods do not eliminate new drivers of change, they may enable managers to retain ecosystem values such as target native species or traditional functions. In the case of using cattle to reduce weed cover, this technique may not reverse non-native species invasion but may allow for the persistence of a diverse native flora by reducing competition with the invading species.

While exact actions may vary depending on management goals, managers might consider the role of adaptive management for determining how best to manage new patterns and processes in a novel ecosystem. Experiments monitored over time provide data on the impacts of chosen manipulations, which can be used to understand if actions result in achievement of management goals. When ecosystems fail to recover, these monitored ecosystems provide data for future management trials. Further, monitoring data can be used to examine species responses to management actions and to background changes leading to novel processes. Identifying which species have the highest resilience to such changes, or which species can adapt to perform an important function, can be paramount in these situations. For example, demonstrating that an endangered species is benefitting from a new species in a system (e.g. southwestern willow flycatcher in tamarisk in the Grand Canyon; see Section 18.5.2.1 for more details on this example) can provide options for species recovery and avoid management mistakes. Chapter 22 provides an example of how adaptive management helped determine a management strategy for the identified novel *Miconia-Cinchona* ecosystem on the Galapagos Islands. Importantly, it also illustrates how novel ecosystem management can result in native species conservation, despite novel conditions.

In all situations, it will likely be critical to identify goals that guide novel ecosystem management. The identification of such goals may highlight particular actions that are more effective than others at managing desired ecosystem attributes and characteristics in these novel ecosystems. For example, an at-risk species might be the focus of management actions at times, while a locally valued function may set the agenda at other times.

18.5.2.1 Goal: Conserve target species or biodiversity

The first goal listed in Figure 18.3 is to conserve native species or biodiversity. A key strategy for conserving a target native species is to remove immediate threats, ensure basic requirements and aim for a novel ecosystem that provides the species with functionally similar habitat to the historical ecosystem. Such habitat would provide the resources needed for the species' survival, but in this case species or ecosystem features not found in the target species' historical or traditional habitat may provide these resources.

The potential role of non-native species in providing resources for rare native species is likely to be particu-larly important in situations when restoration of the native species that formerly provided shelter or an energy source is impractical due to limited economic resources or changes in the physical environment (Schlaepfer et al. 2011). For example, the riparian habitat frequented by southwestern willow flycatchers (*Empidonax traillii extimus*; listed as endangered under the US Endangered Species Act) in the US southwest historically consisted of a mix of native willows (*Salix* spp.), cottonwoods (*Populus* spp.) and other native trees. The non-native invasive tamarisk (*Tamarix* spp.) has replaced much of the native vegetation along riverbanks as a result of human activity and changes in riparian hydrology. Initial reports suggested tamarisk were causing a drop in water table levels and reducing habitat quantity and quality for this native riparian bird species. Nevertheless, results of recent field studies reveal that in some areas up to 75% of flycatchers nest in tamarisk and that fledgling success associated with nests built in tamarisk was indistinguishable from success associated with nests built in native trees (Sogge et al. 2008).

Such strategies may inevitably lead to novel species combinations in these ecosystems, raising questions of the value of allowing non-native species to become naturalized. In the flycatcher example, it may be difficult in many areas to reestablish native woody species that formerly supported the flycatcher because of the extensive modifications to flooding regimes that have occurred. Although removing tamarisk may be a step toward restoring historical vegetation in these regions, doing so may unexpectedly cause direct harm to an endangered native species that now depends, in part, on tamarisk (Schlaepfer et al. 2011). As with many decisions about controlling non-native species, managers familiar with the ecosystem they are trying to manage can couple their knowledge of ecosystem dynamics with knowledge of the demonstrated and potential impacts (both negative and positive) of the non-native on species and ecosystem functions to determine a best plan of action to move forward.

18.5.2.2 Goal: Restore/maintain function or service

The second goal in Figure 18.3 focuses on restoring or maintaining ecosystem functions or services, where services are defined as ecosystem processes that benefit human welfare (Daily 1997). Ecosystem functions can be influenced by both abiotic and biotic components of an ecosystem. For example, resistance of an ecosystem to invasive plants may be influenced by the availability of soil nutrients as well as competition by extant species at a location (Theoharides and Dukes 2007; also see Chapter 3). In novel ecosystems one or both of the abiotic/biotic components may have been altered from a historical state; to restore/maintain a function, management efforts may therefore need to address one or both of these ecosystem components.

Over the past decade, substantial effort has been spent on understanding the role of biota on ecosystem functioning (Loreau et al. 2002; Hooper et al. 2005; Thompson and Starzomski 2007; Naeem et al. 2009). As a result, some general concepts mentioned in Chapter 3 may be key to understanding how altered species composition in novel ecosystems might contribute to target functions. These include: functional groups classified as a set of species with either similar responses to or effects on ecosystem processes (Gitay and Noble 1997); redundancy, the idea that species are at least partially substitutable (Naeem et al. 2002); and biodiversity's contribution to ecosystem multifunctionality: the idea that increased numbers of species or functional groups in an ecosystem result in higher levels of multiple ecosystem functions (Zavaleta et al. 2010). As discussed in Chapter 3, some ecosystem functions and services may only require that a particular functional group be present. In this case, managers may be able to use species in the same functional group as replacements (Parker et al. 2010). It is also possible that supplemental species might be chosen to be more robust to ongoing novel conditions that are not favoring native species (see a more detailed discussion of this topic in Chapter 14). The introduction of non-native tortoises to Mauritius' Round Island provides an example where functionally similar non-natives (*Aldabrachelys gigantean* and *Astrochelys radiata*) fulfilled the role of the extinct native giant tortoise (*Cylindraspis* sp.) (Griffiths et al. 2010). By restoring tortoises to the island, managers expect populations of native palms and other trees whose seeds were spread by tortoise herbivory to increase.

Importantly, although species can be grouped into functional groups based on similar characteristics, not all traits in such similarly grouped species are redundant (Chapter 3; Eviner 2004). Because of this, loss of species from an ecosystem (even if the system is supplemented with other species) may result in an ecosystem that only partially operates like a historical system or a contemporary reference ecosystem. Returning to the case of non-native tortoise introduction on Round Island, while such ecological substitutes have the potential to help managers reach ecosystem recovery goals, careful examination of species interactions will be a part of all such programs to avoid unexpected and negative surprises from the use of non-native species to maintain functioning. Treating these sorts of strategies as adaptive management experiments can enhance the success of the project.

Another goal when managing for function in novel ecosystems might be to ensure ecosystems are able to adapt to ongoing ecosystem changes. Because even species in the same functional group are not truly redundant, managers might be worried that truncated subsets of native species may not provide the same range of ecosystem functions over time (Walker et al. 1999; Winfree and Kremen 2009). In this case, a management tactic may be to preserve high diversity in the ecosystem. There is increasing evidence that such high diversity supports higher levels of multiple ecosystem functions (Zavaleta et al. 2010; Isbell et al. 2011), as well as a more stable function over time (Tilman and Downing 1996). While conserving biodiverse ecosystems is often a common goal for many managers, there may also be a benefit in, at times, focusing conservation efforts on single important species. In particular, some species have been shown to have increased importance (e.g. contribute more to biomass, or have stronger link strength within a food web e.g. Berlow 1999; Hooper et al. 2005) when conditions change. Management efforts that target such important species may help ensure ecosystems can adapt as conditions in novel ecosystems continue to change.

18.5.2.3 Goal: Managing for new species composition or functions

The third goal in Figure 18.3 highlights the management of the new species combinations or functions in novel ecosystems as a management goal. As with managing for native species and traditional functions, managers might need to develop new management methods, possibly through the use of adaptive management. Decisions to manage for new species combinations or functions can at times be at the expense of remaining native species and traditional functions. It is beyond the scope of this chapter to discuss when such options may be preferred or how to make this decision. However, the spreading footprint of people on ecosystems, species and global processes indicates that this will be an issue that is increasingly encountered in ecosystem management. Some have termed ecosystems exhibiting thoroughly novel characteristics as 'designer ecosystems', a term that conveys an uneasy sentiment about human agency (Pimm 1996; MacMahon and Holl 2001). Untethering management goals from traditional conservation and restoration values opens up the prospect of ecosystems managed for purely human interests (see Chapter 37). Assessing the risks of actions that begin to move away from traditional ecosystem management and restoration constraints will be essential. We encourage a cautious approach, so that the development of new approaches for the management of novel ecosystems continues to manifest important features of biodiversity conservation. For an example of the creation of designer ecosystems, see Chapter 23.

18.6 IS THE COST/RISK ACCEPTABLE?

A final decision point when managing novel ecosystems (Fig. 18.1) concerns the cost and/or risk of favored actions. As discussed in the section on social barriers, the success of ecosystem management often relies on adequate funding. One possibility, as managers formulate management strategies based on initial goals, is that solutions are too expensive or possibly the adaptive management needed to figure out optimal solutions is too expensive. There are a variety of commonly used planning tools that can help managers determine if their chosen actions are fiscally feasible, including cost–benefit analysis and many optimization tools (Fischer et al. 2009; Wilson et al. 2009). Optimization tools are commonly used to help managers create reserves that contain the best combination of ecosystem attributes to conserve target species and habitats (systematic conservation planning). More recently, managers are also using optimization tools to understand how the costs of conservation actions can affect the success of conservation planning (Carwardine et al. 2010; Wilson et al. 2011). These tools may be valuable for deciding on priority actions for novel ecosystem management.

A manager's choice of novel ecosystem management actions may also depend on the risks associated with those actions. Risk may be in the form of uncertainty that chosen management actions will have desired outcomes or, alternatively, risk may be in the form of unknown surprises resulting from management actions in novel ecosystems that have few or no analogous ecosystems to refer to for guidance. See Chapter 22 for an example of ecological surprises from novel ecosystem management.

There are a number of tools to help address the risk associated with novel ecosystem management. As discussed earlier in the chapter (Section 18.5.2), adaptive management through careful tracking of different management strategies provides clues to how novel ecosystems might behave in different management regimes. Scenario planning, used to identify a number of possible outcomes, is a second tool that can allow managers to plan for possible surprises from management action (Peterson et al. 2003). For example, urban planners often use scenarios to present their creative yet realistic visions of future cities (e.g. Weller 2009). In the case of novel ecosystems, the process of designing alternative management scenarios may similarly highlight creative options not likely to be encountered in other management approaches. This process may be particularly helpful in engaging stakeholders and building consensus around novel ecosystem management strategies. A final tool used to help define risks associated with management is structured decision making. Here decision-makers use formats such as decision trees to use available information and make optimal choices in the face of uncertainty (Polasky et al. 2011; Gregory et al. 2012). It is possible to make decision trees more transparent and the decisions more deliberate by assigning levels of uncertainty associated with each decision. This method often involves estimating the probability of desired and undesired outcomes at each fork of the tree and, against these estimates, a measure of the likely consequences of each outcome can be determined (Hammond et al. 1999).

When either the cost of management action or risk stemming from novel ecosystem management actions is deemed unacceptably large, managers may have the following options. First, they can change the scale or scope of intervention. Alternatively, ongoing research in the form of adaptive management may highlight additional options for reaching goals as seen in Chapter 22 with the management of invasive species in the Galapagos Islands. Finally, a last choice may be to alter management goals. Where the novel ecosystem management goal was originally maintenance or recovery of a target species or valued ecosystem function, reprioritization may result in shifting management focus to a different species or function of interest. Alternatively, managers might decide to manage for a novel composition of species or functions.

18.7 CONCLUSION: WHY CONSIDER NOVEL ECOSYSTEMS IN MANAGEMENT DECISIONS?

Ecosystems are undergoing intensifying change as a result of anthropogenic drivers at all scales – local, regional and global. Some of these impacts are difficult to reverse, and to try to do so would likely lead to failure of restoration programs. In these cases, where there is evidence that we have moved out of reach of the past, novel ecosystem management offers possibilities for thoughtfully choosing alternative management goals and priorities.

In this chapter we have provided some guidelines for managers faced with the task of successfully managing hybrid and novel ecosystems. Our goal was not to provide a single path for managers to follow because, in reality, there are multiple paths leading to a number of different goals. Rather, our aim was to outline some of the difficult questions and potential pitfalls associated with the management of altered and changing ecosystems and offer guidelines for how managers might proceed. We see this chapter as the beginning of a dialogue focused on hybrid and novel ecosystem management. While we have positioned the manager at the helm, in charge of making critical management decisions, we acknowledge that the most successful road to management of these ecosystems will require the combined efforts of managers, ecologists, policy makers, social scientists and stakeholders.

REFERENCES

Aslan, C., Hufford, M.B., Epanchin-Niell, R., Port, J.D., Sexton, P.J. and Waring, T.M. (2009) Practical challenges in private stewardship of rangeland ecosystems: Yellow starthistle control in Sierra Nevadan foothills. *Rangeland Ecology and Management*, **62**, 28–37.

Bangs, E.E. and Fritts, S.H. (1997) Reintroducing the gray wolf to Central Idaho and Yellowstone National Park. *Wildlife Society Bulletin*, **24**, 402–413.

Berkes, F., Folke, C. and Colding, J. (1998) *Linking Social and Ecological Systems: Management Practices and Social Mechanisms for Building Resilience*. Cambridge University Press.

Berlow, E.L. (1999) Strong effects of weak interactions in ecological communities. *Nature*, **398**, 330–334.

Carwardine, J., Wilson, K.A., Hajkowicz, S.A., Smith, R.J., Klein, C.J., Watts, M. and Possingham, H.P. (2010) Conservation planning when costs are uncertain. *Conservation Biology*, **24**, 1529–1537.

D'Antonio, C.M. and Vitousek, P.M. (1992) Biological invasions by exotic grasses, the grass/fire cycle, and global change. *Annual Review of Ecology, Evolution and Systematics*, **23**, 63–87.

Daily, G.C. (ed.) (1997) *Nature's Services: Societal Dependence on Natural Ecosystems*. Island Press, Washington, DC.

Eviner, V.T. (2004) Plant traits that influence ecosystem processes vary independently among species. *Ecology*, **85**, 2215–2229.

Fischer, J., Peterson, G.D., Gardner, T.A., Gordon, L.J., Fazey, I., Elmqvist, T., Felton, A., Folke, C. and Dovers, C. (2009) Integrating resilience thinking and optimisation for conservation. *Trends in Ecology and Evolution*, **24**, 549–554.

Gardener, M.R., Atkinson, R. and Renteria, J.L. (2010) Eradications and people: lessons from the plant eradication program in Galapagos. *Restoration Ecology*, **18**, 20–29.

Gitay, H. and Noble, I.R. (1997) What are functional types and how should we seek them? in *Plant Functional Types: Their Relevance to Ecosystem Properties and Global Change* (eds T.M. Smith, H.H. Shugart and F.I. Woodward), Cambridge University Press, Cambridge, 3–19.

Gregory, R.L., Failing, L., Harstone, M., Long, G., McDaniels, T. and Ohlson, D. (2012) *Structured Decision Making: A Practical Guide to Environmental Management Choices*. Wiley-Blackwell, New York.

Griffiths, C.J., Jones, C.G., Hansen, D.M., Putto, M., Tatayah, R.V., Müller, C.B. and Harris, S. (2010) The use of extant non-indigenous tortoises as a restoration tool to replace extinct ecosystem engineers. *Restoration Ecology*, **18**, 1–7.

Grumbine, R.E. (1997) Reflections on 'What is ecosystem management?'. *Conservation Biology*, **11**, 41–47.

Hammond, J.S., Keeney, R.L. and Raiffa, H. (1999) *Smart Choices: A Practical Guide to Making Better Decisions*. Broadway Books, New York.

Holl, K.D. and Aide, T.M. (2011) When and where to actively restore ecosystems? *Forest Ecology and Management*, **261**, 1558–1563.

Hooper, D., Chapin, F.S. III, Hector, A. et al. (2005) Effects of biodiversity on ecosystem functioning: a consensus of current knowledge. *Ecological Monographs*, **75**, 3–35.

Isbell, F., Calcagno, V., Hector, A. et al. (2011) High plant diversity is needed to maintain ecosystem services. *Nature*, **477**, 199–202.

Kanowski, J.J., Reis, T.M., Catterall, C.P. and Piper, S.D. (2006) Factors affecting the use of restored sites by reptiles in cleared rainforest landscapes in tropical and subtropical Australia. *Restoration Ecology*, **14**, 67–76.

Keough, H.L. and Blahna, D.J. (2006) Achieving integrative, collaborative ecosystem management. *Conservation Biology*, **20**, 1373–1382.

Levy, S. (2009) The dingo dilemma. *BioScience*, **59**, 465–469.

Loreau, M., Naeem, S. and Inchausti, P. (eds) (2002) *Biodiversity and Ecosystem Functioning: Synthesis and Perspectives*. Oxford University Press, New York.

Mack, R.N. (1981) Invasion of *Bromus tectorum* L. into western North America: An ecological chronicle. *Agro-Ecosystems*, **7**, 145–165.

MacMahon, J.A. and Holl, K.D. (2001) Ecological restoration: A key to conservation biology's future? in *Conservation Biology: Research Priorities for the Next Decade* (eds M.E. Soulé and G.H. Orians), Island Press, Washington, DC, 245–269.

Majer, J.D. (2009) Animals in the restoration process-progressing the trends. *Restoration Ecology*, **17**, 315–319.

Menakis, J.P., Osborne, D., Miller, M., Omi, P.N. and Joyce, L.A. (2002) Mapping the cheatgrass-caused departure from historical natural fire regimes in the Great Basin. USDA Forest Service Rocky Mountain Research Station, Fort Collins.

Naeem, S., Loreau, M. and Inchausti, P. (2002) Biodiversity and ecosystem functioning: the emergence of a synthetic ecological framework in *Biodiversity and Ecosystem Functioning: Synthesis and Perspectives* (eds M. Loreau, S. Naeem and P. Inchausti), Oxford University Press, New York, 3–11.

Naeem, S., Bunker, D.E., Hector, A., Loreau, M. and Perrings, C. (eds) (2009) *Biodiversity, Ecosystem Functioning, and Human Wellbeing: An Ecological and Economic Perspective*. Oxford University Press, New York.

Palmer, M.A., Ambrose, R.F. and Poff, N.L. (1997) Ecological theory and community restoration ecology. *Restoration Ecology*, **5**, 291–300.

Parker, K.A., Seabrook-Davison, M. and Ewen, J.G. (2010) Opportunities for nonnative ecological replacements in ecosystem restoration. *Restoration Ecology*, **18**, 269–273.

Peterson, G.D., Cumming, G.S. and Carpenter, S.R. (2003) Scenario planning: a tool for conservation in an uncertain world. *Conservation Biology*, **17**, 358–366.

Pimm, S.L. (1996) Designer ecosystems. *Nature*, **379**, 217–218.

Polasky, S., Carpenter, S.R., Folke, C. and Keeler, B. (2011) Decision-making under great uncertainty: environmental managment in an era of global change. *Trends in Ecology and Evolution*, **26**, 398–404.

Prach, K. and Hobbs, R.J. (2008) Spontaneous succession versus technical reclamation in the restoration of disturbed sites. *Restoration Ecology*, **16**, 363–366.

Sawtschuk, J., Bioret, F. and Gallet, S. (2010) Spontaneous succession as a restoration tool for maritime cliff-top vegetation in Brittany, France. *Restoration Ecology*, **18**, 273–283.

Scheffer, M. (2009) *Critical Transitions in Nature and Society*. Princeton University Press.

Schlaepfer, M.A., Sax, D.F. and Olden, J.D. (2011) The potential conservation value of non-native species. *Conservation Biology*, **25**, 428–437.

Smith, D.W. and Ferguson, G. (2006) *Decade of the Wolf. Returning the Wild to Yellowstone*. The Lyons Press, Guilford, CT USA.

Sogge, M.K., Sferra, S.J. and Paxton, E.H. (2008) Tamarix as habitat for birds: implications for riparian restoration in the southwestern United States. *Restoration Ecology*, **16**, 146–154.

Standish, R.J., Cramer, V.A., Wild, S.L. and Hobbs, R.J. (2007) Seed dispersal and recruitment limitations are barriers to native recolonisation of old-fields in Western Australia. *Journal of Applied Ecology*, **44**, 435–445.

Suding, K.N. and Hobbs, R.J. (2009) Threshold models in restoration and conservation: a developing framework. *Trends in Ecology and Evolution*, **24**, 271–279.

Suding, K.N., Gross, K.L. and Houseman, G.R. (2004) Alternative states and positive feedbacks in restoration ecology. *Trends in Ecology and Evolution*, **19**, 46–53.

Swetnam, T.W., Allen, C.D. and Betancourt, J.L. (1999) Applied historical ecology: Using the past to manage for the future. *Ecological Applications*, **9**, 1189–1206.

The Nature Conservancy (TNC) (2007) *Conservation Action Planning Handbook: Developing Strategies, Taking Action and Measuring Success at Any Scale*. The Nature Conservancy, Arlington, VA. http://conserveonline.org/workspaces/cbdgateway/cap/practices. Accessed August 2012.

Theoharides, K.A. and Dukes, J.S. (2007) Plant invasion across space and time: factors affecting nonindigenous species success during four stages of invasion. *New Phytologist*, **176**, 256–273.

Thompson, R. and Starzomski, B. (2007) What does biodiversity actually do? A review for managers and policy makers. *Biodiversity and Conservation*, **16**, 1359–1378.

Tilman, D. and Downing, J.A. (1996) Biodiversity and stability in grasslands. *Nature*, **367**, 363–364.

US Fish & Wildlife Service (2002a) FWM#: 400, Series: Habitat Management Plans, Part 620: Habitat Management Practices. Division of Conservation Planning and Policy. http://www.fws.gov/policy/620fw1.html. Accessed August 2012.

US Fish & Wildlife Service (2002b) Exhibit 1, FWM#: 400, Series: Habitat Management Plans, Part 620: Habitat Management Practices. Division of Conservation Planning and Policy. http://www.fws.gov/policy/e1620fw1.html. Accessed August 2012.

Walker, B., Kinzig, A. and Langridge, J. (1999) Plant attribute diversity, resilience, and ecosystem function: the nature and significance of dominant and minor species. *Ecosystems*, **2**, 95–113.

Wasser, S.K., Keim, J.L., Taper, M.L. and Lele, S.R. (2011) The influences of wolf predation, habitat loss, and human activity on caribou and moose in the Alberta oil sands. *Frontiers in Ecology and Evolution*, **9**, 546–551.

Weller, R. (2009) *Boomtown 2050: Scenarios for a Rapidly Growing City Perth, Australia*. University of Western Australia Press, Perth.

Wiens, J.A. (2009) Landscape ecology as a foundation for sustainable conservation. *Landscape Ecology*, **24**, 1053–1065.

Wilson, K.A., Carwardine, J. and Possingham, H.P. (2009) Setting conservation priorities. *The Year in Ecology & Conservation Biology*, **1162**, 237–264.

Wilson, K.A., Lulow, M., Burger, J., Fang, Y.-C., Andersen, C., Olson, D., O'Connell, M. and McBride, M.F. (2011) Optimal restoration: accounting for space, time and uncertainty. *Journal of Applied Ecology*, **48**, 715–725.

Winfree, R. and Kremen, C. (2009) Are ecosystem services stabilized by differences among species? A test using crop pollination. *Proceedings of the Royal Society of London (series B)*, **276**, 229–237.

Zavaleta, E.S., Pasari, J.R., Hulvey, K.B. and Tilman, G.D. (2010) Sustaining multiple ecosystem functions in grassland communities requires higher biodiversity. *Proceedings of the National Academy of Sciences*, **107**, 1443–1446.

Chapter 19

THE MANAGEMENT FRAMEWORK IN PRACTICE – MAKING DECISIONS IN ATLANTIC CANADIAN MEADOWS: CHASING THE ELUSIVE REFERENCE STATE

Stephen D. Murphy

Department of Environment and Resource Studies, University of Waterloo, Canada

19.1 INTRODUCTION

This chapter is a reflection on how the decision framework presented in Chapter 18 can be applied to choices made in interventions and potential restoration of meadow ecosystems in eastern Canada. This case study highlights the difficulty in determining an appropriate reference ecosystem, often due to incomplete historical information, to help identify upfront if there is a problem and inform the subsequent management approach. This example reveals that what was thought to be a correct – almost obvious – approach to restoration is actually much muddier than expected. The decision framework forces a reevaluation of whether the historical reference used was really all that appropriate and, when combined with some compromises made on species selection, results in what is likely a hybrid or novel ecosystem (rather than a historical ecosystem).

19.2 BACKGROUND

As early as in the 1950s, farmers began restoring or rehabilitating agricultural fields to a coastal meadow ecosystem starting with the area around the Bay of Fundy in New Brunswick and Nova Scotia in eastern

Novel Ecosystems: Intervening in the New Ecological World Order, First Edition. Edited by Richard J. Hobbs, Eric S. Higgs, and Carol M. Hall.
© 2013 John Wiley & Sons, Ltd. Published 2013 by John Wiley & Sons, Ltd.

by local (city and county) agencies for conservation purposes often suffer from several constraints: (1) they often involve parcels nested up against urban or suburban areas; (2) they are often fairly limited in area; and (3) due to 'not in my back yard' (NIMBY) attitudes, social issues often constrain the use of apex predators, fire and grazing as management tools. As such, environmental change drivers and land-use legacies have precluded the maintenance of historical conditions. However, by adopting a management mosaic approach, all of the valued aspects of the historical community can be preserved and the historical and non-historical communities combined to maximize native biological diversity.

20.2 MANAGEMENT APPROACH

The city of Boulder, Colorado USA now owns over 18,000 hectares of land that has as its primary purpose the ". . . preservation or restoration of natural areas characterized by or including terrain, geologic formations, flora, or fauna that is unusual, spectacular, historically important, scientifically valuable, or unique, or that represents outstanding or rare examples of native species" (see http://www.bouldercolorado.gov/). Purchases of private lands occurred in the last century and have continued to date, although most recent additions are small due to development within the area. A local land management program evolved along with these purchases and, in addition to conservation and agricultural uses, the public areas now receive over four million visitor days per year. A large fraction of the area has been managed to maintain historical conditions. Conflicts with other uses (e.g. passive recreation and agriculture) and issues related to fragmentation resulted in the production of a 406-page grassland management plan, which was approved in 2010 after many years of work (http://www.bouldercolorado.gov/files/openspace/pdf_grassland_plan/Final_Grassland_Plan_Complete1.pdf).

Preservation of native flora and fauna (and reduction in urban and suburban sprawl) has been the driving force of management. The grassland management plan indentified conservation 'targets' including three types of grassland (mixed-grass prairie mosaic, xeric tallgrass prairie and mesic tallgrass prairie) and a fourth target as 'black-tailed prairie dogs and associates'. Current data indicate that prairie dogs and tallgrass communities cannot coexist, however. Because

the prairie dog is a keystone species of the mixed-grass prairie, a reasonable question to ask is: "Isn't its inclusion within the mixed-grass prairie mosaic necessary to maintain the historical baseline condition?" Vegetation monitoring shows that prairie dogs in the fragmented mixed grassland landscapes nested against the Colorado Front Range transform grasslands into new landscapes, often dominated by non-native plant species. The conservation of one valued native component (the prairie dog) therefore threatened other components. Managers and policy makers realized by 2010 that conservation goals for areas with prairie dogs had to be different than for uncolonized grasslands. A decision to divide the grassland management plan into two targets (mixed grasslands with or without prairie dogs) potentially resolved some conflicts before they could arise.

Of note is that the national/regional concern about sufficient prairie dog numbers has driven local management decisions. At the regional scale, prairie dogs have been considered for listing as endangered species as areas available to prairie dog colonies had been reduced to perhaps 1–2% of their historical range (Kotliar et al. 1999). Accordingly, a subset of local conservationists requested that prairie dogs be given priority status on public lands. Had prairie dog densities and ranges remained more abundant, it could be speculated that the local management emphasis would not favor a substantial area set aside for prairie dog colonies. As long as colonies do not persist beyond decadal scales, such systems are likely reasonably constrained within a successional vegetation sequence that retains native species. However, the persistence of this system remains uncertain and largely contingent upon the periodic die-back of prairie dogs due to the non-native pathogen that causes sylvatic plague (Chapter 15).

In the large area of grasslands where prairie dogs are not to be allowed, the goal is to attempt to maintain historical conditions. The specific focus will be on maintaining native vegetation and a select group of 'indicator species' that argue areas are largely natural. However, these areas are not self-sustaining due to invasive species. As such, the management plan has included 'novel management actions' to maintain desirable species. For example, Seastedt et al. (2008) noted that high-intensity spring grazing by cattle has been used to maintain a relict mesic tallgrass site. Proactive reseeding and planting of native species in grasslands and wetlands have been conducted. The plan also includes an ambitious invasive species control

Box 20.1 Drivers of change, scale and irreversible ecosystem changes

a. Scale of driver vs. scale of management action

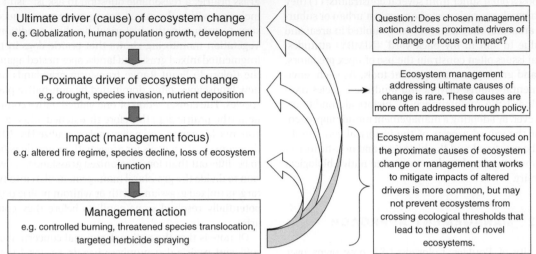

Ultimate driver (cause) of ecosystem change
e.g. Globalization, human population growth, development

Proximate driver of ecosystem change
e.g. drought, species invasion, nutrient deposition

Impact (management focus)
e.g. altered fire regime, species decline, loss of ecosystem function

Management action
e.g. controlled burning, threatened species translocation, targeted herbicide spraying

Question: Does chosen management action address proximate drivers of change or focus on impact?

Ecosystem management addressing ultimate causes of change is rare. These causes are more often addressed through policy.

Ecosystem management focused on the proximate causes of ecosystem change or management that works to mitigate impacts of altered drivers is more common, but may not prevent ecosystems from crossing ecological thresholds that lead to the advent of novel ecosystems.

Novel ecosystems result from changes in biotic and abiotic conditions. The **ultimate cause** of change often stems from factors beyond the scale and scope of individual management action such as increasing human population, increased pollution and expanding development. **Proximate drivers** of change are rooted in these larger-scale ultimate drivers but act at smaller, more localized, scales making them easier targets for management action. Managers will often work not to mitigate drivers of change but rather the **impacts** of such drivers to ecosystems. While such targeted management might mitigate ecosystem changes, without addressing the underlying cause of change managers may need to provide ongoing management to maintain desired species combinations and functions. **Novel ecosystems** may be particularly common when drivers of ecosystem change operate at scales larger than those which can be directly influenced by managers.

b. Example: Prairie Dogs, climate change and management decisions

Ultimate driver (cause) of ecosystem change
Human population growth, development

Proximate driver of ecosystem change
Climate change, altered fire return intervals, arrival of invasive species, C_2O fertilization, increased N deposition, fragmentation, extirpation of specialist predators

Impact (management focus)
New temporal niches in herbaceous communities, species invasion, re-emergence of foothill savannas, 'sesertification' and increase in shrub dominance

Management action
Designate areas acceptable for prairie dog colonies; remove animals from unacceptable areas; weed control in areas without prairie dogs

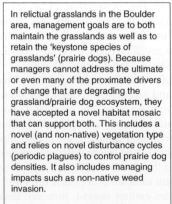

In relictual grasslands in the Boulder area, management goals are to both maintain the grasslands as well as to retain the 'keystone species of grasslands' (prairie dogs). Because managers cannot address the ultimate or even many of the proximate drivers of change that are degrading the grassland/prairie dog ecosystem, they have accepted a novel habitat mosaic that can support both. This includes a novel (and non-native) vegetation type and relies on novel disturbance cycles (periodic plagues) to control prairie dog densities. It also includes managing impacts such as non-native weed invasion.

effort, a series of metrics to assess the status of the landscapes and monitoring is an active part of the project. To date, it could be argued that the vegetation structure and (except for some invasive plants) composition remain within historical ranges of variability.

20.3 SUMMARY

The decision framework (Figure 18.1) asks us to consider the type of ecosystem we are managing. Should the Colorado Front Range grasslands be managed as a possible novel ecosystem? The answer is likely yes, due to multiple ecosystem change drivers occurring at scales beyond the scope of management. As shown in Box 20.1, the management focus is on impacts with ongoing monitoring to assess how fast the ultimate and proximate drivers of change are influencing the status of the lands. The management report acknowledged that climate change drivers and atmospheric chemistry changes were 'unknowns' in terms of management activities. Fire return intervals were appropriately identified and are included as management goals but, with many homes nested within high human occupancy areas, the likelihood of a majority of sites receiving prescribed burns within their historical range of variability is low. The described plan contains the correct science, but the political will to implement that science is uncertain.

Further, no known extirpations of local flora are known, but the removal of apex consumers in the 1800s (wolves, grizzly bears and the prairie-dog specialist, the black-footed ferret) in conjunction with landscape fragmentation has created a trophic food web structure for these areas that likely has no historical precedent. Protection of riparian zones, widespread irrigation ditches and alterations of stream flows have arguably expanded mesic and wetland habitats in public lands. Cattle exclusion to a large expanse of riparian habitat, in conjunction with maintaining stream flows that exhibit patterns unlike native streams, has created a highly desirable species-rich area that arguably never existed during the pre-European Holocene era.

20.4 WHAT ARE THE RISKS OF THE CURRENT MANAGEMENT STRATEGY?

The Boulder Open Space plan is only three years old and the management actions espoused within this effort have not been challenged by extreme events (e.g. extended drought or catastrophic wildfires), the insidious effects of fire suppression, atmospheric chemistry changes or the appearance of a 'super-invader' (a wide-ranging invasive species capable of out-competing natives). The monitoring program built within the plan is sufficient to detect changes, and threshold criteria exist to elicit proactive management activities. To date the stakeholders have, with the possible exception of using frequent fire as a management tool, exhibited a willingness to fund monitoring and proactive management measures designed to create or maintain both prairie dog preserves and a combination of native habitats in historical and non-historical configurations that maximize the native biodiversity potential of the area.

REFERENCES

Kotliar, N.B., Baker, B.W., Wicker, A.D. and Plumb, G. (1999) A critical review of the assumptions about the prairie dog as a keystone species. *Environmental Management*, **24**, 177–192.

Seastedt, T.R., Hobbs, R.J. and Suding, K.N. (2008) Management of novel ecosystems: are novel approaches required? *Frontiers in Ecology and the Environment*, **6**, 547–553.

Chapter 21

THE MANAGEMENT FRAMEWORK IN PRACTICE – HOW SOCIAL BARRIERS CONTRIBUTE TO NOVEL ECOSYSTEM MAINTENANCE: MANAGING REINDEER POPULATIONS ON ST GEORGE ISLAND, PRIBILOF ISLANDS, ALASKA

Kristin B. Hulvey

Ecosystem Restoration and Intervention Ecology (ERIE) Research Group, School of Plant Biology, University of Western Australia, Australia

21.1 BACKGROUND

St George Island is the second largest island in the Pribilof archipelago, with a total land area of 27 square miles. The island has mixed ownership, with some areas federally owned as a part of the Alaska Maritime National Wildlife Refuge (AMNWR) and others owned by the Tanaq Native Corporation, a Native Village Corporation. The island has a history of introduced reindeer grazing, dating to a US government introduction in 1911 (Scheffer 1951). The purpose of the introduction was as a backup food source for the local Aleut population; the herd was little utilized by the locals however, who focused rather

Novel Ecosystems: Intervening in the New Ecological World Order, First Edition. Edited by Richard J. Hobbs, Eric S. Higgs, and Carol M. Hall.
© 2013 John Wiley & Sons, Ltd. Published 2013 by John Wiley & Sons, Ltd.

on fish and seal. As a result the herd grew to around 222 animals before being locally extirpated in 1950 (Scheffer 1951). A second introduction in 1980, this time by the St George residents, resulted in a larger herd that grew to 100 animals by 1991 (Swanson and Barker 1992), c. 430 by 2002 (Sonnen 2004) and around 450–550 animals by 2004 (Hulvey 2004).

The island has little history of reindeer fencing, and the introduced reindeer have been free to move across lands spanning different owners. Five reindeer utilization surveys conducted between 1991 and 2005 show rapid ongoing tundra degradation corresponding to fluctuating herd size. While Swanson and Barker (1992) noted that the lichen ranges (many species) were in excellent condition in 1991, the Natural Resources Conservation Service (NRCS; see Box 21.1) determined in 2002 that the lichen range conditions had deteriorated over part of the island. A graduate student at University of California, Santa Cruz showed that these conditions had degraded further by 2004 (Hulvey 2004).

Continued degradation could have a number of negative ecological and social impacts on the Island. Reindeer may impact nesting habitat and nesting success of the endemic Pribilof Rock Sandpiper (*Calidris ptilocnemis ptilocnemis*), listed as high priority species in Alaska's Comprehensive Wildlife Conservation Plan (Alaska Department of Fish and Game 2006). Continued reductions in the abundance and extent of lichen communities could also be encouraging the spread of nitrogen-fixing lupine that has the potential to alter soil properties. Studies of reindeer-altered lichen communities on St Matthew, another Bering Sea island, indicate that lichens may not recover, even after reindeer are removed for long periods of time (>50 years) (Klein 1987). Finally, members of the St George community including Elders and the Traditional Council have expressed concerns about the negative impacts of the reindeer on the landscape, as well as the possibility that a large herd might face starvation if not properly managed.

21.2 WHAT ARE THE ECOLOGICAL AND SOCIAL BARRIERS INFLUENCING MANAGEMENT?

The creation of a reindeer management plan has been complicated by a number of factors, including a lack of information about the reindeer impact on island ecology and St George's remote location, which makes management logistics and collaborations more difficult. Additionally, the existence of numerous stakeholder groups has contributed to difficulties in coordinating management actions. These stakeholders differ in their ultimate vision of reindeer management because of differing organizational/community/business goals (Box 21.1).

21.3 WHAT MANAGEMENT APPROACH WAS USED?

Currently reindeer management aims to prevent further damage to the St George ecosystem, while allowing the most severely damaged areas to recover. This strategy blends community and other stakeholder interests in conserving the island ecosystem with the community's need to secure income and employment.

Reindeer and ecosystem management evolved as an understanding of reindeer impacts on the island ecosystem increased with focused studies by NRCS and graduate students. Outreach work organized and coordinated by a graduate student, USFWS and NRCS scientists and a local non-profit group resulted in a community-based workshop that included a focus group to discuss reindeer management. Follow-up work included conducting extensive range surveys in collaboration with the Traditional Council's Eco Office, working group discussions including scientists at local universities and ongoing informal networking among all stakeholders.

Using data from surveys, the NRCS determined in 2007 that the tundra could sustainably support c. 100 adult animals without further deterioration. Around the same time, the Tanaq Native Corporation and the Traditional Council entered an agreement to co-manage the herd, thus allowing the Council to have a larger hand in management decisions. The Council sought partnerships to develop a management plan and weighed management options offered by various stakeholders including eliminating all reindeer, increasing the population size or managing the population at a level deemed to be 'sustainable' on the tundra. The group considered the historical ecosystem conditions (i.e. no reindeer) as a reference but also took into consideration the needs of the community, which included the ability of the community to persist on the island when there were minimal economic

Box 21.1 Stakeholders involved in reindeer management decisions

Stakeholder	Connection to reindeer	Initial management goals
Tanaq Native Corporation: Native Corporation that holds the assets of the local Aleut via The Native Claims Settlement Act.	Own the land on the island and most of the island's assets including the reindeer.	Interested in the profits the herd might provide, but also interested in working with other Native groups to protect the island's ecology.
Kayumixtax Traditional Council: Native group that promotes a healthy and financially stable local community.	Invest time, labor and money in island stewardship, including reindeer management.	Interested in ensuring the reindeer do not degrade St George's ecosystem. Interested in securing community profits from the reindeer.
Alaska Maritime National Wildlife Refuge (AMNWR): A government agency that is a branch of the US Fish & Wildlife Service. Portions of St George are part of, and therefore managed by, the AMNWR.	Provide some funding, labor and support for reindeer management. Have agreed to authorize the EQIP fence to cross the Refuge lands as long as reindeer numbers are maintained at those agreed upon in the EQIP contract.	Interested in preserving unique island ecology in its natural diversity including bird life, lichens and native plants. View growing reindeer population as a threat to island ecology due to trampling and overgrazing.
Natural Resources Conservation Service (NRCS): Government agency whose programs aim to help people reduce soil erosion, enhance water supplies, improve water quality, increase wildlife habitat and reduce damage caused by floods and other natural disasters.	Has provided information to AMNWR, Tanaq and the Traditional Council on range conditions via multiple range surveys over the past decade. Has also offered information about how to manage the reindeer population in order to reduce harm to the tundra.	Goal is to provide stakeholders with accurate and relevant information about lichen condition and reindeer population management so that they can make informed decisions about impacts of reindeer on island ecosystems and develop the best reindeer management plan possible.
Private hunting company: Non-local big game hunter company.	Group has provided hired labor for reindeer culls.	Interested in earning money from hunting and therefore in maintaining a larger herd size than NRCS or AMNWR in order to ensure the herd produces the maximum number of trophy animals per year.

opportunities for community members. This made elimination of the reindeer very unfavorable. It also made increasing the herd size to increase income from big game hunting an option; there was a limit to the economic benefit of this operation to the community however, since it was run almost entirely by people not belonging to the community.

A solution involving NRCS's cost-share Environmental Quality Incentives Program (EQIP) provided a workable option. This program combined the community's growing interest in conserving a more historical island ecosystem with its need to support the community. The program required the community to reduce the herd population size to one deemed sustainable, that is, not causing any further permanent change to the lichen ecosystem while letting damaged communities recover. This will be accomplished by building a fence to keep the remaining herd off highly grazed areas. Once the fence is built, the program will provide the community with payments for the 3-year duration of the contract. NRCS is working directly with the Traditional Council to implement the plan. All parties acknowledge that maintenance of any reindeer on the island may result in further changes to the lichen ecosystem, including grazing of the smaller herd on areas of the island that formerly were minimally grazed. There is ongoing monitoring to determine how the reduced herd and fence construction may affect island ecosystems.

21.4   SUMMARY

This example of managing reindeer in the tundra ecosystem on St George Island illustrates the complexity of questions highlighted by the decision framework (Figure 18.1).

21.4.1   What was the management goal?

At the start of this process there was no single management goal for the St George reindeer or tundra ecosystem because of the diverse interests of the multiple stakeholders. After a process that involved developing ecological knowledge, collaboration, information sharing, value identification and sharing, a common workable goal evolved as described.

21.4.2   Will the goal be met without active intervention?

No. The reindeer herd will need to be kept at the level deemed sustainable through ongoing monitoring of the range. Fences will need to be constructed to keep the herd in designated areas for the duration of the management contract with NRCS.

21.4.3   Are ecosystem changes reversible via management?

Perhaps but existing social barriers make management to a historical baseline unlikely. With complete elimination of the reindeer, it is possible that where lichen still dominates mats would increase in depth. It is also possible that some of the area that has crossed from a lichen-dominated ecotype to a vascular-plant-dominated ecotype might recover. Because reindeer prefer some species of lichens over others, the lasting effects of grazing on lichen composition and diversity are uncertain. Additionally, changing climate has been noted by multiple members of the community. A new study on the effects of climate change on lichen growth indicates that, even with reindeer management, the range may not recover to its past state (Klein and Shulski 2011). While it is uncertain how changing climate might alter ecosystem recovery with reindeer removal, restoring to a pre-reindeer state was not an option due to social dynamics including a complex array of stakeholders with different management goals and the economic needs of the local community. These factors suggest the system could be managed as a possible novel ecosystem (Figure 18.2).

21.4.4   What is the management goal (Figure 18.3)?

The current goal is to create and manage an ecosystem in a way that supports community livelihoods and considers historical ecological references. Management therefore seeks to strike a balance where no new degradation is caused by the reindeer, some highly damaged areas are allowed to recover and the community can make some income from reindeer management. To help achieve this balance, local managers collaborated with partners to determine target herd size and lichen range conditions. This included setting up a plan to monitor herd size and range conditions. Based on collected data, decisions will be made about whether and how to adjust herd population size and exclusion zone placement.

21.4.5   Is the cost/risk acceptable?

Yes, but this could change over time. The cost of management, particularly in such a remote area, is large;

this has been a factor that has hindered herd management in the past. Currently costs are covered through a variety of collaborations between the Traditional Council and various partners. For example, the herd size is maintained with help from AMNWR, fencing costs are covered by the NRCS cost-share program and range monitoring is conducted with the help of NRCS specialists. Additionally, some of the local community's management costs (labor) are compensated by NRCS through the cost-share EQIP. This program also compensates for possible lost revenues that could be generated from other activities such as big game hunting, and makes sustainable herd management a plausible economic option for the community.

Risks include the possibility that current solutions may cause deterioration to formerly low-degraded areas because reindeer exclusion from highly degraded areas results in the herd moving into areas that have had less grazing in the past. Continued surveys will hopefully reveal if conditions decline in these areas, allowing managers to adjust management. Additionally, the current management scheme will only provide income for 3 years. The community might not be able to continue current management activities after this without additional funding. This could potentially lead to increases in herd size and degradation from associated rangeland utilization by the reindeer.

REFERENCES

Alaska Department of Fish and Game (2006) Our wealth maintained: A strategy for conserving Alaska's diverse wildlife and fish resources. Alaska Department of Fish and Game, Juneau.

Hulvey, K.B. (2004) St. George Island Reindeer Utilization Assessment Report. Report to the U.S. Fish & Wildlife Service, Homer.

Klein, D.R. (1987) Vegetation recovery patterns following overgrazing by reindeer on St. Matthew Island. *Journal of Range Management*, **40**, 336–338.

Klein, D.R. and Shulski, M. (2011) The role of lichens, reindeer, and climate in ecosystem changes on a Bering Sea island. *Arctic*, **64**, 353–361.

Scheffer, V. (1951) The rise and fall of a reindeer herd. *The Scientific Monthly*, **73**, 356–362.

Sonnen, K. (2004) St. George island Reindeer Utilization Report. USDA-National Resources Conservation Service, Homer.

Swanson, J. and Barker, M. (1992) Assessment of Alaska reindeer populations and range conditions. *Rangifer*, **12**, 33–43.

Chapter 22

THE MANAGEMENT FRAMEWORK IN PRACTICE – CAN'T SEE THE WOOD FOR THE TREES: THE CHANGING MANAGEMENT OF THE NOVEL MICONIA-CINCHONA ECOSYSTEM IN THE HUMID HIGHLANDS OF SANTA CRUZ ISLAND, GALAPAGOS

Mark R. Gardener

Charles Darwin Foundation, Galapagos Islands, Ecuador, and School of Plant Biology, University of Western Australia, Australia

22.1 OVERVIEW

This chapter applies the decision framework in Chapter 18 post hoc to an example in the humid highlands on the island of Santa Cruz in the Galapagos Islands. This example has management goals that are focused at a species level (both threatened native and invasive non-native) and a landscape level. Although the original goal of the project was to return the system to the historical state, the actual goal morphed into manag-

Novel Ecosystems: Intervening in the New Ecological World Order, First Edition. Edited by Richard J. Hobbs, Eric S. Higgs, and Carol M. Hall.
© 2013 John Wiley & Sons, Ltd. Published 2013 by John Wiley & Sons, Ltd.

ing a novel ecosystem. This was done through adaptive management over a period of 20 years. Intervention began without quantitative objectives, effective control methods and an understanding of impacts, thresholds and mechanisms for change. Through a combination of on-the-job learning, scientific knowledge and management evaluation, an effective management mosaic developed that protected biodiversity in priority areas.

22.2   BACKGROUND

The Miconia ecosystem is restricted to the humid highlands of two islands in the Galapagos archipelago and has been impacted by plant and animal invasions, grazing and fire since human colonization in the late 1800s. Its most striking feature is that it is naturally treeless: it was dominated by the endemic shrub *Miconia robinsoniana* and interspersed grass, ferns and bog species. It is habitat for iconic Galapagos animals such as the dark-rumped petrel (*Pterodroma phaeophygi*) and the Galapagos rail (*Laterallus spilonotus*). While exact historical composition of the Miconia ecosystem is unknown, good records exist from surveys in early 1900s (e.g. Svenson 1935) and the pollen record (e.g. Colinvaux and Schofield 1976).

22.3   CHANGES

The proximate driver for change is non-native species invasion; the ecosystem will therefore not recover passively. The invasive tree species red quinine (*Cinchona pubescens*) was introduced to the island of Santa Cruz in the 1940s and began spreading in the 1970s. It now covers more than 11,000ha, reducing indigenous plant species cover and diversity (Jäger et al. 2007, 2009) and is said to inhibit nesting of the dark-rumped petrel and reduce abundance of the Galapagos rail (Shriver et al. 2011). Furthermore, it impacts the unique aesthetics of the previously treeless environment. Black rats (*Rattus rattus*) became established on Santa Cruz during World War Two (Clark 1978). Black rats cause up to 70% reproductive failure in the dark-rumped petrel (Cruz and Cruz 1987a, b). Cattle, horse, donkey and goat grazing has occurred since the early 1900s and intensified in the 1970s with the opening of a good road and pasture development (van der Werff 1983). Several widespread fires between the 1940s and the 1960s also caused change with a reduction of

Miconia robinsoniana and expansion of the fire-tolerant bracken (*Pteridium arachnoides*; Kastdalen 1982).

22.4   MANAGEMENT APPROACH

In light of these changes, management in the National Park area (much of the humid highlands is now dedicated to agriculture) commenced in the late 1970s with the general vision of returning it to a pre-human state. Management focus was to restore endemic Miconia shrublands for two reasons: (1) aesthetic preservation of treeless landscape, in particular, expanding the distribution of *Miconia robinsoniana*; and (2) the protection of dark-rumped petrel. General biodiversity conservation (including rare and threatened species) and ecosystem function have yet to be pragmatically considered as a management objective.

Managers initially deemed ecosystem changes reversible through active management and specifically the eradication of red quinine (MacDonald et al. 1988). Red quinine populations expanded rapidly despite continued efforts, possibly signalling the crossing of an ecological threshold (Fig. 18.2). Social barriers maintain the novel ecosystem state and include the prohibitive cost of the eradication, estimated at $US6 million, (Tye 2006) and red quinine's value as a timber crop (it is harvested both in the agricultural zone and the National Park). Rather than the goal of restoration being to return to a pre-human state, there is now recognition that red quinine is part of the ecosystem. Management control is conducted in priority sites (Buddenhagen and Yanez 2005; Jäger et al. 2007) as part of a novel ecosystem management mosaic.

In terms of the decision framework (Figs 18.1 and 18.3), the goal is now clearly to conserve target species. The methodology to achieve this has been developed through a long-term collaboration between managers and scientists using the adaptive management process. Active management intervention has been: (1) chemical and manual control of red quinine; (2) chemical control of black rats in prime nesting sites of petrel; and (3) hunting of feral animals and exclusion of cattle from National Park areas (which will not be discussed here). Five hundred hectares are currently under management with the objective of killing all adult red quinine with a cost of approximately $US600,000 from 2005 to 2011 (Garcia and Gardener 2012). Control methods have also evolved to become more effective (Buddenhagen et al. 2004) by maximizing tree kill and minimizing off-target damage. The cost of this optimized management ranged from $US14 to $US2225 per ha depending on stem density, treatment type and follow up (Buddenhagen and Yanez 2005). Management has been further optimized by linking the red quinine biology to management (e.g. remove plants before maturation (Rentería 2002).

Unlike red quinine, eradication of black rats has never been considered because of the large size of Santa Cruz Island and the impossibility of preventing reintroduction. Management consists of strategically placed anticoagulant poison baits in special PVC pipes to minimize off-target uptake over several areas totalling about 50 ha, during the petrel breeding season. Once again the permanent presence of rats makes this ecosystem novel (rats prey on and impact a number of vertebrates, invertebrates and plants), but these priority areas have been incorporated into a management mosaic in order to conserve valued biodiversity.

Effective control to below a threshold of impact has been achieved for red quinine and rats over areas of 500 ha and 50 ha, respectively, and now the cost of maintaining these is much reduced (see Fig. 22.1). While control has resulted in increased abundance of

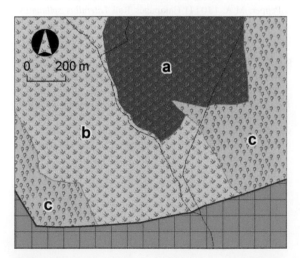

Figure 22.1 Example of a management mosaic in the Miconia–Cinchona ecosystem, Santa Cruz Island, Galapagos National Park: (a) control of rats and red quinine; (b) control of red quinine; and (c) no management with ad hoc harvesting of red quinine. Hashed area is the adjacent agricultural zone. Dotted lines are walking trails. Map by Mandy Trueman.

endemic and iconic species (Jäger and Kowarik 2010; Shriver et al. 2011), there have also been perverse outcomes. Disturbance from control has facilitated new invasions of species such as *Rubus niveus* (Jäger and Kowarik 2010). Hence, increasing resources are being spent on managing *R. niveus* (using different herbicide and application techniques to red quinine) and the system is becoming locked in a feedback cycle where increasing disturbance increases *R. niveus* density and distribution. Ironically, *R. niveus* has a far greater potential to make the system 'more novel' than red quinine by causing up to 60% local extinction of vascular plants (Rentería 2011).

In most cases changes in the Miconia–Cinchona ecosystem have probably not passed an irreversible threshold, but their management in entirety (c. 5000 ha) would require continuous and expensive inputs as reintroduction would occur from adjacent agricultural land. Management therefore cannot address the drivers causing novelty (species invasions) but only focus on mitigation of impacts. While this system has never been managed as a novel ecosystem per se, the size of the management area and limited resources have resulted in a spatial mosaic of different objectives (i.e. intense intervention or do nothing). Another tactic would be to formally recognize the current Miconia–Cinchona ecosystem as novel, and re-adjust management priorities. The same level of resourcing could be used to increase the area of intervention but decrease the intensity (e.g. to 5% cover instead of 0%) such that not all adult trees were removed. Jäger et al. (2009) showed that while a cover of 20% of red quinine results in a loss of abundance of almost all vascular plants, there were no extinctions. Obviously, management costs may rise because of the increased seed bank (from mature plants) and subsequent recruitment. There would therefore be a trade-off between an increasing area of biodiversity protected versus some loss of abundance of native species. There is also a significant social cost; stakeholders would be required to accept the visual impact of trees in a treeless environment. This scenario has yet to be considered by managers but could easily be tested through the adaptive management cycle.

22.5 SUMMARY

The original goal was to return to the pre-human treeless Miconia ecosystem and protect the dark-rumped petrel. This goal could not be met without continuous active intervention. Under current resourcing the ecosystem cannot be restored to historical conditions; in particular, the invasive red quinine tree could not be eradicated and its on-going control is costly. Management has however been adapted such that priority species have been protected in a spatial subset of the entire ecosystem. Adequate knowledge of the reference ecosystem was not a limiting factor in restoration success. The current management strategy satisfies most stakeholders; however, there is potential to optimize protection of more biodiversity over a larger area. Although in the early days of management much of the attention was on specific details, through a process of adaptive management and a change of focus from species to ecosystems we can now clearly see the wood in its entirety instead of just the trees.

REFERENCES

Buddenhagen, C. and Yanez, P. (2005) The cost of quinine *Cinchona pubescens* control on Santa Cruz Island, Galapagos. *Galapagos Research*, **63**, 33–36.

Buddenhagen, C.E., Renteria, J., Gardener, M., Wilkinson, S.R., Soria, M., Yanez, P., Tye, A. and Valle, R. (2004) Control of a highly invasive tree *Cinchona pubescens* in Galapagos. *Weed Technology*, **18**, 1194–1202.

Clark, D.B. (1978) *Population biology of two rodents of the Galapagos Islands.* University of Wisconsin, Madison, USA.

Colinvaux, P.A. and Schofield, E.K. (1976) Historical ecology in the Galapagos Islands: 1. A Holocene pollen record from El Junco Lake, Isla San Cristobal. *Journal of Ecology*, **64**, 989–1012.

Cruz, J.B. and Cruz, F. (1987a) Conservation of the dark-rumped petrel *Pterodroma phaeopygia* in the Galapagos Islands, Ecuador. *Biological Conservation*, **42**, 303–311.

Cruz, R. and Cruz, J.B. (1987b) Control of black rats (*Rattus rattus*) and its effect on nesting dark-rumped petrels in the Galapagos Islands. *Vida Silvestre Neotropical*, **1**, 3–13.

Garcia, G. and Gardener, M.R. (2012) Evaluación de proyectos de control de plantas transformadoras y reforestación de sitios de alta valor en Galápagos. A report by the Galapagos National Park Service and the Charles Darwin Foundation, Puerto Avora, Santa Cruz, Galapagos.

Jäger, H. and Kowarik, I. (2010) Resiliance of native plant community following manual control of invasive *Cinchona pubescens* in Galapagos. *Restoration Ecology*, **18**, 103–112.

Jäger, H., Tye, A. and Kowarik, I. (2007) Tree invasion in naturally treeless environments: Impacts of quinine (*Cinchona pubescens*) trees on native vegetation in Galapagos. *Biological Conservation*, **140**, 297–307.

Jäger, H., Kowarik, I. and Tye, A. (2009) Destruction without extinction: long-term impacts of an invasive tree species on Galapagos highland vegetation. *Journal of Ecology*, **97**, 1252–1263.

Kastdalen, A. (1982) Cambios en la biología de la Isla Santa Cruz entre 1935 y 1965. *Noticias de Galapagos*, **35**, 7–12.

MacDonald, I.A.W., Ortiz, L., Lawesson, J.E. and Nowak, J.B. (1988) The invasion of highlands in Galapagos by the red quinine-tree *Cinchona succirubra*. *Environmental Conservation*, **15**, 215–220.

Rentería, J.L. (2002) Ecología y manejo de la cascarilla (*Cinchona pubescens* Vahl), en Santa Cruz, Galápagos. Área Agropecuaria y de Recursos Naturales Renovables. Undergraduate thesis. Universidad Nacional de Loja, Loja, Ecuador.

Rentería, J.L. (2011) Towards an optimal management of the invasive plant *Rubus niveus* in the Galapagos Islands. PhD thesis. Imperial College, London, UK.

Shriver, W.G., Gibbs, J.P., Woltz, H.W., Schwarz, N.P. and Pepper, M.A. (2011) Galapagos Rail *Laterallus spilanatus* population change associated with habitat invasion by the red-barked quinine tree *Cinchona pubescens*. *Bird Conservation International*, **21**, 221–227.

Svenson, H.K. (1935) Plants of the Astor Expedition, 1930 (Galapagos and Cocos Islands. *American Journal of Botany*, **22**, 208–277.

Tye, A. (2006) Restoration of the vegetation of the dry zone in Galapagos. *Lyonia*, **9**, 29–50.

van der Werff, H. (1983) Effects of feral pigs and donkeys on the distribution of selected food plants. *Noticias de Galapagos*, **36**, 17–18.

Chapter 23

THE MANAGEMENT FRAMEWORK IN PRACTICE – DESIGNER WETLANDS AS NOVEL ECOSYSTEMS

Stephen D. Murphy

Department of Environment and Resource Studies, University of Waterloo, Canada

23.1 OVERVIEW

The concept of a 'designer ecosystem' is inevitably slippery and likely to start fist fights (metaphorically one hopes) but, in practice, sometimes restoration can involve compromises that focus on conserving or restoring as much as possible given extreme anthropogenic changes. In this mix, the suite of approaches to greenroofs, living walls, artificial swales, rehabilitation or remediation of degraded lands and artificial wetlands primarily designed for interdiction and treatment of waste (see also discussion in Chapters 38 and 42) could be included. Each situation may be nothing other than a utilitarian approach to more 'ecological' treatment of pollution or artificial environments; there are however cases where there is more restoration, in the sense of functional or structural fidelity, than appears to exist. One situation can be where wetlands are created, such as the example described here from central Canada.

23.2 BACKGROUND

In central Canada, wetlands have been dewatered/ drained, isolated or otherwise rendered with poor structure (species diversity) and are often nonfunctional or with reduced function in terms of water and nutrient cycling. As with much of the world, the reasons largely relate to agriculture, other resource extractions, urbanization, transportation or often misguided rationales related to control of insect-borne diseases. This can conflict with Canada's Species at Risk Act; from both an ethical perspective (Chapter 31) and a desire to maintain and improve ecosystem services, a loss of wetlands is not acceptable ('no net loss' of wetlands). Entangled in this is the legal suite of how a wetland is defined and what can be done to a wetland. In a review of legal cases in North America over the years, there are federal-state/provincial-municipal-landowner disagreements over what is or is not

Novel Ecosystems: Intervening in the New Ecological World Order, First Edition. Edited by Richard J. Hobbs, Eric S. Higgs, and Carol M. Hall.
© 2013 John Wiley & Sons, Ltd. Published 2013 by John Wiley & Sons, Ltd.

considered a wetland, where a wetland did or did not exist and the ability (or lack thereof) of the Environmental Assessment and Wetlands Act to override grandfathered developments.

Based on a conservation ethic and policy, a goal of some – though, again, not all - landowners and government agencies may be to recover both ecosystem services (such as water filtration) and target species (such as amphibians). To achieve these goals, this is where it is perhaps useful to create new wetlands as 'designer ecosystems'. This might occur in cases where it is difficult to legally establish where wetlands once existed (and what exactly their status was, e.g. deep water swamp or wet meadow) and if it is not feasible economically or ecologically to restore wetlands in situ even if the legal issues were resolved.

The creation of designer wetlands that are not based on some historical baseline or reference ecosystem may also present some difficulties, however. If their creation begins to take us into the realm of 'no net loss' of wetlands, this approach can backfire if the created wetlands are simply glorified stormwater ponds that are poorly designed in terms of wildlife support and other ecological functions. There is an argument that not all stormwater ponds can be multifunctional, an argument supported in areas where the need is driven solely by a demand to reduce stormwater runoff into drains or into natural watercourses. In such cases, especially in Canada, the salt runoff from winter maintenance or high nutrient loads from lawns may not support wildlife. Of course, the crux there is to reduce or find alternatives to the sources of pollution (roadsalt and impervious surfaces, including lawns typical in central Canada) but the interim use of stormwater ponds is likely needed. However, there are cases where additional wetlands can be designed and created to simply act as flood mitigation (as natural wetlands would) and the pollution issue is minimal or has been addressed by other means, such as upstream stormwater ponds.

The creation of designer wetlands can add to a debate on whether one is 'giving up' (see Chapter 37) or enabling poor development practices, familiar to those involved in managing novel and hybrid ecosystems. While it is important to consider these arguments, if considered as an interim step to better ecosystem management then it can be argued that designer ecosystems do have a place. This is especially true if the attempt to restore a complex of ecosystems would take decades or centuries; by that time much of the function and structure to be conserved would probably be extirpated.

23.3 AN EXAMPLE OF CREATING DESIGNER WETLANDS

To illustrate a case where designer wetlands contribute to management goals, we consider an example of the design of small wetland complexes that support keystone species. Function (at least in the way natural wetlands function) is often a secondary or even non-existent concern because the designer wetlands are not always well connected or in the 'right' place. Effectively, we (my graduate students and I as part of teams) use the approach of SLASS (several large and several small) habitats as opposed to SLOSS (single large or several small) habitats. 'Large' here is a relative term because the land available is usually limited. In most cases there is sufficient space to install two $5000\,m^2$ larger wetlands and 5–10 smaller $300\,m^2$ wetlands. If possible, these are placed near adjacent source habitats and populations and are physically connected with small corridors (i.e. 5 m width and several hundred meters length) to create source–sink dynamics that do not also create ecological traps. While there is a risk of disease or other pest propagation via the corridor, the source sites have typically been long subjected to infestations already and the designer systems are being created out of essentially a denuded landscape. Metapopulation dynamics usually carry the designer systems, although there is a risk of elevated inbreeding depression or too-local a set of selection pressures leading to extreme genetic drift. However, if careful about selecting a range of genotypes and phenotypes from regional source populations, these risks and the risk of outbreeding depression can usually be avoided.

These systems do not (and could not) support large herbivores or predators on the scale of moose or bears but these are not found locally (and perhaps were never common). They will support most fish, herptile, bird and mammal populations up to deer and coyotes, with these latter species not always welcomed by nearby human residents. This is also the case with field mice, skunks, raccoons, foxes and even muskrats. Much public consultation is needed for both technical and social reasons.

23.4 THE RESTORATION APPROACH USED

In general, our approach is to plant 6–8 Poaceae, 10–12 Cyperaceae and Juncaceae and 10–15 other key wetland plant species (submergent and emergent).

We add mostly native species of plants but sometimes have to use non-invasive non-regional species if the species pool is lacking. Native species are more tolerant to drought because central Canada experiences this historically. Drought will probably be exacerbated with regional expressions of climate change, such as more intense storm events interspersed with more severe droughts. Into this, we usually introduce the native fish species bass, perch and pumpkinseed (most trout will not tolerate the conditions in wetlands) and most especially native amphibians (leopard frog being the most commonly added). The wetlands are connected via small throughput channels underlain with gravel-mud mixes for oxygenation. The channels meander as they would in natural systems.

23.5 SUMMARY

While designer ecosystems present a different set of issues, the decision framework (Fig. 18.1) raises questions that can inform the overall management approach.

1. Does the ecosystem comprise conditions that prevailed historically (see Fig. 18.1, ecosystem assessment box)? In most cases, there is no known historical reference for designer ecosystems and they will be managed as novel ecosystems (Fig. 18.3).

2. Will the ecosystem recover passively? No, for the reasons outlined earlier, as the wetlands are typically created on highly degraded low-functioning sites where it may not even be possible to determine if they were wetlands historically.

3. Are ecosystem changes reversible? No, again for the reasons outlined earlier. The barriers are ecological and social; a novel ecosystem approach is therefore used.

4. The management goal (Fig. 18.3) is to conserve target species and to restore function/ecosystem services.

5. The costs and risks are usually acceptable, but are determined on a case-by-case basis after extensive public and key informant consultations.

In most cases, management proceeds. The location, methods and mixture of the subset of species and functions are what lend the system its 'novel' label. We use species mostly native to the region and manage for functions that exist in more natural wetlands. However, the system is focused on a smaller subset of structure and function and, again, is usually installed in a non-historical location and therefore would not be a restoration to anything near a historical state or function.

Chapter 24

CHARACTERIZING NOVEL ECOSYSTEMS: CHALLENGES FOR MEASUREMENT

James A. Harris[1], Stephen D. Murphy[2], Cara R. Nelson[3], Michael P. Perring[4] and Pedro M. Tognetti[5]

[1]Environmental Science and Technology Department, Cranfield University, UK

[2]Department of Environment and Resource Studies, University of Waterloo, Canada

[3]Department of Ecosystem and Conservation Sciences, College of Forestry and Conservation, University of Montana, USA

[4]Ecosystem Restoration and Intervention Ecology (ERIE) Research Group, School of Plant Biology, University of Western Australia, Australia

[5]Departamento de Métodos Cuantitativos y Sistemas de Información, Facultad de Agronomia, University of Buenos Aires, Argentina

24.1 INTRODUCTION

Following the definition outlined in Chapters 5 and 6, the essential features of novel ecosystems that distinguish them from unaltered or hybrid systems are:

1. a difference in ecosystem composition, structure or function;

2. thresholds in these attributes that are currently irreversible; and

3. persistence or self-organization.

We need to be able to characterize these three characteristics. Measuring differences, the first step in the process of assessing novelty, is required to understand the degree of dissimilarity between the target and an undegraded system. The types of measurements that might be made to assess differences in composition, structure and function for a potential novel ecosystem are no different from those that would be used to assess these variables in any ecosystem. For a system to be classified as novel however, it must have crossed a threshold (experienced a change that is hard to reverse) and achieved persistence.

The present challenge in writing about identification and measurement in a book like this is that we are writing the manual and guidelines on how to do both right now. This means it is hard to offer much in the way of well-developed examples of identification (see also Chapter 3) and measurement because we do not have the luxury of reviewing extensive and explicit literature. What readers should glean is how we can approach identification and measurement by using or adapting existing approaches; some of these approaches will be familiar in scope and execution. In some cases, we will also be reanalyzing papers in the context of novel ecosystems and discussing where differences between ecosystems becomes a measure of either the drivers of novelty or novelty itself.

24.2 THE CHALLENGE OF MEASURING DIFFERENCES AND MEASURING NOVELTY

24.2.1 The novel ecosystem and measurement framework

The novel ecosystem concept provides an important framework for managers grappling with setting restoration and intervention targets in a changing world. Specifically, it leaps the conceptual boundary of always attempting to return to some historical reference. Nonetheless, the utility of the term hinges on the extent to which degree of novelty can be assessed and whether there are thresholds among ecosystem states that can be measured. There have been several investigations of how much of the world has become novel over time (e.g. Chapter 8; Williams and Jackson 2007; Ellis et al. 2010); these investigations have relied on coarse-scale models of the human footprint and climate change. To date there has not been an examination of the key variables that define novelty at local scales. Understanding novelty at the site-specific level will allow managers to better predict the degree of management intervention necessary to repair degradation and the range of appropriate intervention targets. The most effective measurement variables are ultimately those that managers can use.

24.2.2 Challenges of choosing references for novel ecosystem measurement

The characterization of novelty requires comparing the altered or degraded ecosystem against a reference system. Selection of an appropriate reference from which to measure is not a simple task, however. The fact that ecosystems and their processes are highly dynamic and shifting in response to directional change

renders the use of historical ecosystems as management increasingly difficult (Harris et al. 2006; Thorpe and Stanley 2011). Given continual changes in climate and biophysical conditions at local to global scales, the abiotic and biotic conditions of current ecosystems will be dissimilar to historical ecosystems, regardless of whether there has been extensive anthropogenic disturbance or not. While historical ecosystems may provide critical information for understanding the forces that shaped a particular ecosystem, their appropriateness as a reference model in our changing world is receding; it is not feasible to return an increasing number of systems back to their historical condition because:

• there is a lack of specific information about biotic and abiotic condition at previous points in time;
• there have been alterations in climate and biophysical environment;
• there have been species invasions; and
• there have been species extinctions and extirpations.

If the main criterion for novelty is an inability to return to historical abiotic and biotic conditions in a strict sense, all current systems would therefore be novel by definition, eliminating the utility of the term.

Conceptually, novelty may be measured as the difference in condition between the human-altered ecosystem and what the system condition would have been had the disturbance not occurred. This much is similar to the process of ecological restoration, which has relied traditionally on pre-disturbance reference systems. Of course, with novel ecosystems, historical systems of reference are either not available or helpful only in providing clues rather than more specific guidance. There are two pragmatic approaches to addressing this very situation: (1) using modern reference sites to approximate what conditions at the target site would have been like; and (2) using site conditions from before the degradation happened as a proxy for conditions at the site in the absence of degradation. Each has their own set of limitations; the former introduces spatial bias and the latter temporal bias.

24.2.3 Using modern reference sites as a proxy

Traditionally, the model in ecological restoration has been to find local sites that have not experienced a high degree of degradation and can serve as reference proxies. For example, in the western United States, many river and riparian ecosystems have been degraded by placer and other types of mining which reduce in-channel habitat complexity, species richness and connectivity and alter biogeochemical processes. However, it is possible to find relatively undegraded reaches along many rivers that can serve as models for the undegraded ecosystem trajectory. In this situation, we can derive model conditions for a target site from the mean 'value' of the conditions at a suitable number of references; this is somewhat akin to an 'effect size'. The specific number of reference sites needed to create this model is based on among-site variation in ecosystem properties, with more sites needed where variation is high and fewer where variation is more limited.

The departure from tradition is that we now recognize that using this type of a reference site is not a perfect solution and does not rule out that the comparison and the outcome may involve hybrid or novel ecosystems, as opposed to a more fidelous restoration state. First, all sites have experienced some level of anthropogenic influence even if they have not been directly managed due to factors that operate at large spatial scales such as climate change and acid rain. Second, because ecosystems are complex in multivariate ways, no two are identical; the systems that are being used to quantify reference conditions are therefore unlikely be exact replicates. The degree of difference between two systems may increase with increasing geographic distance or changes in environment. Another problem with using spatial references is that sites that do not appear to have been degraded may in fact have been directly altered by some unknown anthropogenic factor, leading to bias in our understanding of the reference state. Perhaps the most complex aspect of finding suitable references is that some systems can exhibit multiple alternative stable states, even in the absence of degradation. The number of potential states is dependent on the degree of productivity, size of the regional species pool and landscape connectivity (Chase 2003).

24.2.4 Using historical site conditions as a proxy

Although historical conditions per se provide challenges for measuring novelty because of the dynamic nature of ecosystems and climate, in some cases they provide the only available information on the condition of the target site in the absence of degradation.

Dendrochronology, palynology and paleoecology can be used to model pre-Anthropocene climate and biotic composition (Jackson and Hobbs 2009; Ellis et al. 2010), as can traditional knowledge from First Nations (Fletcher and Thomas 2010; Willis et al. 2010). However, using historical conditions as a proxy for a reference is challenging for multiple reasons, not the least of which is that these types of data are available only for a limited number of systems and variables.

Similarly, historical condition varies with spatial scale resulting in variable conditions depending on the scale at which the system is modeled. Using historical information also requires accounting for environmental and demographic determinism and stochasticity, i.e. non-random and random consequences of environmental variation and human intervention. There must, therefore, be a margin of estimated error in any measurements. There are multiple examples of appropriate application of historical references to measure novelty. Nonaka and Spies (2005) determined that the 20th century policy of fire suppression in the western USA has created novel forest ecosystems, though they did not use the term 'novel'. Similarly, Wimberly (2002) and Wimberly et al. (2004) showed that both age and spatial structure of forest ecosystems was novel. Other authors have been more cautious about use of historical models (Hughes et al. 2005; Newson and Large 2006) or use them as a means to help shape decisions about management as opposed to using them as a measurement target (Wohl 2011). If historical information is being considered for use in modeling reference conditions, it is critical to factor into the models differences in climatic conditions between the historical and current period and the effect of this variation in climate on ecological systems. With increasing difference in climate between periods, the value of historical data shifts from providing literal models to providing references or clues about ecosystem composition and function that enables appropriate interventions.

24.2.5 Which approach is best?

Decisions about whether to choose modern reference sites or historical ecosystem conditions as references for the target ecosystems are sometimes driven by availability of data. However, when multiple sources of data are available, the relative importance of temporal versus spatial error in modeling the target ecosystem should be a consideration. For instance, in areas where dramatic ecosystem changes occur in response to small perturbations in climate (e.g. tree-line ecosystems), using historical ecosystem conditions as a reference may introduce much more bias than the use of modern analogs. The spatial error introduced by using local undegraded sites as references – even though environmental conditions may vary in subtle ways from the target system – may be far less than the temporal bias introduced by the historical data. In other cases, high among-site heterogeneity within a landscape may add more error to reference models than the use of data on historical conditions specific to the target site. Utilizing all types of information available, whether temporal or spatial, would seem to be the optimum approach to determining reference and target systems.

24.2.6   Is it possible to measure novelty?

The promising news is that there are many variables that might be used to measure novelty in ecosystems for all three essential features listed earlier: difference in ecosystem composition, structure and function; thresholds in these attributes; and persistence (Table 24.1). We have identified the larger scope of variables that can be measured but there is neither sufficient space nor many worked examples (as yet) to show how all the variables can be applied to novel ecosystems. Consider this a first guideline about what variables should be useful in measuring novel ecosystems. The next two sections focus on the mesoecological and macroecological variables respectively.

24.3   VARIABLES TO MEASURE NOVELTY IN ECOSYSTEM COMPOSITION, STRUCTURE OR FUNCTION

In this section, we describe selected variables from Table 24.1 for measuring and understanding relative novelty of ecosystem states. The variables used to measure novelty are the same variables that would be used to understand the condition of any ecosystem. The mesoecological and macroecological measures presented here represent a jumping-off point for understanding the drivers of novelty and the metrics that managers might consider including in their assessments.

24.3.1   Evolutionary changes

We can examine how natural selection and genetic drift will contribute to or thwart changes to a novel ecosystem state and function. In our experience, it can appear that management of an invasive exotic species has been successful. There is a nuanced effect of selection pressures that may foster an unexpected re-emergence of the exotic species and continued increase in novelty of an ecosystem, however. For example, invasion of North American wetlands by exotic *Lythrum salicaria* had begun creating novel ecosystems. To a great extent, that increase in novelty was mitigated by introduction of natural enemies via biological control programs. Unintentionally, smaller, cryptic, longer-lived genets able to colonize mesic or heterogeneous habitats may be selected for the population that remains. These may be able to colonize related but previously less-disturbed habitats such as wet meadows where there is more environmental heterogeneity (Moloney et al. 2009). Left unmanaged there will be a second wave of increasingly novel habitats, contrary to management expectations based on the apparent success of the natural enemies. The evolutionary mechanisms and rate of change can be used to predict and measure the likelihood of how novel an ecosystem may become, both originally with invasion by *L. salicaria* and then again with the second wave from selection for different genotypes and perhaps even ecotypes.

Intraspecific and interspecific hybridization are even subtler drivers for increasing novelty of ecosystems. For example, intraspecific hybridization within *Phragmites australis* likely facilitated its widespread and increasingly dominant invasion of North America (Morrison and Molofsky 1998; Clevering and Lissner 1999; Warren et al. 2001; Saltonstall 2002). The new hybrid species is dominant and alters the structure and function of wetlands; it has created novel wetlands where it may not be possible to ever restore them (Molofsky et al. 1999; Ailstock et al. 2001). Another example of a hybrid species driving the process of novelty is *Agrostis* in southern Nova Scotia (SD Murphy, personal observation). There, species that are hybrid between commercial crop species and native species have created several genotypes that may be dominating a range of habitats from coastal saltmarshes to upland old-fields. These are so competitive that, again, it is not likely possible to restore the pre-farming landscape. Whether this requires management is an open

Table 24.1 Types and variables of relevance to measuring novelty in ecosystems.

	Type	Variable
MESOECOLOGICAL	Population measures	Evolutionary changes
		Demographic measures
	Community measures	Invasions/new species
		Traits and invasibility
		Diversity and functional groups
		Trophic measures
		New community-scale functions
		Number of new interactions
		Successional processes
	Landscape measures	Connectivity thresholds
	State measures	Equilibrium state:
		■ Biotic coupling
		■ Competition
		■ Saturated
		■ Resource limitation
		■ Density dependence
		■ Optimality
		■ Few stochastic effects
		■ Tight patterns
		Non-equilibrium state:
		■ Biotic decoupling
		■ Species independence
		■ Unsaturated
		■ Abiotic limitation
		■ Density independence
		■ Opportunism
		■ Large stochastic effects
		■ Loose patterns
		Transitional states (rate and amplitude of recovery from disturbance)
		Pools and fluxes of moisture and nutrients
MACRO-ECOLOGICAL		Neutral theory
		Thermodynamic efficiency (energy, emergy, exergy, entropy)
		$1/f$ noise analysis
		Fractal structure analysis
		Complexity and regime shifts
		Thresholds and alternative stable states
		Audits of small : large molecules

question because in the novel state the hybrid species do not seem to alter nutrient or water cycling (at least in the last 20 years of observations). This means that unlike the *Lythrum*- or *Phragmites*-driven examples, the novel ecosystem state may be acceptable to managers. This is consistent with the decision frameworks that were explored in Chapter 18.

24.3.2 Demographic variables

Numbers often indicate how important the ecological influence of populations of species may be (Booth et al. 2010). However, numbers may tell only part of the story. For instance, it may not be the sheer numbers of *Alliaria petiolata* in North America that cause disruption and change ecosystems but rather their impact on soil structure and biota and the cascade effect of expanding openings in the forest horizontal and vertical canopy (Murphy 2005; Burke 2008). Spatial arrangement within populations of any species, especially pre-emption of space, can be a useful measure to show the degree of novelty (Deines et al. 2005; Boulant et al. 2008; Huang et al. 2008). For example, Allison's (2002) study on the long-term prairie restoration near Knox College (USA) can be reassessed in this light because it shows that, even 40 years after an active restoration, the spatial legacy of the monocultures that persisted before the restoration is still evident.

24.3.3 Diversity and functional groups

A useful way to grapple with measuring how diversity might be used in novel ecosystems is provided by Jackson and Sax (2010). Delayed declines in biodiversity (extinction debt) can be measured as a relative indicator on a delayed but inexorable transition to a novel ecosystem. Gains from novel ecosystems that encourage immigration of new species that may compensate for losses in biodiversity (immigration credit) can also be measured. For example, in Great Britain, the species richness of lichens, bryophytes, vascular plants, fungi and songbirds were unaffected or increased in pine plantations, relative to less novel coniferous ecosystems (Quine and Humphrey 2010). The caveat is whether the increase in species richness is due primarily to exotic species and how this may alter ecosystem states or functions (Catterall et al. 2010).

24.3.4 Trophic measures

Trophic measures refer to processes or interactions across trophic levels, such as nutrient cycling, competition and herbivory. Such measures can be adapted for explaining and measuring how novel an ecosystem may become. measuring the impact of grazing can assist in determining degree of novelty in a system. Isbell and Wilsey (2011) measured productivity, light interception, nitrogen uptake and outcomes of experimental manipulation of species richness. From these variables, they could identify that the change from native-dominated meadows to grazed pasturelands changed the ecosystem dynamics and interactions between ecosystem function and species diversity. The exotic-species-dominated ecosystems recovered from grazing faster. Interestingly, increased exotic species richness did not affect ecosystem functions. The opposite was true in native-dominated ecosystems.

A trophic interaction such as a pathogenic invasion can cause a measurable shift to a novel ecosystem (Bishop et al. 2010). *Phytophthora cinnamomi* (a soil pathogen) invaded *Banksia* woodlands in Australia and caused a change in the system's beta-diversity. Some formerly dominant species were highly susceptible. The change in dominant species is probably irreversible once both live plants and then the seedbank of the formerly dominant species declined rapidly; the populations will not be replaced. The pathogen persists across multiple watersheds and thus makes the shift to a novel ecosystem more permanent. There is a twist to this outcome however, because alpha-diversity of native species was maintained. As such, there is still conservation value in the woodlands; it is a reshuffled system but is still dominated by a novel set of native species.

A more complex example is from Lach et al. (2010). They determined that an exotic ant (the white-footed ant, *Technomyrmex albipes*) visited the extrafloral nectaries of an exotic tree (the white leadtree, *Leucaena leucocephala*). The ant hindered the native herbivore (a plant louse, *Heteropsylla cubana*) from harming the exotic white leadtree, the opposite of what a manager would normally desire. Simultaneously, the white-footed ant facilitated bug (Order Hemiptera) damage to another plant, this one a native shrub (beach cabbage, *Scaevola taccada*), which is also the opposite of what a manager would want. The main outcome of interest here is that novelty wrought by one exotic species

(*L. leucocephala*) is being exacerbated by another exotic (*T. albipes*) in a quantifiable manner. This is the sort of measurement that can translate trophic data into degrees of novelty, at least for a subset of the ecosystem.

24.3.5 Number of new interactions

New interactions can be a hallmark of increased novelty. Seastedt et al. (2008) provide a good example from Colorado. To a casual observer, it may seem odd that there is a still a normally fire-driven tallgrass prairie ecosystem in an area where both natural and prescribed burns are prohibited because of a nearby highway. The reason the prairie exists is because of new interactions. Specifically, cattle graze at times when they will eat the exotic cool-season grasses. This removes or reduces the litter layer and the soil is again exposed to light. Concurrently, cattle defecation and urination results in soil fertilization. These new interactions promote and perhaps select in favor of germination of warm-season prairie grasses. In this case, novelty may not actually be measured by the species composition but by the new sets of interactions that maintain the tallgrass prairie. Importantly, similar situations can yield different outcomes where the result is a novel but exotic-dominated and poorly functioning ecosystem (Best 2008). We can therefore use interactions to measure both relative novelty and functional responses.

24.3.6 Successional processes

Successional trajectories can be difficult to predict but Kueffer et al. (2010) obtained sufficient data to analyze the successional trajectories in a tropical secondary forest invaded and dominated by *Cinnamonum verum*. They used a restoration approach of creating small gaps in the canopy and planting native trees using a nodal strategy. They did not recommend trying to extirpate the invasive exotic or restored sites with all native vegetation. Rather, the restoration effort was made under the acknowledgement that some degree of novelty was going to persist. This has ties to the management decisions that were discussed in Chapter 3.

24.4 IDENTIFYING THRESHOLDS IN ECOSYSTEM COMPOSITION, STRUCTURE AND FUNCTION

Thresholds occur when there is a sudden discontinuous change in ecosystem functions and states. The transition can be abrupt, e.g. a few years (Friedel 1991; Carpenter et al. 1999; Scheffer and Carpenter 2003; Suding et al. 2004; Willis et al. 2010). The result is often an alternative stable state where trying to restore or at least change the ecosystem functions and states is very difficult, often because of thermodynamic resistance (Kay 1987; Suding et al. 2004; Suding and Hobbs 2008; Jeffers et al. 2011).

There is a major challenge here for the site manager: identifying thresholds is not a trivial matter, and may involve collecting large datasets over long time series and inevitably involve sophisticated statistical analysis (Suding and Hobbs 2008; Cherwin et al. 2009; Scheffer et al. 2009; Bestelmeyer et al. 2011; Booth et al. 2010; Firn et al. 2010; Murphy 2010). By the time that such a threshold has been identified it may be too late to attempt any restoration to an unaltered system (Bestelmeyer 2006). Invasives are a case in point: by the time the invasion is identified, reversing it may be impossible due to the sheer weight of numbers of individuals and, in particular, propagules. Eradication is only possible if invasives are controlled when population sizes are small. What's more, invasive plants are favored as rare genotypes through the action of the soil biota. It may be that if mycorrhizal symbionts are established within an invasive population, the irreversible threshold has been passed and you have a novel ecosystem. Thresholds also can be crossed because of more direct social pressures. In Kenya, Furukawa et al. (2011) identified a change from a native- to invasive-dominated forest understory as a result of selective fuelwood extraction.

The presence of a threshold means that the ecosystem must have undergone a regime shift from some previous dynamic configuration (Scheffer et al. 2001; Folke et al. 2004; deYoung et al. 2008; Jiao 2009). Regime shifts involve the changing of a dynamic ecosystem with a characteristic structure and function to a different form. Shifts in regime may lead to the formation of novel ecosystems particularly if, once a threshold has been crossed, the system will not easily return to its prior state and the system is composed of unprecedented species or abiotic conditions. These shifts may

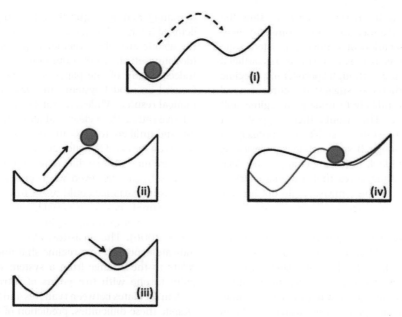

Figure 24.1 Stability and thresholds in two-dimensional (2D) space. A stable area exists in the troughs of a landscape. (i) A change to a different state can be reached either through a change to the variable (ball) pushing it to (ii) and then beyond an unstable threshold point into the new state of (iii) the 'community perspective'; a new state can also be reached by changing the parameters of the landscape (iv) thus forcing the ball to a new position: the 'ecosystem perspective'. The parameters are typically the slow variables of a system and are presumed to not change due to changes in the variable. However, if feedbacks do exist between the environment and the variable, then both can alter. These types of relationship can also be visualized in 3D space. Reproduced from Beisner (2003).

occur through internal factors of the system dynamics (e.g. population variation) and/or through being pushed by changes in an external factor such as nutrient input into a lake – the 'community' and 'ecosystem' perspectives respectively (Beisner et al. 2003). Figure 24.1 explicates this.

Shifts most often involve multiple causative agents but always involve a shift in the feedbacks that maintain a state in one form or another, with feedbacks involving biological, physical and chemical mechanisms (Scheffer et al. 2001). This shift in feedbacks occurs at the threshold, although it may take longer for the result to become apparent in state variables (Walker and Meyers 2004). Measurement of regime shifts likely involves a combination of insights from field data, manipulative experiments and the application of models (Scheffer and Carpenter 2003), although the demonstration of alternate states remains under debate (Beisner et al. 2003). The literature surround-

ing regime shifts and the allied concept of alternative stable states involves much discussion of the concept of resilience. Resilience can be understood with reference to Figure 24.1, and depends upon the topographic characteristics of the basin of attraction that the ball resides in. A system with steeper sides and wider basin will be more resilient than the converse, as it will have quicker return times and be more likely to retain the ball within its sphere of influence following a perturbation.

In recent years, as the appreciation of regime shifts has grown, much work has focused on ways to predict whether a regime shift will occur ahead of time (Fath et al. 2003; Carpenter and Brock 2006; van Nes and Scheffer 2007; Contamin and Ellison 2009; Carpenter et al. 2011). This is because many reported regime shifts involve a shift to a state undesired by humans, such as the change of species-rich coral reefs to algal-dominated reefs (Folke et al. 2004), and an early warning may allow managers to avert or slow the shift in regime. These measures are outlined in more detail later. Identification of a novel ecosystem state could utilize these measures. If the measures outlined later are not behaving as though approaching a regime shift, this would therefore suggest the current novel ecosystem state is unlikely to undergo a regime shift in the near future. This would then suggest that the current system state is stable and unlikely to undergo further change. If such a state is undesirable, management could look to forcing changes that lead to the detection of measures that are characteristic of the regime shift. If these measures do not become apparent, further management actions could be considered.

Measures of approach to regime shifts include rising variance of a 'fast' state variable ahead of the change in state (Carpenter and Brock 2006), such as water column phosphorus in a lake that had much higher variance as it approached an eutrophic turbid condition from an oligotrophic, clear water condition (Carpenter and Brock 2006). Fisher's Information index can be used as a means of tracking when systems are moving between regimes (Fath et al. 2003). Noise detection can be used to determine regime shifts, e.g. measuring when there is 'reddening' in fluctuation patterns (Scheffer and Carpenter 2003). This reddening refers to when there is a low-frequency fluctuation with high amplitude in a population time series, by analogy to light (Petchey 2000). Population variability can be understood ecologically as arising from two contrasting processes: white noise from random environmental fluctuations within set limits and a random walk where populations accrue random, uncorrelated increments of abundances over time (Akcakaya et al. 2003). It is the latter process that is associated with reddened spectra. Slower rates of recovery may also be detected from small perturbations when resilience is likely to be compromised. van Nes and Scheffer (2007) showed that a slower recovery rate was associated with lower resilience. In particular, this measure works even if the perturbation does not necessarily bring a system close to a threshold (van Nes and Scheffer 2007).

Nonetheless, there is a management challenge in using regime shift measures. These measures may be quite generalizable and consistent in the sense that variance of certain state variables increases and variance spectra shift towards longer wavelengths (lower frequencies) just before regime shifts occur (Carpenter and Brock 2006). The challenge is that their relevance

to managers may be questioned, as they often require detailed information about the system in question and the abiotic and biotic relationships between variables (deYoung et al. 2008; Contamin and Ellison 2009). Indeed, many of the papers cited in this subsection focused on model systems and described the mathematical results. While a useful start, the measures are relative rather than clear and absolute, data demanding and unlikely to work in systems with large variation (van Nes and Scheffer 2007).

There may also be difficulties in our interpretation of these measures, even when present. For instance, increasing variance could result from trends in some unmeasured ecosystem variable and thus nothing to do with an impending regime shift (Carpenter and Brock 2006). The measures also do not necessarily indicate approach to a regime shift and could instead relate to the change from a system with one stable point to one with two points of attraction that the system oscillates between (van Nes and Scheffer 2007). Despite these difficulties, prediction of regime shifts is notoriously difficult, so persevering at these endeavors would appear worthwhile (van Nes and Scheffer 2007). In particular, advances in ecological monitoring could lead to the application of more powerful methods such as investigating spectral signatures of, for example, population time series, which appear to be a more powerful way of detecting regime shifts ahead of time (Carpenter and Brock 2006; Contamin and Ellison 2009), although how the frequency of environmental noise affects the likelihood of a regime shift remains an unanswered question (Scheffer and Carpenter 2003). Another unanswered question is how the network typology i.e. the number, types and strengths of interactions among species in an ecosystem, affects the likelihood of a regime shift (Walker and Meyers 2004), although work has progressed in this area (Kondoh 2008).

It may still be possible to consider the converse of these measures: does a lack of variation in measured state variables, despite the noisy environment surrounding ecosystems, suggest that moving to an alternate state is unlikely? In addition to using measures, it is also possible to consider practical examples of irreversible thresholds in regime shifts. One clear example would be extinction of species; presumably extinction is what is generally referred to by Walker and Meyers (2004) when they identified that, of 64 regime shifts identified in their database, 24 were irreversible. Walker and Meyers (2004) include other examples of irreversible shifts beyond species extinction, such as salinization in Australia. Vegetation change in West Africa and Amazon may be another example of an irreversible shift, where a drier climate may have made it impossible for the vegetation to return to the original forest type because the original forests were established during a wetter period in the Earth's history (Walker and Meyers 2004).

Irreversibility will be enhanced by having a wide stability domain (Scheffer et al. 2001), with stability domains typically depending upon slowly changing variables such as land use, nutrient stocks, soil properties and biomass of long-lived organisms. In Figure 24.1, this can be imagined as a wider basin of attraction for the ball. These factors may be predicted, monitored and modified (Scheffer et al. 2001). If the novel ecosystem is appropriate then it may be desirable to enhance resilience by managing these slowly changing variables. This resilience will enhance the capacity of the system to absorb disturbance and reorganize itself while undergoing change, in order to retain essentially the same function, structure, identity and feedback mechanisms. Building resilience into both social and ecological systems through adaptive management should be the ultimate objective of management under regime shifts (deYoung et al. 2008).

24.5 CONCLUSION

We began by stating that there needed to be three essential features to be measured:
1. a difference in ecosystem composition, structure or function;
2. thresholds in these attributes that are currently irreversible; and
3. persistence or self-organization.
We explored how the challenges of measuring differences and novelty are non-trivial and measurement approaches are a work in progress. One issue for managers is how to begin the coarse-scale delineation of novelty in ecosystems. We discussed whether the use of modern references or historical ecosystem proxies was appropriate and which approach is best. This is a common problem in ecology – we do not have a 'control Planet Earth'. The choice of coarse delineation depends largely (but not wholly) on the data available. If there are sufficient data and confidence that modern references exist, then contemporary comparisons are possible. While this engenders more confidence, numerous

caveats mean that this confidence can be misplaced. A more conservative and less costly approach may be to measure current ecosystems against historical ecosystem conditions. This approach lessens but does not avoid the spatial biases characteristic of the modern analogue approach.

We reviewed some of the variables that can be used to measure differences in ecosystem composition, structure and function and used variables more familiar to readers. The examples of how to reinterpret or adapt these variables to measure novel ecosystems are a useful start, but we caution that the examples and publications that will explicitly measure novelty are just now being written. This leaves us with a sense that a gap needs to be filled by testing how measures work and how this might lead to best practices in measuring novel ecosystems by managers.

This is also true of the more complex challenge in measuring novelty via (a) thresholds in ecosystem composition, structure and function and (b) persistence. However, this more macro-ecological approach may be best for managers because the drivers of novelty tend to exist more at this scale. Novelty really implies a regime shift and crossing of ecological thresholds, and it is this process that drives differences in ecosystem composition, structure and function that are tractable at population, community and landscape scales. The conundrum is that while they capture more of what makes an ecosystem novel, the thresholds and persistence measures are even harder to conceptualize, grasp and measure than the familiar techniques already used in population through landscape scales. Population, community and landscape ecology measures have however gone through periods where these were thought too complex to consider or near impossible to measure practically, until computers became common; suddenly, the iterative maths or multiple steps once beyond reach became almost mundane to the experts/ professionals. There will be a learning curve for thresholds and stability measures but, with time and application to examples, they too will become *de rigueur*.

There is a Hellerian (Catch-22) aspect to all of this in the sense that we need to decide and use a decision-making framework before we can quantitatively identify and measure novelty in ecosystems. The catch is that there is a simultaneous need to suggest measures on how to make those decisions. In that context, we view this chapter as an early step in an iterative process. What should happen in the next few years is that the first uses of decision-making management frameworks and guidelines (Chapter 3) will eventually lead to studies where the measures suggested here may be tested and applied. The real test will come as new studies are designed explicitly with the intent of using the measures of drivers of novelty and the state/ function of novelty as outlined here. This chapter is a guideline to the next steps in measuring or assessing how novelty originates, how novelty emerges and how much novelty exists in an ecosystem.

REFERENCES

Ailstock, M.S., Norman, C.M., Bushmann, P.J. (2001) Common reed *Phragmites australis*: Control and effects upon biodiversity in freshwater nontidal wetlands. *Restoration Ecology*, **9**, 49–59.

Akcakaya, H.R., Halley, J.M. and Inchausti, P. (2003) Population-level mechanisms for reddened spectra in ecological time series. *Journal of Animal Ecology*, **72**, 698–702.

Allison, S.K. (2002) When is a restoration successful? Results from a 45-year-old tallgrass prairie restoration. *Ecological Restoration*, **20**, 10–17.

Beisner, B.E., Haydon, D.T. and Cuddington, K. (2003) Alternative stable states in ecology. *Frontiers in Ecology and Environment*, **1**, 376–382.

Best, R.J. (2008) Exotic grasses and feces deposition by an exotic herbivore combine to reduce the relative abundance of native forbs. *Oecologia*, **158**, 319–327.

Bestelmeyer, B.T. (2006) Threshold concepts and their use in rangeland management and restoration: the good, the bad, and the insidious. *Restoration Ecology*, **14**, 325–329.

Bestelmeyer, B.T., Goolsby, D.P. and Archer, S.R. (2011) Spatial perspectives in state-and-transition models: a missing link to land management? *Journal of Applied Ecology*, **48**, 746–757.

Bishop, C.L., Wardell-Johnson, G.W. and Williams, M.R. (2010) Community-level changes in *Banksia* woodland following plant pathogen invasion in the Southwest Australian Floristic Region. *Journal of Vegetation Science*, **21**, 888–898.

Booth B., Murphy, S.D. and Swanton, C.J. (2010) *Invasive Species Ecology*. Second Edition. CABI/Oxford University Press.

Boulant, N., Kunstler, G., Rambal, S. and Lepart, J. (2008) Seed supply, drought, and grazing determine spatio-temporal patterns of recruitment for native and introduced invasive pines in grasslands. *Diversity and Distributions*, **14**, 862–874.

Burke, D.J. (2008) Effects of *Alliaria petiolata* (garlic mustard; Brassicaceae) on mycorrhizal colonization and community structure in three herbaceous plants in a mixed deciduous forest. *American Journal of Botany*, **95**, 1416–1425.

Carpenter, S.R., Ludwig, D. and Brock, W.A. (1999) Management of eutrophication for lakes subject to potentially irreversible change. *Ecological Applications*, **9**, 751–771.

Carpenter, S.R. and Brock, W.A. (2006) Rising variance: a leading indicator of ecological transition. *Ecology Letters*, **9**, 311–318.

Carpenter, S.R., Cole, J.J., Pace, M.L., Batt, R., Brock, W.A., Cline, T., Coloso, J., Hodgson, J.R., Kitchell, J.F., Seekell, D.A., Smith, L. and Weidel, B. (2011) Early warnings of regime shifts: a whole-ecosystem experiment. *Science*, **332**, 1079–1082.

Catterall, C.P., Cousin, J.A., Piper, S. and Johnson, G. (2010) Long-term dynamics of bird diversity in forest and suburb: decay, turnover or homogenization? *Diversity and Distributions*, **16**, 559–570.

Chase, J.M. (2003) Community assembly: when should history matter? *Oecologia*, **136**, 489–498.

Cherwin, K.L., Seastedt, T.R. and Suding, K.N. (2009) Effects of nutrient manipulations and grass removal on cover, species composition, and invasibility of a novel grassland in Colorado. *Restoration Ecology*, **17**, 818–826.

Clevering, O.A. and Lissner, J. (1999) Taxonomy, chromosome numbers, clonal diversity and population dynamics of *Phragmites australis*. *Aquatic Botany*, **64**, 185–208.

Contamin, R. and Ellison, A.M. (2009) Indicators of regime shifts in ecological systems: What do we need to know and when do we need to know it. *Ecological Applications*, **19**, 799–816.

deYoung, B., Barange, M., Beaugrand, G., Harris, R., Perry, R.I., Scheffer, M. and Werner, F. (2008) Regime shifts in marine ecosystems: detection, prediction and management. *Trends in Ecology and Evolution*, **23**, 402–409.

Deines, A.M., Chen, V.C. and Landis, W.G. (2005) Modeling the risks of nonindigenous species introductions using a patch-dynamics approach incorporating contaminant effects as a disturbance. *Risk Analysis*, **25**, 1637–1651.

Ellis, E.C., Goldewijk, K.K., Siebert, S., Lightman, D. and Ramankutty, N. (2010) Anthropogenic transformation of the biomes, 1700 to 2000. *Global Ecology and Biogeography*, **19**, 589–606.

Fath, B.D., Cabezas, H. and Pawlowski, C.W. (2003) Regime changes in ecological systems: an information theory approach. *Journal of Theoretical Biology*, **222**, 517–530.

Firn, J., House, A.P.N. and Buckley, Y.M. (2010) Alternative states models provide an effective framework for invasive species control and restoration of native communities. *Journal of Applied Ecology*, **47**, 96–105.

Fletcher, M.S. and Thomas, I. (2010) The origin and temporal development of an ancient cultural landscape. *Journal of Biogeography*, **37**, 2183–2196.

Folke, C., Carpenter, S., Walker, B., Scheffer, M., Elmqvist, T., Gunderson, L. and Holling, C.S. (2004) Regime shifts, resilience, and biodiversity in ecosystem management. *Annual*

Review of Ecology, Evolution and Systematics, **35**, 557–581.

Friedel, M.H. (1991) Range condition assessment and the concept of thresholds – a viewpoint. *Journal of Range Management*, **44**, 422–426.

Furukawa, T., Fujiwara, K, Kiboi, S. and Mutiso, P.B.C. (2011) Threshold change in forest understory vegetation as a result of selective fuelwood extraction in Nairobi, Kenya. *Forest Ecology and Management*, **262**, 962–969.

Harris, J.A., Hobbs, R.J., Higgs, E. and Aronson, J. (2006) Ecological restoration and global climate change. *Restoration Ecology*, **14**, 170–176.

Huang, H.-M., Zhang, L.-Q., Guan, Y.-J. and Wang, D.-H. (2008) A cellular automata model for population expansion of *Spartina alterniflora* at Jiuduansha Shoals, Shanghai, China. *Estuarine, Coastal and Shelf Science*, **77**, 47–55.

Hughes, F.M.R., Colston, A. and Mountford, J.O. (2005) Restoring riparian ecosystems: The challenge of accommodating variability and designing restoration trajectories. *Ecology and Society*, http://www.ecologyandsociety.org/vol10/iss1/art12/. Accessed August 2012.

Isbell, F.I. and Wilsey, B.J. (2011) Increasing native, but not exotic, biodiversity increases aboveground productivity in ungrazed and intensely grazed grasslands. *Oecologia*, **165**, 771–781.

Jackson, S.T. and Hobbs, R.J. (2009) Ecological restoration in the light of ecological history. *Science*, **325**, 567–569.

Jackson, S.T. and Sax, D.F. (2010) Balancing biodiversity in a changing environment: extinction debt, immigration credit and species turnover. *Trends in Ecology and Evolution*, **25**, 153–160.

Jeffers, E.S., Bonsall, M.B., Brooks, S.J. and Willis, K.J. (2011) Abrupt environmental changes drive shifts in tree–grass interaction outcomes. *Journal of Ecology*, **99**, 1063–1070.

Jiao, Y. (2009) Regime shift in marine ecosystems and implications for fisheries management, a review. *Review of Fish Biology and Fisheries*, **19**, 177–191.

Kay, J.J. (1987) A nonequilibrium thermodynamic framework for discussing ecosystem integrity. *Environmental Management*, **15**, 483–495.

Kondoh, M. (2008) Building trophic modules into a persistent food web. *PNAS*, **105**, 16631–16635.

Kueffer, C., Schumacher, E., Dietz, H., Fleischmann, K. and Edwards, P.J. (2010) Managing successional trajectories in alien-dominated, novel ecosystems by facilitating seedling regeneration: A case study. *Biological Conservation*, **143**, 1792–1802.

Lach, L., Tillberg, C.V. and Suarez, A.V. (2010) Contrasting effects of an invasive ant on a native and an invasive plant. *Biological Invasions*, **12**, 3123–3133.

Molofsky, J., Morrison, S.L. and Goodnight, C.J. (1999) Genetic and environmental controls on the establishment of the invasive grass, *Phalaris arundinacea*. *Biological Invasions*, **1**, 181–188.

Moloney, K.A., Knaus, F. and Dietz, H. (2009) Evidence for a shift in life-history strategy during the secondary phase of a plant invasion. *Biological Invasions*, **11**, 625–634.

Morrison.S.L. and Molofsky, J. (1998) Effects of genotypes, soil moisture, and competition on the growth of an invasive grass, *Phalaris arundinacea* (reed canary grass). *Canadian Journal of Botany*, **76**, 1939–1946.

Murphy, S.D. (2005) Concurrent management of an exotic species and initial restoration efforts in forests. *Restoration Ecology*, **13**, 584–593.

Murphy, S.D. (2010) The planning of ecological restoration projects, in *Restoration Ecology* (Greippson, S. ed.), Kluwer Press, New York, 242–261.

Newson, M.D. and Large, A.R.G. (2006) 'Natural' rivers, 'hydromorphological quality' and river restoration: a challenging new agenda for applied fluvial geomorphology. *Earth Surface Processes and Landforms*, **31**, 1606–1624.

Nonaka, E. and Spies, T.A. (2005) Historical range of variability in landscape structure: a simulation study in Oregon, USA. *Ecological Applications*, **15**, 1727–1746.

Petchey, O.L. (2000) Environmental colour affects aspects of single-species population dynamics. *Proceedings of the Royal Society B: Biological Sciences*, **267**, 747–754.

Quine, C.P. and Humphrey, J.W. (2010) Plantations of exotic tree species in Britain: irrelevant for biodiversity or novel habitat for native species? *Biodiversity and Conservation*, **19**, 1503–1512.

Saltonstall, K. (2002) Cryptic invasion by a non-native genotype of the common reed, Phragmites *australis*, into North America. *Proceedings of the National Academy of Sciences of the United States of America*, **99**, 2445–2449.

Scheffer, M. and Carpenter, S.R. (2003) Catastrophic regime shifts in ecosystems linking theory to observation. *Trends in Ecology and Evolution*, **18**, 648–656.

Scheffer, M., Carpenter, S., Foley, J.A., Folke, C., Walker, B. (2001) Catastrophic shifts in systems. *Nature*, **413**, 591–596.

Scheffer, M., Bascompte, J., Brock, W.A. et al. (2009) Early-warning signals for critical transitions. *Nature*, **461**, 53–59.

Seastedt, T.R., Hobbs, R.J. and Suding, K.N. (2008) Management of novel ecosystems: are novel approaches required? *Frontiers in Ecology and Environment*, **6**, 547–553.

Suding, K.N. and Hobbs, R.J. (2008) Threshold models in restoration and conservation: a developing framework. *Trends in Ecology and Evolution*, **24**, 271–279.

Suding, K.N., Gross, K.L. and Houseman, G.R. (2004) Alternative states and positive feedbacks in restoration ecology. *Trends in Ecology and Evolution*, **19**, 46–53.

Thorpe, A.S. and Stanley, A.G. (2011) Determining appropriate goals for restoration of imperiled communities and species. *Journal of Applied Ecology*, **48**, 275–279.

van Nes, E.H. and Scheffer, M. (2007) Slow recovery from perturbations as a generic indicator of a nearby catastrophic shift. *The American Naturalist*, **169**, 738–747.

Walker, B. and Meyers, J.A. (2004) Thresholds in ecological and social-ecological systems: a developing database. *Ecology and Society*, http://www.ecologyandsociety.org/vol9/iss2/art3/. Accessed August 2012.

Warren, R.S., Fell, P.E., Grimsby, J.L., Buck, E.L., Rilling, G.C. and Fertik, R.A. (2001) Rates, patterns, and impacts of Phragmites *australis* expansion and effects of experimental Phragmites control on vegetation, macroinvertebrates, and fish within tidelands of the lower Connecticut River. *Estuaries*, **24**, 90–107.

Williams, J.W. and Jackson, S.T. (2007) Novel climates, no-analog communities, and ecological surprises. *Frontiers in Ecology and Environment*, **5**, 475–482.

Willis, K.J., Bailey, R.M., Bhagwat, S.A. and Birks, H.J.B. (2010) Biodiversity baselines, thresholds and resilience: testing predictions and assumptions using palaeoecological data. *Trends in Ecology and Evolution*, **25**, 583–591.

Wimberly, M.C. (2002) Spatial simulation of historical landscape patterns in coastal forests of the Pacific Northwest. *Canadian Journal of Forest Research*, **32**, 1316–1328.

Wimberly, M.C., Spies, T.A. and Nonaka, E. (2004) Using natural fire regime-based criteria to evaluate forest management in the Oregon Coast Range, in *Emulating Natural Forest Landscape Disturbances: Concepts and Applications* (A.H. Perera, L.J. Buse and M.G. Weber, eds), Columbia University Press, New York, New York, 146–157.

Wohl, E. (2011) What should these rivers look like? Historical range of variability and human impacts in the Colorado Front Range, USA. *Earth Surface Processes and Landforms*, **36**, 1378–1390.

Chapter 25

CASE STUDY: NOVELTY MEASUREMENT IN PAMPEAN GRASSLANDS

Pedro M. Tognetti

Departamento de Métodos Cuantitativos y Sistemas de Información, Facultad de Agronomia, University of Buenos Aires, Argentina

25.1 INTRODUCTION

Measuring ecosystem novelty seems to be a difficult but necessary challenge. Novelty measurement can be used for either conceptual or management uses. Harris et al. (Chapter 24) have developed a framework to indentify and measure novelty in ecosystems. The authors propose that a novel system is one with different composition, structure and function, that has crossed a potentially irreversible threshold and that shows ecological persistence. Furthermore, they include some approaches and variables to measure novelty, including some possible metrics. An example of novelty description in Pampean old fields in central Argentina is presented here as a case study. A regional database of remnant grasslands (Burkart et al. 2011) and a successional study system spanning more than 30 years (Omacini et al. 1995; Tognetti et al. 2010) have been employed.

Pampean grasslands have been severely modified by human activities since European settlement in the 1500s (León et al. 1984). This region was characterized by a treeless grassland covering more than $700,000\,km^2$ (Soriano 1991; Bilenca and Miñarro 2004). Regional subdivisions were constructed on the basis of hydrogeological features and floristic similarities (Soriano 1991). However, internal variability in floristic composition followed similar patterns in relation to topographic positions (Perelman et al. 2001). Detailed regional characterization can be found elsewhere (Hall et al. 1992; Soriano 1991; Perelman et al. 2001; Chaneton et al. 2002; Bilenca and Miñarro 2004).

Natural and human forces generated and maintained internal heterogeneity (León et al. 1984; Soriano 1991; Hall et al. 1992; Viglizzo et al. 2001). Nowadays, pampean grasslands are mostly converted to agriculture and pasture-based rangelands (Fig. 25.1; Viglizzo et al. 2001; Satorre 2005). Only small corridors of remnant grassland survive within the matrix (Bilenca and Miñarro 2004), but support a high diversity of plants (Burkart et al. 2011) and animals (Comparatore et al. 1996). Hence, grassland

Novel Ecosystems: Intervening in the New Ecological World Order, First Edition. Edited by Richard J. Hobbs, Eric S. Higgs, and Carol M. Hall.
© 2013 John Wiley & Sons, Ltd. Published 2013 by John Wiley & Sons, Ltd.

Figure 25.1 Some components of the Inland Pampa novel landscapes. (a) Remnant corridors dominated by native grasses (*P. quadrifarium*) may be transformed to (b) cattle foraging systems with the introduction of pasture species. Remnants could also be transformed to (c) annual crop fields. Abandonment of agriculture does not permit complete recovery of remnants and results in (d) old-field systems dominated by non-native grasses (*F. arundinacea*, *C. dactylon* and *S. halepense*). Photographs: P. M. Tognetti.

fragmentation (Baldi et al. 2006), plant invasions (Ghersa and León 1999; Chaneton et al. 2001; Poggio et al. 2010) and reductions in area of remnant grassland (Laterra et al. 2003; Perelman et al. 2003; Burkart et al. 2011) may lead to the generation of novel ecosystems (Hobbs et al. 2006, 2009). After agricultural land is abandoned, successional old fields reassemble into exotic-dominated grasslands, with a low community imprint of native species (Facelli et al. 1987; Omacini et al. 1995; Boccanelli et al. 1999; Tognetti et al. 2010). The objective of this chapter is to illustrate differences in community structure and composition for the assessment of novelty in the Inland Pampa, Argentina. As stated in Chapter 24, novelty implies not only differences in composition or structure, but the assessment of thresholds and persistence.

In this chapter the degree of novelty or compositional differences in grasslands reestablishing in ex-arable lands is described.

25.2   FLORISTIC AND STRUCTURAL NOVELTY

One way to construct novelty measures is based on the idea of a distance from historical conditions. In this sense, abiotic and biotic axes from Hobbs et al. (2009) may be measured using multivariate techniques that include different components of each axis. In order to describe degree of novelty in the Inland Pampa old fields, this chapter concentrates only on biotic axes; environmental axes could follow a similar treatment. Different perception levels may also be applied in the analysis. In the most detailed view, species composition and abundance will indicate major changes and the generation of 'novelty'. However, as stated before, in this type of analysis small changes in species composition could indicate the generation of novel or altered ecosystem states (Hobbs et al. 2009). Researchers can therefore also use coarse variables to identify novel assemblies. For example, functional groups or traits in community composition (Diaz and Cabido 1997, 2001) could be used as coarse variables to identify the distance from the original ecosystem state. However, specific or functional composition will refer to different aspects of novelty.

In this sense, research or management objectives can also define distance or novelty of the ecosystem. For example, in biodiversity or species-specific restoration programs, small changes in composition can determine a high degree of novelty. In this sense, Pampean grasslands have been enriched with around 50 species from all around the globe (Cabrera and Zardini 1978; Rapoport 1996; Ghersa and León 1999; Poggio and Mollard 2010). On the other hand, restoration of structural attributes or broad functional properties might be attained with different compositions. This implies large floristic differences but low functional novelty. In this sense, most invading plants in the Pampas were annual forbs and some annual and perennial grasses (Chaneton et al. 2002; Machera 2006). Similarly, functional redundancy of original and newly arrived species could absorb environmental variation independently of species origin, generating stable communities.

25.3   METHODOLOGICAL APPROACH

25.3.1   Objectives and scale of the analysis

The main objective of this analysis was to describe the degree of floristic and functional group novelty in old-field successional systems of the Inland Pampa. This comparison was made at community scale, using a historical database (Facelli et al. 1987; Omacini et al. 1995; Tognetti et al. 2010) and a regional community census (Burkart et al. 2011).

25.3.2   The starting and farthest points

As 'remnant' floristic composition, nine remnant grasslands were selected within the subregion of the Inland Pampa (Soriano 1991; Fig. 25.2; described as segment 'C' in Burkart et al. 2011) in the proximity of an old-field experiment that was used to provide 'novel assemblies' (Fig. 25.1). All of these grasslands correspond to the same general soil and climatic characteristics of the long-term study of secondary succession described by Facelli et al. (1987), Omacini et al. (1995) and Tognetti et al. (2010). For old-field floristic composition, nine samples of age 5–10 years and nine samples of age 15–20 years were randomly selected from a sample pool of c. 200 surveyed annually between 1977 and 2010 in ten adjacent plots (D'Angela et al. 1986; Omacini et al. 1995; Tognetti et al. 2010).

In the remnant grasslands inventory, data consisted of a complete list of vascular plants sampled in a $25\,m^2$ area in each stand (see Burkart et al. 2011 for details). In the successional old-field study, data consisted of a complete vascular plant list obtained from 20 quadrants of $1\,m^2$ randomly located in a $0.5\,ha$ plot. In spite of differences in data acquisition, for this general novelty assessment floristic descriptions are homologous between two study datasets. In both cases, sampling was carried out in late spring/early summer when most species are present and easily identifiable through their reproductive organs.

25.3.3   Selecting variables

To perform the comparison, the same database was examined from two different perspectives. First, species

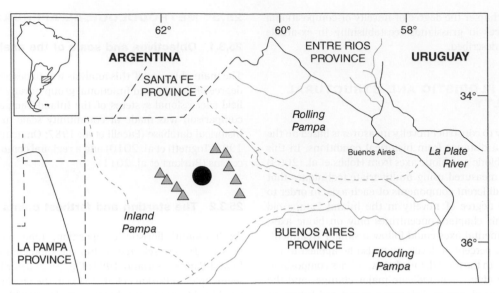

Figure 25.2 Remnant grasslands (triangles) were compared against old-field plots (filled circle) within the Inland Pampa subregion. This comparison was performed using floristic or an a priori functional composition. From Burkart (2011). Reproduced with permission of Springer.

names from the complete floristic list of all samples were considered as descriptive variables. Second, an a priori functional group classification (Diaz and Cabido 1997, 2001) of species by life form (graminoids or forbs, the former including Cyperaceae), life cycle (annual/perennial), growth season (cool and warm season) and origin (natives or non-natives; see Tognetti et al. 2010) was constructed. This produced 16 a priori functional groups.

25.3.4 Distance measurement: Multivariate indices and the current position

Among myriads of procedures, non-metrical multidimensional scaling (NMS) was used to reduce multivariate data into fewer axis coordinates (McCune and Grace 2002). A great advantage of NMS is that it makes few assumptions about the data. In brief, a Euclidean distance matrix is constructed with the original data, and this matrix is contrasted with the n-dimensional arrangement of data in the ordination space. This procedure is repeated until a minimal stress is found, where stress is the distance between original and simulated data. A 1D configuration was chosen because of the simpler graphical approach (for more

details about this ordination technique see Legendre and Legendre 1998; McCune and Mefford 1999; McCune and Grace 2002). All sampling units (each census in remnants and old fields) were included in the analysis without weighting factors. Euclidean distance was preferred because it stresses the differences between samplings (McCune and Mefford 1999). Statistical analysis was performed using a one-way analysis of variance (ANOVA) with position on ordination axis as a dependent variable of grassland type. Planned comparisons were performed considering a priori differences in historical development of remnants and old fields, while unplanned comparisons were carried out using the Tukey test.

25.4 RESULTS

Floristic composition differed between remnant grasslands and old-field communities ($F_{2,24} = 45.8$; $P < 0.001$). While old-field grasslands did not differ in ordination position between mid-successional and later stages (Fig. 25.3), the positions of remnant grasslands on the ordination axes did not overlap at all (Tukey LSD = 0.56 NMS units). On the other hand, ordination of grassland using functional group composition also

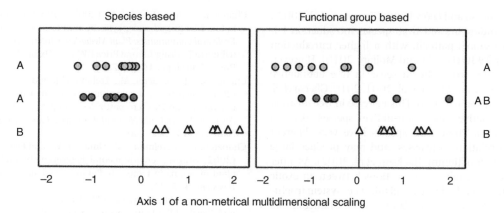

Figure 25.3 Position of grassland communities along ordination axis using species (left) or functional group composition (right) in the analysis. Using species composition, old-field communities (5–10-year-old communities: grey circles/15–20-year-old communities: black circles) were clearly different from remnants (triangles). However, when functional composition was used as a variable, remnant grasslands did not differ from 15–20-year-old old-field plots. Letters on both margins indicate significant differences among grassland type within each panel.

differed between old fields and remnants (planned comparison of old fields versus remnants: $F_{1,24} = 11.5$; $P < 0.005$). However, the overlapping was higher using functional instead of species composition. Moreover, 15–20-year-old old fields did not differ from remnants (Tukey LSD = 1.02 NMS units; Fig. 25.3). Data exploration shows that abundance of annual winter grasses is higher in younger successional plots, while perennial summer and winter grasses prevail in remnants and older plots (15–20 years).

25.5 DISCUSSION

This contribution clearly shows that novelty assessment is dependent upon whether species or functional compositions are used. Based only on floristic data, old-field communities in Inland Pampas clearly established novel assemblies (Omacini et al. 1995; Tognetti et al. 2010). However, considering life form, life span and growing season does not allow a clear differentiation of native-dominated remnant grasslands from non-native-dominated old-field communities.

Comparing this result with Hobbs et al. (2009), the differences between novel and hybrid assemblies are structural (biotic or abiotic) ecosystem features. On the one hand, hybrid systems retain several characteristics of the historical system. Its composition or functioning is no longer within the natural range of variability,

however. On the other hand, novel systems are composed entirely of exotic species from disparate parts of the world. In this work, all three systems are strictly hybrids, containing a mixture of native and exotic species in different proportions (Tognetti et al. 2010; Burkart et al. 2011). This stresses the need to include functional description to the original Hobbs et al. (2009) model, in order to distinguish between novel, hybrid and historical systems which may support the same guilds.

Surprisingly, there was very low congruence between remnants and old fields on the basis of species composition; this was higher when a priori functional groups were considered, however. Despite the fact that non-metrical ordination axis are arbitrary, the Bray–Crutis dissimilarity index with floristic composition indicated a difference close to 60%, while difference decreased using a priori groups. It is possible that the length of the species versus group list influenced dissimilarity measures (McCune and Mefford 1999). However, it also shows that environmental filters might constrain traits from the species pool, generating rules for community assembly (Weiher and Keddy 1999). In some ways, this may suggest that Pampean grasslands have crossed biotic thresholds (Cramer et al. 2008) but have not changed so dramatically that a large shift in ecosystem structure resulted.

The regional species pool was enriched with the introduction of crop weeds and other non-native species (Ghersa and León 1999; Chaneton et al. 2002). Recent studies showed that species introduction followed a sigmoid pattern, with a higher introduction rate near 1930 (Poggio and Mollard 2010). However, other studies stress the increase in tree abundance within the region (Mazía et al. 2001, 2010; Ghersa et al. 2002; Mazía et al. 2010). This increase in tree invasion includes native and non-native species (Ghersa et al. 2002). These new species have very different traits in relation to grasses, and may produce large changes in functioning (Jackson et al. 2002). As other examples, mutualistic interactions between an exotic grass and an endophytic fungi affect ecosystem trophic structure (Omacini et al. 2006, 2009; Chaneton and Omacini 2007). A threshold differentiating hybrid or novel systems might therefore be the inclusion of a trait in the species pool (native or exotic) or new interactions which severely modify ecosystem functioning.

Clearly, this contribution did not entirely resolve the best way to state and measure novelty in ecosystems. In fact, it raised more questions on the assembly of these communities. However, the main value of the novel ecosystems framework is based on the services they provide, despite their novel composition. Novelty description may have theoretical, conceptual or management purposes. In any case, the description will use variables that will inform particular aspects of community. The real distance will emerge from a multifocal approach, considering not only quantitative variables but historical references and current dynamics.

REFERENCES

Baldi, G., Guerschman, J.P. and Paruelo, J.M. (2006) Characterizing fragmentation in temperate South America grasslands. *Agriculture, Ecosystems and Environment*, **116**, 197–208.

Bilenca, D., Miñarro, F. (2004) *Identificación de Áreas Valiosas de Pastizal (AVPs) en las Pampas y Campos de Argentina, Uruguay y sur de Brasil*. Fundación Vida Silvestre Argentina, Buenos Aires.

Boccanelli, S., Pire, E. Torres, P. and Lewis, J.P. (1999) Cambios en la vegetación de un campo abandonado después de un cultivo de trigo. *Pesquisa Agropecuaria Brasilera*, **34**, 151–157.

Burkart, S.E., León, R.J.C, Conde, M.C. and Perelman, S.B. (2011) High native diversity in remnant grasslands on productive soils of the cropping Pampa. *Plant Ecology*, **212**, 1009–1024.

Cabrera, A. and Zardini, E. (1978) *Manual de la flora de los alrededores de Buenos Aires*. ACME, Buenos Aires.

Chaneton E.J. and Omacini, M. (2007) Bottom-up cascades induced by fungal endophytes in multitrophic systems, in *Ecological Communities: Plant Mediation in Indirect Interaction* (Ohgushi T., Craig T.P. and Price P.W. eds), Oxford University Press, Oxford, pp. 164–187.

Chaneton, E.J., Omacini, M., Trebino, H.J. and León, R.J.C. (2001) Disturbios, dominancia y diversidad de especies nativas y exóticas en pastizales pampeanos húmedos. *Anales de la Academia Nacional de Ciencias Exactas, Físicas y Naturales*, **53**, 121–140.

Chaneton, E.J., Perelman, S.B., Omacini, M., and León, R.J.C. (2002) Grazing, environmental heterogeneity, and alien plant invasions in temperate Pampa grasslands. *Biological Invasions*, **4**, 7–24.

Comparatore, V.M., Martínez M.M., Vassallo, A.I., Barg, M.I. and Sacch, J.P. (1996) Abundancia y relaciones con el hábitat de aves y mamíferos en pastizales de *Paspalum quadrifarium* (paja colorada) manejados con fuego (Provincia de Buenos Aires, Argentina). *Interciencia*, **24**, 228–237.

Cramer V., Hobbs, R. and Standish, R. (2008) What's new about old fields? Land abandonment and ecosystem assembly. *Trends in Ecology and Evolution*, **23**, 104–112.

D'Angela, E., Facelli, J.M. and León, R.J.C. (1986) Pioneer stages in a secondary succession of a Pampean subhunid grassland. *Flora*, **178**, 261–270.

Diaz, S. and Cabido, M. (1997) Plant functional types and ecosystem function in relation to global change. *Journal of Vegetation Science*, **8**, 463–474.

Diaz, S. and Cabido, M. (2001) Vive la difference: plant functional diversity matters to ecosys-tem processes. *Trends Ecology and Evolution*, **16**, 646–655.

Facelli, J.M., D'Angela, E. and León, R.J.C. (1987) Diversity changes during pioneer stages in a subhumid Pampean grassland succession. *American Midland Naturalist*, **117**, 17–25.

Ghersa, C.M. and León, R.J.C. (1999) Successional changes in agroecosystems of the Rolling Pampa, in *Ecosystems of Disturbed Ground* (L. Walker, ed.), Elsevier, Amsterdam, pp. 487–502.

Ghersa, C.M., de la Fuente, E.B., Suarez, S. and León, R.J.C. (2002) Woody species invasion in the Rolling Pampa grasslands, Argentina. *Agriculture, Ecosystems and Environment*, **88**, 271–278.

Hall, A., Rebella, C., Ghersa, C. and Culot, P. (1992) Field-crop systems of the Pampas, in *Field Crop Ecosystems* (Pearson, C. ed.), Elsevier, Amsterdam, pp. 413–450.

Hobbs, R.J., Arico, S., Aronson, J. et al. (2006) Novel ecosystems: theoretical and management aspects of the new ecological world order. *Global Ecology & Biogeography*, **15**, 1–7.

Hobbs, R.J., Higgs, E. and Harris, J.A. (2009) Novel ecosystems: implications for conservation and restoration. *Trends in Ecology and Evolution*, **24**, 599–605.

Jackson, R.B., Banner, J.L., Jobbágy, E.G., Pockman, W.T. and Wall, D.H. (2002) Ecosystem carbon loss with woody plant invasion of grasslands. *Nature*, **418**, 623–626.

Laterra, P., Vignolio, O., Linares, M., Giaquinta, A. and Maceira, N. (2003) Cumulative effects of fire on a tussock pampa grassland. *Journal of Vegetation Science*, **14**, 43–54.

Legendre, P. and Legendre, L. (1998) *Numerical Ecology*. Elsevier Science & Technology.

León, R.J.C., Rusch, G.M. and Oesterheld, M. (1984) Pastizales pampeanos – impacto agropecuario. *Phytocoenologia*, **12**, 201–218.

Machera, M. (2006) La invasión de pastizales por especies exóticas: el papel de los disturbios de diferente escala espacial. Doctoral disertation and thesis. Universidad de Buenos Aires.

Mazía, C.N., Chaneton, E.J., Ghersa, C.M. and León, R.J.C. (2001) Limits to tree species invasion in pampean grassland and forest plant communities. *Oecologia*, **128**, 594–602.

Mazía, C.N., Chaneton, E.J., Machera, M., Uchitel, A., Feler, M.V. and Ghersa, C.M. (2010) Antagonistic effects of large- and small-scale disturbances on exotic tree invasion in a native tussock grassland relict. *Biological Invasions*, **12**, 3109–3122.

McCune, B. and Mefford, M.J. (1999) PC-Ord for Windows. MjM Software, Gleneden Beach, Oregon.

McCune, B. and Grace, J.B. (2002) *Analysis of Ecological Communities*. MjM Software, Gleneden Beach, Oregon, USA.

Omacini, M., Chaneton, E.J., León, R.J.C. and Batista, W.B. (1995) Old-field successional dynamics on the Inland Pampa, Argentina. *Journal of Vegetation Science*, **6**, 309–316.

Omacini, M., Eggers, T., Bonkowsky, M., Gange, A. and Jones, T.H. (2006) Leaf endophytes affect mycorrhizal status and growth of co-infected and neighbouring plants. *Functional Ecology*, **20**, 226–232.

Omacini, M., Chaneton, E.J., Bush, L. and Ghersa, C.M. (2009) A fungal endosymbiont affects host plant recruitment through seed and litter mediated mechanisms. *Functional Ecology*, **23**, 1148–1156.

Perelman, S.B., Leon, R.J.C. and Oesterheld, M. (2001) Cross-scale vegetation patterns of Flooding Pampa grasslands. *Journal of Ecology*, **89**, 562–577.

Perelman, S.B., Burkart, S. and León, R.J.C. (2003) The role of a native tussock-grass (Paspalum quadrifarium) in structuring plant communities in the Flooding Pampa grasslands, Argentina. *Biodiversity and Conservation*, **12**, 225–238.

Poggio, S.L. and Mollard, F.P.O. (2010) The alien weed flora of the argentine pampas: disentangling the ecological and historical patterns involved in its formation. *15th European Weed Research Society Symposium*, Kaposvár, Hungría.

Poggio, S.L., Chanenton, E.J. and Ghersa, C.M. (2010) Landscape complexity differentially affects alpha, beta, and gamma diversities of plants occurring in fencerows and crop fields. *Biological Conservation*, **143**, 2477–2486.

Rapoport, E.H. (1996) The flora of Buenos Aires: low richness or mass extinction? *International Journal of Ecology and Environmental Sciences*, **22**, 217–242.

Satorre, E.H. (2005) Cambios tecnológicos en la agricultura argentina. *Ciencia Hoy*, **15**, 24–31.

Soriano, A. (1991) Río de la Plata grasslands, in *Natural Grasslands: Introduction and Western Hemisphere* (Coupland, R.T. ed.). Ecosystems of the World 8A. Elsevier, Amsterdam, pp. 367–407.

Tognetti, P.M., Chaneton, E.J., Omacini, M. and León, R.J.C. (2010) Exotic vs. native plant dominance over 20 years of old-field succession on set-aside farmland in Argentina. *Biological Conservation*, **143**, 2494–2503.

Viglizzo, E.F., Lértora, F., Pordomingo, A.J., Bernardos, J.N., Roberto, Z.E. and Del Valle, H. (2001) Ecological lessons and applications from one century of low external-input farming in the pampas of Argentina. *Agriculture, Ecosystems & Environment*, **83**, 65–81.

Weiher, E. and Keddy, P. (1999) *Ecological Assembly Rules: Perspectives, Advances, Retreats*. Cambridge University Press, Cambridge.

FURTHER READING

Chaneton, E.J., Mazía, C.N., Machera, M., Uchitel, A. and Ghersa, C.M. (2004) Establishment of honey locust (*Gleditsia triacanthos*) in burned pampean grasslands. *Weed Technology*, **18**, 1325–1329.

Darwin, C. (1893) *Journal of Researches into the Geology and Natural History of the Various Countries Visited by the H.M.S. Beagle*. Henry Colburn, London.

Facelli, J.M. and D'Angela, E. (1990) Directionality, convergence, and rate of change during early succession in the Inland Pampa, Argentina. *Journal of Vegetation Science*, **1**, 255–260.

Laterra, P. (1997) Post-burn recovery in the flooding pampa: impact of an invasive legume. *Journal of Range Management*, **50**, 274–277.

Uchitel, A., Chaneton, E.J. and Omacini, M. (2010) Inherited fungal symbionts enhance establishment of an invasive annual grass across successional habitats. *Oecologia*, **165**, 465–475.

Viglizzo, E.F. and Frank, F. (2006) Land-use options for Del Plata Basin in South America: Tradeoffs analysis based on ecosystem service provision. *Ecological Economics*, **57**, 140–151.

Chapter 26

PLANT MATERIALS FOR NOVEL ECOSYSTEMS

Thomas A. Jones

USDA Agricultural Research Service, Logan, Utah, USA

26.1 THE CURRENT STATUS OF RESTORATION PLANT MATERIALS

26.1.1 The appeal of local plant materials

The commonly held preservationist approach to restoration plant materials has the locally adapted population as its centerpiece (Johnson et al. 2010) and is rooted in the conservation biology tradition (Sackville Hamilton 2001). Over forty years ago, Namkoong (1969) stated that the traditional assumption in forest genetics was 'local is best'; over a decade ago Wilkinson (2001) stated that the use of local provenance was the standard approach for restoration plant materials. Part of the emphasis on local populations for restoration relates to the importance placed on the 'precautionary principle' (Broadhurst et al. 2008). This principle encompasses the idea that no risk should be taken in restoration practice that might incur damage to an ecosystem. Furthermore, if action is necessary, it should be as minimally invasive as possible. Hence, the precautionary view is that only 'genetically appropriate' (i.e. local, non-genetically manipulated) materials should be used for ecological restoration (Johnson et al. 2010).

Many justifications have been given for the exclusive use of local plant material in restoration (Montalvo et al. 1997; Endler et al. 2010; Johnson et al. 2010). Features of local populations regarded as critical include adaptation to local environmental conditions and the presence of 'genetic memory', (i.e. adaptation that has been shaped by past selective events) (Montalvo et al. 1997). Undesirable features of non-local germplasm include the potential for: outbreeding depression upon hybridization with remnant local plants; genetic pollution of remnant local populations; inadvertent undesirable selection during commercial propagation (genetic shift); disruption of the natural geographic distribution of genetic arrays; inability to deliver ecosystem services; a compromised ability to respond to climate change; and undesirable cascading effects onto other species (e.g. pollinators) (Montalvo et al. 1997;

Novel Ecosystems: Intervening in the New Ecological World Order, First Edition. Edited by Richard J. Hobbs, Eric S. Higgs, and Carol M. Hall. © 2013 John Wiley & Sons, Ltd. Published 2013 by John Wiley & Sons, Ltd.

Endler et al. 2010; Johnson et al. 2010). Because of the history of invasive species resulting from introductions, non-local natives may arouse the suspicion among preservationists of potential catastrophe (Jones and Robins 2011, section 3.4).

The importance attached to native plants in general and local populations in particular can be partly explained by a 'romance with the familiar' (Marris 2011). Humans value protected areas and naturally wish to preserve natural ecosystems, and restoration with local plant materials is part of this ethos. Not surprisingly, then, ecological restoration has developed as a value-laden applied science (Davis and Slobodkin 2004b; Choi 2007).

26.1.2 The current divergence of professional viewpoints

The preservationist approach to restoration plant materials (described earlier) emphasizes taxonomic and genetic patterns and is predictably widely supported by biological disciplines that emphasize 'patterns' (across space), such as systematists and population biologists. This approach is compatible with the Society for Ecological Restoration's SER1994 definition of ecological restoration: "the process of repairing damage caused by humans to the diversity and dynamics of indigenous ecosystems" (Jackson et al. 1995). This definition has since been replaced by SER with the more flexible and inclusive definition: "the process of assisting the recovery of an ecosystem that has been degraded, damaged, or destroyed" (SER Science & Policy Working Group 2004). This 2004 definition removed the reference to 'indigenous ecosystems' found in the 1994 version. In addition, in its 2004 primer SER recognized that the introduction of diverse genetic stock, as opposed to only local material, is justified when sites display substantial damage and altered physical environments. The primer also states that some non-native species may perform ecological roles formerly played by native species that are no longer prominent at the site.

Relative to the 1994 definition, the 2004 definition of ecological restoration can be seen as more compatible with the interventionist approach. This approach is more willing to take action to ameliorate ecosystem function in order to improve the delivery of the desired ecosystem services. As such, this approach is more likely to be supported by ecosystem-level ecologists who emphasize ecosystem 'processes' (across time) rather than ecosystem patterns (Levin 1992). The debate between the preservationist and interventionist approaches could therefore be succinctly summarized as ecosystem 'patterns versus processes'.

The liberalizing of the definition of ecological restoration has arisen as the restoration ecology profession has come to realize that the challenges of restoration are immense and that 'natural' ecosystems are no longer the norm (Bridgewater et al. 2011). Novel ecosystems may present a nearly intractable situation to the restoration practitioner (Bridgewater et al. 2011). Consequently, based on the novel conditions, it may be desirable to take action to direct a novel ecosystem on a desirable ecological trajectory (Dobson et al. 1997; Vitousek et al. 1997).

The professional divide between the preservationist and interventionist approaches is sizeable and shows few signs of disappearing in the near term. Despite this, individual conservation biologists have begun to advocate for a more flexible approach to biodiversity as a matter of triage or last resort (Parker et al. 2010). The conservation biology profession has come to see that, alone, the traditional wilderness conservation approach is increasingly inadequate to conserve biodiversity (Wilkinson 2004a; Didham 2011). Some proponents of local plant materials acknowledge that performance of local populations may not be optimal for an ecosystem that has been substantially altered (Johnson et al. 2010). Additionally, some restoration ecologists have in recent years recognized the value of non-native species in ecological restoration (Ewel and Putz 2004; Agosta and Klemens 2008; Davies et al. 2010; Goodenough 2010).

The creation of novel ecosystems under triage circumstances is increasingly being considered as a viable option (Bottrill et al. 2008; Mooney 2010). From a conservation viewpoint, novel ecosystems provide new opportunities for land stewardship (Chapin et al. 2009, also see Chapter 39). Because the interventionist approach specifically addresses ecosystems that have been degraded or converted to an alternate state, it does not necessarily conflict with the preservationist movement which has always been primarily concerned with maintenance of relatively pristine lands. Nevertheless, Didham (2011) expresses the legitimate concern that natural wildlands could be negatively affected if a new emphasis on novel ecosystems diverts attention and funding away from traditional preservation efforts.

26.2 RECONCILIATION OF THE TWO APPROACHES

Reconciliation of the two approaches may come from examining Whisenant's (2002) model of the three stages of ecosystem degradation (Fig. 26.1). The preservationist approach may have the greatest potential for success in Whisenant's (2002) Stage A or early in Stage B, prior to the point where ecological function becomes seriously compromised (Fig. 26.1). The preservationist approach relies on the restoration of specific genotypes (Wilkinson 2004a) and seems to rely on these pre-disturbance genotypes to regenerate the functional properties of the pre-disturbance ecosystem. However, the preservationist approach is less likely to be effective as ecosystem function becomes increasingly compromised. This occurs following the crossing of a biological threshold into Stage B, and especially once an abiotic threshold has been crossed into Stage C (Fig. 26.1). It is likely that, under these conditions, the interventionist approach may result in the greatest opportunity to return to a more desirable level of ecosystem function and to increase the delivery of desired ecosystem services.

The Restoration Gene Pool (RGP) concept (Jones 2003) has been proposed in an attempt: (1) to reconcile the preservationist and interventionist points of view for the application of restoration plant materials, and; (2) to provide a comprehensive plant materials framework for restoration. The RGP concept serves as a restoration analog to the gene pool concept developed for germplasm classification outlined by Harlan and deWet (1971). The RGP concept consists of four RGPs, ordered from greatest to least genetic similarity (and greatest to least preference to the restoration practitioner) relative to the target (local) population, termed primary (1°), secondary (2°), tertiary (3°) and quaternary (4°). The RGP concept shows respect for the preservationist viewpoint by recognizing the precautionary principle, as evidenced by the placement of local material of the target taxon in the 1° (most preferred) RGP. Nevertheless, the 1° RGP, the restoration practitioner's preferred choice, may no longer be sufficiently adapted, not economically feasible or not logistically possible. In such

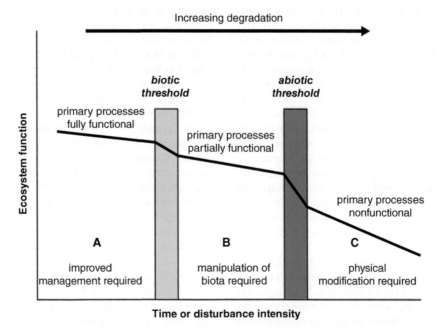

Figure 26.1 Three stages of degradation of ecosystem function (A, B, and C) separated by biotic (A→B) and abiotic (B→C) thresholds. With permission from Steve Whisenant.

cases, the RGP concept recognizes the interventionist viewpoint, as represented by the 2°, 3° and 4° RGPs.

Because maximizing genetic identity by choosing the 1° RGP does not necessarily maximize ecological adaptation, particularly when the restoration site has been modified, a higher-order RGP may be substituted. In contrast to the 1° RGP, the 2° RGP consists of non-local material of the target taxon. In the 3° RGP, which is really something of a special case unavailable for many species (Jones and Monaco 2007), material is developed by hybridization of a closely related taxon with the target taxon in order to introduce desirable traits. An example would be the development of blight-resistant American chestnut (*Castanea dentata*). Because blight resistance could not be found in the

American chestnut, it was hybridized with Chinese chestnut (*C. mollissima*) and the hybrid was back-crossed three times to American chestnut while artificially selecting for blight resistance (Hebard 2006). In contrast to the 3° RGP, the 4° RGP consists of distinct taxa of the same functional guild as the target taxon, but that can be more effective in the restoration environment than the target taxon.

Jones and Monaco (2007) provided a flowchart for use as a systematic tool to choose the most appropriate RGP and to justify the choice made by the restoration practitioner under the given circumstances (Fig. 26.2). By choosing the appropriate RGP based on cumulative experience, the RGP concept incorporates the principle of adaptive management (Holling 1978).

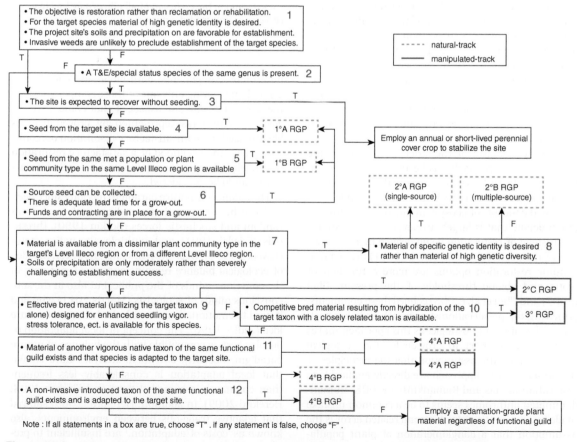

Figure 26.2 A decision-making flowchart for the Restoration Gene Pool concept (Jones and Monaco 2007).

26.3 ISSUES RELATED TO THE PRESERVATIONIST APPROACH FOR NOVEL ECOSYSTEMS

26.3.1 Ecological issues

A commonly held view by the general public is that 'wild' environments still exist in nature (Hobbs et al. 2010) and, when they have been altered, ecological restoration may return them to the pre-disturbance state (Hobbs et al. 2011). In contrast, restoration ecologists have been more circumspect regarding their ability to achieve such ambitious societal expectations (Zedler and Callaway 1999; Ashley et al. 2003; Hobbs 2004). The reality is that 37% of the Earth's terrestrial surface is occupied by novel ecosystems, and the Earth's remaining natural ecosystems are highly fragmented and embedded in areas modified by human activity, termed anthropological biomes or 'anthromes' (Ellis and Ramankutty 2008). Furthermore, anthropogenically driven effects such as climate change, ozone depletion and nitrogen deposition affect ecosystems worldwide (Arneth et al. 2010). Hobbs et al. (2011) stated that the term 'restoration' can be a false promise because it conveys the idea that a return to the original state is actually possible. They further contend that, when policymakers and the public embrace restoration without understanding its limitations, perverse policy outcomes may result.

Recently, criticism of the strict application of the preservationist approach has become more frequent. Hobbs and Harris (2001) pointed out that many restoration projects are rooted in outdated and now-discarded concepts of ecosystem function, resulting in much debate that is largely irrelevant or unanswerable. Instead of restoration being defined by an ecosystem of the past, Hobbs and Harris (2001) argued that genuine restoration options are more a function of biotic and abiotic thresholds of the present. Choi (2007) and Choi et al. (2008) echoed these comments, criticizing the preservationist approach as being past-oriented, static and nostalgic, as this approach makes little conscious effort to provide for the ecological future. Making the assumption that the principles of the preservationist approach are directly applicable to restoration in Ellis and Ramankutty's (2008) human-dominated anthromes seems to be a common mistake made by preservationists. A second related error is the assumption that a conglomeration of plant populations constitutes an ecosystem. Instead, ecosystems

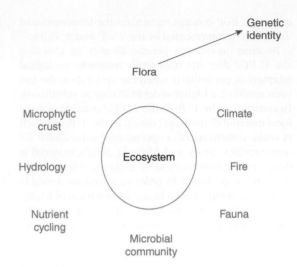

Figure 26.3 The ecosystem has many interactive components, but an emphasis on a single component (see red lettering) to the exclusion of all others may lead to a perverse or unintended outcome.

have many components, both biotic and abiotic (Fig. 26.3). Emphasizing only one of those components (e.g. the flora) and only one aspect of that component, its genetic identity (as represented by provenance), is likely to lead to an unsatisfactory restoration outcome. To increase the likelihood of a satisfactory restoration result, all ecosystem components and their interactions should be considered.

The balance-of-nature paradigm has long since been replaced by a contrasting paradigm that emphasizes random and stochastic forces (Botkin 1990), thresholds (Friedel 1991) and multiple alternative states (Laycock 1991). Over 20 years later, however, the idea of ecological balance remains engrained, not only in the conscious minds of the public, but also in the subconscious minds of ecologists (Marris 2011) who find it hard to emotionally part with the balance-of-nature idea (Zedler 1996). The concept of local adaptation or 'local is best' is compatible with the balance-of-nature paradigm. Recently, two meta-analyses have shown that local adaptation is considerably less frequent than commonly assumed (Leimu and Fischer 2008; Hereford 2009). In addition, Hereford (2009) showed that adaptive trade-offs between environments, also known as 'costs of adaptation,' are insufficient to preclude simultaneous adaptation to multiple environments. This suggests that general adaptation across a range of environments is more common than generally believed.

Because novel ecosystems are becoming the norm, multiple authors have contended that ecological *processes* must trump ecological *patterns* in their importance in restoration, making dynamic systems themselves the target (Wilkinson 2004a; Choi 2007; Jackson and Hobbs 2009). Likewise, it is likely that empirical performance of a plant material will ultimately trump its provenance as the primary consideration for the choice of plant materials for restoration of novel ecosystems (Jones et al. 2010). In this vein, Handel et al. (1994) suggested a search for genotypes adapted to various environmental stresses that may be able to solve particular and severe restoration problems.

Montalvo et al. (1997) have argued that non-local genotypes are not sustainable as part of a restoration strategy. However, it follows from this logic that when an environment becomes novel, its local genotypes are no longer sustainable as they are not evolutionary products of the novel environment. To enhance sustainability in the novel environment, humans may therefore need to assist nature by developing new genotypes specifically adapted to the novel conditions (Jones and Monaco 2009; Prévot-Julliard et al. 2011). As the frequency of novel ecosystems increases (Ellis et al. 2010, also see Chapter 8), cultural values, sentimentalities and priorities are likely to shift as the belief in retaining ecosystems as particular assemblages increasingly shackles restoration efforts to unrealistic, unsustainable and anachronistic objectives (Choi 2004; Hobbs et al. 2009; Jackson and Hobbs 2009).

26.3.2 Evolutionary and genetic issues

Clewell (2000) regarded the concept of historical authenticity as a target for ecological restoration to be more of a burden than a benefit. As noted by Seastedt et al. (2008), historical coevolutionary relationships are becoming increasingly uncommon as ecosystems are increasingly modified, yet documentation of both micro-evolutionary responses and novel biotic environments is increasing (Ashley et al. 2003). Clewell (2000) believes that genuine restoration is necessarily *natural* because of its emphasis on natural autogenic processes. He goes on to argue that because genuine restoration is *authentic* (i.e. it is ultimately directed by

nature and not by the practitioner) it cannot guarantee a fixed endpoint, as pursued by preservationists. If his argument is correct then the preservationist approach is fundamentally non-Darwinian, as it seeks to control the evolutionary outcome rather than allowing natural selection to take its course. Another non-Darwinian tendency of preservationists is to place faith in natural selection as an optimizing agent; this results in local genotypes being regarded as optimal, an attribute of natural selection that Darwin ardently denied (Gould and Lewontin 1979; Gould 1998).

Some evolutionary biologists such as Broadhurst et al. (2008) are recognizing that strict adherence to a local-only policy may be detrimental. These authors advise against using wildland seed collections from fragmented habitats because of insufficient evolutionary potential due to inbreeding or low genetic variation. They argue that these features compromise a germplasm's ability to meet new environmental challenges. In addition, they see the merit in specific genotypes (2°, 3° RGPs) or functionally similar non-native species (4° RGP) designed to overcome these new challenges. Broadhurst et al. (2008) contend that the benefits of collecting for genetic diversity may be compromised by strict adherence to the 'local is best' paradigm. While they agree with Johnson et al. (2010) that the local gene pool is an excellent starting point for a restoration population, Broadhurst et al. (2008) believe that local seed sources are inherently limited for ecological restoration, particularly across broad geographic scales. In addition, they reject the argument that remnant populations may be negatively affected by contact with non-local material (e.g. outbreeding depression) on the basis that this problem is minor relative to the severity of the land-degradation problem.

Broadhurst et al. (2008) argue that maintaining genetic *processes* by redistributing genes across the landscape is much more important than concentrating on reinstating local genetic *patterns* in unstable environments. Genotypes are products of biological history, contingency and stochastic change (Gould 1998) and, as a consequence of the continuing influence of man, the pace of environmental change is increasing (Kareiva et al. 2007). Harris et al. (2006) warned that local populations may be a 'genetic dead-end', with a potentially limited ability for evolutionary response as novel ecosystems with heightened environmental challenges come to predominate. Nevertheless contemporary evolution (Stockwell et al. 2003), an increase in ecological fitness over relatively short timeframes, may

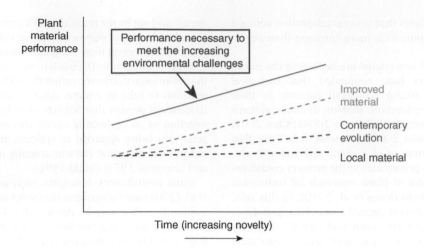

Figure 26.4 As novel ecosystems become entrenched over time and concomitant environmental challenges increase, the gap between the performance necessary to meet these challenges (blue) and the performance of local material (dashed red) increases unless the difference is made up by contemporary evolution (dashed orange). Improved material specifically designed to be adapted to the modified conditions (dashed green) may offer the best potential to narrow this gap.

offer potential for adaptation of local populations to novel conditions if suitable and sufficient genetic variation is present (Fig. 26.4).

26.3.3 Practical issues

Minteer and Collins (2010) have encapsulated an increasingly widely held sentiment, namely that our options for saving threatened organisms are dwindling, time is running out and we must be willing to search for new solutions. They stated that the challenge of the future will be to create new species combinations in novel ecosystems (restoration) without disrupting the historical species in traditional ecosystems (preservation). Increasingly, restoration professionals are coming to the realization that, overall, an overly prescriptive approach that limits restoration options is not in the best interests of the planet (Davis and Slobodkin 2004a). Restoration is rarely easy and is becoming increasingly challenging (Zedler and Callaway 1999; Suding et al. 2004). While idealism is laudable, preservationist attitudes tend to give way to more realistic contemporary solutions when failures are experienced (Hobbs 2004; Wilkinson 2004b).

Rational arguments can be made both for and against a local-only policy, as reflected in a wide range of attitudes on the part of on-the-ground suppliers of native seeds (Smith et al. 2007). For example, Smith et al.'s (2007) survey found that, while some businesses promote local plant materials to maximize adaptation, maintain genetic integrity and minimize the potential for outbreeding depression, one seed industry professional argued that the presumed benefits of local seed often did not apply in practice. Half of the survey respondents indicated that the technical and financial challenges of collecting wildland seeds deter their use. They stated that the cost of harvesting wildland seeds or producing limited quantities of seed of local populations under cultivation usually far outweighs the benefit. Theoretical issues surrounding restoration are therefore sometimes deemphasized because of pervasive practical deterrents.

26.4 OBJECTIVES FOR PLANT MATERIALS FOR NOVEL ECOSYSTEMS

26.4.1 The utility of novel ecosystems

A unifying concept for novel ecosystems is that the transformation of ecosystems as a result of intentional or inadvertent human activities is increasing, and this transformation is to some extent irreversible (Hobbs et al. 2009). Ecosystems should therefore be managed to stimulate an ecological trajectory that results in inherent value (Hobbs and Harris 2001). Kareiva et al. (2007) of The Nature Conservancy (USA) called for the wise domestication of nature, i.e. the selection of desirable ecosystem attributes and the thoughtful management of trade-offs among ecosystem services so that both man and nature might thrive. Wilkinson (2004a) came to the conclusion that, because of the extent of ecosystem modification on a worldwide scale, an approach based on entities (e.g. local genotypes) would have to be at least partly abandoned in favor of an ecosystem process-based approach emphasizing ecosystem services. Harris et al. (2006) regard many restoration goals of the past to be 'overly prescriptive', and they anticipate a new emphasis on generating functional, emergent and designer ecosystems for the future. While admitting that designed ecosystems would be anathema to some, Palmer et al. (2004) have argued that they will be part of the future sustainable world and will be managed for provision of ecosystem services. Prévot-Julliard et al. (2011) advocated assisting nature to increase sustainability and stated that non-native species may be important in restoration when they contribute to ecosystem function, particularly if they confer social or economic advantages. Vitousek et al. (1997) decried the loss of local populations to land transformation, recognizing that increasing levels of human investment will be necessary to save the Earth's remaining wildlands. In other words, sustaining pristine wildlands will likely become increasingly problematic, and options in addition to preservation must be found (Hobbs et al. 2010).

26.4.2 Critical functional traits

Plant materials that display desired functional traits are critical for the development of functioning ecosystems. However, within the discipline of community ecology, traits have long been overshadowed by an emphasis on taxonomic entities, resulting in a loss of general ecological mechanistic understanding (McGill et al. 2006). Jones et al. (2010) assembled a list of functional traits for desirable perennials that may stimulate ecological processes that in turn may overcome obstacles associated with annual-grass invasion of sagebrush-steppe rangelands. In a study of 28 species used in prairie restoration in western Oregon, Roberts et al. (2010) identified traits and sets of traits associated with performance variables such as establishment frequency and plant cover. They also found that these relationships varied among the three functional groups in the study (annual forb, perennial forb and perennial grass). These researchers hope to create trait-based models to predict species performance in restoration, thus improving the efficacy of restoration seed mixes.

Wilkinson's (2004b) parable of the assembly of a novel ecosystem, the Green Mountain cloud forest on Ascension Island in the south Atlantic, is most relevant for ecological restoration today. Over the course of several decades following Joseph Hooker's visit in 1843, this desert island, devoid of trees and cursed with minimal soil resources, was transformed by numerous introductions into a luxuriant tropical ecosystem with ample ecological function (Wilkinson 2004b). Because the non-native plants originated from a variety of locales, the novel flora of Ascension Island self-organized by the mechanism of ecological fitting (Janzen 1985) with little reliance on previously acquired coevolutionary relationships. Ecological fitting is the consequence of the placement of disjunct species in a common environment with the subsequent persistence of some species and loss of others based on their ecological fitness, leading to a functional novel ecosystem. The Ascension Island example has not been without some negative consequences for its meager endemic flora (Gray 2004). Nevertheless, it demonstrates what ecological restoration can accomplish when a variety of species with a variety of ecological functions come together and generate a myriad of novel ecological interactions, as Janzen (1985) described for naturally coexisting species in Santa Rosa National Park in Costa Rica.

26.4.3 Ecosystem services

The delivery of desired ecosystem services is of particular importance in the design or direction of novel ecosystems (Kareiva et al. 2007). Here there is an opportunity to encourage or insert populations, species or groups of species that confer desirable properties. Services that may be conferred by desirable plants include: aesthetic beauty; weed suppression and enhanced plant-community richness; water-quality improvement; water storage and flood abatement; enhanced plant cover and soil stability; soil genesis; carbon sequestration; 'tight' nutrient cycling; waste decomposition; wildfire suppression and air-quality improvement; climate regulation; biological nitrogen fixation; and, finally, provision of habitat, food and shelter for wildlife, pollinators and livestock (Daily et al. 1997; Zedler 2003; Gleason et al. 2008; Hardegree et al. 2011).

It has been claimed that only non-genetically manipulated, local, native (i.e. genetically appropriate) plant materials are sufficiently diverse and resilient to adapt to novel environments and to generate healthy self-sustaining landscapes that produce diverse products and provide ecosystem services (Johnson et al. 2010). However, for novel ecosystems it stands to reason that plant populations that possess functional traits that confer adaptation to the novel environment have the greatest potential to overcome obstacles that impede ecosystem recovery (Jones et al. 2010). Consequently, populations adapted to the novel environment may have the greatest potential to deliver ecosystem services. In addition, selection of plant materials for the ability to provide key ecosystem services offers an opportunity for the design of novel ecosystems, and this would seem to contribute to the wise domestication of nature *sensu* Kareiva et al. (2007).

26.4.4 Adaptation and fitness

In order for a plant to be a successful component of a novel ecosystem, it must be adapted to the environmental conditions of that ecosystem. From an evolutionary perspective, Sgrò et al. (2011) emphasized that highly adapted genotypes are critical in novel ecosystems, and they will not necessarily be found locally. While this concept has long been applied in crop improvement for agricultural environments (Sgrò et al. 2011), these authors argue that it is just as applicable for natural populations. For natural ecosystems, adaptation is something that has been presumed based on the mere presence of the plant at the site, although this may not always be the case (White and Walker 1997). On the other hand, because the populations of a novel ecosystem have not been exposed to a long history of natural selection at a site, adaptation to the novel environmental conditions must be verified by direct experimentation.

A key feature of an adapted population is ecological fitness, the ability of a population to reproduce and regenerate itself on a site. A population is less likely to express fitness in the restoration environment if its environment-of-origin is quite ecologically different (Raabová et al. 2007). In other words, it is unwise to translocate edaphic or climatic ecotypes or subspecies to ecologically distinct sites (van Andel 1998). This generalization highlights the importance of a thorough understanding of the biology and ecology of the species of interest when developing plant materials of that species. If an environmental stress characterizes the novel environment, an understanding of how the stress affects fitness (Goldberg and Novoplansky 1997) can assist in developing stress-tolerant populations. As we shall see later, as fitness and stress tolerance are under genetic control, it is reasonable to apply artificial selection to increase the prospects for restoration success in challenging novel environments.

In novel ecosystems, an important component of the fitness of populations of desired species will often be their ability to compete with undesirable species that are invasive (Jones and Monaco 2009). As environments change, adaptation of resident species may decline and they may be displaced by newcomers with superior fitness (Walther et al. 2009). This may be the case particularly if the newcomer displays a profound effect on native biodiversity in the given environment, qualifying it as one of the approximately 10% of invasive species known as 'transformer species' (Richardson et al. 2000; see also Chapter 11).

26.4.5 Sustainability and resilience

Novel ecosystems are self-maintaining by definition (Hobbs et al. 2006), and their value as such stems partly from the fact that ongoing human maintenance is not required (see Chapters 5 and 6). To be sustainable, populations must be able to survive occasional extreme environmental events (Brown and Amacher 1999). Local populations generally possess such genetic memory due their presumed long-term residence at the site (Montalvo et al. 1997). However, there may be little reason to believe that local populations are necessarily superior to non-local populations when environmental conditions have been modified by human activity and have moved outside their historical range of variability (Hobbs et al. 2009). On the other hand, selection in the novel environment may provide for adaptation to that environment via contemporary evolution. For example, Leger (2008) reported that remnant big squirreltail (*Elymus multisetus*) plants that persisted following invasion by the annual grass, *Bromus tectorum*, displayed enhanced competitive ability relative to the unselected (control) population.

Kessler and Thomas (2006) suggested managing for resilience in changing environments, and they recognized that this might require a modified biological community. Briske et al. (2008) advocated resilience management to limit the probability that an at-risk system might cross a threshold to a less desirable state. To direct the system along a restoration pathway, residual properties of the affected ecosystem must be enhanced before they decay over time (Briske et al. 2008). To buttress resilience, Seastedt et al. (2008) suggested enhancing genetic diversity, species diversity and functional diversity, which may be accomplished by the seeding of effective plant materials.

Because some ecosystems have been irreversibly transformed, Hobbs et al. (2006) argued that we have little choice but to learn how to turn lemons into lemonade. To do this, we must begin by considering how the current state has arisen, appreciate the new realities, be willing to experiment and make bold choices and then apply an adaptive management strategy to learn from our mistakes and gradually increase our rate of success. Restoration has been described as a gamble that may be straightforward but often does not go as planned, resulting in unintended consequences (Suding 2011).

26.5 DEVELOPING PLANT MATERIALS FOR NOVEL ECOSYSTEMS

26.5.1 Contemporary evolution and assisted evolution

Contemporary evolution, that which occurs over relatively short timeframes, is now understood to be commonplace, particularly under conditions of anthropogenic change (Stockwell et al. 2003). However, most conservation programs pay little attention to evolutionary mechanisms, and this inattention may be detrimental over time (Kinnison et al. 2007). Ashley et al. (2003) called for 'evolutionarily enlightened management' that considers both evolutionary and ecological consequences of management decisions. They suggested that genetic changes due to selective forces associated with human activity may be desirable if fitness is increased. Sgrò et al. (2011) concurred and further argued that valuable genetic material is not limited to local provenance, as supported by the meta-analysis of

Hereford (2009). Weeks et al. (2011) provided a rebuttal to arguments opposing translocation of populations. From an evolutionary genetics perspective, they argue that the risks of translocation are generally overemphasized and its benefits are generally underappreciated.

Jones and Monaco (2009) proposed 'assisted evolution' as a paradigm for plant materials for restoration of altered ecosystems. As they describe it, assisted evolution is based on four principles regarding plant materials: (1) they reflect general historical evolutionary patterns; (2) they are particularly suited to the modified environment; (3) they are able to adapt to contemporary selection pressures; and (4) they contribute to the restoration of ecosystem structure and function. The advantage of this paradigm is that it recognizes that 'local has value' (point 1), yet it does not compromise the importance of plant material performance as measured empirically (points 2–4). The tools of hybridization and artificial selection (discussed later) may be combined with elements of the preservationist viewpoint in a 'restoration-appropriate' fashion to improve the success of restoration projects (Jones 2009; Jones and Robins 2011).

26.5.2 Augmenting genetic variation

Numerous researchers have suggested broadening the genetic base of plant materials to enhance the potential for adaptation to novel conditions (van Andel 1998; Jones 2003; Rice and Emery 2003; Broadhurst et al. 2008; Lankau et al. 2011; Sgrò et al. 2011; Weeks et al. 2011). For cross-pollinating species, this may be easily accomplished by passive hybridization among multiple populations in isolation via wind or insect pollinators, as appropriate. For self-pollinating or apomictic (propagating asexually by seed) species, this would typically be accomplished without hybridization by simple mechanical mixing of seed lots. All available populations from a specific geographical area or habitat type may be included or, alternatively, populations could be included based on performance as determined beforehand by field evaluations.

In this vein, Rice and Emery (2003) called for 'coarse selective tuning' that would involve assembling regional seed mixes collected from a variety of microclimates within a climatic zone and/or augmentation from the edges of the species' distribution to capture novel genetic variation for environmental extremes.

Once this material is seeded, they anticipate that it will be 'fine-tuned' locally by the removal of non-optimal genotypes by the forces of stabilizing selection. Guided by this principle, Jones (2003) released the P-7 germplasm of bluebunch wheatgrass (*Pseudoroegneria spicata*), a wind-pollinated species, for use in restoration of semi-arid rangelands of the northwestern USA. This 'multiple-origin polycross' (Jones 2003) was created by passively hybridizing populations from multiple sites into a new (polycross) population. Broadhurst et al. (2008) advocated a similar approach, which they termed 'composite provenancing'. Weeks et al. (2011) also recommended this approach, referring to the 'strategic mixing of populations'. Moreover, this approach has been recommended by SERI in its primer on ecological restoration (SERI 2004) which refers to the introduction of 'diverse genetic stock' for the development of novel, more adaptive ecotypes.

A contrary view is offered by Johnson et al. (2010). Because its constituent populations originate from different localities, they argue a multiple-origin polycross likely consists of large amounts of non-adapted genetic variation (Rogers and Montalvo 2004, chapter 8; Johnson et al. 2010). Thus, outbreeding depression could result upon hybridization of the multiple-origin polycross with remnant local plants (Rogers and Montalvo 2004, chapter 8). However, Carney et al. (2000) have demonstrated that natural selection can overcome initial outbreeding depression, as hybrids of sunflower species *Helianthus annuus* and *H. bolanderi* that initially displayed low fertility became genetically stable over a 50-year period. Erickson and Fenster (2006) reported similar results after six generations following hybridization in the legume *Chamaecrista fasciculata*. However, Endler et al. (2010) have argued that outbreeding depression may lead to extirpation of the population before it can rebound.

26.5.3 Artificial selection

Lankau et al.'s (2011) call for 'pre-adapted' seeds raises the topic of artificial selection, the common technique of the plant and animal breeding disciplines. Since ancient times, breeding has led to the domestication of many plant species, and it predates the advent of the science of genetics by millennia (Kingsbury 2009). However, since the beginning of the 20th century, the development of breeding practices and procedures has closely followed its companion discipline of genetics

as this powerful and diverse discipline has evolved (Kingsbury 2009). Consequently, breeding may be thought of as applied genetics. In the forestry discipline, for example, Namkoong (1969), Ledig and Kitzmiller (1992) and O'Neill and Yanchuk (2005) mention many timber species that have potential for selection using non-local provenances, as they are 'generalists' that tend to feature populations that are fairly widely adapted.

For restoration species, traits related to ecological fitness or successful seed production capability in a cultivated setting are the most obvious choices for artificial selection. Traits associated with ecological fitness might include tolerance to various biotic or abiotic stresses, seedling establishment, competitive ability against invasive species or seed production ability. Traits desirable for commercial seed production might include improved germinability and seedling establishment, increased seed yield and reduced seed shattering. Traits chosen for selection would typically be those that most limit restoration success. As an example of a trait associated with ecological fitness, James et al. (2011) determined that seedling recruitment of bluebunch wheatgrass on rangelands is limited primarily by poor seedling emergence rather than by germination or seedling persistence. As an example of a seed-production trait, Whalley et al. (1990) demonstrated that a more acute angle between the opened glumes, the two appendages surrounding the seed, is associated with seed shattering resistance in Indian ricegrass (*Oryzopsis hymenoides* = *Achnatherum hymenoides*). The subsequent release of 'Rimrock' Indian ricegrass in 1997 led to dramatically increased seed yields, resulting in lower seed prices. In fact, the increased affordability of Indian ricegrass seed has greatly stimulated the use of this species in rangeland restoration in the northwestern USA.

Traits that contribute to seedling establishment success are likely particularly important in novel ecosystems where invasive plants are present. In rangelands of the western USA, for example, environmental requirements for seedling establishment are much more restrictive than those for mature-plant growth (Hardegree et al. 2011), and establishment may occur only in the occasional favorable year (Harrington 1991). Even in humid systems, traits relating to establishment have been shown to be more important than dispersal-related traits for determining colonization ability (Pywell et al. 2003). While genetic memory (Montalvo et al. 1997) derived from past natural selective events on mature-plant traits is important, for restoration applications in novel ecosystems, establishment traits are therefore likely to be even more critical. Jones et al. (2010) detailed how functional traits (e.g. seed size, specific leaf area, C : N ratio and root length) affect specific ecological processes (e.g. seed dispersal, rate of development, soil nutrient dynamics and nitrogen capture, respectively). Presumably, all such traits are subject to genetic improvement via artificial selection. As the targeted traits become better understood and selection protocols are improved, artificial selection may become more refined and effective. Accomplishing this task may require interaction among scientific disciplines such as plant ecology, physiology and genetics (Jones et al. 2010).

In the case of cross-pollinated species, selection may be applied between populations (i.e. choosing one population for restoration instead of another) and/or within populations (i.e. genetic manipulation of a population). For self-pollinated or apomictic species, however, selection would typically be applied only between populations (Jones and Robins 2011, see section 2). This is because hybridization does not take place each generation for self-pollinated or apomictic species. On the other hand, predominantly cross-pollinated species are often self-incompatible and are therefore typically obligated to cross (Jones 2005, see question 3).

Selection within cross-pollinated populations may be practiced generation after generation (Dudley 2007), a process termed recurrent selection. The intent of this procedure is to effect a continuous increase in the frequency of desirable alleles in the population (Dudley 2007). Recurrent selection is particularly effective for quantitatively inherited traits (i.e. those controlled by alleles at many genetic loci, each with a relatively small effect). Under such circumstances, progress from selection may continue even after many generations of selection (Dudley 2007). In the process, care should however be taken to maintain an effective population size (N_e) that is sufficiently high to limit inbreeding to a minimal level (Jones and Robins 2011, see section 3.7).

Within the realm of native plant material development for restoration applications, artificial selection has been controversial (Jones and Monaco 2009). Nevertheless, the disciplines of conservation biology and plant breeding have much to offer one another (Jones 2009). However, misconceptions must be overcome before the potential of this interaction can be realized. Jones and Robins (2011) have offered suggestions as to how artificial selection and other forms of genetic manipulation can be implemented in ways that are most appropriate for the development of plant materials for restoration of disturbed landscapes. Genetically improved materials specifically developed for enhanced adaptation to increasing environmental challenges can be developed, while plant materials based strictly on local provenance are genetically static (with the exception of increased adaptation via contemporary evolution; Fig. 26.4).

26.5.4   Release and environmental evaluation

The formal decision whether to 'release' a plant material that previously had been experimental is made by the sponsoring organization(s) (Jones and Young 2005, question 14). The decision is based primarily on the empirical data presented by the developer that compares the candidate material to previously released materials of the same species. Prior to release, the plant material is considered to be experimental and is not available to the general public. Upon release, however, seed is made available to the seed industry for multiplication for sale for restoration practice. Unless the plant material is legally protected as intellectual property, release also means that the material can be freely used by other plant material developers to develop future plant materials.

The USDA-Natural Resources Conservation Service (NRCS) (2010) has developed a procedure to evaluate prospective releases for potential environmental impact. Each prospective release is scored in four categories: (1) impact on habitats, ecosystems and land use (six criteria); (2) ease of management (six criteria); (3) conservation need and plant use (three criteria); and (4) biological characteristics (11 criteria). Depending on the score for each category, the outcome of the evaluation may be: (1) to release; (2) not to release; or (3) not to release without higher-level approval pending an environmental assessment or environmental impact statement. To date, USDA-NRCS has discontinued 25 plant materials, all of non-native origin and all released before 1990. Each plant material scored medium or high for spread by seed, vegetative spread, potential invasiveness and/or impact on habitat (personal communication J. Englert 2012, USDA-NRCS, Washington DC). Some examples of plant materials discontinued on this basis are releases of *Bromus inermis, Coronilla varia, Elaeagnus umbellata, Lespedeza thunbergii* and *Lonicera maackii.*

26.6 CONCLUSION

Local materials have been championed in cases when the target environment remains within the normal environmental 'adaptive envelope' (Johnson et al. 2010) and when the management objective is restoring natural levels of genetic diversity (Rogers and Montalvo 2004, chapter 7). However, novel approaches are needed for management of novel ecosystems (Seastedt et al. 2008) for which the emphasis on natural genetic diversity may be secondary to the need to regenerate ecosystem function (Wilkinson 2004a; Jones and Monaco 2009). For these reasons, and because novel ecosystems are quickly becoming the rule rather than the exception, the application of 'intelligent effort' in the development of native plant materials (Booth and Vogel 2006) has considerable merit. This empirical approach emphasizes testing of prospective plant materials for use in restoration environments. One of its greatest assets is its potential to usher in an extraordinary era of interdisciplinary interaction among ecologists, geneticists, physiologists and evolutionary biologists for the development of native plant materials that can more effectively restore damaged landscapes.

REFERENCES

Agosta, S.J. and Klemens, J.A. (2008) Ecological fitting by phenotypically flexible genotypes: implications for species associations, community assembly and evolution. *Ecology Letters*, **11**, 1123–1134.

Arneth, A., Harrison, S.P., Zaehle, S. et al. (2010) Terrestrial biogeochemical feedbacks in the climate system. *Nature Geoscience*, **3**, 525–532.

Ashley, M.V., Wilson, M.F., Pergams, O.R.W., O'Dowd, D.J., Gende, S.M. and Brown, J.S. (2003) Evolutionarily enlightened management. *Biological Conservation*, **111**, 115–123.

Booth, D.T. and Vogel, K.P. (2006) Revegetation priorities. *Rangelands*, **28**, 24–30.

Botkin, D.B. (1990) *Discordant Harmonies: A New Ecology for the Twenty-First Century.* Oxford University Press, New York.

Bottrill, M.C., Joseph, L.N., Cawardine, J. et al. (2008) Is conservation triage just smart decision making? *Trends in Ecology and Evolution*, **23**, 649–654.

Bridgewater, P., Higgs, E.S., Hobbs, R.J. and Jackson, S.T. (2011) Engaging with novel ecosystems. *Frontiers in Ecology and the Environment*, **9**, 423.

Briske, D.D., Bestelmeyer, B.T., Stringham, T.K. and Shaver, P.L. (2008) Recommendations for development of resilience-based state-and-transition models. *Rangeland Ecology and Management*, **61**, 359–367.

Broadhurst, L.M., Lowe, A., Coates, D.J., Cunningham, S.A., McDonald, M., Vesk, P.A. and Yates, C. (2008) Seed supply for broadscale restoration: maximizing evolutionary potential. *Evolutionary Applications*, **1**, 587–597.

Brown, R.W. and Amacher, M.C. (1999) Selecting plant species for ecological restoration: a perspective for land managers, in *Revegetation With Native Species: Proceedings. 1997 Society For Ecological Restoration Annual Meeting.* L.K. Holzworth and R.W. Brown, compilers. 1997 November 12–15. Forest Service, Rocky Mountain Research Station, Ogden, Utah, pp. 1–16.

Carney, S.E., Gardner, K.A. and Riesberg, L.H. (2000) Evolutionary changes over the fifty-year history of a hybrid population of sunflowers (*Helianthus*). *Evolution*, **54**, 462–474.

Chapin, F.S., Kofinas, G.P. and Folke, C. (2009) *Principles of Ecosystem Stewardship: Resilience-Based Natural Resource Management in a Changing World.* Springer-Verlag, New York.

Choi, Y.D. (2004) Theories for ecological restoration in changing environment: toward 'futuristic' restoration. *Ecological Research*, **19**, 75–81.

Choi, Y.D. (2007) Restoration ecology to the future: a call for new paradigm. *Restoration Ecology*, **15**, 351–353.

Choi, Y.D., Temperton, V.M., Allen, E.D., Grootjans, A.P., Halassy, M., Hobbs, R.J., Naeth, M.A. and Torok, K. (2008) Ecological restoration for future sustainability in a changing environment. *Ecoscience*, **15**, 53–64.

Clewell, A.F. (2000) Restoring for natural authenticity. *Ecological Restoration*, **18**, 216–217.

Daily, G.C., Alexander, S., Ehrlich, P.R. et al. (1997) Ecosystem services: benefits supplied to human societies by natural ecosystems. *Issues in Ecology*, **2**, 1–16.

Davies, K.W., Nafus, A.M. and Sheley, R.L. (2010) Non-native competitive perennial grass impedes the spread of an invasive annual grass. *Biological Invasions*, **12**, 3187–3194.

Davis, M.A. and Slobodkin, L.B. (2004a) Restoration ecology: the challenge of social values and expectations. *Frontiers in Ecology and the Environment*, **2**, 44–45.

Davis, M.A. and Slobodkin, L.B. (2004b) The science and values of restoration ecology. *Restoration Ecology*, **12**, 1–3.

Didham, R. (2011) Life after logging: strategic withdrawal from the Garden of Eden or tactical error for wilderness conservation? *Biotropica*, **43**, 393–395.

Dobson, A.P., Bradshaw, A.D. and Baker, A.J.M. (1997) Hopes for the future: restoration ecology and conservation biology. *Science*, **277**, 515–522.

Dudley, J.W. (2007) From means to QTL: the long-term selection experiment as a case study in quantitative genetics. *Crop Science*, **47**(S3), S20–S31.

Ellis, E.C. and Ramankutty, N. (2008) Putting people in the map: anthropogenic biomes of the world. *Frontiers in Ecology and the Environment*, **6**, 439–447.

Ellis, E.C., Goldewijk, K.K., Siebert, S., Lightman, D. and Ramankutty, N. (2010) Anthropogenic transformation of the biomes, 1700 to 1900. *Global Ecology and Biogeography*, **19**, 589–606.

Endler, J., Mazer, S., Williams, M., Sandoval, C. and Ferren, W. (2010) Problems associated with the introduction of non-native genotypes on NRS reserves. http://nrs.ucop.edu/research/guidelines/non_native_genotypes.htm. Accessed August 2012.

Erickson, D.L. and Fenster, C.B. (2006) Intraspecific hybridization and the recovery of fitness in the native legume *Chamaecrista fasciculata*. *Evolution*, **60**, 225–233.

Ewel, J.J. and Putz, F.E. (2004) A place for alien species in ecosystem restoration. *Frontiers in Ecology and the Environment*, **2**, 354–360.

Friedel, M.H. (1991) Range condition assessment and the concept of thresholds: a viewpoint. *Journal of Range Management*, **44**, 422–426.

Gleason, R.A., Laubhan, M.K. and Euliss, N.J. Jr. (eds) (2008) Ecosystem services derived from wetland conservation practices in the United States Prairie Pothole Region with an emphasis on the US Department of Agriculture Conservation Reserve and Wetlands Reserve Programs. US Geological Professional Paper 1745, 58 p.

Goldberg, D. and Novoplansky, A. (1997) On the relative importance of competition in unproductive environments. *Journal of Ecology*, **85**, 409–418.

Goodenough, A.E. (2010) Are the ecological impacts of alien species misrepresented? A review of the 'native good, alien bad' philosophy. *Community Ecology*, **11**, 13–21.

Gould, S.J. (1998) An evolutionary perspective on strengths, fallacies, and confusions in the concept of native plants. *Arnoldia*, **58**, 11–19.

Gould, S.J. and Lewontin, R.C. (1979) The spandrels of San Marco and the Panglossian paradigm: a critique of the adaptationist programme. *Proceedings of the Royal Society of London B*, **205**, 581–598.

Gray, A. (2004) The parable of Green Mountain: massaging the message. *Journal of Biogeography*, **31**, 1549–1550.

Handel, S.N., Robinson, G.R. and Beattie, A.J. (1994) Biodiversity resources for restoration ecology. *Restoration Ecology*, **2**, 230–241.

Hardegree, S.P., Jones, T.A., Roundy, B.A., Shaw, N.L. and Monaco, T.A. (2011) Assessment of range planting as a conservation practice, in *Conservation Benefits of Rangeland Practices* (D.D. Briske, ed.) USDA-Natural Resources Conservation Service, pp 171–212.

Harlan, J.R. and deWet, J.M.J. (1971) Toward a rational classification of cultivated plants. *Taxon*, **20**, 509–517.

Harrington, G.N. (1991) Effects of soil moisture on shrub seedling survival in a semi-arid grassland. *Ecology*, **72**, 1138–1149.

Harris, J.A., Hobbs, R.J., Higgs, E. and Aronson, J. (2006) Ecological restoration and global climate change. *Restoration Ecology*, **14**, 170–176.

Hebard, F.V. (2006) The backcross breeding program of the American Chestnut Foundation. *Journal of the American Chestnut Foundation*, **19**, 55–77.

Hereford, J. (2009) A quantitative survey of local adaptation and fitness trade-offs. *American Naturalist*, **173**, 579–588.

Hobbs, R.J. (2004) Restoration ecology: the challenge of social values and expectations. *Frontiers in Ecology and the Environment*, **2**, 43–44.

Hobbs, R.J. and Harris, J.A. (2001) Restoration ecology: repairing the Earth's ecosystems in the new millennium. *Restoration Ecology*, **9**, 239–246.

Hobbs, R.J., Arico, S., Aronson, J. et al. (2006) Novel ecosystems: theoretical and management aspects of the new ecological world order. *Global Ecology and Biogeography*, **15**, 1–7.

Hobbs, R.J., Higgs, E. and Harris, J.A. (2009) Novel ecosystems: implications for conservation and restoration. *Trends in Ecology and Evolution*, **24**, 599–605.

Hobbs, R.J., Cole, D.N., Yung, L. et al. (2010) Guiding concepts for park and wilderness stewardship in an era of global environmental change. *Frontiers in Ecology and the Environment*, **8**, 483–490.

Hobbs, R.J., Hallett, L.M., Ehrlich, P.R. and Mooney, H.A. (2011) Intervention ecology: applying ecological science in the 21st century. *Bioscience*, **61**, 442–450.

Holling, C.S. (1978) *Adaptive Environmental Assessment and Management*. Wiley, Chichester, UK.

Jackson, L.L., Lopoukhine, N. and Hillyard, D. (1995) Ecological restoration: a definition and comments. *Restoration Ecology*, **3**, 71–75.

Jackson, S.T. and Hobbs, R.J. (2009) Ecological restoration in the light of ecological history. *Science*, **325**, 567–569.

James, J.J. Svejcar, T.J. and Rinella, M.J. (2011) Demographic processes limiting seedling recruitment in arid grassland restoration. *Journal of Applied Ecology*, **48**, 961–969.

Janzen, D.H. (1985) On ecological fitting. *Oikos*, **45**, 308–310.

Johnson, R., Stritch, L., Olwell, P., Lambert, S., Horning, M.E. and Cronn, R. (2010) What are the best seed sources for ecosystem restoration on BLM and USFS lands? *Native Plants Journal*, **11**, 117–131.

Jones, T.A. (2003) The Restoration Gene Pool concept: beyond the native vs. non-native debate. *Restoration Ecology*, **11**, 281–290.

Jones, T.A. (2005) Genetic principles and the use of native seeds. *Native Plants Journal*, **6**, 14–24.

Jones, T.A. (2009) Conservation biology and plant breeding: special considerations for development of native plant materials for use in restoration. *Ecological Restoration*, **27**, 8–11.

Jones, T.A. and Young, S.A. (2005) Native seeds in commerce. *Native Plants Journal*, **6**, 286–293.

Jones, T.A. and Monaco, T.A. (2007) A restoration practitioner's guide to the Restoration Gene Pool concept. *Ecological Restoration*, **25**, 12–19.

Jones, T.A. and Monaco, T.A. (2009) A role for assisted evolution in designing native plant materials for domesticated landscapes. *Frontiers in Ecology and the Environment*, **7**, 541–547.

Jones, T.A., and Robins, J.G. (2011) Appropriate use of genetic manipulation for the development of restoration plant materials. *Progress in Botany*, **72**, 249–264.

Jones, T.A., Monaco, T.A. and James, J.J. (2010) Launching the counterattack: interdisciplinary deployment of native-plant functional traits for repair of rangelands dominated by invasive annual grasses. *Rangelands*, **32**, 38–42.

Kareiva, P., Watts, S., McDonald, R. and Boucher, T. (2007) Domesticated nature: shaping landscapes and ecosystems for human welfare. *Science*, **316**, 1866–1869.

Kessler, W.B. and Thomas, J.W. (2006) Conservation biology from the perspective of natural resource management disciplines. *Conservation Biology*, **20**, 670–673.

Kingsbury, N. (2009) *Hybrid – The History of Science and Plant Breeding*. The University of Chicago Press, Chicago, IL.

Kinnison, M.T., Hendry, A.P. and Stockwell, C.A. (2007) Contemporary evolution meets conservation biology II: impediments to integration and application. *Ecological Research*, **22**, 947–954.

Lankau R., Jorgensen, P.S., Harris, D.J. and Sih, A. (2011) Incorporating evolutionary principles into environmental management and policy. *Evolutionary Applications*, **4**, 312–325.

Laycock, W.A. (1991) Stable states and thresholds of range condition on North American rangelands: a viewpoint. *Journal of Range Management*, **44**, 427–433.

Ledig, F.T. and Kitzmiller, J.H. (1992) Genetic strategies for reforestation in the face of global climate change. *Forest Ecology and Management*, **50**, 153–169.

Leger, E.A. (2008) The adaptive value of remnant native plants in invaded communities: an example from the Great Basin. *Ecological Applications*, **18**, 1226–1235.

Leimu, R. and Fischer, M. (2008) A meta-analysis of local adaptation in plants. *PLoS ONE*, **3**(12), e4010.

Levin, S.A. (1992) The problem of pattern and scale in ecology. *Ecology*, **73**, 1943–1967.

Marris, E. (2011) *Rambunctious Earth: Saving Nature in a Post-Wild World*. Bloomsbury, New York.

McGill, B.J., Enquist, B.J., Weiher, E. and Westoby, M. (2006) Rebuilding community ecology from functional traits. *Trends in Ecology and Evolution*, **21**, 178–184.

Minteer, B.A. and Collins, J.P. (2010) Move it or lose it? The ecological ethics of relocating species under climate change. *Ecological Applications*, **20**, 1801–1804.

Montalvo, A.M., Williams, S.L., Rice, K.J., Buchmann, S.L., Cory, C., Handel, S.N., Nabhan, G.P., Primack, R. and

Robichaux, R.H. (1997) Restoration biology: a population biology perspective. *Restoration Ecology*, **5**, 277–290.

Mooney, H.A. (2010) The ecosystem-service chain and the biological diversity crisis. *Philosophical Transactions of the Royal Society B*, **365**, 31–39.

Namkoong, G. (1969) Nonoptimality of local races, in *Proceedings of the 10th Southern Conference on Forest Tree Improvement*. Houston, Texas, pp. 149–153.

National Resources Conservation Service (2010) Worksheet for documenting an environmental evaluation of NRCS plant releases. Section 540.83 in *National Plant Materials Manual* (4th edition) www.nrcs.usda.gov/Internet/FSE_DOCUMENTS/stelprdb1042145.pdf. Accessed August 2012.

O'Neill, G. and Yanchuk, A. (2005) A primer on seed transfer for compliance and enforcement in BC. http://www.for.gov.bc.ca/hre/forgen/seedtransfer/SeedTransferPrimer20.pdf. Accessed August 2012.

Palmer, M., Bernhardt, E., Chornesky, E. et al. (2004) Ecology for a crowded planet. *Science*, **304**, 1251–1252.

Parker, K.A., Seabrook-Davison, M. and Ewen, J.G. (2010) Opportunities for nonnative ecological replacements in ecosystem restoration. *Restoration Ecology*, **18**, 269–273.

Prévot-Julliard, A.-C., Clavel, J., Teillac-Deschamps, P. and Julliard, R. (2011) The need for flexibility in conservation practices: exotic species as an example. *Environmental Management*, **47**, 315–321.

Pywell, R.F., Bullock, J.M., Roy, D.B., Warman, L., Walker, K.J. and Rothery, P. (2003) Plant traits as predictors of performance in ecological restoration. *Journal of Applied Ecology*, **40**, 65–77.

Raabová, J., Münzbergová, Z. and Fischer, M. (2007) Ecological rather than geographic or genetic distance affects local adaptation of the rare perennial herb, *Aster amellus*. *Biological Conservation*, **139**, 348–357.

Rice, K.J. and Emery, N.C. (2003) Managing microevolution: restoration in the face of global change. *Frontiers in Ecology and the Environment*, **1**, 469–478.

Richardson, D.M., Pyšek, P., Rejmánek, M., Barbour, M.G., Panetta, F.D. and West, C.J. (2000) Naturalization and invasion of alien plants: concepts and definitions. *Diversity and Distributions*, **6**, 93–107.

Roberts, R.E., Clark, D.L. and Wilson, M.V. (2010) Traits, neighbors, and species performance in prairie restoration. *Applied Vegetation Science*, **13**, 270–279.

Rogers, D.L. and Montalvo, A.M. (2004) Genetically appropriate choices for plant materials to maintain biological diversity. University of California. Report to the USDA Forest Service, Rocky Mountain Region, Lakewood, Colorado, USA. http://www.fs.fed.us/r2/publications/botany/plantgenetics.pdf. Accessed August 2012.

Sackville Hamilton, N.R. (2001) Is local provenance important in habitat creation? A reply. *Journal of Applied Ecology*, **38**, 1374–1376.

Seastedt, T.R., Hobbs, R.J. and Suding, K.N. (2008) Management of novel ecosystems: are novel approaches required? *Frontiers in Ecology and the Environment*, **6**, 547–553.

Sgrò, C.M., Lowe, A.J. and Hofmann, A.A. (2011) Building evolutionary resilience for conserving biodiversity under climate change. *Evolutionary Applications*, **4**, 326–337.

Smith, S.L., Sher, A.A. and Grant III, T.A. (2007) Genetic diversity in restoration materials and the impacts of seed collection in Colorado's restoration plant production industry. *Restoration Ecology*, **15**, 369–374.

Society for Ecological Restoration International (SERI) Science & Policy Working Group (2004) *The SER Primer on Ecological Restoration.* Society for Ecological Restoration International, Tucson, Arizona.

Stockwell, C.A., Hendry, A.P. and Kinnison, M.T. (2003) Contemporary evolution meets conservation biology. *Trends in Ecology and Evolution*, **18**, 94–101.

Suding, K.N. (2011) Toward an era of restoration in ecology: successes, failures, and opportunities ahead. *Annual Review of Ecology, Evolution, and Systematics*, **42**, 465–487.

Suding, K.N., Gross, K.L. and Houseman, G.R. (2004) Alternative states and positive feedbacks in restoration ecology. *Trends in Ecology and Evolution*, **19**, 46–53.

van Andel, J. (1998) Intraspecific variability in the context of ecological restoration projects. *Perspectives in Plant Ecology, Evolution and Systematics*, **1/2**, 221–237.

Vitousek, P.M., Mooney, H.A., Lubchenco, J. and Melillo, J.M. (1997) Human domination of Earth's ecosystems. *Science*, **277**, 494–499.

Walther, G.-R., Roqyes, A., Hulme, P.E. et al. (2009) Alien species in a warmer world: risks and opportunities. *Trends in Ecology and Evolution*, **24**, 686–693.

Weeks, A.R., Sgrò, C.M., Young, A.G. et al. (2011) Assessing the benefits and risks of translocations in changing envi-

ronments: a genetic perspective. *Evolutionary Applications*, **6**, 709–725.

Whalley, R.D.B., Jones, T.A., Nielson, D.C. and Mueller, R.J. (1990) Seed abscission and retention in Indian ricegrass. *Journal of Range Management*, **43**, 291–294.

Whisenant, S.G. (2002) Terrestrial systems, in *Handbook of Ecological Restoration. Volume 1: Principles of Restoration* (M.R. Perrow and A.J. Davy, eds), Cambridge University Press, Cambridge, pp. 83–105.

White, P.S. and Walker, J.L. (1997) Approximating nature's variation: selecting and using reference information in restoration ecology. *Restoration Ecology*, **5**, 338–349.

Wilkinson, D.M. (2001) Is local provenance important in habitat creation? *Journal of Applied Ecology*, **38**, 1371–1373.

Wilkinson, D.M. (2004a) Do we need a process-based approach to nature conservation? Continuing the parable of Green Mountain, Ascension Island. *Journal of Biogeography*, **31**, 2041–2042.

Wilkinson, D.M. (2004b) The parable of Green Mountain: Ascension Island, ecosystem construction and ecological fitting. *Journal of Biogeography*, **31**, 1–4.

Zedler, J.B. (1996) Ecological issues in wetland mitigation: an introduction to the forum. *Ecological Applications*, **6**, 33–37.

Zedler, J.B. (2003) Wetlands at your service: reduced impacts of agriculture at the watershed scale. *Frontiers in Ecology and the Environment*, **1**, 65–72.

Zedler, J.B. and Callaway, J.C. (1999) Tracking wetland restoration: do mitigation sites follow desired trajectories? *Restoration Ecology*, **7**, 69–73.

Chapter 27

CASE STUDY: MANAGEMENT OF NOVEL ECOSYSTEMS IN THE SEYCHELLES

Christoph Kueffer[1], Katy Beaver[2] and James Mougal[3]

[1]Plant Ecology, Institute of Integrative Biology, Swiss Federal Institute of Technology (ETH), Zurich, Switzerland

[2]Plant Conservation Action group (PCA), Victoria, Mahé, Seychelles

[3]National Park Authority, Victoria, Mahé, Seychelles

27.1 INTRODUCTION

The granitic islands of the Seychelles are continental fragments that formed part of the former Gondwana supercontinent until they became isolated in the middle of the ocean between Africa and India some 65 million years ago. Before a permanent French colony was established towards the end of the 18th century, these islands were probably only occasionally visited by ships for short periods. Within less than 100 years most of the larger islands were almost completely deforested, from the coasts to the mountainous interiors, and transformed into a highly degraded landscape dominated by non-native species. Another 100 years of opportunistic and changing management for various crops and forest products followed until, with the rise of tourism in the 1970s, most land use of inland habitats was abandoned. Apart from the devel-oped coastal areas, lowland plateau and lower mountain slopes, the granitic Seychelles islands are today once again fully covered in forest. But now these forests are composed of a mixture of ancient endemic species that evolved in isolation for millions of years and globe-trotting non-native species that arrived on the islands over the past c. 200 years.

Much of the higher inland forest is now inside protected areas and mainly managed for conserving native biodiversity and for water catchment. For much of the past two decades, invasive non-native species have been the priority nature conservation management concern. The presence of non-native predators such as rats, non-native creepers and the most prevalent non-native tree, *Cinnamomum verum*, are seen as especially significant management challenges. In the past the focus has been on preserving individual species and a vision of restoring an ecosystem back to a historical

Novel Ecosystems: Intervening in the New Ecological World Order, First Edition. Edited by Richard J. Hobbs, Eric S. Higgs, and Carol M. Hall.
© 2013 John Wiley & Sons, Ltd. Published 2013 by John Wiley & Sons, Ltd.

state. Here we discuss how the novel ecosystem concept can provide new perspectives for conservation in Seychelles, which are well worth exploring as a means to maintain native biodiversity in a significantly changed ecological context.

27.2 THE GRANITIC ISLANDS OF THE REPUBLIC OF SEYCHELLES

To appreciate the story of Seychelles' novel ecosystems, some information on the biogeography and environmental history of the granitic Seychelles islands is needed; this is summarized in this section. More information and detailed reference lists can be found elsewhere (Sauer 1967; Stoddart 1984; Kueffer and Vos 2004).

27.2.1 A small but unique island biota

The Republic of Seychelles in the Western Indian Ocean encompasses two types of islands: c. 70 coralline and 40 granitic islands. The granitic islands (4–5° S, 55–56° E) are characterized by a tropical maritime

climate and great isolation from continents (over 1500 km from both Africa and India). The total land area is c. 240 km² with the largest island Mahé comprising an area of 155 km² and rising to about 900 m a.s.l. (Fig. 27.1). The other three main islands are Praslin (38 km²), Silhouette (20 km²) and La Digue (10 km²). A further six islands are between 100 and 300 ha in size, and the remaining c. 30 islands range from 100 to <1 ha.

The granitic islands are of continental origin, i.e. they formed part of the ancient Gondwana supercontinent. Some 130 million years ago Madagascar, India and the granitic Seychelles broke away from Africa. India and Seychelles separated from Madagascar c. 85 million years ago, and c. 65 million years ago India and Seychelles broke apart. The continental origin of the granitic Seychelles has two important implications. First, in contrast to the comparatively young volcanic or coralline rocks of true oceanic islands, the granitic Seychelles are mostly formed from ancient granite. Consequently soils are very nutrient poor, especially in terms of phosphorus (Kueffer et al. 2008; Kueffer 2010). Second, the species of the Seychelles evolved from an established fauna and flora. This led to a harmonic flora and fauna, meaning that examples of

Figure 27.1 View of the southern part of Mahé, the largest of the Seychelles granitic islands. It is characterized by a rugged topography with exposed granitic outcrops (inselbergs) intermixed with closed-canopy forest. Photograph courtesy of Katy Beaver.

prolific adaptive radiation are rare and different guilds and taxonomic families are relatively homogeneously present, in contrast to true oceanic islands. The angiosperm flora includes some 200 species of which c. 70 are endemic, including an endemic family (Medusagynaceae) and enigmatic species such as the coco-de-mer palm (*Lodoicea maldivica*) with the largest seed of the plant kingdom. The endemic vertebrate fauna includes some 12 birds, 1 fruit-bat, 1 insectivorous bat, several species of reptiles (including giant tortoises), freshwater fishes and amphibians.

27.2.2   250 years of trial-and-error cultivation

Four main periods of human land use can be distinguished: (1) early colonization (first permanent settlements, subsistence agriculture and exploitation of timber and giant tortoises c. 1770–1800); (2) colonial exploitation (rise of export agriculture and deforestation, mainly for timber, c. 1800–1900); (3) colonial administration (independent crown colony, coconut and cinnamon industry, reforestation, c. 1900–1976); and (4) post-colonial era (independence, reorientation from primary to tertiary sector, rise of nature conservation; 1976 to present).

The Seychelles islands were not permanently settled until 1770; this was first as a French colony until it was officially turned over to Britain in 1815. During the early settlement phase, the population size increased from a few hundred to a few thousand people who made a living from exploitation of timber and giant tortoises, small-scale farming and the experimental introduction of some spice trees such as cloves, nutmeg and cinnamon (*Cinnamomum verum*). As early as the 1820s, most of Mahé was cleared of its original forest. By the 1870s, the native vegetation remained only as small patches in inaccessible parts of the mountains (Baker 1877). As well as tree cutting, deliberate and accidental forest fires played an important role in the degradation of the forests (Senterre 2009).

Most of the agricultural products of the 19th century lost their importance for the economy early in the 20th century with the exception of coconut, while the turn of the century was the start of the cinnamon industry. The history of cinnamon in Seychelles is a curious one, as the plant was introduced deliberately at an early stage but exploitation was long delayed; it then involved naturalized rather than cultivated

stands. By the late 19th century, cinnamon commonly dominated the secondary vegetation at mid to high elevations. Small quantities of cinnamon products had been exported in the early 19th century, but the cinnamon cultivar on Mahé was believed to be of inferior quality and was consequently rarely exploited until 1907. Thereafter cinnamon became the main export after copra. With the beginning of the cinnamon industry, an intensive and extraordinarily rapid second phase of forest destruction began. Cinnamon was cropped for its bark and leaves and the other woody plants were felled to provide additional firewood to operate the cinnamon oil distilleries.

The end of the cinnamon industry as a major economic activity coincided with the opening of the international airport in 1971 and the subsequent development of the tourist industry, which was strongly promoted by the Seychelles government after independence in 1976. In parallel with the reorientation of the economy from the primary to tertiary sector, nature conservation (a concern among experts since the 1960s) rapidly increased in importance. Today, fisheries and the service sector (tourism, offshore financial services) are the most important sectors of the Seychelles economy. Despite population growth (some 86,500 inhabitants in 2010) and housing development, about 70% of the total land area is still covered by vegetation, although almost exclusively secondary.

27.2.3   *Cinnamomum verum*: keystone species of Seychelles novel landscapes

Cinnamomum verum Presl (syn. *C. zeylanicum*; the true cinnamon) is a native of lowland evergreen climax forests in the Western Ghats of India and Sri Lanka. Today, cinnamon is by far the most abundant tree in the Seychelles and is a common to dominant species in all except coastal habitat (Kueffer and Vos 2004). Across many square kilometers, cinnamon makes up between 50% and over 90% of adult trees. These are exceptionally high abundances of a non-native species on a habitat, and especially island, scale.

Paradoxically, cinnamon does not apparently behave like a particularly aggressive invader at present in Seychelles. Fragments of relatively undisturbed native inselberg vegetation, palm forest or montane cloud forest vegetation are only weakly to moderately invaded by cinnamon despite the fact that they have been small habitat fragments in a sea of cinnamon forests for over

100 years. Even these moderate levels of invasion can however be a major threat to the remaining patches of native vegetation as we will discuss later, but it appears that cinnamon is not a particularly aggressive invader of undisturbed native vegetation. Under its own canopy in cinnamon-dominated forests, growth rates of cinnamon juveniles are particularly low (Kueffer et al. 2007, 2010). Cinnamon is also slower growing than most of the other invasive trees present in Seychelles under full sunlight, and its shade or drought tolerance is not outstanding compared to co-occurring native and non-native woody species (Fleischmann 1999; Kueffer et al. 2007, 2010; Schumacher et al. 2008, 2009). Why then has cinnamon managed to conquer Seychelles in such a dramatic way?

It appears that cinnamon profited from a short window of opportunity of a few decades in the early 19th century. Among the non-native and native plants present at the time, cinnamon was the one species that had the right traits to profit from the rapid large-scale deforestation. It colonized much deforested and eroded land and became so abundant that it persisted for the following 200 years. Had *Psidium cattleianum* or *Syzygium jambos* been introduced to the Seychelles a few decades earlier in the early rather than late 19th century (Kueffer and Vos 2004), one of these non-native species might have become the most abundant tree of Seychelles. Indeed, both species were already present in the 18th century in the French colonies of the Mascarenes (Mauritius, Rodrigues, La Réunion) – the main trading partner of Seychelles at the time – and in fact much of Mauritius is today covered in *Psidium* thickets while large areas of Rodrigues have been invaded by *Syzygium* (Kueffer and Mauremootoo 2004). A historical coincidence may therefore have determined the current ecology of the Seychelles. Possibly, this was partly a stroke of luck for biodiversity of Seychelles: *Psidium* invasions in Mauritius are impenetrable, species-poor thickets where native biodiversity recovers only after intensive weeding (Baider and Florens 2011) while, as we discuss in Section 27.3.2, native biodiversity thrives rather well in Seychelles cinnamon forests. Cinnamon may have acted as a barrier against such a dramatic invasion by *Psidium* in Seychelles, although other possible explanations such as differences in soil properties between the granitic Seychelles and the volcanic Mascarenes cannot be excluded.

Two unique traits of cinnamon were probably important for its successful establishment and later persistence. First, cinnamon forms a very dense topsoil root mat that apparently allows it to strongly impede juvenile regeneration of other species and efficiently take up scarce nutrients (Kueffer et al. 2007). This may have facilitated the initial establishment of the species on eroded land and later may have ensured a high resistance of cinnamon stands against colonization of other species. The second important trait of cinnamon is its prolific production of very nutritious fruits that are preferentially eaten and dispersed by native frugivorous birds and fruit-bats (Kueffer et al. 2009). Efficient seed dispersal by native and non-native frugivores probably made possible the initial rapid colonization of deforested land. Thus, the combination of a historical coincidence (the early introduction of cinnamon), an ecologically impoverished native island flora and two unique traits of cinnamon – strong belowground competition and prolific production of nutritious fruits – apparently contributed to the outstanding success of this non-native species and led to a switch to an alternative vegetation state dominated by a non-native tree.

27.3 A FRESH PERSPECTIVE ON HABITAT MANAGEMENT IN SEYCHELLES

In this section we introduce three habitat types that are of particular importance for nature conservation in Seychelles, and discuss for each of them how the novel ecosystem concept can provide fresh perspectives for habitat management. We take two main messages from the novel ecosystem discussion as a starting point. First, in the case of strongly anthropogenically altered ecosystems, restoration back to a historical 'pre-human' or 'natural' state is not necessarily the best management strategy for conserving native biodiversity or maintaining other ecosystem services. A historical state is often no longer a realistic target and, even if it could be achieved, may be less resilient to unavoidable anthropogenic stressors than an ecosystem that is specifically designed to resist such pressures. Second, on islands in particular this means, among other things, that both negative impacts and potential benefiting roles of non-native species should be considered in ecosystem management (Kueffer and Daehler 2009).

The first example that we discuss is nature conservation management on small islands that are free of non-native mammalian predators and therefore of particular value to nature conservation management. The second example is cinnamon-dominated secondary forest at mid-elevation on the largest island, Mahé, where large-scale control of cinnamon is not feasible and alternative management strategies must be explored. The third example is isolated inselberg vegetation, which opens the discussion as to whether those remnants of native vegetation that are apparently relatively resistant to ongoing environmental change should be 'artificially' enriched with rare native species.

27.3.1 Small islands

Several small islands of some 20–200 hectares are today managed primarily for, or with high priority given to, nature conservation purposes (e.g. Aride, Conception, Cousin, Cousine, Frégate, North). While some of these islands are strict nature reserves, others support a luxury hotel surrounded by land which is managed as a natural area. In the past these islands were heavily exploited, sometimes first through the extraction of guano that accumulated as a result of former seabird colonies, and then as coconut plantations. Thanks to restoration activities, native-dominated coastal forests have reestablished or are in the process of reemerging (Fig. 27.2).

The unique value of these islands for nature conservation however is that they are the only natural areas left that are free of non-native predators such as cats and rats (in most cases due to successful eradication).

Figure 27.2 Restored native lowland forest on a small rat-free island that now provides habitat for threatened endemic species such as the Seychelles Magpie Robin, *Copsychus sechellarum*, shown in the photograph (courtesy of Eva Schumacher).

Consequently, some of these islands once again host large seabird colonies; three endemic land bird species occur exclusively on these islands (mostly through reintroductions or species management) and the reintroduction of three more endangered land bird species is currently being considered or implemented. A number of endemic plant, reptile and invertebrate species also occur exclusively or in greater numbers on these small islands and the introduction of other plant and animal species, including terrapins and certain insects, is currently under discussion or implementation. On some of the islands, giant tortoises from Aldabra (a coralline island of the Republic of Seychelles) have been introduced and are roaming freely, thus reintroducing an animal which had become extinct on the granitic islands due to over-exploitation and which can play an important role in ecosystem functioning, e.g. in seed dispersal.

These small islands are sometimes presented to tourists as a resurrection of the pre-human state of Seychelles, and the idea that such habitats are being restored to a natural state also prevails in discussions among nature conservation managers. However, upon closer inspection, species assemblies on offshore islands are fundamentally novel. The future will show how well these novel communities function and how much native biodiversity can be maintained in them over the long term. Some conservation management may be constantly necessary (e.g. supplementary feeding of some species, translocation of individuals between islands to ensure gene flow, invasive species control), but thanks to tourism (entry fees in nature reserves, hotel guests on islands with a tourism resort) financial resources for management are partially secured at least at the moment.

The example of small islands illustrates two important aspects related to the novel ecosystem concept. First, while some major anthropogenic disturbances (in this example, the past transformation of the habitat through intensive land use) do not necessarily represent an irreversible hurdle for restoring native biodiversity, other anthropogenic stressors (such as the presence of non-native predators) are fundamental barriers that hinder recovery of native biodiversity unless removed and contained. This was possible in these islands and makes them offshore havens of native biodiversity, despite otherwise much novelty.

Second, in particular on oceanic islands, the design and management of novel species assemblies for nature conservation is already regularly happening and first experiences are promising. These small islands may offer an alternative to *ex situ* conservation for species that would go extinct in habitats with a presence of non-native predators. In contrast to *ex situ* conservation, the preserved species are not completely isolated from ecological interactions with other organisms. The working of ecological interactions however also means that different conservation targets, e.g. two rare species, will sometimes conflict, as is already becoming apparent. At the moment, the reduction of the abundance of the currently dominant native tree, *Pisonia grandis*, is for instance discussed because its sticky seeds can incur mortality in seabirds and some rare land birds.

27.3.2   Mid-elevation cinnamon forests

On Mahé at mid-elevations between c. 400 and 600 m, a novel forest ecosystem dominated by non-native cinnamon formed (Fig. 27.3). The lower end of this habitat zone is limited by abandoned or neglected former timber plantations, while the upper end is determined by the beginning of the montane cloud forest zone. Although these mid-elevation forests are dominated by non-native trees, many threatened native plant and animal species occur in this habitat and much of this habitat is inside protected areas.

Typically, 70–90% of adult trees are non-native cinnamon; cinnamon has therefore long been considered the most problematic plant invader of the Seychelles.

Figure 27.3 Novel mid-elevation forest dominated by the non-native tree *Cinnamomum verum* and characterized by a very dense seedling carpet of mostly cinnamon seedlings. Photograph courtesy of Eva Schumacher.

The canopy is only some 10–15 m tall, with formerly coppiced multi-stem cinnamon trees relatively widely spaced and almost no second understory layer below the canopy. Light penetration to the ground is consequently relatively high and ranges between c. 5 and 20% of total light above the canopy. Apart from scattered native trees, old and large trees are lacking. It is difficult to reconstruct historical forest structure but the canopy was probably 15–20 m high and, although not comparable to mainland tropical rainforests, vegetation must have been denser with a well-developed understory layer and in particular old and slowly dying and decomposing adult trees (Vesey-Fitzgerald 1940; Stoddart 1984; Kueffer 2006).

The novel biotic characteristics of the resulting ecosystem are particularly evident in the case of the regeneration niche. In the former native forest, the ground was probably covered with a thick layer of slowly decomposing leaf litter of endemic species such as the tree *Northea hornei* (Sapotaceae) and several palms and pandans. Seedlings (especially of small-seeded species) probably established mostly in microhabitats such as on decaying wood, with seedling densities of some 10 individuals per square meter; these are conditions that now only occur in pockets of relatively undisturbed native forest. In contrast, cinnamon litter decomposes quickly producing a thin litter layer and a soil with a low organic material content (Kueffer et al. 2008; Kueffer 2010), and densities of 100 seedlings per square meter are common (Fig. 27.3; Kueffer 2006). In contrast, soil nutrient availabilities seem to be relatively resilient to vegetation changes and soils in cinnamon-dominated stands remain as nutrient poor as in remnant stands of native trees (Kueffer et al. 2008, 2010; Kueffer 2010). The effects of the changed litter and soil environment on the native biota, for instance native litter-dwelling amphibians or invertebrates, are poorly understood.

A second novel factor that strongly shapes regeneration is root competition by adult cinnamon trees, which produce a dense root mat just below the soil surface (Kueffer et al. 2007). Whether allelopathy is involved in these belowground interactions of cinnamon is not known. It appears that the strong belowground root competition of cinnamon adult trees functions as an effective barrier against the invasion by other non-native invaders, while native regeneration from the few deposited seeds is apparently relatively good under a cinnamon canopy (Kueffer et al. 2007). This is in contrast to many other invasive non-native woody species (e.g. *Psidium cattleianum*) which often form very dense aboveground thickets, strongly suppressing native plants mainly through competition for light. In the case of cinnamon, such dense thickets can only very occasionally be found.

Thirdly, regeneration dynamics have fundamentally changed due to the dominance of non-native species and rarity of native species on a landscape scale, which results in a seed rain that is composed almost exclusively of non-native species (mostly cinnamon) and seed limitation of native trees (Kueffer 2006; Kueffer et al. 2007). In an enumeration of some 60,000 seedlings, only 1.5% were from native species and about half of these were of the common endemic palm, *Phoenicophorium borsigianum* (Kueffer 2006). Seed limitation is an important reason for the poor regeneration of many native species (Kueffer 2006), and unless action is taken their populations will decline further.

As well as the high belowground root competition that characterizes cinnamon and leads to a substantial change in the regeneration niche, fruit characteristics of the species are also novel (Kueffer et al. 2009). Cinnamon produces fruits containing 55 times more lipid than the median values of the studied native species, and they are also high in protein and total energy content but low in sugars (Kueffer et al. 2009). These characteristics are typical of Lauraceae fruits, and the lack of such nutritious and lipid-rich fruits from the native flora can likely be explained by a lack of specialized frugivory on a small island (Kueffer et al. 2009). In combination with non-native fruit trees from nearby gardens and plantations, lipid-rich cinnamon fruits and sugar-rich fruits from other co-occurring invasive plants such as *Clidemia hirta* provide a fundamentally different food source to frugivorous animals than pre-human forests. This improved food source has probably contributed to the current high population densities of native frugivorous birds and fruit-bats, and these densities may be higher than under historical conditions. This novelty may mean that cinnamon contributes to a high-quality habitat for frugivorous animals, but it may also mean that cinnamon competes with native plant species for seed dispersal services and thereby negatively impacts native plant regeneration. Indeed, in a food preference experiment with the most common native frugivorous bird, the Seychelles bulbul (*Hypsipetes crassirostris*) cinnamon fruits were preferred to all other tested non-native and native fruits except one (Kueffer et al. 2009).

In light of the novel ecosystem concept, the management challenge in the case of mid-elevation forests in Seychelles is to build on the positive services provided by cinnamon while mitigating negative impacts (Kueffer 2003; Kueffer et al. 2010). The cinnamon-dominated vegetation matrix may for instance help to build the resilience of a mixed non-native/native forest against invasions of more problematic species (e.g. *Alstonia macrophylla, Falcataria moluccana, Psidium cattleianum*) in a novel situation where propagule pressure of functionally novel non-native species is very high. Removal of whole patches of cinnamon forests has been tried in the past but led to a rapid reinvasion by other non-native species and increased erosion (Fleischer-Dogley 2004). For instance, the light-demanding invasive legume tree *Falcataria moluccana* regenerates vigorously from the seed bank in forest gaps, producing densities of several 100 seedlings per square meter within weeks after gap formation (Schumacher 2007). On the other hand, cinnamon probably also has many negative effects on native biodiversity. For instance, the impacts on invertebrates, whether in the litter layer or those depending on trees as food source, are poorly understood. Juveniles of native species regularly die suddenly in cinnamon forests, and it is not known whether cinnamon may act as a host species for a disease. Despite problematic aspects of the large-scale presence of cinnamon, complete removal of cinnamon is unlikely to be more effective than mitigating current and emerging future negative side-effects of the species.

An emerging idea aiming at increasing the proportion of native biodiversity without losing the resistance to invasions provided by a cinnamon-dominated vegetation is to establish small patches of completely native vegetation stands of a few meters in diameter in a matrix of cinnamon forest (Kueffer et al. 2010). These stands can be established by either weeding remnant stands of mostly native trees or by forming artificial gaps through the felling of a few cinnamon trees and replanting them with native plants (Fig. 27.4). Remnant stands of native vegetation seem to be relatively resistant to the invasion by cinnamon and other non-native plants and there is therefore hope that these native patches could be maintained with relatively little management costs. In fact, pilot experiments with native and non-native seedlings have shown that in small forest gaps a combination of belowground competition (from surrounding cinnamon adult trees that penetrate into the gaps) and

Figure 27.4 A promising restoration strategy for novel mid-elevation forests is to create small patches of native vegetation in a matrix of non-native forest. A few cinnamon trees are felled to form an artificial gap which is then replanted with native plants. The photograph shows the cut stump of a cinnamon tree in the foreground, replanted saplings of different endemic palm species in the center and non-native vegetation composed mainly of *Cinnamomum verum* and some *Psidium cattleianum* in the background. Photograph courtesy of Eva Schumacher.

Figure 27.5 Inselbergs are granite rock outcrops that are covered by patchy and stunted vegetation growing in small pockets of soil. Endemic pitcher plants (*Nepenthes pervillei*) cover the granite in the foreground, behind which is a mixture of endemic species including various shrubs, a pandan species and a tall sedge. Photograph courtesy of Eva Schumacher.

aboveground shade (that may be provided by a native tree canopy) could shift the competitive advantage from invasive non-native to native species (Kueffer et al. 2010).

A possible explanation is that non-native and native woody species in Seychelles tend to rely on a different growth strategy for dealing with low resource availabilities (Schumacher et al. 2008; Kueffer et al. 2010). While non-native species tend to depend on high phenotypic plasticity and the ability to allocate biomass to below- and aboveground parts depending on whether below- or aboveground resources are most limiting, many native (and in particular endemic) species in contrast use a slow-growth resource-efficient growth strategy that minimizes resource use rather than maximizes resource uptake. In consequence, native species tend to have an advantage compared to non-native species in situations where both belowground and aboveground resources are in short supply and the plastic resource-foraging strategy of non-native species does not work well (Schumacher et al. 2008, 2009; Kueffer et al. 2010).

If a management strategy that aims at establishing and maintaining small native-dominated vegetation patches in a matrix dominated by cinnamon works, then this may help to overcome some of the problems

of a cinnamon-dominated landscape. These native vegetation patches could for instance function as sources of native seed, which may promote a scattered establishment of native species in the cinnamon-matrix forest. Alternatively, the deeper litter layer in the native patches may provide habitat for native litter-dwelling animals (possibly after reintroduction). A forest of cinnamon that keeps out invaders and maintains high levels of native frugivorous birds and fruit-bats with interspersed native vegetation patches may therefore be a novel forest ecosystem that can maintain high levels of native biodiversity in a highly anthropogenic landscape. Whether and how such a management vision may play out positively for the native biodiversity of Seychelles will however depend on long-term adaptive management. Many surprises and new hurdles to overcome must be expected.

27.3.3 Inselberg vegetation

Inselbergs are more or less extensive granite rock outcrops that are covered by a patchy and stunted vegetation (Fig. 27.5). Plants grow in very shallow accumulations of soil in crevices and cavities. Such inselberg vegetation, characterized by climatic extremes (dry and sun-exposed microclimate, no shelter from strong rainfall or wind), occurs throughout the elevation gradient but primarily from c. 200–700 m. Inselbergs are the most resistant habitat to non-native invasions, i.e. only 2–10% of the adult layer are non-native plants (Fleischmann 1997). About 50% of the endemic tree and shrub species of the Seychelles can be found on inselbergs and some threatened species are almost exclusively found on inselbergs (e.g. *Glionnetia sericea*, *Medusagyne oppositifolia*; Huber and Ismail 2006).

While some of the inselberg flora is adapted to this extreme habitat, for other species it represents a very marginal habitat. For instance, *Mimusops sechellarum* is a Sapotaceae tree that grew historically in low- to mid-elevation rainforests and reached a height of 30 m, but is now often found as a stunted tree of a few meters on inselbergs. Although the inselberg vegetation is still mostly composed of native species, many of these species grow in a microhabitat that is untypical for them and take on an extreme phenotype (e.g. as a 5 m instead of 30 m tree).

Inselbergs provide a unique opportunity for biodiversity conservation in Seychelles, however. The habitat is mostly resistant to non-native invasions (although some invasive plant species such as *Alstonia macrophylla* are increasingly a problem) and anthropogenic land use; this allows a high level of native biodiversity to remain scattered in the middle of forests at all elevations, and management costs would be relatively low. At present, the plant composition of inselbergs is probably not yet novel in a strict sense. For example, the occasional *Mimusops* tree must have grown on inselbergs also in pre-human times. If these habitats are increasingly managed for native biodiversity and certain rare native species are planted that did not originally belong to the inselberg habitat (or only as occasional rare individuals), then they may increasingly be considered novel. As topographic features that stand out from the surrounding forest they can serve as seed sources of native species in a landscape dominated by non-native species.

It is tempting to consider the deliberate introduction of threatened native plant and animal species to inselbergs even if inselbergs were naturally only a marginal habitat for some of these species. Species conservation on inselbergs could then become a case of *ex situ* species conservation done in the midst of the former habitat of the species. Thanks to the special conditions on inselbergs (especially a certain resistance to invasion) this could be carried out in a cost-efficient manner. The added benefits are that at least some ecological interactions would be maintained in the wild,

and there is the possibility that an occasional individual of these rare species will establish in the surroundings of the inselberg. The practical implementation of such conservation will face important problems: for instance climate change may make dry inselberg habitat less suitable in the future, and ecotourism increasingly puts pressure on some inselbergs. As with small islands and cinnamon-dominated mid-elevation forests, the novel ecosystem idea is a source of new ideas that are urgently needed.

27.4 CONCLUSIONS

We have discussed three habitat types that are of particular importance for nature conservation management in Seychelles. For each of the three habitats, the novel ecosystem concept can provide fresh perspectives for devising effective and long-term management strategies aimed at saving the unique endemic biota of Seychelles. Importantly, the novel ecosystem concept does not mean that the conservation of native biodiversity or the active management of critically endangered species are abandoned as central goals of natural area management. Rather, it takes the changing management challenges into account in an increasingly anthropogenic world. These emerging ideas do not negate or make obsolete old conservation strategies, such as preventing and containing non-native invasions, and they should not be seen as a panacea for all conservation problems.

On islands in particular, where anthropogenic zones and different natural areas are compressed into a very small area, a major challenge will be to manage emerging mosaics of different novel ecosystems at a landscape scale (e.g. Kueffer 2003). For instance, while cinnamon may be beneficial for biodiversity management in mid-elevation forests, it can be a problematic invader of montane cloud forests (Kueffer 2006). These cloud forests are often only a few hundred meters away from mid-elevation forests, and it is not obvious how cinnamon-free cloud forests can be maintained next to vast areas of cinnamon-dominated forest. As another example, the close distance between natural areas and anthropogenic zones provides endemic frugivores with the opportunity to forage for additional food from introduced fruit trees in gardens. This high connectivity however makes the prevention of the spread of invasive pests and diseases very challenging.

Ultimately, maintaining native biodiversity in Seychelles will require the design of novel landscapes that interweave sustainable land use at low to mid-elevations with different types of ecologically novel biodiversity conservation areas in the inland (Kueffer 2003). This will ideally allow synergies among activities, e.g. using native tree species for forestry or managing production land also as buffer zones against invasive species around biodiversity areas.

The current landscapes of Seychelles are the result of a changing land-use history characterized by a number of abrupt changes and by historical contingencies such as the spread of one non-native tree species, cinnamon, across the landscape. These historical legacies of the past 200 years cannot be erased, but that does not mean that the endemic biota of the Seychelles cannot be saved, at least partly, in the wild. However, this will require envisioning novel future landscapes rather than lamenting over a lost past paradise.

REFERENCES

Baider, C. and Florens, F.B.V. (2011) Control of invasive alien weeds averts imminent plant extinction. *Biological Invasions*, **13**, 2641–2646.

Baker, J.G. (1877) *Flora of Mauritius and The Seychelles. A Description of the Flowering Plants and Ferns of Those Islands.* C Reeve & Co, London.

Fleischer-Dogley, F. (2004) Habitat restoration in the Morne Seychellois National Park, in *Case Studies on the Status of Invasive Woody Plant Species in the Western Indian Ocean: 5. Seychelles* (C. Kueffer and P. Vos, eds), Forestry Department, Food and Agriculture Organization of the United Nations, Rome, Italy, pp. 30–31.

Fleischmann, K. (1997) Invasion of alien woody plants on the islands of Mahé and Silhouette, Seychelles. *Journal of Vegetation Science*, **8**, 5–12.

Fleischmann, K. (1999) Relations between the invasive *Cinnamomum verum* and the endemic *Phoenicophorium borsigianum* on Mahé island, Seychelles. *Applied Vegetation Science*, **2**, 37–46.

Huber, M. and Ismail, S. (2006) *Suggested IUCN red list status of the endemic woody plants of the Inner Seychelles.* Institute of Integrative Biology, ETH Zurich, Zurich, Switzerland.

Kueffer, C. (2003) Habitat Restoration of Mid-altitude Secondary Cinnamon Forests in the Seychelles, in *Proceedings of the Regional Workshop on Invasive Alien Species and Terrestrial Ecosystem Rehabilitation in Western Indian*

Ocean Island States. Sharing Experience, Identifying Priorities and Defining Joint Action. (J.R. Mauremootoo, ed.) Indian Ocean Commission, Quatre Bornes, Mauritius, pp. 147–155.

Kueffer, C. (2006) *Impacts of woody invasive species on tropical forests of the Seychelles.* Dissertation ETH No. 16602, Department of Environmental Sciences. ETH Zurich, Zurich.

Kueffer, C. (2010) Reduced risk for positive soil-feedback on seedling regeneration by invasive trees on a very nutrient-poor soil in Seychelles. *Biological Invasions*, **12**, 97–102.

Kueffer, C. and Mauremootoo, J. (2004) *Case Studies on the Status of Invasive Woody Plant Species in the Western Indian Ocean. 3. Mauritius (Islands of Mauritius and Rodrigues).* Forestry Department, Food and Agriculture Organization of the United Nations, Rome, Italy.

Kueffer, C. and Vos, P. (2004) *Case Studies on the Status of Invasive Woody Plant Species in the Western Indian Ocean: 5. Seychelles.* Forestry Department, Food and Agriculture Organization of the United Nations, Rome, Italy.

Kueffer, C. and Daehler, C. (2009) A habitat-classification framework and typology for understanding, valuing and managing invasive species impacts, in *Management of Invasive Weeds* (Inderjit, ed.) Springer, Berlin, pp. 77–101.

Kueffer, C., Schumacher, E., Fleischmann, K., Edwards, P.J. and Dietz, H. (2007) Strong belowground competition shapes tree regeneration in invasive *Cinnamomum verum* forests. *Journal of Ecology*, **95**, 273–282.

Kueffer, C., Klingler, G., Zirfass, K., Schumacher, E., Edwards, P. and Güsewell, S. (2008) Invasive trees show only weak potential to impact nutrient dynamics in phosphorus-poor tropical forests in the Seychelles. *Functional Ecology*, **22**, 359–366.

Kueffer, C., Kronauer, L. and Edwards, P.J. (2009) Wider spectrum of fruit traits in invasive than native floras may increase the vulnerability of oceanic islands to plant invasions. *Oikos*, **118**, 1327–1334.

Kueffer, C., Schumacher, E., Dietz, H., Fleischmann, K. and Edwards, P.J. (2010) Managing successional trajectories in alien-dominated, novel ecosystems by facilitating seedling regeneration: a case study. *Biological Conservation*, **143**, 1792–1802.

Sauer, J.D. (1967) *Plants and Man on the Seychelles Coast. A Study in Historical Biogeography.* The University of Wisconsin Press, Madison, London.

Schumacher, E. (2007) *Variation in growth responses among and within native and invasive juvenile trees in Seychelles.* Diss. No. 16988, Department of Environmental Sciences. ETH Zurich, Zurich.

Schumacher, E., Kueffer, C., Tobler, M., Gmür, V., Edwards, P.J. and Dietz, H. (2008) Influence of drought and shade on seedling growth of native and invasive trees in the Seychelles. *Biotropica*, **40**, 543–549.

Schumacher, E., Kueffer, C., Edwards, P.J. and Dietz, H. (2009) Influence of light and nutrient conditions on seedling growth of native and invasive trees in the Seychelles. *Biological Invasions*, **11**, 1941–1954.

Senterre, B. (2009) *Forest fire risk assessment on the Seychelles main granitic islands. Consultancy report*. Ministry of Environment-UNDP-GEF Project, Victoria, Seychelles.

Stoddart, D.R. (ed.) (1984) *Biogeography and Ecology of the Seychelles Islands*. DR W. Junk Publishers, The Hague, Boston, Lancaster.

Vesey-Fitzgerald, D. (1940) On the vegetation of the Seychelles. *Journal of Ecology*, **28**, 465–483.

Chapter 28

PERSPECTIVE: MOVING TO THE DARK SIDE

Patricia L. Kennedy

Department of Fisheries and Wildlife & Eastern Oregon Agricultural & Natural Resource Program, Oregon State University, USA

Novel ecosystems first entered my consciousness in the early 1990s when I was asked by Russ Graham, a very insightful and thoughtful silviculturist, to define 'desired future conditions' of a landscape. We were serving on a scientific panel charged with developing management guidelines for the northern goshawk (*Accipiter gentilis*; Fig. 28.1), a North American bird of prey. This raptor was proposed several times for listing as threatened under the US Endangered Species Act because of its association with old growth forests. The panel's mission was to develop management guidelines that would halt future declines. I had spent my graduate years trapping, tagging, chasing and observing this critter and was therefore one of the 'species experts' on this inter-disciplinary panel.

The northern spotted owl (*Strix occidentalis caurina*) had recently been listed and in this era the common approach to managing nesting habitat of raptors of conservation concern was to establish buffer zones around nest sites or to establish preserves (areas that were off-limits to human activities). The panel was trying to think outside of the box; Russ was therefore suggesting that if we, the biologists, could define north-ern goshawk nesting habitat the foresters could create these conditions on the landscape using a wide variety of forest management tools including (horrors) timber harvest. At first I reacted with tremendous resistance. How could we dare try to approximate Mother Nature? Why not determine how much of Mother Nature needs to be set aside for conservation of this species? My compromising side thought was that we could certainly do this without putting the timber industry out of business.

During our weeks of discussion, Russ taught me a lot about the dynamic nature of forest landscapes (I had been well schooled in the equilibrium theory of forest succession). Reserves would not stay as old growth forests perpetually; the landscapes needed to contain all seral stages so that we were constantly growing these forests. In this fire-suppressed landscape, timber harvest was a good surrogate. After much debate and soul searching the panel developed a management plan that proposed using forest management practices to create heterogeneous landscapes that would be suitable for northern goshawks. Although we received national awards for this plan

Novel Ecosystems: Intervening in the New Ecological World Order, First Edition. Edited by Richard J. Hobbs, Eric S. Higgs, and Carol M. Hall.
© 2013 John Wiley & Sons, Ltd. Published 2013 by John Wiley & Sons, Ltd.

and it is well cited, the true test of its merit lay in the fact that it was hated by both the conservation community and the timber industry. It did not place buffers around all known nest sites, allowed for timber harvest near nest sites and increased harvest rotation periods so that more large trees could grow in this landscape. I had now entered the slippery slope of promoting managed landscapes for conservation of wildlife habitat.

Fast forward 10 years and, with colleagues and students, I am studying the response of grassland songbirds to non-native plants on one of the largest remnants (c. 65,000 ha) of the Pacific Northwest Bunchgrass Prairie (erroneously referred to as the Palouse Prairie; Fig. 28.2). Unlike most grassland remnants this prairie is amazingly clean from the standpoint of non-native plants. Patches do exist, however, and lo and behold these little guys (which are all ground nesters) like to place their nests in clumps of non-native vegetation (Fig. 28.3). Over time I realized these birds had a more flexible view of non-native plants than the reigning paradigm of non-native plant management: destroy at all cost. They use non-native vegetation if it provides enough structure to hide their nests and plenty of bugs to eat. These non-native sites were not ecological traps either, and reproductive success was comparable to native sites. I was now moving farther down the slippery slope by contemplating the value of non-native plants as wildlife habitat.

My most recent step towards the dark side occurred in 2008 and was the fault of John Wiens. It was time

Figure 28.1 Northern Goshawk (*Accipiter gentilis*). Photograph: Pat Kennedy.

Figure 28.2 Searching for grassland bird nests in the Zumwalt Prairie, 2005. Photograph courtesy of Andrea Lueders.

Figure 28.3 The object of our searches: a nest of horned larks on the Zumwalt prairie. Photograph courtesy of Andrea Lueders.

know him personally. He was a plant ecologist and, although a man of many talents, he would not be described as a fire ecologist and was publishing papers in weird journals (*Global Ecology and Biogeography*) on novel ecosystems. Why in the world did I contact him and, more interestingly, why did he agree to host me? There is no concrete answer to this question, but as a result of that terrific year in Perth I am now co-authoring papers suggesting desired future conditions might contain non-native plants and be suitable wildlife habitat.

What sacrilegious intellectual leap is next for this wildlife biologist? Will I soon be promoting tolerance of feral pigs, cats and foxes? I shudder at the possibilities and hope my wildlife professional societies will continue to honor my memberships.

for a sabbatical and I had a grant to write synthesis documents on the effects of fire management on wildlife so I could locate anywhere. John and I shared a love of Australia so he suggested I contact Richard Hobbs. I knew Richard's reputation of course, but did not know him personally. He was a plant ecologist and, although a man of many talents, he would not be described as a fire ecologist and yes, publishing papers in varied journals (*Global Ecology and Biogeography* on novel ecosystems). Why in the world did I contact him and more interestingly why did he agree to have me? There is no concrete answer to this question, but as a result of that terrific year in Perth I am now co-authoring papers suggesting desired future conditions might contain non-native plants and be suitable wildlife habitat.

What meritorious intellectual idea is next for this wildlife biologist? Will I soon be promoting tolerance of feral pigs...? and foxes? Shudder at the possibilities and hope the wildlife professional societies will continue to honor my memberships.

Figure 2.3.2. Top object is an scientist's idea of horned larks on the Panmunt prairie. Photograph courtesy of Arthur Landers.

...on a sabbatical and I had a grant to write synthesis documents on the effects of fire management on wildlife. I could locate anywhere. John and I sabbatical above of Australia say he suggested I contact Richard Hobbs. I knew Richard's reputation of course, but did not

Part V

How Do We Appreciate Novel Ecosystems?

Chapter 29

PERSPECTIVE: COMING OF AGE IN A TRASH FOREST

Emma Marris

Columbia, Missouri, USA

My friend Taya and I were out at her parents' country place, about twelve acres in the western foothills of the Cascades. I was maybe eight, visiting for the first time. Taya was taking me on a tour. We were struggling along, as short-legged people do through dense, early successional Northwest forest. She stopped and took hold of a small sapling. "This", she said, "is the difference between our land and a park". And then, shockingly, she stepped on the sapling until it was bowed in two and then snapped it with her boot, killing it dead. Or maybe she ripped it out of the ground with her two hands – she was a very strong girl, I remember. I don't remember the details of the act. But I do remember that she killed a tree and also the sensation of my mind being blown right out my ears. (Taya's childhood arbor-cide didn't presage sociopathy or anything close to it. She's now a vet.)

I was a city kid, so well schooled in the leave-no-trace ethos of wilderness preservation by school and camp that the idea of killing a tree . . . it wasn't that it was wrong. It was that I had never even considered the possibility. Nature was, to me, inviolate, unchanging, ancient and pure. *Pristine.* It was better than God – less judgmental, more fun to play in, but just as serious and Big.

For a while, I went up with Taya and her parents almost every weekend (or so it seems to me now). As we played in those woods, it ceased to be the Big Pristine to me. It became, instead, a familiar friend. Taya and I sought out springy tree limbs to use as bouncing seats, made expeditions to nearby notable rocks and built tiny worlds out of sticks and plants in the suggestive architectural spaces of rotting stumps. I particularly relished making trails. Taya's dad lent me a pint-sized machete and taught me how to wield it.

I know now that the forest on the property was decidedly not old growth; it had been logged as recently as the 1950s. In amongst the slender alders, thorny salmon berries and sword ferns dusty with golden spores were massive stumps of giant cedars that had once grown there. I recall that some of them had notches in their great grey sides where the springboards had been fitted. I played, then, in a trash forest, a weedy place left behind after the good stuff was gone, no doubt rife with scotch broom and other invasive species.

Ecologists and conservationists have long ignored such forests. In thrall to the lure of 'pristine wilderness', they shunned second growth. But that's beginning to change. People like the UCLA's Susanna Hecht

Novel Ecosystems: Intervening in the New Ecological World Order, First Edition. Edited by Richard J. Hobbs, Eric S. Higgs, and Carol M. Hall.
© 2011 Emma Marris. Published 2013 by John Wiley & Sons, Ltd.

and the Smithsonian's Joe Wright have published work suggesting that secondary forests in the tropics are more like old growth than we used to think, that they have a surprising number of plant and animal species in them and that they therefore have pretty decent conservation value. (Although some other researchers think the value may be oversold due to the occasional appearance in secondary forest of a species that is mostly restricted to old growth.) Less work has been done in temperate forests, but the shift in thinking seems to be reaching all corners of the conservation world; I think that's an excellent thing. Not only are many secondary forests real steals when it comes to conservation bang for buck, but I hope that if the pros start valuing them, the rest of us will learn to value them too.

After half a dozen years reporting on conservation, mostly for *Nature*, I have learned that The Big Pristine doesn't exist. Doesn't now; didn't ever. Around the world, prehistoric humans have had far greater impact on ecosystems than we ever knew. Climate change also means that every place, old growth and trash alike, is now changed by humanity's presence. Conservation approaches that move beyond fealty to the Big Pristine are starting to overtake older approaches centered on putting walls around protected areas and keeping hands off. More and more you hear talk of moving species to deal with climate change, or letting them move around without automatically freaking out about invasive species. More and more conservation projects are set in cities or on farms. I am by no means recommending large-scale killing of saplings by children as a conservation tool, but you never know. Nature has become less sacrosanct, and conservation is learning to become more interventionist. It is a bona fide paradigm shift, and it is fascinating to watch it happen. I like to think that the time I had as a child to pay such close attention to the vibrancy and beauty of that second-growth forest prepared me to document it.

My unscientific guess is that upwards of 80% of ecologists and conservationists first learned to love nature in places that weren't pure, pristine or wild: culverts in cul de sacs, empty lots, summer camps on formerly logged land, hedgerows and the weedy margins of agricultural fields. We loved them as kids, and now we are learning to love them again as scientists and environmentalists. It is a kind of homecoming.

Chapter 30

ENGAGING THE PUBLIC IN NOVEL ECOSYSTEMS

Laurie Yung[1], Steve Schwarze[2], Wylie Carr[3], F. Stuart Chapin III[4] and Emma Marris[5]

[1]Resource Conservation Program, College of Forestry and Conservation, University of Montana, USA
[2]Communication Studies, University of Montana, USA
[3]Society and Conservation, University of Montana, USA
[4]Institute of Arctic Biology, University of Alaska Fairbanks, Fairbanks, USA
[5]Columbia, Missouri, USA

30.1 INTRODUCTION

Novel ecosystems are such a departure from familiar ecosystem categories that they are simultaneously a difficult conceptual shift and a potentially transformative opportunity. The concept of novel ecosystems could reshape the way we think about conservation, our interactions with nature and the public dialogue about ecosystem management. Novel ecosystems are fundamentally social *and* ecological. They emerge from the intersection of self-willed nature and human intervention and impact. In addition to being, in part, social creations, novel ecosystems provide many important goods and services including places to connect with nature, sources of clean drinking water, resources for local livelihoods and refuges for diverse species. Thus far, the concept of novel ecosystems has been restricted to scientific circles, and the idea of novelty and the potential of such places has not yet become part of a broader public dialogue. In this chapter, we advocate for public engagement in novel ecosystems and suggest strategies for facilitating this broader dialogue. As Dietz and Stern (2008: vii) point out, "the environmental problems of the 21st century can be effectively addressed only by processes that link sound scientific analysis with effective public deliberation".

Careful deliberation about the values of novel ecosystems (see Chapter 31) and how to manage them into the future is necessary if we wish to identify and prioritize the types of public goods and services and ecological functions provided by such systems. In Chapter 41, Marris and others suggest that specific management strategies for novel ecosystems "will now ideally be decided on within a specific landscape in proportion to their ability to offer local, regional and global stakeholders the opportunity to add to the values and services of multifunctional landscapes". To do so, we need to engage these stakeholders through dynamic, transparent, meaningful public involvement processes. Engagement processes can also build a richer research agenda, inform policy-making and otherwise ensure that public values and interests are integrated into decisions about novel ecosystems.

We envision that scientists, managers, policy-makers, non-governmental organizations, and many others have a role to play in communication and engagement. Public engagement can be a formal process that feeds into official decision-making. Public

Novel Ecosystems: Intervening in the New Ecological World Order, First Edition. Edited by Richard J. Hobbs, Eric S. Higgs, and Carol M. Hall.
© 2013 John Wiley & Sons, Ltd. Published 2013 by John Wiley & Sons, Ltd.

engagement can also emerge from community conversations and informal alliances. Effective communication and engagement require involvement of scientists (to ensure dialogue with scientific knowledge) and decision-makers (to ensure that public views influence policy-making) as well as professional facilitators, communication experts and other specialists. Most importantly, a diversity of stakeholders representing different values, interests and ideologies need to be included in such efforts. This chapter examines strategies and methods for communicating about the novel ecosystems concept and for engaging the public in thinking about and planning for such places.

30.2 CHALLENGING CONVENTIONAL CATEGORIES

Most landscape and seascape descriptors, such as *wilderness, park, wasteland* and *brownfield*, come with normative connotations. Novel ecosystems defy easy categorization, however, because they mix and match traditionally positive and negative properties: they are diverse but invaded, neglected but resilient, new but natural, anthropogenic but wild. Novel ecosystems thus confront the simple binaries that permeate conservation discourse. As Hobbs et al. (2006: 2) suggest, novel ecosystems are defined by two key characteristics: "(1) novelty: new species combinations, with the potential for changes in ecosystem functioning; and (2) human agency: ecosystems that are the result of deliberate or inadvertent human action, but do not depend on continued human intervention for their maintenance". Chapter 6 expands the definition of a novel ecosystem to include the abiotic, biotic, and social components of the system. Recognizing the centrality of human actions to shaping modern ecosystems is critical because it has the potential to move us beyond discourses that dichotomize nature and humans. Because novel ecosystems so completely resist the typical human/nature divide, they may be able to facilitate a new public dialogue grounded in an appreciation of the inextricable links between social and natural systems.

For instance, the concept of novel ecosystems could replace perceptions that nature only exists 'out there' in national parks and wilderness areas with recognition that ecological processes occur all around us everyday. Few people in the world live more than a half mile from a city park, an overgrown culvert, an empty lot, a wild field margin or a backyard gone to seed.

Interestingly though, one typical response to the widespread fear that fewer and fewer people in the affluent west make and maintain connections with nature is to encourage national park visitation. Another response is to nurture nearby nature. For example, the US National Wildlife Federation's Garden for Wildlife program teaches people to landscape their backyards for the benefit of wildlife, including edible plants, water sources, cover and breeding locations. Similarly, outdoor classrooms such as those promoted by the Nature Explore Program take schoolchildren outside but not off school grounds, to learn about nature with materials and spaces close at hand. The novel ecosystems concept may enable people to mentally recategorize nearby 'natural' places as quiet, beautiful nature, as opposed to messy, neglected wrecks, opening up social and political space for people to value and engage with the different types of ecosystems that exist all around them.

Novel ecosystems also challenge conventional conservation strategies. The decades-long movement to create parks and wilderness has, in many senses, created a kind of blindness to the conservation opportunities that exist outside of those protected areas that are close simulacrums of historical ecosystems, absent of permanent human habitation and use. For instance, since 1988, over 20 former US military bases have been integrated into the national wildlife refuge system. These sites frequently suffer from severe toxic chemical contamination and simultaneously boast significant biological diversity (Havlick 2007). Marris (2011) however points out that, in North America's conservation culture (and the bulk of international conservation), nature without human presence or influence has long been privileged over nature that bears the fingerprints of humanity. The result of these and other cultural threads is a widespread trope of humanity as the opposite and enemy of nature. When we see lands that don't look historically correct, or bear obvious signs of human use, we either reject them out of hand as being useless for conservation, or we roll up our sleeves and begin planning how best to restore them to their former glory. Given the great expense and, in many cases, the near impossibility of restoring altered systems to a semblance of their historical condition, we would be foolish not to ask whether something else of value can be made out of such places. The novel ecosystems concept starts that conversation. When a place is categorized as irredeemably degraded, it is dismissed. When it is categorized as a novel ecosystem, the next questions become "what values can it provide?" and "what roles can it play?" These are key questions about which public goods novel ecosystems should provide, questions that public engagement processes can and should help to answer.

30.3 CONSIDERING THE AUDIENCE IN COMMUNICATIONS

To initiate a public dialogue about which public goods might be provided by novel ecosystems, we need to effectively communicate with a diverse citizenry. Whether communication is intended to be informative or persuasive, any strategic approach to public dialogue must start with a consideration of the audience. This is easier said than done, especially in diverse, multicultural societies. Differing levels of knowledge, conflicting attitudes and dispositions toward action and diverse values and social identities confound blanket assumptions and stereotypes. While such diversity can pose real challenges, it can also present opportunities for innovative communication strategies.

On any conservation issue, level of knowledge needs to be a primary consideration. For scientists and managers this is especially critical, as the general public's scientific knowledge may lag behind their own. In some cases, members of the public may possess experiential knowledge that confirms, complements or even contradicts traditional scientific knowledge. For example, in some remote Alaskan indigenous communities, local residents may have observed new species or ecological patterns of which scientists and managers are unaware, as in the appearance of salmon in rivers of arctic Alaska in response to regional warming and sea ice retreat. Where level of knowledge is a concern, communication needs to include basic scientific information about ecosystem processes, species migration and other topics that undergird the concept of novel ecosystems. Even the distinction between native and non-native species may not be a salient idea for some audiences. At the same time, communicators need to be aware of the risks of the 'information-deficit' model of scientific communication, in which audiences are presumed to have a knowledge deficit that can only be overcome through the transmission of scientific expertise (Ziman 1991; Wynne 1995; Irwin and Wynne 1996). This model inhibits possibilities for collaborative learning and contributions based on local and experiential knowledge. In this context, novelty needs to be treated less as a concept that communicators must 'get across' to others, and more as a feature of shared inquiry in particular locations. For example, local knowledge among hunters has improved scientific understanding of the emergent habitat for deer hunting in Alaska (Brinkman et al. 2009).

Beyond knowledge, communicators must consider audience attitudes and dispositions toward action. Even for communication that is largely informative or educational, an audience's attitude about a topic can shape the kinds of information that might be perceived as relevant and thus lead to greater engagement. For example, a birder's desire to see new species might shape informative communication about ecosystem change. Likewise, an activist steeped in US-style preservationism and a timber industry worker may have different attitudes about pine beetle infestation; in turn, they may be more or less receptive to different kinds of information. Moreover, persuasive communication intended to influence personal or political action requires careful analysis of underlying attitudes. A useful example of such analysis can be found in the 'Six Americas' studies related to climate change (Maibach et al. 2010). These studies do not simply measure beliefs; instead, they categorize audience types based on broader dispositions toward personal and political action that are shaped by factual beliefs, deeper value commitments and social and political identities.

Social values and identities provide yet another layer when considering the audience. Even audience members who share a broadly pro-environmental attitude may be inspired by different core values: a spiritual reverence for non-human nature, a concern for human health, a commitment to future generations or an opposition to industrial progress. These values feed into one's social and political identity and one's sense of self in relation to others, providing a backdrop for decision-making (Crompton and Kasser 2010). The stuff of identity is far more rich and variegated than the 'blank slate' that is often presumed by information-deficit models of scientific communication. Effective communication about novel ecosystems must then go beyond information transfer; it must engage these identities in order to demonstrate *why novel ecosystems should matter to people in different walks of life*. For example, Coulthard et al. (2011) argue that effective engagement with the global fisheries crisis must include attention to social well-being, a much more complex way of thinking about people's values, motivations and aspirations than one narrowly rooted in economic self-interest. Without a clear connection to the every-day concerns that shape the identities of ordinary citizens, the idea of novel ecosystems risks becoming a relic of intellectual history rather than a beneficial concept for practical decision-making.

The above ideas point to a more dynamic way of considering audience than merely identifying demographic features and then tailoring a message accordingly. In fact, effective communication is not simply about adapting to the audience, but *constituting* the audience. Instead of treating the audience's identity as a fixed characteristic, communicators can help audience members see themselves as a certain kind of person: an active, empowered citizen who has a stake in the long-term health of their socio-ecological community. As well as providing information and tapping into the specific values and identities of audience members, engagement with novel ecosystems also requires a conscious effort to encourage a broader ecological identity among citizens (Thomashow 2002).

30.4 FRAMING THE MESSAGE ON NOVEL ECOSYSTEMS

Developing a consistent message about novel ecosystems may also involve deliberate circulation of one or more frames. While the concept of a frame is defined differently across disciplines, a communicative approach considers a frame as an organizing theme or storyline that provides meaning to events (Entman 1993). Frames give meaning to events by structuring those events as consistent with broader patterns of meaning that are shared in a culture. In this view, frames are effective not because they are new ways of defining a situation, but because they are mundane; audiences have already heard this kind of story before in a variety of topics and contexts.

Importantly, in this view the idea of novel ecosystems is not itself a frame. Instead, the strategic or inadvertent use of frames can shape understanding of novel ecosystems and the possibilities for engagement and action. It is easy to see, for example, how the very notion of novel ecosystems might be introduced to a broader public through an 'end of nature' frame that pits human activity as inherently at odds with the rest of the natural world, or as a tale of scientific breakthrough by intrepid researchers operating on the frontiers of knowledge. Likewise, human interventions to deal with non-native species could be characterized

as a surgical strike on those species, as opening a Pandora's Box of unintended consequences or as a conservation of resources. Every frame directs attention to some aspects of reality while deflecting others, and in doing so promotes particular ways of seeing problems, causes and solutions.

The challenge is to generate frames that will attract public attention, concern and action on the issues that are most important for novel ecosystems at the appropriate scale. Because of the way that frames shape perception, communicators need to be mindful of how frames can implicitly privilege certain kinds of values and outcomes. The metaphor of invasive species, while not precisely a frame, illustrates how an ostensibly descriptive characterization contains within it an implicit attitude and course of action. Consider too how a degraded park, vacant lot or untended school playground might be treated differently if it was framed as a 'living laboratory' versus depicted as a 'community revitalization' effort.

The point here is not to search for the one perfect frame. Indeed, what seems to make identifying an ideal frame so difficult is the fundamentally *mixed* character of novel ecosystems, as discussed earlier: diverse yet invaded, anthropogenic yet wild. We might instead seek a multiplicity of frames that express *discursive hybridity* (not to be confused with hybrid ecosystems, a term used throughout this book) that Marafiote and Plec (2006) found in colloquial, everyday environmental talk. The messy, ambiguous, contradictory character of novel ecosystems may pose a category problem for some, but similarly discordant views and beliefs often characterize human relationships with the rest of the natural world. Frames such as 'living laboratory', 'garden' and 'adventure', for example, each conceive of nature as a site of human intervention while also acknowledging nature's autonomy. The character of novel ecosystems also opens the door to multiple frames that appeal to different stakeholder groups who may value the same ecosystem for different reasons. A California grassland savanna invaded by European annuals may appeal to some as pasture for cattle, to others as habitat for native mammals or butterflies and to others as a 'natural' ecosystem typical of the region.

Resistance to a one-size-fits-all frame for communicating about novel ecosystems therefore reflects an openness to the inherent complexity of novel ecosystems themselves, but also to the complexity and contradictions that characterize human relationships to the natural world. Moreover, communication about novel ecosystems should not rely too heavily on getting the frame 'right' for other reasons: frames facilitate top-down forms of communication and governance that are at odds with public engagement (Brulle 2010); frames are always susceptible to co-optation and colonization by competing voices, as in the cases of sustainable development and ecosystem services (Peterson et al. 2005, 2010); over time, frames can become a kind of cultural common sense that blind their users to their limitations. While useful, frames need to be seen as just one dimension of effective communication. Furthermore, communication processes need to be open to emergent frames and challenges to existing frames that surface during public engagement processes (Walmsley 2009).

30.5 GROUNDING COMMUNICATION IN SPECIFIC PLACES

Communication about novel ecosystems should be less about abstract scientific concepts and more about local, on-the-ground changes. Research on climate change communication has revealed that place-based information about concrete changes to local landscapes resonates with the public. A place-based approach, whereby climate change impacts are discussed in terms of changes in specific communities and local examples, has been widely advocated (Thompson and Schweizer 2008; Shome and Marx 2009). For example, Grossman (2005) suggests that effective communication of scientific research requires situating findings in specific places where changes are occurring. Thomashow (2002) similarly recommends that by linking people to tangible and accessible changes on the land. To the extent that communication about novel ecosystems is tangible and germane to people's daily lives, they will be more engaged and invested in understanding ecosystem change and contemplating responses to such change. Discussing changes to local ecosystem services and locally iconic species may be one way to make such changes more relevant. People have significantly modified 75% of the terrestrial ice-free surface (Ellis and Ramankutty 2008; see also Fig. 8.4), with the greatest degree of modification in those regions where population density is highest and where people inhabit and utilize working landscapes (note that only 27–35% of the earth's surface would be considered novel, see Chapter 8). Thus, most people probably have greatest familiarity with ecosystems that

exhibit substantial novelty. It is therefore important that communication recognize the ecological and cultural value of modified or novel ecosystems as touchstones to nature for a broad spectrum of society. For example, most British ecosystems are the product of a long history of human habitation. Many ecosystems of Scotland are valued today for their moorlands rather than for the extensive forests that disappeared long ago.

30.6 EMPOWERING PEOPLE THROUGH A FOCUS ON SOLUTIONS

Communication about ecosystem change is most effective if it is empowering and solutions-based. Information that focuses on problems and blames society for their occurrence makes people feel guilty, angry or afraid and generally leads to public disengagement from the issues (Baumeister et al. 2001). Widespread disengagement of society from climate-change issues, for example, may substantially reflect the presentation of gloom-and-doom information and scenarios by climate-change scientists (Maibach et al. 2010). In contrast, positive messaging that focuses on solutions tends to engage the public and engender creativity and a search for coherent long-term strategies for action (Fredrickson and Branigan 2005). Discussions of tangible solutions make communication empowering and build people's sense of efficacy (Schweizer et al. 2009; Pike et al. 2010). Most ecological issues are like a glass that is both half-empty and half-full: half a problem and half an opportunity for a solution. Clear identification of the nature of problems as a context for discussing potential solutions is a strategy that is most likely to engage society and engender action.

In the context of novel ecosystems, it may be more constructive to focus less on the species that have been lost with land-use change and more on ecosystem properties such as connectivity that can enhance the probability that species that are currently rare or endangered can migrate or adjust to changes in climate or land use. For example, the Greater Yellowstone Ecosystem is a corridor of managed and unmanaged habitat within the Rocky Mountains that allows wolves and other species to migrate latitudinally in response to expected climate warming or to recolonize areas of local population extinction.

Similarly, a focus on the functional properties of novel ecosystems that sustain a breadth of ecosystem services is more helpful than simply documenting the compositional or structural changes that have occurred. This is not to say that ecologists should downplay the seriousness of past and current changes but rather that the discussion of problems should be coupled with strategies to identify and explore potential solutions. Declines in forest health, for example, reflect in part a failure of long-lived trees to adapt or migrate as rapidly as climate is changing. Focusing on potential solutions such as assisted migration or reseeding with geographically diverse (rather than locally adapted) seed sources moves the debate from discussion of attribution to discussion of the risks and benefits of alternative adaptation options (Millar et al. 2007).

A shift in focus from problems to solutions also implies a shift in scientific goals from research that is motivated primarily by a search for basic understanding of patterns and processes (fundamental research) to science that also seeks to understand the dynamics of changes that society might seek to promote (use-inspired fundamental research) (Stokes 1997). This suggests that communication about solutions is a two-way endeavor, as the public has a critical role to play in the development of research programs that focus on responses to emerging ecosystem novelty.

30.7 INTEGRATING PUBLIC INTERESTS AND CONCERNS INTO DECISION-MAKING

The goal of effective communication should be to engage people in a broad public dialogue that informs policy decisions. Public engagement in public land management and public policy formation is increasingly seen as a democratic imperative, and the benefits of an engaged civil society are widely touted. In the case of novel ecosystems, public participation in decision-making for public lands is a legal requirement in many countries. For example, public participation has been mandated for federal land management agencies in the United States since passage of the Administrative Procedures Act (1944) and was greatly enhanced by subsequent legislation, including the National Environmental Policy Act (1969) and the National Forest Management Act (1976). In many countries, legislation mandating citizen involvement in decision-making emerged in the 1970s and 1980s in response to public concerns about mismanagement of public lands and declining environmental quality. Public participation was established as an accountabil-

ity mechanism to ensure that agency decisions reflected the diverse interests of society. It also served as an explicit acknowledgement that environmental policy-making was a value-laden exercise that should be open to broad public input. Emerging models of shared governance similarly involve the public in decisions that reach far beyond publicly owned lands as ecosystem management moves beyond public lands to landscape-scale conservation efforts that incorporate multiple ownerships and agencies.

To the extent that public participation effectively incorporates public views, it is an important mechanism for ensuring that both public values and concerns about novel ecosystems are integrated into decision-making. Such processes enable the public to articulate the goods and services they wish to prioritize in such ecosystems, and allow the persistent worries outlined in Chapter 37 to be expressed, debated and negotiated. As such, public concerns that novel ecosystems might mean giving up on cherished conservation goals, pave the way for unfettered development or inspire weakening of current environmental regulations can be addressed through effective public engagement processes. As described later, effective public dialogue provides opportunities to sort through trade-offs and carefully consider the potential consequences of different courses of action.

Public engagement works well when it is connected to both formal and informal decision-making processes. While public engagements do not make policy decisions, they are also not simply thought experiments with no real world relevance. The purpose of such engagements is to inform policy-making, which requires that policy-makers be receptive to the insights and recommendations that flow from public dialogue. Similarly, scientists and managers need to be open to the ways that citizens conceptualize and value different ecosystems and willing to consider non-traditional approaches to ecosystem management.

An example of the role of public engagement in novel ecosystems is the transition of the Rocky Mountain Arsenal (RMA), located just ten miles from downtown Denver, from military complex to wildlife refuge. After this 17,000 acre facility was declared a superfund site, site surveys revealed impressive wildlife populations and surprisingly healthy habitat in some locations. To date, more than 12,000 acres of the RMA has been transferred to the National Wildlife Refuge System (Havlick 2007). Starting in the 1990s, public participation in decision-making processes at the site have informed clean-up plans, health monitoring of residents near the facility and community education. Public participation has been facilitated through community advisory boards, surveys, in-depth interviews, public meetings and public information campaigns. While not without controversy, these participatory mechanisms have enhanced the ability of neighboring communities to determine how the clean-up of the RMA would be handled and provided opportunities for members of the public to voice concerns and seek redress.

30.8 ENGAGING THE PUBLIC TO IMPROVE NOVEL ECOSYSTEM MANAGEMENT

Public participation can improve the management of novel ecosystems in a number of ways. First, public participation has the potential to make novel ecosystem management more democratic by improving the inclusivity and transparency of decision-making processes. Democratic rationales for public participation maintain that people have an inherent right to participate in decision-making processes that will affect their lives. The management of novel ecosystems certainly has the potential to affect peoples' lives and livelihoods. Allowing potentially affected publics to participate lends legitimacy to novel ecosystem management decisions by ensuring that public concerns are heard and considered. Participation can also improve transparency by allowing for explicit discussion of the various values and considerations that factor into management decisions – particularly opening up the often taken for granted assumptions inherent in the study and management of ecosystems.

Along these lines, public participation has the potential to substantively improve management decisions from both a social and an ecological perspective. It can encourage the inclusion of broader social knowledge and active deliberation between publics, experts and policy makers. Substantive motivations for public participation encourage the incorporation of diverse perspectives into processes and expose the often unstated values that lie behind management decisions. By broadening the value considerations, public participation can make management decisions more socially robust or inclusive of broader perspectives that improve satisfaction and public buy-in to management schemes. Chapin et al. (2010) argue

that stakeholder involvement is required for building political capital and socially acceptable plans. They suggest that: "Planning processes must engage stakeholders as co-problem-solvers at all stages, from initial framing and goal setting to evaluation of monitoring results and adjustment of management actions. Engagement of diverse stakeholders increases the likelihood that the full breadth of synergies and tradeoffs will be considered and that there will be broad support for policies and actions that emerge. This process can draw on diverse sources of local, traditional, and formal knowledge systems that widen the consideration of goals and approaches available to address complex problems."

Additionally, public participation can help define the role of an ecosystem in a particular community as well as the desired role of society in that ecosystem. Dialogue between concerned publics, experts and decision-makers can help define how an ecosystem could be managed to meet the goals of citizenry and managers, whether for biodiversity, livelihood, species preservation, carbon sink or recreation. This discussion can lead to more ecologically robust management decisions by inclusion of broader knowledge of how the ecosystem functions in relation to the communities that surround and use it.

What types of participatory processes could make novel ecosystem management more democratic and robust? Drawing upon similar democratic and substantive motivations for public participation, scholars in science and technology studies have begun to advocate for what they describe as "upstream public engagement." Rogers-Hayden and Pidgeon (2007: 346) describe upstream engagement as "dialog and deliberation amongst affected parties about a potentially controversial technological issue at an early stage of the research and development process and in advance of significant applications or social controversy". In other words, 'upstream' public engagement is a tool for opening up the value-laden assumptions inherent in emerging science and technology by involving the public early and "throughout the entire process of scientific research and development" (Corner and Pidgeon 2010: 33). We see significant potential for upstream engagement in the area of novel ecosystems. As discussed previously, the concept of novel ecosystems is based on a recognition that ecosystems are shaped by social processes as well as ecological processes. Ecosystem management decisions are thus inherently value laden. Recognizing the inevitable place that social processes and values have in contemporary ecosystems, upstream public engagement provides an early opportunity for scientists, publics and decision-makers to discuss the values associated with particular ecosystems and approaches towards management. Such discussions make the role of society in ecosystems more explicit and encourage public recognition of the ties between natural and social systems and processes.

In a comprehensive assessment of public participation, Dietz and Stern (2008: 226) conclude that: "When done well, public participation improves the quality and legitimacy of a decision and builds the capacity of all involved to engage in the policy process. It can lead to better results in terms of environmental quality and other social objectives. It also can enhance trust and understanding among parties."

Contrary to Seastedt et al.'s (2008: 552) concern that, in the context of novel ecosystems, "managers need protection from the public and policy makers, who are quick to condemn when activities designed to produce long-term results do not produce short-term benefits," we think that the inclusion of publics and policy makers alongside scientists and managers has the potential to improve the management of novel ecosystems in both the short and long term.

30.9   NEGOTIATING TRADE-OFFS

Public engagement offers an important opportunity for dialogue and negotiation of trade-offs, and careful consideration of which public goods should be prioritized in which places. The ecological novelty that characterizes many ecosystems provides an opportunity for dialogue about dynamics and change that is often missing from discourses about 'pristine nature'. If people perceive novelty and change as acceptable attributes of ecosystems, this provides an entry point for discussions of causes of change (e.g. stochastic and directional changes in climate and land use) and the ecological and societal consequences of these changes. This may, for example, foster discussions of trade-offs among ecosystem services associated with resource extraction (e.g. logging), maintenance of cultural landscapes (e.g. small-scale farms) or the inclusion or exclusion of fire control from designated wilderness. Such discussions can be contentious because people often differ in the values they place on alternative uses of the ecosystems

they inhabit or use (Yung et al. 2003; Ardoin 2006). Nonetheless, recognition of novelty as a frequent and acceptable state of ecosystems allows the dialogue to focus on the consequences that come from novelty rather than on an artificial construct of 'naturalness' (Cole and Yung 2010). In this way, people can make choices about trade-offs in extant ecosystems in the context of their goals and values.

30.10   PLANNING STRATEGIES AND CAUTIONS

Engagement processes should be inclusive and transparent, and promote collective problem definition, mutual learning, trust and dialogue. Dietz and Stern (2008: 3) outline the following principles for effectively melding scientific analysis and public participation:
"**1.** ensuring transparency of decision-relevant information and analysis;
2. paying explicit attention to both facts and values;
3. promoting explicitness about assumptions and uncertainties;
4. including independent review of official analysis and/or engaging in a process of collaborative inquiry with interested and affected parties; and
5. allowing for iteration to reconsider past conclusions on the basis of new information."
As Welp et al. (2006) point out, stakeholders can also help identify research questions that are socially relevant and advance scientific understanding. Detailed information on how to organize and manage participation and integrate science into engagement processes can be found in Dietz and Stern (2008).

It is important to also recognize the significant potential of group process in public engagement. When people work together in a group, they often utilize multiple sources of information and develop a more sophisticated understanding of the problem at hand.

Public participation in scenario planning can be a particularly effective way to engage stakeholders in joint discovery and mutual learning, and to integrate both expert and non-expert knowledge into decisions (Chapin et al. 2010). Because the specific trajectory of novel ecosystems is uncertain, scenario planning with the public can enable engagements to become relatively specific in terms of problem definition and possible paths forward while still acknowledging multiple futures (Peterson et al. 2003).

While there are many reasons to engage the public (e.g. democratic values and improved outcomes), we want to caution against a naïve or overly optimistic view of public involvement. Environmental issues are highly politicized and engagement processes bring different values and interests, thus different ideas, about how ecosystems should be managed to the forefront. Public engagement can, at times, increase conflict and controversy, and delay already difficult decisions. Deliberation does not always lead to consensus (in fact, it rarely leads to widespread agreement), but it can reveal public views, advance understanding and inform those decisions that would otherwise be made largely by experts.

30.11 COMMUNICATION AS ENGAGEMENT; ENGAGEMENT AS COMMUNICATION

There is an important synergy between communication and engagement, especially if we set aside assumptions about one-way expert communication filling gaps in public knowledge and move toward a vision of communication and engagement as two dimensions of broader public dialogue. To state it simply, the nature of communication influences the process and outcomes of public engagement. Similarly, public engagement processes can help us understand the outcomes of particular communication strategies. Communication can and should be structured to encourage engagement and dialogue, rather than privileging certain ways of thinking and acting (Brulle 2010). In this context, communication invites members of the public to participate in a broader dialogue in which they are active and vital participants in decisions, and bring relevant knowledge and experience to the discussion. Scientists, citizens and decision-makers together develop the discourse and practice that surround novel ecosystems and, in doing so, influence the social and ecological trajectory of such places. Because the concept of novel ecosystems does not assume one correct or better way of managing such systems, instead demonstrating a greater openness in appropriate management goals (Chapter 18), there is tremendous political space for deliberation regarding which public goods they will provide. Engagement provides a mechanism for the public to claim novel ecosystems and to decide together how to manage them.

REFERENCES

Administrative Procedures Act (1944), Pub.L. 79-404, 60 Stat. 237.

Ardoin, N.M. (2006) Toward an interdisciplinary understanding of place: lessons for environmental education. *Canadian Journal of Environmental Education*, **11**, 112–126.

Baumeister, R.F., Bratslavsky, E., Finkenauer, C. and Vohs, K.D. (2001) Bad is stronger than good. *Review of General Psychology*, **5**, 323–370.

Brinkman, T.J., Chapin, T., Kofinas, G. and Person, D.K. (2009) Linking hunter knowledge with forest change to understand changing deer harvest opportunities in intensively logged landscapes. *Ecology and Society*, **14**(1), 36. http://www.ecologyandsociety.org/vol14/iss1/art36/ Accessed August 2012.

Brulle, R.J. (2010) From environmental campaigns to advancing the public dialog: environmental communication for civic engagement. *Environmental Communication*, **4**(1), 82–98.

Chapin, F.S., Zavaleta, E.S., Welling, L.A., Deprey, P. and Yung, L. (2010) Planning in the context of uncertainty: Flexibility for adapting to change, in *Beyond Naturalness: Rethinking Park and Wilderness Stewardship in an Era of Rapid Change* (Cole, D.N. and Yung, L., eds), Island Press, Washington, DC, pp. 216–233.

Cole, D.N. and Yung, L. (eds) (2010) *Beyond Naturalness: Rethinking Park and Wilderness Stewardship in an Era of Rapid Change*. Island Press, Washington, DC.

Corner, A. and Pidgeon, N. (2010) Geoengineering the climate: the social and ethical implications. *Environment*, **52**(1), 24–37.

Coulthard, S., Johnson, D. and McGregor, J.A. (2011) Poverty, sustainability and human wellbeing: a social wellbeing approach to the global fisheries crisis. *Global Environmental Change*. http://eprints.ulster.ac.uk/17469/1/Coulthard_et_al_2011_GEC.pdf. Accessed August 2012.

Crompton, T. and Kasser, T. (2010) Human identity: a missing link in environmental campaigning. *Environment: Science and Policy for Sustainable Development*, **52**(4), 23–33.

Dietz, T. and Stern, P.C. (2008) *Public participation in environmental assessment and decision making*. National Research Council, National Academies Press, Washington DC.

Ellis, E.C. and Ramankutty, N. (2008) Putting people on the map: anthropogenic biomes of the world. *Frontiers in Ecology and the Environment*, **6**(8), 439–447.

Entman, R. (1993) Framing: Toward clarification of a fractured paradigm. *Journal of Communication*, **43**(4), 51–58.

Fredrickson, B.L. and Branigan, C. (2005) Positive emotions broaden the scope of attention and thought-action repertoires. *Cognition and Emotion*, **19**, 313–332.

Grossman, D. (2005) Observing those who observe. *Nieman Reports*, **59**(4), 80–85.

Havlick, D. (2007) Logics of change from military-to-wildlife conversions in the United States. *Geojournal*, **69**, 151–164.

Hobbs, R.J., Arico, S., Aronson, J. et al. (2006) Novel ecosystems: theoretical and management aspects of the new ecological world order. *Global Ecology and Biogeography*, **15**, 1–7.

Irwin, A. and Wynne, B. (1996) *Misunderstanding Science? The Public Reconstruction of Science and Technology*. Cambridge University Press, New York.

Maibach, E., Roser-Renouf, C. and Leiserowitz, A. (2010) Global warming's six Americas: an audience segmentation analysis. Yale Project on Climate Change and the George Mason University Center for Climate Change Communication.

Marafiote, T. and Plec, E. (2006) From dualisms to dialogism: hybridity in discourse about the natural world, in *The Environmental Communication Yearbook* (S. Depoe, ed.), Lawrence Erlbaum, Mahwah, NJ, pp. 49–75.

Marris, E. (2011) *Rambunctious Garden: Saving Nature in a Post-Wild World*. Bloomsbury, New York.

Millar, C.I., Stephenson, N.L. and Stephens, S.L. (2007) Climate change and forests of the future: managing in the face of uncertainty. *Ecological Applications*, **17**, 2145–2151.

National Environmental Policy Act (1969) 42 USC 4321-4347.

National Forest Management Act (1976) 16 USC 1600-1614.

Peterson, G.D., Cumming, G.S. and Carpenter, S.R. (2003) Scenario planning: a tool for conservatrion in an uncertain world. *Conservation Biology*, **17**(2), 358–366.

Peterson, M.N, Peterson, M.J. and Peterson, T.R. (2005) Conservation and the Myth of Consensus. *Conservation Biology*, **19**(3), 762–767.

Peterson, M.J., Hall, D.M., Feldpausch, A.M. and Peterson, T.R. (2010) Obscuring ecosystem function with application of the ecosystem services concept. *Conservation Biology*, **24**(1), 113–119.

Pike, C, Doppelt, B. and Herr, M. (2010) *Climate communications and behavior change: a guide for practitioners*. The Climate Leadership Initiative, University of Oregon.

Rogers-Hayden, T. and Pidgeon, N. (2007) Moving engagement 'upstream'? Nanotechnologies and the Royal Society and Royal Academy of Engineering's inquiry. *Public Understanding of Science*, **16**(3), 345–364.

Schweizer, S., Thompson, J.L., Teel, T. and Bruyere, B. (2009) Strategies for communicating about climate change impacts on public lands. *Science Communication*, **31**, 266–274.

Seastedt, T.R., Hobbs, R.J. and Suding, K.N. (2008) Management of novel ecosystems: Are novel approaches required? *Frontiers in Ecology and the Environment*, **6**(10), 547–553.

Shome, D. and Marx, S. (2009) *The Psychology of Climate Change: A Guide for Scientists, Journalists, Educators, Political Aides, and the Interested Public*. Center for Research on Environmental Decisions, Columbia University.

Stokes, D.E. (1997) *Pasteur's Quadrant: Basic Science and Technological Innovation*. Brookings Institution Press, Washington.

Thomashow, M. (2002) *Bringing the Biosphere Home: Learning to Perceive Global Environmental Change*. MIT Press, Cambridge, MA.

Thompson, J.L. and Schweizer, S.E. (2008) The conventions of climate change communication. National Communication Association convention, San Diego, CA. http://www.earthtosky.org/data/climate/PDF_Resources/thompson%20%20schweizer%20nca%202008.pdf Accessed August 2012.

Walmsley, H.L. (2009) Mad scientists bend the frame of biobank governance in British Columbia. *Journal of Public Deliberation*, **5**(1), 1–26.

Welp, M., Vega-Leinert, A., Stoll-Kleemann, S. and Jaeger, C.C. (2006) Science-based stakeholder dialogues: theories and tools. *Global Environmental Change*, **16**, 170–181.

Wynne, B. (1995) Public understanding of science, in *Handbook of Science and Technology Studies* (S. Jasanoff, ed.), Sage Publications, Thousand Oaks, CA.

Yung, L., Freimund, W. and Belsky, J. (2003) The politics of place: understanding meaning, common ground, and political difference on the Rocky Mountain Front. *Forest Science*, **49**(6), 855–866.

Ziman, J. (1991) Public understanding of science. *Science, Technology and Human Values*, **16**(1), 99–105.

Chapter 31

VALUING NOVEL ECOSYSTEMS

Andrew Light[1], Allen Thompson[2] and Eric S. Higgs[3]

[1]Institute for Philosophy and Public Policy, George Mason University & Center for American Progress, Washington, D.C., USA

[2]School of History, Philosophy, and Religion, Oregon State University, USA

[3]School of Environmental Studies, University of Victoria, Canada

31.1 INTRODUCTION

Learning to appreciate novel ecosystems will take practice and, as we argue in this chapter, perhaps some new approaches to valuing them. The challenge is not in learning to value novel ecosystems for the first time. After all, as Jackson (Chapter 7) notes, novel ecosystems are nothing new. Significant new ecological assemblies have arisen throughout the past. What is different now is the pace and extent of such changes.

Perring and Ellis (Chapter 8) argue that novel ecosystems comprise between 28% and 32% of the ice-free land surface. Some of these ecosystems have been with us for millennia (e.g. agricultural and urban transformations); those anthropogenic landscapes have been valued by people for a variety of reasons. Think of the Italian countryside: verdant hills dotted with vineyards, small farms, winding narrow roads and modest towns. This is a landscape celebrated for its cultural history, aesthetic qualities and the services procured by anthropogenic ecosystems. There are no shortage of values that people attach to such landscapes and the ecosystems within, and these values in turn shape how people restore and intervene in ecosystems as well (Hall 2005). Novel ecosystems, as defined in this book, are appreciated by many and largely elicit little controversy.

Not so for most ecologists, natural historians and lovers of wild nature. They worry about irreversible biodiversity losses, uncontrolled species invasions, unpredictable effects of rapid climate change and continuing land conversion from wild to productive ecosystems. Focusing on novel ecosystems, some fear, will shift an emphasis to ecosystem function and services perhaps at the expense of critical species, and is seen as tantamount to giving up the good fight for conservation values. Standish et al. (Chapter 37) chronicle some of these misapprehensions and persistent concerns, some of which will remain intractable issues well into the future. How will we come to terms with greater complexity in how we manage ecosystems, increased momentum for novelty for the sake of novelty, shifting cultural values about natural heritage and the hubris that may accompany an increased confidence in our ability to appropriately intervene in novel ecosystems?

Ecologists are especially good at focusing on subtle changes in species' relationships, and it is in these subtleties that a new order of complexity emerges not only in our understanding of ecosystems but also how we appreciate them. For example, Stone et al. (2008) studied ancient multitrophic interactions in ecosystems centered on the Turkey oak (*Quercus cerris*). The Turkey oak was native to the British Isles before the last glaciation (more than 115,000 years before present). It did not naturally return to the UK, but was planted for horticultural purposes over the last 300 years. As an exotic species in the context of the modern era, the Turkey oak now serves as a reservoir food source for two bird species with significant conservation value (blue and great tits; *Cyanistes caeruleus* and *Parus major*). The two bird species are laying eggs earlier in response to climatic changes, which interferes in the synchrony with typical food sources (notably native gall wasps). However, invasive gall wasps migrating northwards have filled the critical food gaps for the great and blue tits, and have created novel interactions with native gall wasps. These subtle interactions are an example of what is expected under conditions of rapid environmental, ecological and land conversation changes, and raise thorny questions (Hobbs et al. 2009). Is the Turkey oak ecosystem native or novel? It appears to be either one, depending on temporal range. What conservation and restoration values should take priority? How do we decide on a way forward in the face of a future that promises even more uncertainties? Ecologists might be perceived as dragging their feet into the new ecological world order, but their caution helps shape a measured response to valuing novel ecosystems.

The purpose of this chapter is to reconcile the complex ethical dynamics of novel ecosystems with the twin facts that novel ecosystems are with us and will likely be more appreciated in coming years, given the critical role they will likely play in adapting to a warmer world. How best then to value them? We turn first to the articulation of some of the ecological values present in novel ecosystems, and provide an initial taxonomy. Next we turn to the field of environmental ethics, which focuses on the moral status of environments. While our view is that more traditionally oriented environmental ethicists are not obliged to value novel ecosystems less than historical ecosystems we argue that some likely will, given other debates that have evolved in the field over the moral status of anthropogenic environments. We conclude with a

brief look at a new promising approach – environmental virtue ethics – which could offer an alternative framework for understanding our moral relations with novel ecosystems.

31.2   WAYS OF VALUING NOVEL ECOSYSTEMS

There are suggestions in preceding chapters for why we should value novel ecosystems when compared with historical reference or humanly manipulated ecosystems. They provide us with a foundation for how novel ecosystems should be valued if we already value the more traditional ends of ecosystem management such as preserving biodiversity, encouraging persistence or enhancing resilience. Consider the following eight aspects of novel ecosystems.

1. Some novel ecosystems provide a **continuation of benefits** to people. Hallett et al. (Chapter 3) give an account of shifting fortunes of the Rodrigues fody (*Foudia flavicans*). Dependent on mature forest stands on the smallest of the Mascarene Islands, the species went into sharp decline with the conversion of forests to agricultural fields in the 1960s. The fody was saved partly by fast-growing exotic trees that satisfied the requirements of this habitat specialist. Thus, a novel ecosystem stepped in to provide the continuation of a species with high conservation value.

2. Some novel ecosystems may provide **new and improved benefits** to people. There are two versions of this prospect. The first is inadvertent new benefits afforded by, for example, the creation of new habitats. On the northwest coast of Wales is Cei Balast, or Ballast Island, which was created by the dumping of widely sourced ballast rock from a global shipping trade in Welsh slate and copper. The vegetation communities that assembled from native and introduced species in the wake of industrial activities now have significant conservation status (see Chapter 27 for more information). In these cases, the new value of the ecosystem is managed to maintain particular services. The second version comprises 'designer ecosystems', which are explicitly designed ecosystems aimed at delivering new functions previously not provided by those ecosystems or optimized functions, or both. Examples are biofiltration ecosystems designed for effluent treatment, or green roofs that provide an improved building envelope. Such living machines are highly contrived and regulated, and taken as such may make significant contributions to sustainable social systems. However, engineered solutions will almost certainly gather more concerns when ambitions turn to mimicking or replacing or improving upon existing ecosystems.

3. Novel ecosystems may have **higher diversity** than historical ecosystem. This is frequently the case with urban ecosystems, which are often hybrid ecosystems comprising native and non-native vegetation. Where species richness is higher, it may contribute to increased resilience in the face of rapid environmental and ecological changes.

4. Novel ecosystems may require **innovative management approaches** and so will call for controlled experimentation (see point seven), providing a knowledge base that will be necessary in a rapidly changing world. Such bold experiments would be perceived as difficult or inappropriate in historical ecosystems. Seastedt et al.'s account (Chapter 15) of intensive grazing by native black-tailed prairie dogs in urban and suburban landscapes of the Colorado Front Range is a good example. Prairie dog grazing is compounded by the effects of climate change, nutrient deposition and non-native plant invasion, and is transforming the vegetation and soils into emerging novel communities. Such complex system interactions are compelling new approaches to management for resiliency.

5. In the face of significant changes, for example when climate change forecloses on historical species, allowing or even encouraging the formation of novel ecosystems will **increase resilience** to further changes. At the same time, they provide additional evolutionary resources for the development of new species characteristics and behaviors. Ecosystems that are prevented from changing are brittle and break easily.

6. Novel ecosystems, when embraced for conversation values, will provide **additional conservation capacity**. In large network efforts, such as Gondwanalink in Western Australia or Yellowstone-to-Yukon in western North America, the incorporation of hybrid and novel ecosystems opens the potential for more significant conservation connectivity and buffer zones.

7. Unmanaged novel ecosystems constitute **uncontrolled experiments** that will inform management in a rapidly changing world. Careful scrutiny of these experiments will build understanding of how best to intervene in novel and hybrid ecosystems. It is likely also that hybrid and novel ecosystems will be incorporated in long-term ecological studies and experiments.

8. Novel ecosystems manifest key characteristics of **wildness** including ongoing change, uninhibited growth and free-flowing evolutionary processes. Over time, novel ecosystems may become the new normal (see Marris et al. Chapter 41), and as such become representative of new values of wildness. As such, the new wildness will be distinct from traditional values of wilderness determined by the degree of human influence.

These eight initial ecological values of novel ecosystems, while certainly not exhaustive, offer a starting point for investigation into how novel ecosystems could be valued in a robust moral framework. As suggested at the start of this chapter however, these characteristics of novel ecosystems are only morally valuable if we already have reason to value ecosystems, resilience and wildness. Before turning to what we think is an alternative framework for understanding and appreciating the value of novel ecosystems, we will first look at how those holding any of several traditional views in the field of environmental ethics would likely approach questions of the value of novel ecosystems.

31.3 NOVEL ECOSYSTEMS THROUGH THE LENS OF TRADITIONAL ENVIRONMENTAL ETHICS

Setting aside issues of the economic value of novel ecosystems (which is beyond the scope of this chapter) we can divide between two distinct (although overlapping) projects when it comes to the general question of the moral or ethical value of novel ecosystems: describing the ways in which people actually do value (or do not value, or hold in diminished value) novel ecosystems, or describing the ways in which novel ecosystems should be valued. The former project would be largely descriptive and empirical. It would best be accomplished through anthropological and sociological surveys of people's attitudes of how they do or do not value novel ecosystems. The latter project would in contrast be normative.

We will focus on the latter project here, although investigations into people's actual beliefs about the value of novel ecosystems are certainly worth doing. Even if these were done as well as it could be, a normative conception of these systems would still be required in order to have a full account of the value of novel ecosystems. A fully normative account of the value of anything simply is what that value should be understood to be, rather than what anyone claims is the value of the thing in question. One important reason to have such a sense of an item's value in this case is that any survey of the value of novel ecosystems, no matter how comprehensive, will reveal many disagreements as to how people think these systems should be valued. These disagreements would point us in different directions for the purposes of making law or policy and so need to be informed by the best normative assessments of the relative merits of these competing assessments of value.

Environmental ethics is the field of philosophy that has focused on providing this kind of assessment of the value of non-human entities ranging from individual non-human animals to whole ecosystems, including characteristics of non-human natural entities such as wildness. In general, since its formal origins in the early 1970s, the principle question that has occupied the time of most philosophers working in environmental ethics is how the value of nature could best be described so that it is recognized as directly morally considerable in and of itself, rather than only indirectly morally considerable because it is appreciated or needed by humans (see Light 2002).

Pursuit of this question has led many, if not most, theorists in the field to embrace one or another version of the claim that non-human natural entities or systems possess some kind of special intrinsic or inherent value based on some kind of property that they have. If certain natural entities possess this kind of value then they are directly morally considerable, similar to how most moral theories find humans to be directly morally considerable and not necessarily dependent for their moral value on some use that they have for other persons. For example, on a theory that original, non-anthropogenic ecosystems have some kind of intrinsic value, then a moral claim to preserve or protect a particular ecosystem from development would not be contingent on describing some value that this system has for humans.

As might be expected, this picture gets much more complicated as there are many varieties of environmental ethics and abundant disagreements about why non-human entities might possess some form of non-instrumental value that should be respected in a moral sense. This array of views further complicates answering the question of how traditional approaches to environmental ethics value novel ecosystems. Unfortunately, there are as yet no systematic works by any environmental ethicists on novel ecosystems which would give

us any insight into how even a particular school of thought would value them. Still, we can make some progress here by taking a representative sampling of different schools of thought in the field and trying to discern how they would value novel ecosystems based on how they have discussed relevant comparison cases.

There are many ways that have proven useful for categorizing and distinguishing among the different schools of thought in environmental ethics. For example, an early and still pervasive starting place for many in the field is between human-centered views (anthropocentric) and non-human-centered views (non-anthropocentric) to the question of what is morally considerable, followed by further subdivision of schools of thought within each of these approaches. We will combine that basic distinction with another on whether a school of thought places direct value on ecosystems or if it values ecosystems only for indirect reasons. What we are looking for here is whether the adherents to a given school of thought value ecosystems for some quality they have in and of themselves, or for some role ecosystems play in the health or well being of some other entity which is found to be valuable in and of themselves. This will help us to more quickly hone in on the possible pitfalls that an assessment of novel ecosystems could have in some of the most predominant theories in the field.

We consider three types of an environmental ethic, divided by whether they attribute direct or indirect value to ecosystems: individualistic non-anthropocentrists, holistic non-anthropocentrists and holistic anthropocentrists.

Both individualistic and holistic non-anthropocentrists share the view that there are things that are not human which are properly the objects of direct moral consideration. The main point of difference between them is generally that individualists believe that the only entities which we owe moral obligations directly to are those with individual interests of some kind – such as the interest in avoiding pain (in the case of utilitarians such as Singer 1990) – or being the subject of an individual life with a unique trajectory (in the case of Kantians such as Regan 1983). On these accounts, entities which do not have minds – and so cannot have interests – are not directly morally considerable. This would therefore exclude what we might call 'collective entities' such as species and ecosystems from direct moral consideration.

In contrast, holistic non-anthropocentrists (e.g. Rolston 1988; Callicott 1989) value ecosystems for direct reasons. Ecosystems themselves are the proper subjects of moral concern. The value that ecosystems have is not generally due to any interest that can be attributed to them but rather to some intrinsic property that they have which generates a reason for holding them the proper subjects of moral consideration.

Finally, holistic anthropocentrists (e.g. Norton 1984) do not extend moral consideration beyond humans and so do not directly value either non-human individuals or non-human collective entities such as species or ecosystems. The 'holistic' aspect comes in their insistence that questions of natural value should be holistic – inclusive of and appreciative of the role that ecosystems play in the health and flourishing of humans – at the level of individual and collective decision-making. But again, holistic anthropocentrists can value ecosystems indirectly insofar as they are useful for humans or for things other than humans.

Again, this tripartite division of views in the field is not intended to be comprehensive or complete as a description of the variety of views in environmental ethics. This division is however illuminating for our discussion of how current work in environmental ethics could be applied to consider the value of novel ecosystems, because it gives us a way of focusing on the kinds of views in the field which we expect will be the biggest challenge to claims that novel ecosystems have positive value when compared to historical or reference ecosystems. Although some representatives of all three views may find fault with novel ecosystems – especially in trade-off scenarios where a decision has to be made to allow or encourage a novel ecosystem in place of a historical or hybrid ecosystem – the camp most likely to discount them is holistic non-anthropocentrism.

Indeed, holistic non-anthropocentrists, who arguably represent the bulk of views in environmental ethics to date, are the only of these three camps who would claim that ecosystems writ large are the proper objects of moral consideration. To put it crudely, in this respect they have the most at stake on the question of how to value novel ecosystems. Any theory that we could fairly categorize as belonging in this camp would, of necessity, have to have some kind of account of the specific qualities or characteristics of ecosystems that make them the proper objects of direct moral consideration. Novel ecosystems may or may not share those qualities and so may or may not fare well when assessed by one of these views. For example, if

a version of holistic non-anthropocentrism claimed that ecosystems were intrinsically valuable (or relatively more valuable) when they were non-anthropogenic in origin, say, based on a claim that things that evolved separately from humans have a unique kind of value that humanly produced environments do not have, then novel ecosystems would either not be intrinsically valuable on that account or would at least be less valuable than original non-anthropogenic systems.

In contrast, because individualist non-anthropocentrists and holistic anthropocentrists do not value ecosystems directly, they do not have a necessary starting point from which they must come to a moral assessment of novel ecosystems. They can still certainly assess the relative value of novel ecosystems with respect to the sorts of beings that they find directly morally valuable (all individual sentient entities in the case of Singer and Regan and all humans in the case of Norton), but novel ecosystems in themselves do not serve as a test case for their theory of why ecosystems are valuable because they do not hold ecosystems to be intrinsically valuable. In this respect, our claim is that figures from these camps should not have a reason to necessarily discount the moral value of novel ecosystems or otherwise discriminate against them since there is nothing morally at stake for them concerning the value of ecosystems in general. Ecosystems are only derivatively valuable on these accounts and so, at best, what could be claimed is that novel ecosystems were lacking in some quality that non-novel ecosystems have. Whether novel ecosystems lacked that quality would likely be in dispute. Certainly, we could have an active debate for any particular example of these views on the value of novel ecosystems. Even if the argument we have made here is rejected however, it ought to be conceded that novel ecosystems are potentially more controversial for holistic anthropocentrists.

Holistic non-anthropocentrists have a direct moral stake in the question of how to value ecosystems and hence must confront the question of the value of novel ecosystems in moral terms. A precedent in the literature that could give us insight on how this camp of environmental ethicists would approach the value of novel ecosystems is the existing criticisms that have emerged from holistic non-anthropocentrists concerning restored ecosystems. We will consider these objections and whether they offer a serious threat to finding a positive value in novel ecosystems. Our conclusion will be that while there may eventually emerge individual accounts that disagree, again, there is no necessary reason why all holistic non-anthropocentric views would have to find novel ecosystems of lesser value than non-novel systems. We will make this conclusion sharper by claiming that holistic non-anthropocentrists should not disvalue novel ecosystems even if they find them of comparatively lesser value than original non-anthropogenic systems. Partly, our confidence in this conclusion is that holistic non-anthropocentrists are not of one voice about the relevant comparison cases represented by restored ecosystems.

31.4 A LESSON FROM RESTORATION ECOLOGY

To understand the critical literature in environmental ethics on restoration ecology we have to back up slightly and recognize that holistic non-anthropocentrism inherits a particular kind of ontological problem (meaning, a problem about the metaphysical nature of a thing or type of thing) that is not entailed by the other two views. While individualistic non-anthropocentrists and holistic anthropocentrists can more or less easily identify examples of the sorts of things they have determined are proper objects of direct moral consideration, this is less easily accomplished with holistic non-anthropocentrism.

For example, holistic anthropocentrists claim that only humans are, properly speaking, moral subjects. While there are active debates on the ends, as it were, of moral considerability for this camp – on issues of whether human fetuses are persons and on whether brain-dead humans are persons and deserving of direct moral consideration – it is relatively easy to identify human persons out of a crowd of other things that are the proper objects of direct moral consideration. In comparison, holistic non-anthropocentrists have a harder time identifying their objects of direct moral consideration. An environmental ethic of this sort for example cannot be properly said to find the same value indiscriminately in all environments, especially if we were to include any human-made interior environment (like a dining room) in a range of comparison cases. This problem has created an active body of literature on marginal cases among holistic non-anthropocentrists with a focus on both non-human species and ecosystems which, although apparently 'natural', are dependent on humans for their existence.

For example, there have been active debates among holistic non-anthropocentrists on whether species of domesticated animals are natural or artifactual, and hence properly speaking the kinds of collective entities who should be directly valued by this kind of view (Katz 1983). Similarly, these sorts of concerns have also been raised about humanly produced, exterior environments, partly producing the critical literature on restoration ecology from this field.

The two most prominent critics of restoration ecology from a non-anthropocentric holist position are Robert Elliot (1997) and Eric Katz (1997). Both have criticized restored ecosystems for different but overlapping reasons. They have at least two shared concerns that should be addressed here: (1) that restored environments – insofar as they are attempts to copy environments that came before them – are offered by their proponents as a fair substitution for original environments; and (2) that ecological restoration will be used to justify destruction of original environments. Various examples are offered by both to demonstrate this worry and generally involve actual examples of proposals by some party to develop a 'pristine' or generally untouched area on the condition that it be restored later.

Both Elliot and Katz resist the positive claims of these examples by arguing that because nature has an 'originary value' – that is, its value derives in part from the fact that its origins are not dependent on humans – a substitute restoration can never count the same as the original. Elliot, for example, argues that an analogy can be drawn between the value of an ecological restoration and the value of an artistic fake. Just as the faked art work (no matter how perfect) can never replicate the causal origins of the original and the fake (no matter how good) can never be that same object that was painted by person A at time T producing object X, a restored environment can never replicate the evolutionary origins of a non-anthropogenic natural environment. The upshot is that, on this view, restored ecosystems cannot have the same value as original ecosystems and, by extension, that the possibility of restoration is an insufficient justification for the destruction of a non-anthropogenic natural environment, given Elliot's account of the value of original, non-anthropogenic ecosystems. Through these arguments, both philosophers make a contribution to the necessary metaphysical burden inherited by holistic non-anthropocentrism to accurately describe the sorts of collective entities that are the proper objects of direct moral consideration and the sorts that are not. Restored environments are 'tainted' due to their origins and hence, at least, can never equal the value of a non-anthropogenic system.

Both of these core concerns about restoration ecology have been answered in various ways in the literature. One illustrative example for the purposes of this discussion is offered by Light (2000) who distinguishes between two types of restorations: 'malicious' and 'benevolent'.

On Light's account, malicious restorations are those that are offered as substitutes for original systems and the possibility of their creation is supposed to justify the destruction of the original system. In contrast, benevolent restorations are those that are undertaken to remedy some kind of intentional or unintentional destruction of a natural system and not offered as a prior justification for a destructive act. One of the primary problems in the cases offered by Elliot and Katz is that the act of restoration is being used to justify destruction of some arguably original non-anthropogenic system. However, in the absence of such duplicitous motivations, what is really wrong with restored environments? Most restoration projects are not malicious, as described by Elliot and Katz, but rather benevolent insofar as they aim to make up for a 'harm', given that non-anthropocentric holists believe that non-humans are the kinds of things in the world that can be harmed in a moral sense.

If we were to apply this reasoning to novel ecosystems, then we should be able to resolve any similar difficulties. For one, novel ecosystems are not generally offered by an agent as a substitute for anything. Novel ecosystems are a description of a state of the world that has evolved as a result of human changes to the environment either through action or inaction. In general, humans have not set out to create novel ecosystems but rather they are more accurately described as an unintended byproduct of our engagement with the world. Insofar as humans have not intentionally created novel ecosystems, there should be no need even for a distinction between malicious and benevolent novel ecosystems (compare the discussion of hubris in Chapter 37).

Elliot and Katz do however diverge in several places. It is easier to show the compatibility of Elliot's critique of restoration with a positive valuation of novel ecosystems since he is generally more willing to accept a range of different forms of restoration as producing different kinds of value, some of which are very close to the value he finds in original ecosystems. For example, while Elliot thinks a naturally regenerated forest has less value than an original forest (because he holds that natural value is quantitative across time and that the regenerated forest has also lost continuity with the forest that stood there before), he admits we could "accept that the value of the regenerated forest is roughly equal to the value of the original forest" (1997; 92). Elliot also, "with some hesitation", accepts some human intervention to accelerate regenerative processes because they have the ability to restore natural values more quickly in a degraded area than "development through natural forces of ecological complexity" (1997; 93). While Elliot might discount the value of novel ecosystems as compared to original untouched systems (if they can be found today), there seems little room for him to discount them completely given his approach to these other activities.

In contrast, Katz is more concerned that a commitment to restoration also risks certain psychological flaws that have moral implications. A faith in restoration for Katz is a form of hubris. Acceptance of restored environments is a slippery slope to acceptance of a completely artifactual world, one that is entirely humanly made and which has lost its non-anthropogenic value. When combined with his holistic non-anthropocentrism, these psychological flaws become condemnable moral practices representing a form of unjust domination of humans over other entities (Katz 1997; 113): "The attempt to redesign, recreate and restore natural areas and objects is a radical intervention in natural processes. Although there is an obvious spectrum of possible restoration[s] . . . all of these projects involve the manipulation and domination of natural areas. All of these projects involve the creation of artifactual realities, the imposition of anthropocentric interest on the processes and objects of value. Nature is not permitted to be free, to pursue its own independent course of development."

Unlike Elliot, Katz condemns all forms of restoration, no matter their motivation and no matter the reason for their engagement. This sort of objection presents a more difficult challenge to novel ecosystems as surely any appreciation of their potential positive value is in part dependent on an acceptance that humans have reordered the world through both the things we have done to change it and acceptance of what comes of the world after we abandon our day-to-day management of it. If the act of restoring an ecosystem, no matter what the reasons, is a form of domination then novel ecosystems may also be tarred by extension of this reasoning.

To respond to such a line of reasoning let's first distinguish between two kinds of contexts in which we might be asked to assess the value of novel ecosystems: (1) the act of 'discovery', where we come to learn that an ecosystem that we either did or did not previously recognize as resulting from a causal chain involving human intervention meets the definition of a novel ecosystem; and (2) in the 'response' to novel ecosystems, where we are asked to make a choice as to whether to allow or even encourage their emergence unimpeded or try to prevent them from emerging by managing for historical fidelity. A view like Katz's should not prove a problem for either contexts of engagement with novel ecosystems, but for different reasons in each case.

Regarding the context of discovery, we return to the intentional origins of Katz's objections to restoration. Katz's argument is that restorations are not valuable (or not as valuable as original natural systems) because the humanly replicated system can never replace the original. An assessment of the value of novel ecosystems in the context of discovery does not begin in the same place. Rather than proactively arguing for the value of novel ecosystems, the sober study of them starts only with an acknowledgment of their reality. At present, about 30% of the Earth may arguably be constituted by such systems (Chapter 8). An appreciation of the value of these systems is not an exercise in a project necessarily intending to encourage their expansion – which might be subject to Katzian criticisms of "domination of natural areas" – but rather a project of understanding the role and function they can play in the flourishing of human and non-human entities which presently inhabit them.

At first glance, however, the context of response might be more difficult and suggest a reason that non-anthropocentric holists critical of practices such as restoration ecology would be forced to disvalue novel ecosystems. At least a view structured like Katz's would not necessarily lead to this conclusion, however. On Katz's view, every form of intervention is a form of domination. If the choice is between allowing for the spread of novel ecosystems, tainted as they likely are on his view by the hand of humans, or managing systems for historical fidelity through restoration, then both represent a form of unjust domination of humans over nature. While someone defending this view might try to justify a choice of one over another, there seems

to be no clear reason why restoration for fidelity would win out over an embrace of the positive value of novel ecosystems. In fact, from the way in which Katz makes his argument, a revision of his account which would be critical of novel ecosystems would be question-begging given that the problem with restoration is that it restricts nature's freedom, not allowing it to "pursue its own independent course of development". The emergence of novel ecosystems would seem to be an arguable case of nature doing exactly that, pursuing its own trajectory of change, whereas management for historical fidelity would be the imposition of our will upon it.

In the next section we will offer an example of a new and emerging view in environmental ethics which does not share the same general structure of the sorts of views we have discussed so far; this new view offers a different approach to understanding the moral value of novel ecosystems. Before closing this section however, we return to the three general camps in traditional environmental ethics and summarize how they could embrace the value of novel ecosystems for the reasons we identified in Section 31.3.

Since individualistic non-anthropocentrists (particularly sentientists such as Singer and Regan) draw the line of direct value of ecosystems at individual animals, then demonstrating the indirect value of novel ecosystems should be as relatively straightforward as it is with any other kind of habitat. Insofar as it can be demonstrated that novel ecosystems expand or protect the habitats of other animals, then these systems are valuable. From this perspective, there are no general reasons why any kind of ecosystem is better than another so long as they are equally compatible with the flourishing of the forms of life that inhabit them and not, for example, confined systems in zoos or aquariums which can be independently criticized on these views.

Since holistic anthropocentrists draw the line of direct value of ecosystems at humans, then demonstrating the indirect value of novel ecosystems will be a matter of demonstrating their utility for maximizing human flourishing and improving those characteristics of non-human systems which we can make broader anthropocentric claims for valuing such as biodiversity (see Norton 1987).

This leaves us with the question of whether holistic non-anthropocentrists can or should value novel ecosystems. Insofar as it can be demonstrated that they can add value to ecosystems and ecological communities, then given our previous answers against those who would criticize the human dimensions of other kinds of environments they should be found to be valuable for all the reasons offered earlier. It is also clear that novel ecosystems will play an important role in a changing climate. Even under the most optimistic greenhouse gas mitigation scenarios, given the duration of CO_2 in the atmosphere the warming forced by historical emissions to date will continue to cause changes outside the historical range of different ecosystems well into the future. Given the current low ambition of mitigation efforts, we can expect the pace and duration of ecosystemic changes wrought by climate change to be even more dramatic in the future. In this respect, novel ecosystems could play a very important role in adaptation to a warmer world. Novel ecosystems will provide habitat for species that are shifting their ranges due to climate change.

31.5 AN ALTERNATIVE APPROACH TO VALUING NOVEL ECOSYSTEMS: ENVIRONMENTAL VIRTUE ETHICS

Fortunately several alternative views to the dominant approaches in environmental ethics exist which could provide a more interesting platform from which to assess the value of novel ecosystems, and which do not get stuck in the same kind of metaphysical puzzles. We will focus here on environmental virtue ethics, not by way of necessarily endorsing this view (although one of us has) but by way of providing an example that focuses on other characteristics of novel ecosystems and deepens our appreciation of the value questions that surround them. Further, we explore possible application of how an appeal to virtue can help answer questions about managing ecosystems with no historical analog.

To understand the way that a virtue ethics could provide a distinctive approach to novel ecosystems, we first need to distinguish virtue theory from the dominant strains of modern ethical theory in general, and then distinguish environmental virtue ethics from the three types of traditional environmental ethics discussed earlier.

What is a virtue? According to adherents to the work of the Ancient Greek philosopher Aristotle, who was the first to systematically investigate this topic, a virtue is a character trait, i.e. a developed and relatively stable behavioral disposition that a person needs to live well in the sense that possession of the virtues is constitutive of human flourishing. "To possess or exhibit a par-

ticular character trait is to standardly take certain types of considerations (under certain types of circumstances) as reasons (or motivations) for acting, feeling, or desiring in certain ways" (Sandler 2012, p. 68). Thus, the subset of particularly environmental virtues can be understood as forms of human character excellence regarding various relations to the organisms and ecosystems of the biosphere where our specifically human lives unfold.

Two features are central to making virtue theory distinct from other ethical theories: the account it offers of what makes a particular action morally right and the explanation it provides of the ultimate ground of moral normativity. First, whereas more recent ethical theories, such as utilitarianism or a Kantian deontology, attempt directly to explain what makes an action morally right, virtue theory aims instead to describe and defend particular traits of character as genuine virtues, that is, as substantive components of a well-lived human life. On this view, the notion of right action is dependent upon a more primary conception of the virtuous person; the morally right thing to do is whatever the virtuous person would do in those circumstances when acting in her characteristic virtue. (This approach to normative guidance may be useful for managing novel ecosystems as we discuss later.)

Second, both traditional anthropocentric and non-anthropocentric theories of right action are structured to attribute some form of intrinsic value to human beings or parts of the non-human world, respectively, and then it is this value that grounds any moral duties we may owe directly to them. Alternatively, according to virtue theory, human beings ought to first cultivate and then to exercise good traits of character just because these virtues are substantively part of the well-lived human life. One's moral obligations, then, are not grounded on the intrinsic value of some human or other entity, but instead are derived from the demands of a virtue, such as justice, that constitute a well-lived life. The upshot is that a virtue theory approach to environmental ethics does not have to involve any direct appeal to the value of organisms, species or ecosystems but instead must articulate how particular attitudes, behaviors and emotive responses toward such entities exemplify various forms of human excellence.

With this general characterization, we can further illustrate how an environmental virtue ethic (EVE) is distinct by contrasting it briefly with the three traditional forms of environmental ethics discussed in the previous sections. First, compared to an individualist non-anthropocentric theory, which finds individual sentient organisms to bear an intrinsic value, EVE seeks to articulate and defend the other regarding virtues, such as compassion, as extending to non-human animals. Second, compared to a holist non-anthropocentric theory, EVE strives to identify particular character traits relevant to our relations with species, ecosystems or places and then defend some of these traits as genuine virtues. For example, we may cast virtues such as care, compassion, non-maleficence or even reciprocity and justice more broadly, recognizing them to include various collective entities. Finally, it is often thought that EVE most closely resembles a form of holist anthropocentrism, since both focus on how our relations with ecosystems, places and kinds of organisms are intimately connected with the conditions of human flourishing. What is however distinctive about EVE in this context is that it is not limited to a conception of flourishing in terms of subjective well-being; it does not have to value the non-human world in terms of either the formation or mere satisfaction of human preferences. Instead, it can offer an account of how ecosystems fit within an objective needs-based account of what it is to live a human life with befitting dignity. This, too, may turn out to be an important feature of the virtue approach because while many people today express preferences for historical ecosystems, novel ecosystems may still provide many necessary services or otherwise meet the conditions we need to flourish.

We may expect that many traditional virtues will also be specifically environmental virtues (e.g. frugality, temperance) and some traditional vices will be environmental vices (e.g. greed). But as we face a future of rapid and unprecedented anthropogenic environmental changes at local, regional and global scales, it remains an open question if new environmental conditions will call for correspondingly new environmental virtues. Consider the case of restoration again, especially given the now accepted fact that ongoing environmental change threatens the viability of using historical references to set goals for ecological restoration (Hobbs et al. 2009).

In a recent collection of essays devoted in part to an investigation of the role that an understanding of virtues could play in thinking about the future of restoration, we see a variety of appeals that could provide some insight. Throop (2012) argues that if restoration is conceived metaphorically as a kind of healing, then humility, sensitivity, self-restraint and respect for others will be key virtues for restorationists whose practice can still exhibit a high degree of historical fidelity. More directly, Higgs (2012) suggests that a sophisticated and varied kind of sensitivity to history itself may be a novel virtue for restoration practitioners. Alternatively, Sandler (2012) maintains that global warming will weaken the virtue of historical fidelity associated with forms of assisted ecosystem recovery while at the same time raising the salience of other virtues such as openness, accommodation and, in particular, a new virtue of reconciliation as "a disposition to accept and respond appropriately to ecological changes that, though unwanted or undesirable, are not preventable or ought not be actively resisted" (Sandler 2012; 76).

By now it should be clear that environmental virtue ethics offers a distinct approach to environmental ethics and has application in contexts of significant and sometimes irreversible environmental change. What remains to be addressed is how a virtue ethic approach would apply to novel ecosystems. Here we merely sketch two such applications: (1) virtues associated with the general acknowledgement of novel ecosystems as a category; and (2) how an appeal to virtues may help answer questions about managing ecosystems that have no historical analog.

How should good human beings respond to novel ecosystems? It seems that *honesty* demands we acknowledge the existence and prevalence of ecosystems without historical analog, that *prudence* requires rebranding what were once thought of only as degraded landscapes and that *courage* may be required to confront these emerging realities. On the other hand, the *vice of hubris*, a perennial worry concerning ecosystem intervention, initially appears misplaced in the context of novel ecosystems to the extent, as we argued earlier, that most novel ecosystems are discovered rather than intentionally created. Finally, many people are emotionally attached to the idea of conserving or restoring historical ecosystems because they attribute intrinsic value to them (see Chapter 37). However, it seems more and more important that we become able to reconcile ourselves to the consequences of environmental change that we cannot avoid (Sandler 2012). At the same time, since novel ecosystems by definition arise from human activity, acknowledging that they deserve our attention would seem to exhibit a virtue of environmental responsibility (Thompson 2012).

In closing, we explain how virtue theory can provide guidance, albeit indirectly, concerning the management and stewardship of novel ecosystems. Typically, normative questions about management are directed by the goal of conserving or recreating historical conditions. Ethically, historical-based goals have been seen as at least innocuous, if not fully laudable. In cases where protecting or recreating past ecological conditions are not feasible however, what new goals ought to guide ecosystem intervention?

One option is to aim for new goals such as ecosystem services, biological diversity or resilience. This plurality however brings with it a new problem: which end, or combination of ends, should be pursued in any particular case? Virtue theory offers another option by encouraging us to ask about excellence in landscape, wild land and natural resource management practices by focusing on virtuous practitioners themselves. Instead of judging someone to be a good ecosystem manager, due to the fact that she aims at the right goals and undertakes suitable means, we could reverse the order of explanation and evaluate management goals and methodologies by appeal to what good or virtuous managers would try to do.

If we follow this strategy, the next step would be to establish some criteria by which we could identify particularly good ecosystem managers; one place to start is by filling out our description of the social role occupied by these professionals. Higgs and Hobbs (2010) offer some starting points. Good ecosystem managers have traits that encourage public and practitioner engagement, exhibiting suitable humility, flexibility and restraint, exercising creativity instead of a superficial formalism, with an orientation toward long-term resilience for a site and in relation to the human communities impacted by management decisions. More broadly, we might think the practice of ecosystem management as playing a role in assisting economic, social and environmental sustainability. Developing clearer ideas about how ecosystem managers may meet the plurality of demands that confront them in their practice could help us identify who among us manage ecosystems well.

31.6 CONCLUSION: A PLEA FOR PRAGMATISM IN PRACTICE

While this chapter has surveyed an array of approaches to valuing novel ecosystems from a number of different theories in environmental ethics, we have not made a strong claim for preferring one view over another. It was not our intention to do so.

Fortunately, it is not the case that all theories of environmental values force a decision to use only one or another school of thought when morally assessing a given natural entity or deciding on one's duties to it in a given situation. So-called 'pluralists' and some 'pragmatists' in environmental ethics make a compelling case that, under certain circumstances, we should instead seek to find the greatest overlapping consensus of views from a variety of approaches on why any given thing in nature has value, and then appreciate the array of values that should be operative in our decisions with respect to that thing.

Particularly at the level of practice and policy, this pragmatic methodology asks us to set aside theoretical debates in environmental ethics and focus on morally responsible solutions to environmental problems which reflect the ends that are converged upon by a variety of stakeholders who have competing accounts of why things in the world are morally valuable (see Light 2010a). While anthropocentists, non-anthropocentrists and virtue ethicists may value novel and non-novel ecosystems for different reasons, they may all agree on common decisions and policies concerning them.

In this respect, a decision to allow the spread of a novel ecosystem in any given place and an attempt at recreation of an original system could be governed by a pragmatic assessment of which choice garners greater gains on some measure: biodiversity, biomass, ecosystem services, etc. Just as valuing original or reference ecosystems should not necessarily entail disvaluing novel ecosystems, valuing novel ecosystems should not necessarily entail failing to value some other kind of ecosystem which bears less of a human imprint. Valuing novel ecosystems, like valuing many kinds of landscapes, is necessarily part of the larger task of forming a comprehensive environmental ethic which can do more than celebrate the few remaining isolated bits of wilderness in the world (Light 2010b). As with all of these different kinds of environments, novel ecosystems deserve our attention because there are better and worse forms of them.

Similarly, from the other side of a potential comparison, it is also worth pointing out that valuing less humanly impacted (even original) environments is not at odds with valuing novel ecosystems. Particularly from the perspective of a holistic form of non-anthropocentrism, which attempts to articulate the full value of individual and collective non-human entities which make up the world, there will often be ecosystems which are lacking in one or more characteristics that we would find valuable for some reason or another. Just because we might value the relative non-human origins of an ecosystem in one place, this does not mean that we would necessarily be compelled to discount the overall value of another ecosystem that had many of the same components of the reference system but with a more obvious trace of human involvement.

Nonetheless, even more pluralistic and pragmatic accounts of natural value that embraced novel ecosystems will confront challenges. As we continue to force the world to change so quickly, pressures on species extinction and the disappearance of historical ecosystems will surely escalate. We are confronted with a difficult future as we decide what we can plausibly save, given limitations of time and money. No single theory of environmental value will provide all the answers on what we should do in the face of scarce resources and mounting uncertainty. An account that appreciates the full range of ecosystems that in fact make up the world is however more likely to be in a position to make meaningful contributions to the decisions we face today than one that neglects systems only because they more clearly bear the trace of human contact.

REFERENCES

Callicott, J.B. (1989) *In Defense of the Land Ethic*. State University of New York Press, Albany.

Elliot, R. (1997) *Faking Nature*. Routledge Publishers, London.

Hall, M. (2005) *Earth Repair: A Transatlantic History of Environmental Restoration*. University of Virginia Press, Charlottesville.

Higgs, E. (2012) History, novelty and virtue in ecological restoration, in *Ethical Adaptation to Climate Change* (A. Thompson and J. Bendik-Keymer, eds), The MIT Press, Cambridge, MA, pp. 81–102.

Higgs, E.S. and Hobbs, R.J. (2010) Wild design: Interventions and ethics in protected areas, in *Beyond Naturalness: Rethinking Park and Wilderness Stewardship in an Era of Rapid Change* (D. Cole and L. Yung, eds), Island Press, Washington, DC, pp. 234–251.

Hobbs, R., Higgs, E., and Harris, J. (2009) Novel ecosystems: implications for conservation and restoration. *Trends in Ecology & Evolution*, **24**, 599–605.

Katz, E. (1983) Is there a place for animals in the moral consideration of nature? *Ethics and Animals*, **4**, 74–87.

Katz, E. (1997) *Nature as Subject*. Rowman and Littlefield, Lanham, MD.

Light, A. (2000) Ecological restoration and the culture of nature: A pragmatic perspective, in *Restoring Nature: Perspectives from the Social Sciences and Humanities* (P. Gobster and B. Hull, eds), Island Press, Washington, DC, pp. 49–70.

Light, A. (2002) Contemporary environmental ethics: From metaethics to public philosophy. *Metaphilosophy*, **33**, 426–449.

Light, A. (2010a) Methodological pragmatism, pluralism, and environmental ethics, in *Environmental Ethics: The Big Questions* (D. Keller, ed.), Blackwell Publishers, Cambridge, MA, pp. 318–326.

Light, A. (2010b) The moral journey of environmentalism: From wilderness to place, in *Pragmatic Sustainability: Theoretical and Practical Tools* (S. Moore, ed.), Routledge Press, London, pp. 136–148.

Norton, B.G. (1984) Environmental ethics and weak anthropocentrism. *Environmental Ethics*, **6**, 131–148.

Norton, B.G. (1987) *Why Preserve Natural Variety?* Princeton University Press, Princeton.

Regan, T. (1983) *The Case For Animal Rights*. University of California Press, Berkeley.

Rolston, H. (1988) *Environmental Ethics*. Temple University Press, Philadelphia.

Sandler, R. (2012) Global Warming and the Virtues of Ecological Restoration, in *Ethical Adaptation to Climate Change* (A. Thompson and J. Bendik-Keymer, eds), The MIT Press, Cambridge, MA, pp. 63–80.

Singer, P. (1990) *Animal Liberation*, 2nd edition. Avon Books, New York.

Stone, G., van der Ham, R. and Brewer, J. (2008) Fossil oak galls preserve ancient multitrophic interactions. *Proceedings of the Royal Society B*, **275**, 2213–2219.

Thompson, A. (2012) The virtue of responsibility for the global climate, in *Ethical Adaptation to Climate Change* (A. Thompson and J. Bendik-Keymer, eds), The MIT Press, Cambridge, MA, pp. 203–222.

Throop, W. (2012) Environmental virtues and the aims of restoration, in *Ethical Adaptation to Climate Change* (A. Thompson and J. Bendik-Keymer, eds), The MIT Press, Cambridge, MA, pp. 47–62.

Chapter 32

CASE STUDY: A ROCKY NOVEL ECOSYSTEM: INDUSTRIAL ORIGINS TO CONSERVATION CONCERN

Michael P. Perring

Ecosystem Restoration and Intervention Ecology (ERIE) Research Group, School of Plant Biology, University of Western Australia, Australia

Porthmadog, on the northwest coast of Wales, United Kingdom, is a small town that has relatively recent origins having been set up in the 1820s by an enterprising businessman, William Maddocks (Elias 2010). Prior to obtaining permission for the building of the town, he built a sea wall in 1811 (The Cob) to allow reclamation of land for agriculture from the Glaslyn Estuary. The reclamation changed the landscape to such an extent that it led a local poet (Dafydd Sion James) to comment (in Welsh) in 1813: "Potatoes instead of cockles, ponies instead of seals" (Elias 2010). While some minds were distracted by the changes to the land (or sea, depending upon your perspective), other minds were directed to the still extant sea since the diversion of the Afon (River) Glaslyn led to the scouring out of a viable harbor. This harbor serviced the burgeoning trade for slate and also copper that was being quarried and mined from the surrounding area, in what is now Snowdonia National Park. Famous three-masted sail boats (Porthmadog schooners) carried the slate around the rest of the UK and Ireland before expanding into the world market with the completion of the Ffestiniog Railway in 1830. This world market included the Netherlands, Belgium, France, Germany, Scandinavia, the eastern seaboard of the United States and latterly San Francisco via Cape Horn, as well as special building projects associated with the British Empire around southeast Asia and perhaps Australia (T. Elias, pers. comm. 2012). At its height, over 116,000 tons of slate was exported on over 1000 vessels in 1873 (http://www.porthmadog.co.uk/townguide/html/page_5.html, accessed August 2012).

The vessels' seaworthiness depended upon the heavy load of ballast provided by the slate. Once unloaded at their destination, alternative ballast was required from the local area where they were docked. In addition, other cargoes may have been loaded for alternative ports whereupon new ballast was taken on board (T. Elias, pers. comm. 2012). Upon the schooners'

Novel Ecosystems: Intervening in the New Ecological World Order, First Edition. Edited by Richard J. Hobbs, Eric S. Higgs, and Carol M. Hall.
© 2013 John Wiley & Sons, Ltd. Published 2013 by John Wiley & Sons, Ltd.

eventual return to Porthmadog, ballast was initially unloaded to create Cei Newydd (South Snowdon Wharf). Once this wharf was completed, an alternative site needed to be found (see aqua tint photo by C.F. Williams 1850, record XS4302/17/9 in Caernarfon Archives). A trio of brothers (Joseph, John and Evan Lewis), who were also ships' carpenters and block makers, created the timber foundations of a new island on what was previously a sand bank where ballast was then unloaded; the island was referred to as Lewis Island on early charts, but is now known as Cei Balast (Ballast Island; see newspaper cutting dated June 5th 1970, record XS4302/2/36 in Caernarfon Archives; Fig. 32.1). After less than a century of the slate trade, the First World War intervened and trade ceased. Tourist trains now journey the route from Porthmadog to the old slate quarries at Blaenau Ffestiniog to remind people of the industrial heritage of this area. However, another reminder is just over the sea wall in Porthmadog itself; this reminder goes mainly unremarked except upon Ordnance Survey Maps and maritime charts.

Rocks from around the world – granites from Scandinavia, limestone and sandstone from unknown but not local origins and metamorphic and igneous stones (T. Elias, pers. comm. 2012) – now form the rocky foundations of a woodland novel ecosystem on Cei Balast.

The habitat on Cei Balast is described as A.2.1 (Dense continuous scrub) in the Phase 1 National Vegetation Classification system (COFNOD 2012). This dry description belies the conservation interest that is found within the site and its surrounding area, an interest that arguably only comes about from the industrial heritage and relatively short history of intensive human land use. As well as some widespread native species that have spontaneously colonized – including blackthorn (*Prunus spinosa*), hawthorn (*Crataegus monogyna*), willow species (*Salix spp.*) and western gorse (*Ulex gallii*) – the unloading of ballast led to the introduction of species such as wild chamomile (*Matricaria disciodea*), evening primrose (*Oenothera agg.*) and perhaps cotoneaster (*Cotoneaster spp.*) (Elias

Figure 32.1 Looking south to Cei Balast from Porthmadog outer harbor wall. Photograph © Twm Elias 2012. Reproduced with permission.

2010; T. Elias, pers. comm. 2012). Of most conservation interest is a diminutive plant commonly known as Welsh mudwort (*Limosella australis*) which is found on adjacent saltmarsh but was most likely introduced in the ballast that constitutes Cei Balast (Jones 1991).

The Latin moniker for Welsh mudwort belies its non-native origins, and the fact that the type specimen was described in Australia almost a century before it was discovered in South Wales in 1897 (Jones 1991). Just over 20 years later, this plant was also described in the Glaslyn estuary and this population, together with one in the Dysynni estuary (in mid-north Wales), now make up the only known populations in the UK, having become locally extinct in south Wales (Jones 1991). Indeed, only three populations exist throughout Europe (Elias 2010). *Limosella australis* is therefore on Schedule 8 of the Wildlife and Countryside Act (1981) and is Nationally Scarce and Rare, as well as being on the IUCN threat listing of Welsh vascular plants and part of Gwynedd's Biodiversity Action Plan (COFNOD 2012).

Despite all these designations however, many would regard it as an introduced species given its widespread distribution in the southern hemisphere and its presence on the northeast coast of the United States (from whence it likely arrived). However, does this mean it is not deserving of conservation, particularly given its common-name connection to Wales? Is it, indeed, regarded as a Welsh species? Has the rocky novel ecosystem of Cei Balast given this species a means to continue its genetic lineage? As Jones (1991; p. 8) pointed out: "But perhaps it's worth considering this one: that when Columbus hailed far horizons, the species of the Americas discovered their New World as well, and for *Limosella australis*, Wales could have been that place".

The conservation status of *Limosella australis* highlights the complexities surrounding novel ecosystems and the influence of humanity on nature. However, the wider area of the Glaslyn estuary also highlights that human enterprise is not necessarily universally negative, an impression that sometimes appears the case from the reporting of environmental issues in the popular press. Not only is the estuary home to some rare and endangered native animal species (e.g. the otter; COFNOD 2012), the marsh that is now found behind The Cob is home to a number of rare plants, including Welsh broad-leaved marsh orchid (*Dactylorhiza majalis ssp. cambrensis*) and the dwarf spike-rush (*Eleocharis parvula*) (Elias 2010). More broadly, the demand for timber to service the burgeoning shipping industry in the 19th century led to the planting of oak woodlands. These woodlands now form a Local Nature Reserve (Coed-y-Borth) with 120-year-old upright native oaks (Elias 2010) likely providing habitat for birds such as the greater spotted woodpecker (COFNOD 2012). None of these areas, or the habitat that is now present on them, would exist without the industrial heritage of Porthmadog and the enterprising vision of William Maddocks. From industrial origins, a rocky novel ecosystem has arisen that has provided a species of conservation concern, while the industrial origins themselves have provided habitat and safe haven for a number of rare animal and plant species.

ACKNOWLEDGEMENTS

Thank you to John Healey (Bangor University), Bleddyn Williams (North Wales Wildlife Trust), Lucy Kay (Countryside Council for Wales), John Ratcliffe (CCW), Angharad Jones (Caernarfon Archives Office), Richard Gallon (COFNOD), Barry Davies (Gwynedd County Council), Twm Elias (Snowdonia National Park Study Centre), Polly Spencer-Vellacott (BSBI), Gwen Roberts (Porthmadog Library) and Wendy McCarthy (BSBI) for dealing with my many queries in the research for this chapter, initially prompted by Nicholas Crane and the BBC team behind Coast.

REFERENCES

COFNOD (2012) Biodiversity Information Search Report E02693 (Public). North Wales Environmental Service, Bangor, North Wales.

Elias, T. (2010) Glaslyn Marsh – 200 years since the construction of the Cob (Translation). *The Nature of Wales*, **37**, Winter 2010. http://www.naturcymru.org.uk/past-issues/issue-37/translation-morfa-glaslyn.aspx. Accessed August 2012.

Jones, A. (1991) Welsh mudwort? *Botanical Society of the British Isles, Welsh Bulletin*, **52**, 6–8 (Winter 1991).

Chapter 33

THE POLICY CONTEXT: BUILDING LAWS AND RULES THAT EMBRACE NOVELTY

Peter Bridgewater[1] and Laurie Yung[2]

[1]Global Garden Consulting, UK
[2]Resource Conservation Program, College of Forestry and Conservation, University of Montana, USA

33.1 INTRODUCTION

Novel ecosystems pose important challenges and critical questions to policy at multiple scales. The author of Chapter 10, for example, asks: "How do we adapt to and manage ecosystems we have never seen before? And what reference states do we use for restoring ecosystems?" While these questions are directed towards management issues, they are also germane in directing policy development. Developing policy and legislating for the unknown and unseen itself requires a novel approach. For example, in Chapter 14, novel plant communities are described that provide habitat features for native fauna, and introduced animals can act as ecological engineers, creating ecosystem novelty (see Bridgewater et al. 2012). How we handle such challenges in policy will have far-reaching consequences for our ability to take advantage of the opportunities and benefits that result from novelty.

Novel ecosystem change also has important implications for human communities. Such changes require careful thinking about what kinds of policies allow for sustainable use of local resources and continued provision of ecosystem services in the context of rapid change. Chapter 3 proposed that management moves toward intervention to maintain or enhance ecosystem function, arguing that "a novel ecosystem paradigm shifts management concerns from the specific goal of maintaining historical ecosystems toward an admittedly more qualitative consideration of how the ecosystem functions to provide species habitat and ecosystem services". This shift in management goals also needs to be reflected in policy changes that focus on ecosystem functions and the ecosystem services that flow from those functions. Novelty may alter access to important food resources, change ecosystem function or result in positive or negative changes to ecosystem service delivery.

Similar questions are raised in the management chapters of this book on the desirability of restoration, or when management for persistence of novel ecosystems makes more sense than management for restoration, and is likely to result in more stable land- and seascapes than pursuing restoration to historical

Novel Ecosystems: Intervening in the New Ecological World Order, First Edition. Edited by Richard J. Hobbs, Eric S. Higgs, and Carol M. Hall.
© 2013 John Wiley & Sons, Ltd. Published 2013 by John Wiley & Sons, Ltd.

states. To address these issues of increasing change and dynamism in ecosystems, policy development also needs to be more dynamic and legislation more flexible (not normally a characteristic of legislation). There is often a tension between the dynamism of ecological systems and the predictability we desire in our legal systems (Craig 2010). But as we reshape the world, deliberately or accidently, and create a dynamic, even rambunctious (Marris 2011) biosphere, our approach to policy (and ultimately legislation) must be similarly rambunctious (i.e. dynamic and capable of dealing with unpredictable change).

International and national biodiversity policy comes from a historical focus on conservation of existing species and ecosystems – largely species, in fact. This is despite the fact that the Convention on Biological Diversity (CBD) urges that conservation, sustainable use and equitable sharing of benefits should be tackled equally (CBD 1992). Similarly, in many countries domestic environmental legislation tends to focus on species conservation, conceptualizes nature as stable and static and articulates success in terms of the 'preservation' of historical conditions and species compositions (Doremus 2010; Camacho 2011). In the United States, for example, the canon of natural resource and environmental law, while visionary at the time of passage, is largely founded on an equilibrium paradigm that assumes enduring, unchanging ecosystems (Thrower 2006). These laws assume that ecosystems are inherently stable, and changes that do occur are largely predictable and often reversible (Craig 2010). As such, they are often not well-suited to management and conservation of dynamic, changing systems and the uncertainty and unpredictability of ecological change (Karkkainen 2002). Furthermore, natural resource law often prioritizes historical preservation, a goal that will be increasingly difficult to meet both financially and practically (Camacho 2011). Because novel ecosystems are both dynamic and a departure from historical conditions, such ecosystems are either ignored or deliberately excluded from such laws.

Importantly, novelty does not suggest abandonment of efforts or policies to conserve biodiversity. Novelty will however likely require a different sort of prioritization, thinking outside the confines of historical conditions, and consideration of the values and services we desire. For example, in the context of rapid environmental change, extinctions and range changes are highly likely; such dramatic shifts may require policies that enable prioritizing some species over others (so-called triage). In some systems, change might alter historical species assemblages but maintain or even 'improve' ecosystem service delivery.

Because ecosystem novelty is a relatively new concept and/or because some view novelty as harmful to conservation goals, most law and policy says little about the conservation of novel ecosystems and does little to guide management of such novelty as it emerges. Yet the challenge of novel ecosystems is that not only are they not static, they are highly dynamic. Not only is policy largely silent about the changes represented by novel ecosystems, some policies actively resist such change by requiring the maintenance of historical species assemblages or habitat. Instead of offering direction regarding novelty, most policy ignores or is unable to address rapid environmental change and novel ecosystems or provide mechanisms to address or conserve valued functions and services from such ecosystems. Lack of policy attention to novel ecosystems is particularly notable given how widespread these systems are (the extent of novel ecosystems is explored in Chapter 8). Much of this is due to the current corpus of environmental law and policy being derived from mid-20th century thinking when environmental stability seemed more assured, especially measured and perceived against human life-spans.

In this chapter, we examine how novelty and novel ecosystem changes are addressed (or not) in existing domestic and international policy, and what types of changes might improve policy guidance. Rather than a comprehensive review of relevant policies, we consider a range of policies related to biodiversity and endangered species, invasive species, protected areas and ecosystem services. In addition to international examples, we look at domestic policy from the United States, the United Kingdom, Canada and Australia. While limited in regional scope and cultural diversity, these domestic examples hopefully provide broader insights. There are also policies at local, regional and provincial/state levels that require a similar rigorous examination; these help drive bottom-up change, which is nourished by broader national and international policies (some are already addressing novel ecosystems; the Municipality of Saanich, British Columbia, Canada, for example, makes explicit mention of novel ecosystems in its Invasive Species Management Strategy). However, we argue that significant gaps remain and outdated policy and legislative frameworks continue to limit our ability to respond to the challenges and opportunities posed by the emergence of novel ecosystems. However, we also find some key areas of innovation and some policy space to address novelty. Wherever possible, we suggest how policy might be amended and improved to both acknowledge the existence of novel ecosystems and provide mechanisms for management (and even conservation) of such systems. This chapter is written in the context of a 'Call to action on understanding and managing novel ecosystems' issued by the 2011 workshop which gave birth to this present volume (see Bridgewater et al. 2011).

33.2 WHAT IS POLICY?

While policy is often taken for granted as something that emerges in response to new challenges and is relatively easy to craft, the reality of policy-making is quite different. Policy can be understood as the formulation of political views based on evidence (e.g. observational and experimental science) and the subsequent development of legal frameworks, responding to the problems and questions posed by the science, and shaped by the politics. Robust policy on a topic emerges from a sound evidentiary base and from public dialogue and deliberation. Typically environmental policy not only considers the science relevant to the question at hand, but also includes assumptions about the other major issues impacting on government at any one time. Koetz et al. (2011) note that "not only science and politics itself, also their interactions, which are particularly intensely and inextricably interwoven when it comes to the identification, formulation, implementation and analysis of environmental policies (Nowotny et al. 2001; Young 2004; van den Hove 2007), can be understood and discussed in institutional terms."

There is a broadly linear, but also highly complex, relationship between the framing of questions in science, development of scientific explanations for real-world events and policy development from that science. Policy does not however simply flow from science; policy is developed through political processes that integrate social values and economic interests, and typically reflect the conflict and controversy that surround most environmental problems. This political process eventually leads to the development of new laws or changes to existing legal frameworks. Good politics and good policy require public engagement, as discussed in detail in Chapter 30. As ecosystem management moves away from traditional goals such as preservation of historical species assemblages/habitats to the delivery of ecosystem services, public policy will need to set priorities and allow for trade-offs through democratic processes (Camacho 2011).

In the context of rapid ecosystem change, including the likelihood of abrupt change and surprises, our policy and legal institutions (as part of the socio-ecological system in which we live) must become better at adapting to changes and allowing for a more dynamic approach to policy making and the creation of innovative and flexible legal frameworks.

Environmental policies at all scales date largely from the late 1960s; before that time almost no national governments had separate Ministries or Agencies dealing specifically with environmental matters or well-developed laws governing the management of public lands, and the environment was not discussed at all at international fora. Following the global meeting in 1972 on Environment and Human Development, many international developments in environmental policy and law occurred. The United Nations Environmental Programme (UNEP) was born in 1974, and a plethora of conventions dealing with environmental issues arose at the same time or during the following decade. Some 20 years after the Stockholm summit, a second summit occurred in Rio de Janeiro, Brazil. This 1992 summit is referred to as the World Conference on Environment and Development (WCED). There were a number of achievements flowing from this summit, chief among which (for our purposes) were the signing into force of three conventions: Convention on Biological Diversity (CBD 1992); UN Framework Convention on Climate Change (UNFCCC 1992); and the UN Convention on Combating Desertification (UNGA 1994). Some of these international conventions required domestic legislation to be passed to reflect the concerns of the relevant convention, while others are silent on such requirements; many countries have done so to a greater or lesser extent.

Of particular relevance to novel ecosystems are the conventions dealing with biodiversity which arose in this period (see also Jóhannsdóttir et al. 2010), including: Convention on Wetlands (Ramsar, Iran, 1971), Convention Concerning the Protection of the World Cultural and Natural Heritage, Convention on International Trade in species of Wild Flora and Fauna (CITES) and Convention on Migratory Species. While all of these conventions have scope and legal context to deal with novelty, in most cases this has not been addressed by decisions of the conventions' governing bodies.

33.3 NATIONAL BIODIVERSITY POLICIES

Biodiversity policies, developed since the 1990s in most countries, respond to the threat of extinction through efforts to conserve and manage (although often expressed as preserve) species in their historical range, rather than recognizing the more dynamic (and real) situation of change. Novel ecosystems are one aspect of an emerging paradigm that recognizes non-equilibrium ecosystem dynamics, and yet policies that relate to novel ecosystems are largely non-existent while many extant 'biodiversity polices' actually militate against recognition, let alone prescribing management for, novel ecosystems. The reasons for this lie chiefly among the lack (to date) of an effective scientific evidence base for the status of novel ecosystems and a vigorous policy setting dealing with species loss and ecosystem protection, rather than biodiversity change. Such policy directions do chime with prevailing public views and opinion, yet miss the point that some changes in species may not presage ecosystem destruction and instead promote ecosystem survival. These challenges are complicated by the fact that other key changes (e.g. invasive species) are often seen to be a major cause of biodiversity loss rather than, in many cases, being elements of biodiversity change as earlier chapters have discussed. Furthermore, where a nation's political structure has a federal system of government there is often vigorous policy and political discussion over which level (local, state, federal) of government should have which responsibilities for biodiversity management and conservation (Nie 2008).

Protected area policy in particular reflects outmoded ideas about ecosystems and often resists novelty and change. Protected area policies often emphasize the protection or restoration of historical ecosystem components (Cole and Yung 2010). For example, the Leopold Report (Leopold et al. 1963), whose recommendations formed the basis for United States national park policy for many decades, simultaneously acknowledges the dynamism of ecosystems and advocates that parks protect a "vignette of primitive America" and "the condition that prevailed when the area was first visited by the white man" (pg. 33). To the extent that conservation goals are based on incorrect assumptions about ecosystem persistence, decision-makers may be unable to formulate effective goals for biodiversity change in protected areas given the context of rapid change (Cole and Yung 2010). The ecological integrity framework utilized by Parks Canada incorporates but also moves beyond historical conditions by recognizing some forms of ecological change (Woodley 2010). Application of this ecological integrity framework, especially as represented in the Parks Canada Principles and Guidelines for Ecological Restoration in Canada's Protected Areas (Parks Canada and the Canadian Parks Council, 2008), to novel ecosystem change represents an important policy experiment.

Despite the preponderance of policies that either ignore or resist ecosystem change, regulations that aid implementation of legislation can be amended to accommodate new understandings of nature and the values of novel ecosystems. In many countries, the body of existing law is relatively broad and vague and could provide the flexibility to incorporate novel ecosystems without the need to redefine statutory language. As Kareiva et al. (2008) argue in the context of the United States, existing laws can be used opportunistically to acknowledge and address environmental change. As directives and administrative policies emerge that focus on resilience and adaptive capacity (as opposed to historical conditions), opportunities open up to explicitly consider novelty. For example, the US Forest Service's 2008 Interim Directive on Ecological Restoration and Resilience emphasizes maintenance of "the adaptive capacity of ecosystems", "current and likely future ecological capabilities" and "future changes in environmental conditions". Such language moves away from conservation goals that focus on historical assemblages of species and toward ecosystem function and process, thus providing political space for embracing and valuing novelty.

There are a number of specific areas of US domestic policy that intersect with novel ecosystems, including policies related to invasive species and endangered species. As Camacho (2011; p. 1427) points out, invasive species law is "premised on the distinction between native and non-native". Policies that determine which species are considered invasive where can both constrain and guide the types of novelty that emerge and how that novelty is managed. Invasive species policy essentially determines which species are considered acceptable in which locations, and thus influences how resources and energy are allocated to control (or attempt to control) invasion, migration and range shifts. In Europe, with successive waves of human interaction with landscapes, ecosystems and species over millennia, the concept of nativeness is even more problematic and often defined by reference to arbitrary dates (e.g. 1500 AD in the UK, which means all the species introduced by the Romans are considered native).

Invasive species policy often fails to recognize that, in many cases, non-native species provide benefits such as supporting rare native species or serving as a functional substitute for extinct species (Schlaepfer et al. 2011). These policies can be problematic if they rigidly define nativeness in terms of historical range. Writing about Canadian protected areas, Scott and Lemieux (2005) point out that existing policies on invasive species may not easily accommodate the range shifts that will likely occur in the context of climate change. In other words, native species that move as a result of changing conditions may be considered invasive and undesirable in new locations and steps might be taken to eradicate them. Invasive species policy can also be problematic to the extent that non-natives are blindly resisted and, in many cases, demonized, even if they play important roles in novel ecosystem function. More recent thinking around invasive species policy recognizes ecosystem change and suggests that the focus should be on outcomes and the impacts of species on biodiversity, human health and ecological services rather than whether the species in question is native or non-native (Davis et al. 2011).

Further, if domestic policies differ between countries, species may be prevented from moving across international boundaries. Policy related to invasive species therefore needs to be flexible enough to differentiate between potentially harmful non-natives and migration of species that should be embraced to ensure that species are able to adapt and persist in the face of change. At a regional level, the directives of the European Union (EU) contribute to management of invasives at a continental scale, where EU member nations are in a political union. However, much of the effects of theses directives also flow to, or are adopted by, countries that are not yet included into the political framework of the EU (e.g. Norway, Switzerland, Serbia, Turkey).

In relation to biodiversity, two key directives have been adopted:
• Directive 2009/147/EC on the Conservation of Wild Birds (Birds Directive) (EC 2010); Directive 92/43/EEC of 21 May 1992 on the conservation of natural habi-

tats and of wild fauna and flora (Habitats Directive) (EC 1992).

These directives provide for the protection of animal and plant species of European importance and the habitats that support them, particularly through the establishment of a network of protected sites called Natura 2000. Further relevant legislation includes a Water Framework Directive, under which Member States are required to protect and improve their inland and coastal waters, and a Marine Strategy Framework Directive to achieve good environmental status in their marine environment by 2020. These four directives, of varying maturity, have placed great stress on the use of protected areas as cornerstones for conservation, and strongly emphasize the perception of a constant and unchanging natural world, presenting a somewhat flawed approach. The developing work on Green Infrastructure, which emphasizes a strategic approach to support protected areas with networks of open spaces set within a multi-functional landscape, may help remedy these flaws. Much of the focus is still on ensuring support for protected areas and endangered species however, with zero reference to emerging novelty in European landscapes (e.g. Kettunen et al. 2007). This mix of policy prescription through guidelines and frameworks, with degrees of legal suasion, has been changing and expanding as more countries join the European Union and the need to strive for pan-European consistency increases. One area yet to be developed is the incorporation of ecosystem services into Commission directives, as discussed by Cliquet et al. (2009) and in Section 33.4.

Endangered species policy also presents important challenges and opportunities for novel ecosystems. To the extent that endangered species policy is based on the assumption that species should be restored only to the locations where they previously existed, such policies may prove problematic. As conditions change, species may be better adapted to new locations, and thus merit recovery and protection in areas outside of their historical range. Interestingly, the United States Endangered Species Act (Endangered Species Act 1973) allows for the designation of critical habitat outside the geographic area occupied by the species, if essential to conservation (Ruhl 2008). In many countries, managing agencies may also have the authority to relocate endangered species to habitat outside of their historical range. For example, under the United States Endangered Species Act, managing agencies can release species "to areas not formerly occupied by the species" if habitat within their historical range has been destroyed (Ruhl 2008). In some cases, domestic policy may therefore allow for managed relocation or the intentional establishment of novelty in particular places for the purpose of conserving species under conditions of change.

Policy for novel ecosystems may also require the prioritization of conservation resources. Are there types of novelty we wish to maintain or nurture? In which cases should we intervene? Answering such questions requires some degree of prioritization or triage since, given the extent of global climate change, many species are likely to go extinct or change their range and many ecosystems will suffer resorting or recombination (Soulé 1986). One of the challenges associated with triage is developing policy mechanisms that enable investment in protection and recovery of some species, while allowing for decisions not to assist other species. Ruhl (2008) argues that managing agencies in the United States need to "differentiate between species that are unlikely to survive climate change under any circumstances and those that are likely to benefit from assistance in their home ecosystems" to avoid wasting resources on species that cannot be recovered (p. 60). In other words, from a policy perspective, novelty does not suggest abandonment of efforts to conserve biodiversity. It may however indicate a need for prioritization of particular species in particular places (as opposed to conservation of all species in the locations where they historically occurred). Some endangered species statutes allow managing agencies to prioritize based on the likelihood of success.

To fully embrace novelty and ensure that efforts to conserve species are effective in the context of novel ecosystems, domestic endangered species policy therefore needs to be responsive to change, able to designate habitat outside of historical range and protect species in places where they have not previously occurred (and consider managed relocation in certain instances). More broadly, novelty requires us to consider species recovery at the continental scale, and to define success in terms of native species persisting and thriving somewhere as opposed to their historical habitat.

Domestic and international policies determine if and how we resist, ignore or guide novel ecosystem change; they also reflect whether and how we, as societies, embrace novelty. The extent to which the scientific community and civil society resists or questions the reality and relevance of novel ecosystems complicates how novel ecosystems are perceived by, and then incor-

porated in, policy frameworks. For example, in many nations novel ecosystems could have important roles in both climate change adaptation and mitigation, on land and in marine environments. As they are however ignored in current policy, their potential role in adapting to and mitigating against climate change is lost. Ecosystem restoration is also an area of increasing interest, with many nations now embracing restoration of or compensation for habitat loss and restoration as a tool for addressing broader development policy goals (Nelleman and Corcoran 2010). Again, the emphasis is all too frequently on restoration to a historical state and not restoration to, or creation of, functional land and seascapes that may include novel ecosystems as components.

33.4 ECOSYSTEM SERVICES: THE NEW WAVE

Ecosystem services and their delivery represent a new policy focus after the completion of the Millennium Ecosystem Assessment (MA 2005) and various regional and national level initiatives, e.g. the recently completed UK National Ecosystem Assessment (UNEP-WCMC 2011). Services can be derived from ecosystems with different structural and content settings than 'natural' ecosystems. Neither the Millennium Assessment nor the UK National Ecosystem Assessment had any focus or mention of novel ecosystems and their actual or potential role in delivery of ecosystem services. This is partly due to a lack of an effective science base; in the policy frameworks developed around ecosystem services concepts, such assessments have therefore not been focused on the kind of change novelty is bringing. Ecosystem services are seen as having a clear economic flavor and a key role in the delivery of potential 'biodiversity offsets' from development.

The prevailing view in both the science community and civil society is that we have to maximize and maintain the extent of 'natural' ecosystems and thus their ecosystem services. This view is inevitably translated by policy makers and politicians into a stance that ignores or sidelines ecosystem change and novel ecosystems. And yet, almost without realizing the consequences, a focus on ecosystem service delivery as a sole or major policy objective means that the components and processes of the ecosystems could conceivably change, provided the services were continually delivered. Understanding the balance between ecosystem components processes and service delivery from ecosystems means taking account not only of natural systems but semi-natural ('hybrid') and certainly novel ecosystems. The MA provided a means to redefine the Ramsar Convention's wise use concept in terms of sustainability, especially the capacity of the ecosystem to continue to deliver the services on which other ecosystems and people depend (Bridgewater 2008). The messages from the MA also forced a re-think of the Ramsar Convention's concept of maintenance of ecological character, and thus the list of Wetlands of International Importance (Ramsar sites). "Change in ecological character", which is a concept important in measuring Ramsar site management effectiveness, can also be resolved in terms of ecosystem service delivery. The criterion for being and continuing to be a Ramsar Site therefore essentially becomes an issue of the continuing capacity to deliver ecosystem services.

Several conventions, including the Convention on Biological Diversity, the Ramsar Convention on Wetlands and the UN Convention on Combating Desertification, took note of the MA ecosystem services framework and have incorporated it into their rubric. Such international acceptance flowed into national activity in several countries including the UK, Australia, Germany, the Netherlands, Sweden and others. The concept and value of ecosystem services now pervades much discussion and debate (e.g. TEEB 2011; UKNEWP 2011).

Ecosystem services provided by biodiversity can serve as a buffer for people against the negative impacts of climate change (Cliquet et al. 2009). While Cliquet et al. 2009 do not deal specifically with novelty and ecosystem services, they develop an interesting concept in the context of EU policies: the concept of an 'Ecosystem Framework Directive'. However, they are arguing for a less legalistic approach than the birds and habitat directives discussed earlier. Rather, they suggest an ecosystem services approach that should have as its main purpose the provision of "a framework that (a) raised awareness and then (b) moved operators towards an ecosystems approach, focusing on biodiversity indices, ecosystem functions, goods and services rather than individual habitats and species".

They further argue that an ecosystem services framework should "provide a safety net to catch those sites which would no longer be protected if particular species, species' assemblages or habitat types can no longer persist under changing bioclimatic conditions. It should encourage and facilitate the use of ecological restoration to produce new areas capable of delivering valued ecosystem services in a way which is congruent with what the area would support biophysically."

This is much closer to ideas and concepts of novel ecosystems, except possibly for the last point, where nature gardening (managing and conserving novel ecosystems) may well be the most appropriate approach to ensuring continual delivery of ecosystem services. In conclusion, Cliquet et al. 2009 ask "whether a directive aiming at protecting and developing ecosystem services could substantially amend the difficulties which we encounter with current policy and legislation." This question identifies the problems at European scale, but is equally applicable at national or international scale, i.e. what is the right focus for policy intervention, and what legal mechanisms need to be in place to back those policies?

33.5 INTERNATIONAL BIODIVERSITY POLICY

Like much national policy, international policy on biodiversity, through the Convention on Biological Diversity (CBD) as *primus inter pares* among the biodiversity-related group of conventions, has largely ignored novel ecosystems or dismissed them as potentially undesirable. For example, the Report of The Second Meeting of the Ad Hoc Technical Expert Group on Biodiversity and Climate Change (CBD 2010) included the following: "Some of the more controversial techniques that could be used include assisted migration of regional tree species, the importation of invasive alien tree species or the use of genetically modified tree stock. These latter techniques should take into account risks from the development of novel ecosystems, which may have impacts on the endemic species of the area. On the other hand, when used in forest areas that are already managed, such approaches may have some potential to ease the pressure on natural forests."

While this document is not a formal policy statement of the CBD, it does suggest that some parties to the CBD see novel ecosystems as threatening to 'endemic biodiversity'. While novelty can threaten endemic species in some cases, this narrow view prevents broader thinking about what specifically is threatened under what circumstances, and thus overlooks ways to incorporate novel ecosystems into broader biodiversity management and, inter alia, help ameliorate the problems of global environmental change.

Recent discussions on adaptation to climate change nationally (and internationally through the UNFCCC) have had considerable focus on preventing deforestation, especially in the tropics, rather than on embracing novelty where it is occurring. This process, initiated at the 2009 Conference of the Parties to the UNFCCC and known as Reduction in Degradation and Deforestation (REDD), has come to include conservation and sustainable management (the so-called REDD+), opening the possibility for more active intervention (Alexander et al. 2011). At the Conference of the Parties to the UNFCCC in 2011 in Durban, South Africa this process was advanced through 'ecosystem-based adaptation' (EBA) which is defined (UNFCCC 2011) as "how ecosystems can play a role in helping people adapt to climate change". The CBD has also recently considered ecosystem-based adaptation including examining where it could be a cost-effective strategy to address the impacts of climate change, particularly in vulnerable areas where adaptive capacity is low. In some cases, ecosystem-based adaptation has been linked to local community-based adaptation and assumptions that local communities can improve their adaptive capacity through certain natural resource management strategies (Huq et al. 2005). For example, coastal ecosystems can play a role in coastal protection and buffer the impacts of storms while maintaining fish supplies and wetlands and rivers that protect, manage and produce water.

Although they do not reference novel ecosystems explicitly, EBA provides an example that shows how the concepts are being utilized (even if not by name): "For example, managing wetlands to provide coastal protection may require emphasis on silt accumulation and stabilization, possibly at the cost of reduced wildlife and possibilities for recreation. It is therefore important that decisions to implement ecosystem-based adaptation are subject to risk assessment, scenario planning and adaptive management approaches that recognize and incorporate these potential trade-offs. It is also worth noting that broad stakeholder participation can be a measure of successful implementation, since local users manage many ecosystems and ecosystem services."

This statement points to the trade-offs inherent in decisions about novelty and the ways in which we acknowledge, embrace and guide novel ecosystem changes. To the extent that policy recognizes that historical ecosystems are not always 'better' and more adaptive to current and future conditions, that policy opens the door to careful consideration of trade-offs. It acknowledges that, under certain circumstances, retention of ecosystem function or key livelihood resources might be prioritized over historical species compositions. This passage also raises an important question about who decides and which stakeholders are involved in decisions about trade-offs in the context of emerging novelty. One challenge under REDD+ is to secure land rights for local and indigenous communities to ensure they can benefit from restoration activities. Similarly, as policies to address novelty struggle with important trade-offs, questions about who decides and who benefits will certainly arise.

In the context of ecological (also termed green) infrastructure, novel ecosystems are dealt with in The Economics of Ecosystems and Biodiversity (TEEB 2011), an international effort to assess and communicate the economic benefits of global biodiversity. Chapter 9 of TEEB (2011) has the following: "True restoration to prior states is rarely possible, especially at large scales, given the array of global changes affecting biota everywhere and that 'novel' ecosystems with unprecedented assemblages of organisms are increasingly prevalent (see Hobbs et al. 2006; Seastedt et al. 2008)."

Nevertheless, the growing body of available experience on the restoration and rehabilitation of degraded ecosystems suggests that this is a viable and important direction in which to work, provided that the goals set are pragmatic and realistic (Jackson and Hobbs 2009). They state, for example, "Restoration efforts might aim for mosaics of historic and engineered ecosystems, ensuring that if some ecosystems collapse, other functioning ecosystems will remain to build on. In the meantime, we can continue to develop an understanding of how novel and engineered ecosystems function, what goods and services they provide, how they respond to various perturbations, and the range of environmental circumstances in which they are sustainable."

This passage shows an appreciation of the existence and role of novel ecosystems, yet still expresses some hesitancy and fails to delve into the policy implications of managing novel ecosystems or put the positive or negative effects of such management into an economic context. At least the topic was raised!

At their fourth meeting in 2011 the Chairs of the Scientific Advisory Bodies of Biodiversity-Related Conventions (CSAB4 2011) noted the following with reference to a discussion on restoration: "[There was a] need for careful communication on ecosystem restoration with decision makers prioritizing the reduction of direct and indirect pressures, and also the need to consider how long a restored habitat/ecosystem might remain in its restored condition, or become degraded again. There were opportunities for cross-convention collaboration to address multiple benefits relevant to more than one convention. The Convention on Migratory Species (CMS) agreed that care needed to be taken with communicating what could, and what could not be restored and at what cost, and to what extent? The Ramsar Convention on Wetlands highlighted the way in which climate change was rapidly causing changes that required consideration of novel ecosystems and how to address these in policy, and the fact that no ecosystem was truly static over hundreds of years. CMS highlighted the need to consider ecological networks and barriers to movement (linear development – roads, power-lines, etc.) rather than single, stand-alone sites."

For the first time in CBD deliberations, there was an appreciation that novel ecosystems should be a focus of policy. This may indeed be an important topic for future policy negotiations.

The CBD and related conventions also address invasive species, that broad spectrum of biodiversity activity that has its roots in the area of biosecurity which developed in international policy in the 1990s. Similar to national policies, the general direction for international invasive species policy has been to assume eradication is the most effective way to manage invasive species. Other options, including management of novel species mixes and with ecosystem services they often continue to offer, are not considered.

The CBD subsidiary body (SBSTTA, Subsidiary Body on Scientific, Technical and Technological Advice) debated restoration at its 15th meeting in November 2011. Regrettably, novel ecosystems and their scientific, existential and policy role were not given much attention in these discussions and deliberations and, for the moment, they remain largely unmentioned in the biodiversity and climate change debates.

33.6 EMERGING POLICY EXAMPLE: THE UK GOVERNMENT'S POLICY POSITION

The UK Government published a Natural Environment White Paper in 2011 (UKNEWP 2011). Although a UK

government document, the way responsibility for environmental issues is divided among the devolved administrations for Scotland, Wales and Northern Ireland meant that this paper was focused on England. Like many such documents it covers a breadth of issues. Although the topic of novel ecosystems per se is not covered, there are some very interesting pointers regarding the role of novel ecosystems in policy development.

In the case of woodlands and forests the following is included (our emphasis): "Our ambition is for a major increase in the area of woodland in England, better management of existing woodlands and a renewed commitment to conserving and restoring ancient woodlands. Forests and woodlands must play a full part in achieving *a resilient and coherent ecological network* across England. We want to create more opportunities for planting productive and native woodlands; more trees in our towns, cities and villages; and a much larger proportion of existing woodlands brought into active management. We also want to increase the use of sustainably grown and harvested wood products. Together, this will enhance the wide range of benefits that woodlands provide, including renewable energy and timber, *new wildlife habitats and green space for people to use and enjoy, helping us to mitigate and adapt to the future changing climate*. It will also *increase resilience to climate change, pests and diseases, and help to halt the loss of biodiversity*."

This is explicitly about managing novel systems, indeed even developing novel systems to cope with a stressed landscape, which continues to have further stressors such as climate change, threats to wildlife health and biodiversity loss. Yet the idea of novel ecosystems is still not specifically developed and drawn out. The policy challenges which novel ecosystems can help address are clear, however, including climate change mitigation/adaptation, threats to wildlife health, increasing urbanization and loss of resilience in landscapes. The paper continues: "Our landscapes have never been static and they will continue to evolve, reflecting the choices we make as a society. These choices sometimes lead to conflicting demands on our landscapes: for new housing and transport infrastructure, growing food and energy needs, improved flood protection, or increases in forestry. Climate change will also affect how landscapes evolve, both directly and through the land use planning choices we make in response."

Again with no explicit mention of novel ecosystems, the paper builds exactly the criteria that generate novel ecosystems and how that process may be managed through effective planning processes. It also suggests some means of managing and molding novel ecosystems:

• "We will work with local communities in a number of areas throughout England to support local engagement in landscape planning;

• "We will work with civil society to update and improve the consistency of the national landscape character area profiles and integrate information on the ecosystem goods and services that they provide;

• "Gardeners can do more for nature by adopting environmentally friendly practices . . .

• "By co-ordinating gardening efforts to create a network of interlinking habitats across neighbourhoods, biodiversity can be greatly enhanced in towns and cities."

These suggestions offer ways to engage the public in emerging novel ecosystems at three scales: civil society, local communities and individual gardeners. Mirroring the recommendations outlined in Chapter 30 on public engagement, the UK white paper proposes that members of the public have critical roles to play in landscape planning and monitoring, identification of goods and services and enhancement of biodiversity. The final suggestion, that gardeners have a role in biodiversity conservation, implies that tomorrow's ecosystem management may be more like gardening than currently accepted management-by-protection (Bridgewater 1997; Marris 2011). The gardening metaphor has important policy implications, as it endorses active intervention in ecosystems (as opposed to passive protection) and explicit decisions regarding what kinds of goods and services we wish to prioritize as ecosystems change. Both active intervention and choices about trade-offs require policy guidance, again suggesting a real need for policies specific to novel ecosystems (see Chapter 37 on concerns about novel ecosystems). As this policy paper develops and turns to hard policy, and then perhaps legislative changes, it will be important to have novel ecosystems seen, recognized, understood and valued throughout that process.

33.7 CONCLUSIONS

Novel ecosystems present important challenges to environmental policy at local, national and international scales. Novelty flies in the face of the current policy focus on retention and restoration of historical conditions and species assemblages. Current policies are largely silent regarding the existence and management of novel ecosystems and many policies are rooted in outmoded notions of static, stable ecosystems. Developing policy and legislating for the unknown and unseen requires us to tackle a number of new questions. How do we adapt to and manage ecosystems that we have never seen before? What goals do we adopt for restoration in the absence of historical targets? How do we make decisions at different scales about novelty and change?

To address these questions in a better way, research on the origins, persistence and maintenance of novel ecosystems, including how novel assemblages positively or negatively affect ecosystem function, is needed. While the scientific community is often accused of simply asking for more research, in this case we are dealing with new and unpredictable situations and ecosystems; the research, properly framed, can significantly contribute directly to the policy-making process. Such research results can also help inform public dialogue and deliberation and the ways in which benefits, trade-offs or negative impacts are considered in democratic decision-making about novel ecosystems.

While specific policy on novel ecosystems is notably absent, ecosystem change and climate change in particular have inspired policy-makers to begin to consider novel ecosystems in emerging policy documents. As we move forward, future policy discussions need to:

• recognize the existence and value of novel ecosystems and the many benefits they can provide, including ecosystem function, resources for livelihoods, ecosystem services and adaptive capacity in the face of climate change;

• acknowledge and establish mechanisms for navigating the trade-offs inherent in decisions about maintaining historical conditions versus embracing novel elements and prioritizing goals in the context of competing demands; and

• develop processes to make policy more flexible and dynamic in the face of novel and often rapid ecosystem change to enable policy to keep apace of scientific knowledge and changing social values.

In the absence of policy that explicitly addresses novel ecosystems, we cannot acknowledge, account for, or conserve the types of novelty we (or importantly, future generations) might wish to value. Ecosystems will go 'novel' whether or not we have developed the policy to guide their transformation. Even good policy will not enable us to determine what types of novelty occurs where. Policy on novel ecosystems will however allow us to consider the benefits of emerging ecosystem configurations and move beyond knee-jerk reactions to maintain existing ecosystems at all costs. Such policies will also enable us to acknowledge and value novel ecosystems without compromising efforts for overall biodiversity conservation, management and restoration at larger scales.

REFERENCES

Alexander, S., Nelson, C.R., Aronson, J. et al. (2011) Opportunities and challenges for ecological restoration within REDD+. *Restoration Ecology*, **19**, 683–689.

Bridgewater, P.B. (1997) The Global Garden: Eden revisited, in *Conservation Outside of Reserves* (P. Hale and D. Lamb, eds), Surrey Beatty, Sydney, pp. 35–38.

Bridgewater, P. (2008) A new context for the Ramsar Convention: Wetlands in a Changing World. *RECIEL*, **17**, 100–106.

Bridgewater, P., Higgs, E.S., Hobbs, R.J. and Jackson, S.T. (2011) Engaging with novel ecosystems. *Frontiers in Ecology and the Environment*, **9**, 423–423.

Bridgewater, P., Lei, G. and Lv, C. (2012) From Stockholm to Rio II: the natural and institutional landscapes through which rivers flow, in *River Conservation and Management* (P.J. Boon and P.J. Raven, eds), John Wiley, Oxford, pp. 295–311.

Camacho, A.E. (2011) Transforming the means and ends of natural resources management. *North Carolina Law Review*, **89**, 1405–1450.

CBD (1992) Convention on Biological Diversity. United Nations. http://www.cbd.int/doc/legal/cbd-en.pdf. Accessed August 2012.

CBD (2010) AHTEG on Climate Change. http://www.cbd.int/doc/meetings/sbstta/sbstta-14/information/sbstta-14-inf-21-en.pdf. Accessed August 2012.

Cliquet, A., Backes, C., Harris J. and Howsam, P. (2009) Adaptation to climate change: Legal challenges for protected areas. *Utrecht Law Review*, **5**, 158–175.

Cole, D.N. and Yung, L. (eds) (2010) *Beyond Naturalness: Rethinking Park and Wilderness Stewardship in an Era of Rapid Change*. Island Press, Washington, DC.

Craig, R.K. (2010) "Stationarity is dead" – long live transformation: five principles for climate change adaptation law. *Harvard Environmental Law Review*, **34**, 10–73.

CSAB4 (2011) http://www.cbd.int/doc/legal/cbd-en.pdf. Accessed August 2012.

Davis, M.A., Chew, M.K., Hobbs, R.J. et al. (2011) Don't judge species on their origins. *Nature*, **474**, 153–154.

Doremus, H. (2010) The Endangered Species Act: static law meets dynamic world. *Journal of Law and Policy*, **32**, 175–235.

EC (European Commission) (1992) Council Directive 92/43/EEC of 21 May 1992 on the conservation of natural habitats and of wild fauna and flora. *Official Journal of the European Commission L*, **206**, 22/07/1992, 7–50.

EC (European Commission) (2010) Directive 2009/147/EC of the European Parliament and of the Council of 30 November 2009 on the conservation of wild birds. *Official Journal of the European Commission L*, **20**, 26.1.2010, 7–25.

Endangered Species Act (1973) United States Public Law, 93–205.

Hobbs, R.J., Arico, S., Aronson, J. et al. (2006) Novel ecosystems: theoretical and management aspects of the new ecological world order. *Global Ecology and Biogeography*, **15**, 1–7.

Huq, S., Yamin, F., Rahman, A., Chatterjee, A., Yang, X., Wade, S., Orindi, V. and Chigwada, J. (2005) Linking climate adaptation and development: A synthesis of six case studies from Asia and Africa. *Ids Bulletin-Institute of Development Studies*, **36**, 117–122.

Jackson, S.T. and Hobbs, R. (2009) Ecological restoration in the light of ecological history. *Science*, **325**, 567–569.

Jóhannsdóttir, A., Cresswell, I. and Bridgewater, P. (2010) The current framework for international governance of biodiversity – is it doing more harm than good? *RECIEL*, **19**, 139–149.

Kareiva, P., Enquist, C., Johnson, A., Julius, S.H., Lawler, J., Petersen, B., Pitelka, L., Shaw, R. and West, J.M. (2008) Synthesis and conclusions, in *Preliminary review of adaptation options for climate-sensitive ecosystems and resources: A report by the U.S. Climate Change Science Program and the Subcommittee on Global Change Research* (S.H. Julius and J.M. West, eds), US Environmental Protection Agency, Washington, DC, pp. 9.1–9.66.

Karkkainen, B. (2002) Collaborative ecosystem governance: Scale, complexity, and dynamism. *Virginia Environmental Law Journal*, **21**, 189–243.

Kettunen, M., Terry, A., Tucker, G. and Jones, A. (2007) Guidance on the maintenance of landscape features of major importance for wild flora and fauna – Guidance on the implementation of Article 3 of the Birds Directive (79/409/EEC) and Article 10 of the Habitats Directive (92/43/EEC). Institute for European Environmental Policy (IEEP), Brussels.

Koetz, T., Farrell, K.N. and Bridgewater, P. (2011) Building better science-policy interfaces for international environmental governance: assessing potential within the Intergovernmental Platform for Biodiversity and Ecosystem Services. *International Environmental Agreements: Politics, Law and Economics*, **11**, 1–21.

Leopold, A.S., Cain, S.A., Cottam, D.M., Gabrielson, I.N. and Kimball, T.L. (1963) Wildlife management in the national parks, in *Transactions of the 28th North American Wildlife and Natural Resources Conference*, 28–45.

MA (Millennium Ecosystem Assessment) (2005) *Ecosystems and Human Well-being: Synthesis*. Island Press, Washington DC.

Marris, E. (2011) *Rambunctious Garden: Saving Nature in a Post-Wild World*. Bloomsbury, New York.

Nellemann, C. and Corcoran, E. (eds) (2010) Dead Planet, Living Planet – Biodiversity and Ecosystem Restoration for Sustainable Development. A Rapid Response Assessment. United Nations Environment Programme, GRID-Arendal. www.grida.no.

Nie, M. (2008) *The Governance of Western Public Lands: Mapping Its Present & Future*. University Press of Kansas, Lawrence.

Nowotny, H., Scott, P. and Gibbons, M. (2001) *Re-Thinking Science: Knowledge and the Public in an Age of Uncertainty*. Blackwell, Cambridge.

Parks Canada and the Canadian Parks Council (2008) Principles and Guidelines for Ecological Restoration in Canada's Protected Natural Areas. http://www.pc.gc.ca/docs/pc/guide/resteco/index_e.asp. Accessed August 2012.

Ruhl, J.B. (2008) Climate change and the endangered species act: Building bridges to the no-analog future. *Boston University Law Review*, **88**, 1–62.

Schlaepfer, M.A., Sax, D.F. and Olden, J.D. (2011) The potential conservation value of non-native species. *Conservation Biology*, **25**, 428–437.

Scott, D. and Lemieux, C. (2005) Climate change and protected area policy and planning in Canada. *The Forestry Chronicle*, **83**, 696–703.

Seastedt, T.R., Hobbs, R.J. and Suding, K.N. (2008) Management of novel ecosystems: are novel approaches required? *Frontiers in Ecology and the Environment*, **6**, 547–553.

Soulé, M.E. (1986) Conservation biology and the real world, in *Conservation Biology: The Science of Scarcity and Diversity* (M.E. Soulé, ed.), Sinauer Associates, Sunderland, MA, pp. 1–12.

TEEB (2011) *The Economics of Ecosystems and Biodiversity in National and International Policy Making* (P. ten Brink, ed.), Earthscan, London and Washington.

Thrower, J. (2006) Adaptive management and NEPA: How a non-equilibrium view of ecosystems mandates flexible regulation. *Ecological Law Quarterly*, **33**, 871–895.

UKNEWP (2011) *The Natural Choice: Securing the Value of Nature*. The Stationery Office, London.

UNEP-WCMC (2011) UK National Ecosystem Assessment: understanding nature's value to society: synthesis of the key findings. http://uknea.unep-wcmc.org. Accessed August 2012.

UNFCCC (1992) United Nations Framework Convention on Climate Change. United Nations, http://unfccc.int/resource/docs/convkp/conveng.pdf. Accessed August 2012.

UNFCCC (2011) Ecosystem-based approaches to adaptation: compilation of information. FCCC /SBSTA/2011/INF.8. http://unfccc.int/resource/docs/2011/sbsta/eng/inf08.pdf. Accessed August 2012.

UNGA (1994) Elaboration of an International Convention to Combat Desertification in Countries Experiencing Serious Drought and/or Desertification, Particularly in Africa. United Nations General Assembly A/AC.241/27. http://www.unccd.int/en/about-the-convention/Pages/Text-overview.aspx. Accessed August 2012.

van den Hove, S. (2007) A rationale for science-policy interfaces. *Futures*, **39**, 807–826.

Young, O.R. (2004) Institutions and the growth of knowledge: Evidence from international environmental regimes. *International Environmental Agreements: Politics, Law and Economics*, **4**, 215–228.

Woodley, S. (2010) Ecological integrity: A framework for ecosystem-based management, in *Beyond Naturalness: Rethinking Park and Wilderness Stewardship in an Era of Rapid Change* (D.N. Cole and L. Yung, eds), Island Press, Washington DC.

Chapter 34

PERSPECTIVE: LAKE BURLEY GRIFFIN

Peter Bridgewater

Global Garden Consulting, UK

The edges of many parts of Lake Burley Griffin in Canberra, Australia, make a very good example (in which I must declare a personal involvement) of a novel ecosystem. Lake Burley Griffin is an artificially created lake from the 1960s. A mixture of planting and adventive seeding establishment has created emergent edge vegetation of *Phragmites australis*, *Typha domingensis* with *Lythrum salicaria*, backed by a shrub/tree vegetation with *inter alia Salix* species, *Alnus glutinosa*, *Rubus fruticosus* agg., *Fraxinus excelsior* and a range of associated shrubs and forbs. If a person from western Europe was blindfolded and transported to the lake edge from the edge of a lake in Europe, they would be hard put to notice they were on a different continent! These vegetation communities, acting as buffers to a lake ecosystem (including being home to a range of native and introduced fauna) existed in stable conditions for decades. Public perceptions however have increasingly moved to see European species in the 'wild' as negative, with a desire to see them replaced by 'native' species.

In May 2011 it became apparent that the local Government (the Australian Capital Territory Legislative Assembly) were intent on removing some of this vegetation along part of the lake shoreline, partly to create better views across the lake and partly because of concerns over the undesirable nature of the vegetation. I wrote to the ACT Minister for the Environment and Sustainable Development urging that the system be left intact and subjected to appropriate management, citing it as an important novel ecosystem delivering ecosystem benefits.

The Minister replied as follows: "The plant species being removed along the foreshore of Lake Burley Griffin are declared as noxious weeds in the ACT (Australian Capital Territory) under the *Pest Plants and Animals (Pest Plants) Declaration 2009*. Active management of these species is in line with the Australian Government's Australian weed strategy which forms part of an integrated approach to national bio-security.

"The woody and tree species removal targets *Salix* species, plus *Populus alba*, *Alnus glutinosa* and *Rubus fruticosus* (aggregate) species. Such invasive species require extensive management due to their vigorous and smothering nature.

Novel Ecosystems: Intervening in the New Ecological World Order, First Edition. Edited by Richard J. Hobbs, Eric S. Higgs, and Carol M. Hall.
© 2013 John Wiley & Sons, Ltd. Published 2013 by John Wiley & Sons, Ltd.

"Following the weed removals (*sic*) the Lake Burley Griffin shoreline will be revegetated with native species that will assist many positive ecological outcomes, such as increased fish habitat, reduced turbidity and nutrient control. Cleared areas where vistas are to be maintained will be revegetated with grasses, ground-covers and emergent plants. Plant selection has been considered for ecological values, environmental conditions, aesthetic values and maintenance requirements. The works along the foreshore of Lake Burley Griffin will compliment (*sic*) works on a regional scale in the Molongolo and Murrumbidgee catchments."

On the face of it, this was a perfectly reasonable response as it drew on existing legislation to defend an action. Yet the fact that this ecosystem has been stable, and enabled a protective function along the lake, was ignored by both the Minister and presumably his advisors, political and expert.

I tried once more in September of the same year after the vegetation had been rather brutally, yet incompletely, cleared. A few *Allocasuarina* tree seedlings had been planted but nothing else was in place, and the lake edge was now exposed to erosion.

The Minister replied further: "The ACT government recognises the values of both native and non-native aspects of the natural environment in which Canberrans live. The recently completed community consultation *Time to Talk* identified the community's appreciation of these values."

In February 2012 I was again able to visit the site. It had, as I feared, moved from 'clearance' to 'destruction'. Lake-edge erosion had set in with a vengeance since the *Phragmites/Typha* swampy edge to the lake was no longer well protected or integrated with the extensive near-surface root systems of the *Salix*, *Populus* and *Alnus*, and large chunks had been washed away and lay dying. There was little evidence of any new plantings of native species. An expensive and difficult

management situation had therefore been created, contrasting with the still well-vegetated edges of Sullivans Creek, a creek flowing from the grounds of the Australian National University to the lake.

Canberra (possibly one of the most artificial cities on earth) should offer chances to use its many novel ecosystems in balance with restoration and management of native ecosystems, to truly create the bush capital image Canberra tries to project. Yet here we have an example of the problems novel ecosystems have, and will have, in the face of public perception and legal constructs that have been badly conceived, and are now difficult to unravel. Public servants and politicians often try to deliver what they see as good outcomes, but they are always under public pressure from civil society in general and non-governmental organizations in particular. These groups hold earnest but not necessarily well-informed views, often advocating return to Arcadian conditions for the planet that almost certainly never existed.

While this vignette refers solely to a localized space, this situation is repeated every day across the world. Legislation from local to global is almost all derived from past environmental settings and perceptions, yet flexibility and nimbleness are now the key requirements to manage our planet. Leadership can and must come from the global legal instruments dealing with biodiversity, yet they too are reluctant to grasp this nettle. We therefore return to the message of this perspective: public perception, and thus pressure on the political pulse, must move from the past to the future, embracing and managing change while attempting restoration and conservation where this is possible. I experienced disappointment and dismay on my last visit, and I admit to feeling it was not worth pursuing this issue, yet I know I should have. We all have responsibility to keep these issues, the opportunities, problems and solutions in front of our political leaders.

Chapter 35

CASE STUDY: SHALE BINGS IN CENTRAL SCOTLAND: FROM UGLY BLOTS ON THE LANDSCAPE TO CULTURAL AND BIOLOGICAL HERITAGE

Barbra A. Harvie[1] *and Richard J. Hobbs*[2]

[1]School of GeoSciences, Institute of Geography and the Lived Environment, University of Edinburgh, UK

[2]Ecosystem Restoration and Intervention Ecology (ERIE) Research Group, School of Plant Biology, University of Western Australia, Australia

Mining for mineral or energy resources occurs worldwide and many mining enterprises result in the production of waste or residues that require storage and/or treatment. In central Scotland, mining for coal and oil shale was widespread. The extraction of oil from shales began in the 1850s and developed into an important industry, supplying 25% of the city of London's lamp oil. The arrival of the internal combustion engine and the motor car led to the production of motor spirit (petrol) from shale oil that was sold throughout the UK, and the development of the diesel engine in 1938 led to further refinement processes. The maximum output from the industry in Scotland was in 1913 when 27.5 million barrels of crude oil were produced. However, production declined due to competition from oil produced elsewhere and ceased altogether in 1962.

The process of retorting crude oil from deep-mined oil-bearing shale resulted in 7 tonnes of burnt shale waste for every 10 barrels of oil produced. The waste was left lying around West Lothian (west of Edinburgh) in large heaps, creating the area's unique red shale bings or spoil heaps. These tower above the naturally low-lying landscape of West Lothian. For some, they represented a major blot on the landscape, a reminder of the large-scale exploitation of resources and result-

Figure 35.1 Naturally occurring species-poor calcareous grassland community on the plateau of Greendykes. Photographs: B.A. Harvie.

ant waste products and an intrusion on the natural geomorphology of the area. However, the burnt shale (blaes) was also recognized as a useful resource: a source of landfill for roads and other construction. Changes to UK legislation after the Aberfan mining disaster in Wales in 1966 resulted in many small bings being reclaimed as green space from the late 1960s onwards. The remaining sites, mostly privately owned, were retained for their commercial value.

However, those bings that remained in the landscape soon came to the attention of botanists and other naturalists as they were gradually colonized by plants and animals. The physical and chemical structure of burnt shale is unlike coal spoil or any other type of industrial waste. The substrate varies in physical and chemical characteristics and thus provides a variety of different, locally distinct habitats. As a consequence of their additional industrial heritage they are a unique habitat, not found anywhere else in Britain or Western Europe. The bings are now home to several nationally (UK) rare and protected plant and animal species from

Stereocaulon saxatile (an inconspicuous lichen) to badgers (*Meles meles*). The sites form refugia for many other locally rare fauna and flora, and more than 350 plant species have been recorded on them. Some of the bings have remained unmanaged since shale extraction ceased (Fig. 35.1), while others have been reshaped, seeded and planted in various ways during the 1970s and 1980s when land reclamation was a national political priority. The resulting vegetation, in all instances, is so unusual within the region that the oil-shale bings constitute one of the eight main habitats in West Lothian's Biodiversity Action Plan. At the base of one bing, a genetically distinct birch woodland has established naturally and contains many of the associated ground flora and bryophyte species of long-established native woodlands. Another bing is designated as a Scottish Wildlife Trust nature reserve because of its diversity of habitats and species.

Several garden escapes have been recorded on the bings and are well established on many sites. These now pose an interesting dilemma: should they be encouraged as an integral part of a novel vegetation type or should they be removed for fear that they spread uncontrollably throughout the surrounding landscape? Species like these may become a permanent component of plant assemblages on the bings and form part of established novel community types. Many environmentalists and ecological managers may consider that they are not natural and should therefore be discouraged, but what is natural on an entirely human-created habitat?

The bings are now also of considerable social and historical importance; two are scheduled as historical industrial monuments. They are a focus of community identity in a population whose common culture of mining is slowly being eradicated by families of non-West Lothian origin taking up residence in the many new housing developments in the county. As a consequence, the bings have potential as an education resource at all levels because of the historical importance of the shale-oil industry that created them (global engineering expertise and products ranging from paraffin to detergent), the ecological importance of their extensive flora and fauna (nature reserves and primary succession) and the geological importance of the sedimentary rocks that they were mined from (Carboniferous limestone series). Socially, they provide much used public open space. Several bings, such as the Five Sisters (Fig. 35.2), are seen as important landmarks with social, cultural and historical significance in addition to their biological values.

Restoration and management of spoil waste is common in many parts of the world, but it has been argued that it is unnecessary on the shale bings. Restoration has largely followed 50-year-old policies with a standard recipe of: reducing the height and gradient of the heap; rounding peaks and ridges; covering with topsoil; applying fertilizers (liberally); and sowing with commercial grass mix. On low-lying heaps, trees (usually birch and alder) were planted directly into the spoil at the bottom of the heap without any amelioration. Often the only purpose was to obtain a satisfactory visual effect. Management decisions on the restoration of the bings and similar sites are affected by an unrealistic public perception of what post industrial and other waste sites should be restored to, and how quickly they should be restored. Scottish Natural Heritage Information and Advisory Note Number 50 (http://www.snh.org.uk/publications/on-line/advisorynotes/50/50.htm) emphasizes the "need for careful consideration to be given to any bing reclamation proposals to ensure that the distinctive and potentially unique natural heritage interest is adequately considered". Recent changes to local conservation policy have now ensured that many of the remaining bings are safe from demolition, reshaping, reclamation and restoration.

Figure 35.2 Five Sisters, near West Calder, West Lothian, Scotland. Photograph: R.J. Hobbs.

The shale bings of central Scotland represent an interesting case of novel ecosystems: their novelty arises because of the deposition on the landscape of new structures comprising physically and chemically distinct substrates that are subsequently colonized in an idiosyncratic way by both native and non-native species. They have also gone from being perceived as undesirable features to valuable social and cultural components of the local landscape. Is this the future for other types of novel ecosystem too?

The material in this chapter is based on Harvie (2005, 2007, 2011) and Harvie and Russell (2007), which provide more detailed information on the subject.

REFERENCES

Harvie, B.A. (2005) West Lothian Biodiversity Action Plan: Oil Shale Bings. West Lothian Council, Linlithgow.

Harvie, B.A. (2007) The importance of the oil-shale bings of West Lothian, Scotland to local and national biodiversity. *Botanical Journal of Scotland*, **58**, 35–47.

Harvie, B.A. (2011) To tree or not to tree? *Reforesting Scotland*, **44**, 20–21.

Harvie, B.A. and Russell, G. (2007) Vegetation dynamics on oil-shale bings; implications for management of post-industrial sites. *Aspects of Applied Biology*, **82**, 57–64.

Part VI

What's Next?

Chapter 36

PERSPECTIVE: A TALE OF TWO NATURES

Eric S. Higgs

School of Environmental Studies, University of Victoria, Canada

I gaze through the window and see the sloping muscle-brown trunk of an arbutus tree, and just beyond the furrowed silver gray bark of Douglas fir. The ground-storey vegetation is suppressed by the thin soils of a sandstone ridgetop but moss, exuberant moss, is everywhere. In the distant background is a glimpse of the ocean and Vancouver Island. This is a scene etched in my imagination. When I am away from this place I can easily conjure up the image, and it is the subject of many photographs. Changing the composition – a tree falling, others dying because of heat or water stress – would be a source of lament, a minor tragedy in my world of things. I am sensitive to such change.

Years ago, a property management company on orders from the absentee landlord cut down a massive deciduous tree under some vague complaint that its leaves were clogging the downspout. The tree likely predated the original farmhouse that was razed to make way for the housing development in which I lived. The summer shade the tree provided was gone in the blink of an eye, and so was the habitat provided by that singular tree. It took me a while to accommodate the alteration, and in some ways I never did; that I

remember a distant and seemingly obscure event is testament to its influence.

I wonder about adjustment when it comes to novel ecosystems. Will the change from historical to hybrid, or hybrid to novel ecosystems push us deeper into lament by the loss of familiar settings for daily life? Will the changes be sufficiently gradual (years? decades?) that we acclimatize? Or, will we adjust to new cultural views of nature that will make it easier to love novel ecosystems? There is no longer a single type of nature rooted in historical continuity and changed through direct human agency. The new natures are driven often by diffuse indirect effects (climate change, nitrogen deposition) that are far beyond our immediate control, and certainly by complicated interactions of these with new and changing values about ecosystems (ecosystem services).

The window that frames my reflections belongs to a house my family and I own at the north end of Galiano Island, one of the southern Gulf Islands archipelago in the waters of the Salish Sea between the cities of Vancouver and Victoria (Canada). The place found us as much as we found it. It was a moment in life when

Novel Ecosystems: Intervening in the New Ecological World Order, First Edition. Edited by Richard J. Hobbs, Eric S. Higgs, and Carol M. Hall.
© 2013 John Wiley & Sons, Ltd. Published 2013 by John Wiley & Sons, Ltd.

caution evaporated and the choice just felt right. I visit often, typically every two or three weeks and for longer stretches during holidays. There's a house and a small studio on the six hectares, a large vegetable garden and orchard, a treehouse, a meadow with a small pond and, most remarkable, a labyrinth replicated from the original at Chartes Cathedral. The former owners loved this place very much. They selectively felled and milled the trees that became the studio and home, and did most of the planning and building. People who watched the construction report they had never seen a building site with so little impact on the surroundings. Only a few trees were cut, and the home is sited to be virtually invisible from a distance. It simply blends in. Every window frames an exquisite wildness.

But just how wild is it? The property is determined by a large north–south ridge. The house sits atop the ridge, the meadow 30 m below a precipitous sandstone cliff and the studio and garden down a gentler slope in the opposite direction from the ridge. The property is adjoined on two sides by the 142 ha Dionisio Provincial Park, and on the third by Penalakut First Nation reserve lands. These two lightly traveled lands are between us and the ocean, which is why there is a feeling of solitude and quiet. The land was logged early in the first half of the 20th century, and subsistence activities by early settlers created or at least maintained the meadow below the home. A few old-growth Douglas fir (*Pseudotsuga menzeisii*) are evident to the trained eye on the dry, thin soil ridge (these are not the grand tall fir trees associated with this species, but gnarled, tough survivors). Some magnificent and likely very old arbutus boles dot the ridge. Less than 100 m off the ridge down in the valleys that run parallel are western red cedar, sitka spruce, grand fir and western hemlock that require slightly deeper soils and more moisture. On the other side of the boundary in the Provincial Park are some very large old-growth trees (although this area was also logged earlier in the 20th century).

From where I write, the ridgetop comprises wind-hardened trees, a mossy understory with some shrubs and grasses. When I first set eyes on the site, I appreciated its aesthetic qualities – the way the house nestles on the ridge just so – but presumed the small diameter trees represented a recent forest community, the open mossy understory an anomaly of human management, and the grasses, as is often the case in this part of the world, agronomic. The more time I spend here the more I realize this ecosystem is intact and historically

continuous. The trees I cored to ascertain age turned out to be much older than they appear; this is a likely consequence of thin soils, high winds and hot dry summers. The moss communities are unusual but apparently naturally selected for the site. I was struck one day that many of the bunchgrasses I had assumed were agronomic turned out to be Roemer's fescue (*Festuca roemeri*), a native species. I came for the solitude and quiet, and found an intact ecosystem, too.

The agencies of novelty are close by. Climate change looms over the southern Gulf Islands, as it does everywhere in the region: increasing sea level, earlier springs, extreme temperatures, less predictability. Land conversion remains a possibility, but perhaps less likely now with much of the north end of Galiano in parkland. Invasive species are a significant driver. Most visible is Scotch broom (*Cytisus scoparius*), which is evident along roadsides and disturbed sites on Galiano. It is clearly expanding, but stays out of deeper forested sites. There's a healthy patch growing along the road near the lower meadow. I pull it out, but it will be a sustained struggle to hold the line. Scotch broom is everywhere in the region. It is a scourge for restorationists, but for many its showy yellow flowers turn swaths of landscape along the west coast of North America yellow in the spring. Broom is a bully plant, displacing native species, and its flammability alters the fire regime. The seedbank is long lived, as much as 40 years. Short of some unexpected miracle, Scotch broom is here to stay. If it is, then how do we come to terms with it? Do we revile it and wage a steady campaign against it? Or, might we learn to appreciate its qualities, hold it back where needed, and not give up on native species? How will generations in the future regard it?

Ten years ago, I wrote 'A tale of two wildernesses' in *Nature By Design*. My perspective was formed by field research in Jasper National Park, an icon of Canadian wilderness, and the recently completed Wilderness Lodge at Disney World in Orlando, Florida. I wondered what these two different views of nature had in common, and what the perils might be in a world where we could specify the kind of nature we would enjoy. My impulse was to anchor the practice and theory of ecological restoration in keystone concepts such as ecological integrity and historical fidelity. I reasoned then that historical references and continuity were vital in tempering our distinctly human ambitions; they anchor restoration goals to something other than us. The problem with the Wilderness Lodge was that it lacked either integrity or fidelity. It is what literary theorists term a simulacrum, a copy with no true original. Floating free from ordinary constraints of natural process, this alternate icon of wilderness offers us little in terms of moral guidance; it is a technological artifact. The story is complicated by the fact that Jasper had turned its back on parts of its history, reinforcing William Cronon's maxim that wilderness is a manufactured idea. More so than a decade ago, the idea of nature expressed by the Wilderness Lodge and other manufactured landscapes is today shaping and reshaping the idea of wilderness in Jasper and other wild places.

I worry that the spread of technologically shaped ideas about nature will collide with the idea of novel ecosystems (have they already?) to produce ecosystems that are serviceable commodities. This ascendancy of human intervention fulfills an ideal of nature that has more to do with industrial calculus and colonial authority than it does with the aspirations of restoration ecologists. The work of restorationists and conservationists over the last two decades has emphasized careful study of ecosystem characteristics including historical continuity (or lack thereof), and modest engagement by people in recovering damage. Will this work be undone as people rush to the opportunities afforded by novel ecosystem management that emphasizes function and pragmatics, a bright future unhinged from the tedious constraints of history?

On Galiano, these questions seem distant. Working with an historical ecosystem that is tending towards hybrid, there are firm moral handholds in scaling the future. Perhaps in 50 years a warmer world will have changed the conditions that my son will experience, or perhaps this warmth might fuel an epidemic that races through the arbutus trees. There will be new invasive species to contend with, but also a powerful resolve to address these issues head on. If Scotch broom maintains a similar invasion dynamic, then people on Galiano may learn to accept it without having to give in. As Gary Snyder (1990) proposes in his *Practice of the Wild*, "We can chase off mosquitoes and fence out varmints without hating them. No expectations, alert and sufficient, grateful and careful, generous and direct."

When familiar ecosystems pass over the threshold where returns to history are no longer practicable, then we need a moral basis for action. These ideas are just emerging and will, I hope, be based on something more than 'anything goes'. Indeed, historical knowledge may prove more, not less, important in the future. In the meantime, maintaining a commitment to restoration of historical and hybrid ecosystems will help us form an enduring and appropriate response to novel ecosystems. Restoration brings us to the brink of hope, and it isn't clear quite yet what hope looks like in a world replete with novel ecosystems.

REFERENCE

Snyder, G. (1990) *The Practice of the Wild*. North Point, San Francisco.

Chapter 37

CONCERNS ABOUT NOVEL ECOSYSTEMS

Rachel J. Standish[1], Allen Thompson[2], Eric S. Higgs[3] and Stephen D. Murphy[4]

[1]Ecosystem Restoration and Intervention Ecology (ERIE) Research Group, School of Plant Biology, University of Western Australia, Australia

[2]School of History, Philosophy, and Religion, Oregon State University, USA

[3]School of Environmental Studies, University of Victoria, Canada

[4]Department of Environment and Resource Studies, University of Waterloo, Canada

37.1 INTRODUCTION

Over 20 years ago, Michael Soulé, one of the leaders of the conservation movement, anticipated the great challenge that novel ecosystems would pose to conservation scientists and managers seeking outcomes for nature conservation (Soulé 1990). In his presidential address to the then recently formed Society for Conservation Biology he acknowledged the new combinations of species emerging in our wake and stressed the need for a new ecological discipline to make sense of these assemblages. Soulé called the new discipline 'mixoecology'. His concluding plea for society members to think deeply about what is right – *how* people should intervene in the new world and what *goals* should be pursued – will resonate with anyone who has considered whether or not the existence of novel ecosystems should be acknowledged in conservation practices.

The concept of novel ecosystems is unfamiliar and unsettling for many people, especially those schooled in historically rooted conceptions of nature: nature free from human intervention. Novel ecosystems challenge conventional conservation and restoration practices, yet they are not immune to these practices. Novel ecosystems can be managed to fulfill traditional goals that people value such as biodiversity conservation and watershed protection (Chapter 18). Nevertheless, there are reasons to be cautious with widespread adoption of the novel ecosystems concept and associated management practices. Firstly, it is important that people fully understand the concept and its implications for management before either embracing it wholeheartedly or rejecting it outright. Secondly, as ecologists and philosophers we share concerns about novel ecosystems and think it is wise to continue to question the utility of the concept and the level of intervention that might be required to achieve desirable outcomes, such as biodiversity conservation and human well-being, in our rapidly changing world.

In the 20 years since Soulé's address to the Society for Conservation Biology, there has been increasing recognition of the dilemma that novel ecosystems pose to modern conservation practice (e.g. Ehrenfeld 2004; Chazdon 2008; Kowarik 2011; Morris et al. 2011). Yet few authors have aired strong concerns about the concept or the practices being suggested for management of novel ecosystems (but see Keulartz 2012). The concerns raised here are therefore anticipatory but informed by observations of how people tend to respond to change (including novelty) and other key elements of novel ecosystems (e.g. invasive species). We do not write as champions of novel ecosystems: our purpose in exposing them to greater scrutiny is to find appropriate ways to understand and manage them as they progressively occupy a larger proportion of the biosphere. At times humanity may lament such change, but denial is not a helpful response.

The purpose of this chapter is to articulate and begin to discuss some of the concerns about novel ecosystems. Some of these concerns can be resolved through clarification of the concept, while others are persistent and challenge our best moral, political, economic and cultural wisdom. Giving voice to such concerns illustrates how complicated the idea of novel ecosystems is and indicates there may be a need for new systems of belief and behavior to support relationships of respect and humility between people and ecosystems in a rapidly changing world.

We will focus on two types of concerns. The first are connected with *misapprehensions* that arise, mostly from lack of information or misunderstanding of the implications of managing novel ecosystems. We think these concerns can be alleviated with clarification of the concept. These concerns should not be ignored or underestimated because they can easily obstruct an informed and constructive discussion about novel ecosystems and their management. The second type includes more *persistent concerns* about novel ecosystems. These are much more difficult concerns to work through because they require revisiting and possibly altering systemic patterns of social tradition and moral beliefs that pertain to nature and its conservation. We consider each type of concern in Sections 37.2 and 37.3, respectively.

37.2 MISAPPREHENSIONS ABOUT NOVEL ECOSYSTEMS

Some of the concerns that novel ecosystems generate can be adequately addressed by clearing away mistaken beliefs or other misconceptions. We cover three misapprehensions in this section: those regarding management of invasive species; second, concern that

acceptance of novel ecosystems will result in the replacement of traditional conservation and restoration practice; and lastly, momentum: the concern that novel ecosystems will be valued simply because they are new.

37.2.1 Invasive species

The first misapprehension is that *accepting or acknowledging novel ecosystems implies that managers will surrender any attempt to control invasive species*. This particular concern has been the topic of a recent polarized debate in the literature (Davis et al. 2011; Simberloff et al. 2011). Authors of this book suggest that accepting novel ecosystems into management frameworks does not necessarily mean that invasive species will be left unmanaged. The attempt to control invasive species that interfere with the attainment of management goals will remain a critical intervention (Chapter 3). Similarly, security measures aimed at preventing the spread of potentially invasive species into new regions will remain important. What the novel ecosystems concept does demand is a more sophisticated approach to the management of non-native species, firstly by distinguishing between invasive species (or potentially invasive) and species that are not invasive and not likely to become so in the future (Chapters 3 and 11). Increasingly, the presence of non-native species in ecosystems is unavoidable and so it is critical that managers distinguish between those non-native species that are likely to foreclose options for management and those that are not (Allison 2011; Davis et al. 2011). In the context of the novel ecosystems conceptual framework, invasive species that drive systems across ecological thresholds into novelty or are critical to its maintenance as a novel ecosystem are likely to foreclose options for management. For example, invasive mammalian predators have caused ecosystems in New Zealand to cross biotic thresholds; in this case maintaining predator-free zones is essential for the persistence of some species (Norton 2009). By distinguishing between non-native species on the basis of the likely consequences of their invasion, effective solutions can follow. These could include the option of allowing non-native species to persist if they are 'playing nicely' with the resident native species (Carroll 2011). Additionally, people can attach many values to novel ecosystems and this includes the possibility of valuing – and so deliberately maintaining – particular non-native species that fulfill desired functions and/or services (Chapter 3; Schlaepfer et al. 2011).

Of course there is a risk that benign or even desired non-native species become invasive or otherwise undesirable as environmental conditions continue to change (Simberloff et al. 2011). Predicting which plant species are likely to become invasive, where invasion is likely to occur and what impacts might result from plant invasions are research topics that have received much attention in recent decades (e.g. Rejmánek and Richardson 1996; Ehrenfeld 2010; Vilà et al. 2011). Recent research suggests that scientists are getting closer to being able to predict the identity and context of plant species invasions (Catford et al. 2011; Pyšek et al. 2012). Progress on this front has also been made for other groups of organisms (e.g. insects; Worner and Gevrey 2006). Such research is critical to the development of effective decision-making tools that enable managers to choose which species to actively manage and where to invest their control efforts. Importantly, managers can decide to change their strategy if monitoring reveals that any species – native or not – has become invasive or if the goals change and it is no longer wanted (Section 3.2). Another more persistent concern regards people's sentiments against non-native species. We cover this concern in Section 37.3, the section on persistent concerns.

37.2.2 Replacement of traditional management practices

The second misapprehension we address is that *acceptance of novel ecosystems will result in the replacement of traditional conservation and restoration practice*. By 'traditional' we mean any management that has been practiced in the past, including by those with traditional ecological knowledge. Instead, rather than replacing traditional practices, the novel ecosystems management framework offers additional, and potentially more effective, options for managing ecosystems that are resistant to more traditional practices (Chapter 3). Where ecological or social thresholds prevent the conservation or restoration of historical ecosystems, and failed attempts attest to this fact, managers therefore have the option to aim instead for goals related to conservation of the individual species, functions or services presently existing in situ. The incorporation of novel ecosystems into *established* management frameworks (Chapter 18) calls for a broader framework for traditional goals and practice rather than suggesting

their replacement (see also Hobbs et al. 2010, 2011). Indeed, the common and traditional practice of biodiversity conservation remains a possible and perhaps overlooked utility of novel ecosystems. Basing conservation and restoration purely on history or habitual practice will become increasingly unsuccessful: such practices are likely to create something like an ecological zoo that will ultimately disappear as environments continue to change at unprecedented rates (Jackson and Hobbs 2009). Nonetheless, conservation and restoration, as conventionally or traditionally understood, will remain critical practices for some time to come because people value these practices and value historical landscapes. In this context, the value of historical knowledge may become more, not less, important to people even as the priority given to specific historical reference conditions fade (Higgs 2012).

It is an open question as to whether novel ecosystems will promote the practice of restoration in a more flexible form or merely engender a malaise of defeatism. It seems to us that people who are willing to accept that certain ecological thresholds are likely to be irreversible are more willing to recognize the benefits of novel ecosystems in management frameworks. On the other hand, people who are unwilling to accept the possibility of irreversible ecological thresholds may see little benefit and, perhaps based on various misapprehensions or persistent concerns, will actively resist accepting novel ecosystems. Failing to acknowledge the existence of an ecological threshold could result in costly failure to meet management goals; for this reason alone, it may be worth considering the novel ecosystem management framework. If, however, there is uncertainty as to whether an ecological threshold exists, then there is the option to implement management actions that can be easily reversed or to wait until more information is available (Section 3.2). Critically, the decision to manage an ecosystem as if it were novel is reversible if, for example, funding becomes available to intervene to overcome an ecological threshold and push the system towards a hybrid or historical ecosystem. Our purpose, then, in addressing misapprehensions and concerns, is to help make room for a more practical response towards novel ecosystems.

37.2.3 Momentum

In today's predominant consumer culture there is a social value that ascribes worth to novelty: the idea that something is valuable simply because it is new. *The concern, then, is that people will value novel ecosystems simply because they are new, without regard to ecosystem function or biodiversity conservation.* Rapid obsolescence and fast-evolving technologies create a positive social response to novel consumer products. There is therefore a worry that the value people attach to simply being new will transfer from the marketplace of consumer products to unfolding phenomena in landscape ecology, that ecosystems with no historical analog will be valued simply because they have no precedent. The worry unfolds that the very term 'novel ecosystem' gives privilege to the value of novelty. This pattern is confounded by the emergence over the last decade of ecosystem services as an approach to valuing ecosystems. Widely lauded as an approach to generating new support for conservation and restoration through the explicit valuation of ecosystem services (Millennium Ecosystem Assessment 2005; Aronson et al. 2006), it often emphasizes the monetization of ecological processes. The pattern of monetization can lead to viewing ecosystems as commodities delivering specific benefits (Kosoy and Corbera 2010), which when combined with a positive value of novelty could amplify the preference given to new ecosystems over historical ecosystems. That such change in valuation is happening very quickly – both novel ecosystems and ecosystem services were largely unnoticed two decades ago – places a premium on anticipating these patterns.

On the other hand, the reasons novelty has value in a consumer marketplace depend upon conditions of gaining advantage in a competitive arena (for the producer) or for increasing perceived status (for the consumer), conditions that may not accurately transfer to the domain of decisions regarding ecosystem management. Second, those people who might view novel ecosystems as preferable *simply because they are novel* have the misapprehension that novel ecosystems are indeed new. The generation of new species assemblages in response to changing environmental and ecological conditions has long been recognized by paleoecologists (Jackson and Williams 2004; see also Chapter 8). What is new is the pace of change, and the concern that ecological novelty will swamp historical and hybrid ecosystems. To address any misapprehensions that novel ecosystems will find greater cultural purchase as a consequence of our contemporary fascination with novelty per se, we simply need to be reminded of the long history of novel ecosystems. In contrast, as we will discuss in Section 37.3.2, there is at present widespread positive evaluation of particular ecosystems founded on the belief that they have not been disturbed

by human activity, or are 'old'. We only mean to point out that valuing ecosystems simply on the grounds that they are 'new' or 'old' may be far too simplistic in light of the open ecological future facing humanity.

There is, however, a sense in which the idea of novelty may become a more persistent concern, and therefore immune to simple clarification. The spread of novel ecosystems by human conversion of ecosystems to productive purposes, as well as by rapid ecological and environmental change, has enlarged, and will continue to enlarge, their presence in our lives. The fact of their spreading existence coupled with the patterns of consumer fascination with novelty and ecosystem services might create cultural momentum for the idea of novel ecosystems. There may well be a shift from trepidation about novel ecosystems to acceptance and then to embracing them, if not celebrating them. The concern with such momentum is the authority associated with this transition in belief. Taken on their own terms and (as we argue later in Section 37.3.2) situated within a place-based view of ecosystems, acceptance of novel ecosystems may lead to more effective intervention. Beyond this argument, however, are worries that embracing novel ecosystems will authorize boundless management intervention under a pretext of sheer human creativity, leading eventually to those that result from and manifest distinctly human technologically motivated ambitions (Keulartz 2012). This worry is consistent with the argument made by some historians of technology that attribute a distinctive type of determinism to technological advance (Roe Smith and Marx 1994). On this view, technological progress can develop a cultural momentum such that it becomes difficult to imagine a world not given over to a greater technological character, and that such change is more-or-less inevitable: this cultural determinism invests technology with sufficient agency to effect change. Pushed far enough, determinism can develop significant momentum in its own right; determinism becomes "the human tendency to create the kind of society that invests technologies with enough power to drive history" (Roe Smith 1994). In addressing novelty, it may be the case that such determinism will develop and become the human tendency to create the kind of ecosystems that invests novelty with enough power to drive our ecological thinking. This view – that novelty will become such a prevalent cultural ideal and novel ecosystems so ubiquitous that people will undertake interventions that invest ecosystems with novelty and nothing else – is rather extreme but does indicate the trajectory of concern. This worry certainly fits the next category of persistent concerns.

37.3 PERSISTENT CONCERNS

Persistent concerns are not dispensed with merely by clarification. They demand our attention because they render an impression, considered against a background of ways people traditionally value nature, of novel ecosystems as troublesome. There are no easy solutions here.

37.3.1 Complexity

Accepting novel ecosystems into management frameworks will add complexity. There is no doubt that this is true. Intelligent and effective solutions for the management of novel ecosystems require managers to embrace the reality that socio-ecological systems are extraordinarily complex. From a social perspective, there are questions about what objectives should guide the management of novel ecosystems and how and for whom novel ecosystems ought to be managed. There will be much qualitative analysis and reflective dialogue about the values people might call on to defend specific goals, in specific contexts, as desirable. Emerging discourse about values in restoration will inform conversations about the merits of alternative goals for novel ecosystems and their desirability (e.g. Nassauer 2004; Ostrom and Nagendra 2006; Higgs and Hobbs 2010; Zweig and Kitchens 2010). Perhaps the best answers will arise from the detailed process of stakeholder participation, which necessarily becomes more complex as the scale of projects increase (Chapters 18 and 39). Once a set of objectives are agreed upon, then it may be challenging to accomplish them as old and new tools (e.g. non-native species) are tested for their ability to do the job under novel ecological conditions. Hopefully, this process will lead to mutually desirable and practical solutions for the management of novel ecosystems. However, objectives are likely to be questioned if these cannot be met in timeframes that are acceptable to managers or if the context for goal setting shifts due to factors such as environmental change, funding availability or the membership or values embraced by groups of stakeholders.

From an ecological perspective, rapidly changing systems comprising new species assemblages under-

score the challenges of ecology and bring attention to Homer-Dixon's claim that "Ecology will be the master science of the 21st century" (Homer-Dixon 2009) and Egler's earlier maxim that "ecosystems are not only more complex than we think, but more complex than we can think" (Egler 1977). Ecologists have long realized that ecosystems are dynamic, yet ecosystem management is often aimed at a static goal: the historical ecosystem (Hobbs and Harris 2001). There are valid reasons for this. One is that monitoring the fate of individual species is much easier than undertaking measures to assess ecosystem properties that implicitly deal with change, such as whether or not ecosystems are self-sustaining or resilient to disturbance (Ruiz-Jaen and Aide 2005). Of course, the assumption has been that these properties would emerge from a system that contained its historical assemblage of species.

A distinct advantage of developing a conceptual framework for the management of novel ecosystems is that such a framework must acknowledge that ecosystems are dynamic and complex (Chapter 3). While other models for management have incorporated multiple ecosystem states into their frameworks (e.g. state-and-transition models), the historical ecosystem is generally regarded as the ideal goal for management (e.g. Standish et al. 2009). Within the framework for the management of novel ecosystems, on the other hand, there is acknowledgement that some change is likely to be irreversible and so it is not possible to go back to a historical ecosystem. Management then can focus on the maintenance of (dynamic) functions rather than a particular (static) species composition (Chapter 3). Managers can therefore aim to achieve a functional goal with any one of a number of species compositions depending on the desired function (e.g. habitat provision for an endangered species). Prioritizing the restoration or maintenance of function over species composition is different than, but not a necessarily a replacement of, the emphasis of traditional approaches which focused instead on restoring or conserving historical species compositions. Consequently, this aspect of novel-ecosystems thinking represents a significant shift in the way people can approach ecosystem management, but one that is consistent with current understanding of ecosystem dynamics (Wallington et al. 2005; Hobbs and Suding 2009).

It may be that certainty of outcome in ecosystem management is an illusion, but a sense of certainty about the goals of management was and is still important to citizens, NGOs, the private sector and government. In this usage, 'certainty' really means general agreement about which goal ought to be desired. As we have discussed, it also helps if managers are confident of the means to achieve this shared goal. Throw novel ecosystems into the mix and the door is opened to more possibilities, both in terms of what goals ought to be desired and the exact means by which managers achieve these goals. It will be challenging to implement new methods if these push against traditional values (Box 37.1). We can therefore see that managing novel ecosystems will add complexity not only to the techni-

cal aspects of intervention, but also to the process of first identifying and then justifying various goals, as managers will have to navigate different social belief systems, including both traditional and emerging values for nature.

37.3.2 Will people care about novel ecosystems?

This concern relates to the natural heritage value of novel ecosystems or lack thereof. Some of the most deeply held

---

Box 37.1 What is the right thing to do? Conflicting values over novel means to achieve management goals in Canada

The overarching mandate of Parks Canada is to maintain 'ecological integrity' of the parks under their management (Fig. 37.1). Ecological integrity is defined as "a condition that is determined to be characteristic of its natural region and likely to persist, including abiotic components and the composition and abundance of native species and biological communities, rates of change and supporting processes" (Woodley 1993; Parks Canada Agency 2005; Parks Canada and Canadian Parks Council 2008). How managers might achieve this mandate is an open question. From the perspective of high-ranking Parks Canada staff and politicians responding to the public it is deemed important to maintain the status quo, especially charismatic species, and ensure the parks generate revenue. While few would argue that conserving rare charismatic species is not important, there is conflict emerging over the ways to achieve this goal. One option is to restore or create habitat to maintain rare species as has been initiated for rare birds in Point Pelee National Park. Another option would be to log large areas of the park and so promote rare birds with an affinity for habitat edges. However, logging in national parks would be considered morally wrong by many people, particularly conservationists. These same people are concerned that the novel ecosystems concept may be used to justify logging of the parks. Ultimately, we have to decide among conflicting values. The shared values that emerge

from these discussions will set the context for judging the morality of the means of achieving them. For example, the stakeholders may decide that the priority is to save rare birds from extinction and that selective logging is the best way to achieve this goal. Further, stakeholders may decide that selective logging is unlikely to undermine ecological integrity at the landscape scale; it is therefore possible to maintain multiple values within the park network. Under such a scenario, logging in parks may be judged as being a morally acceptable thing to do.

Figure 37.1 Pine-dominated forest in Ontario, Canada. Photograph courtesy of Stephen Murphy.

ethical and aesthetic values, and in many cases economic values, center on natural heritage. Perception of natural heritage encompasses the flora and fauna of a region as well as the geological forms on land or sea *as people understand themselves to have received it*. What people value as part of their natural heritage is therefore set within a narrative structure: for all of us, an understanding of where we live is informed by the cultural narratives we use to identify it as a particular place (Higgs 2003; O'Neill et al. 2008). Non-native species seem to threaten some things people value as part of their natural heritage. Indeed, non-native species often lie at the root of strong opposition to the concept of novel ecosystems; the possibility that non-native species might fulfill an ecosystem function that people actually value is quickly dismissed by those with a strong preference, or bias, for native species. Perhaps then, much of what we humans value as part of our natural heritage is structural (compositional) rather than functional. Can people continue to leverage the social and emotional power of natural heritage while recognizing that the world has changed and will continue to change?

The optimistic answer is yes. Novel ecosystems may not classify as part of how people commonly think of our natural heritage in the 'new worlds' of North America and Australia (i.e. what many people think of as untouched wilderness) but in fact some people value, as part of their natural heritage, what are in fact older novel ecosystems (e.g. in Europe), even if they do not always perceive them as such (Box 37.2). This observation stands as a testament to the possibility that newly emerging novel ecosystems may gain a heritage value in the future. The form that these heritage values take will be context specific, depending on the specific mix of natural and cultural features. This is because heritage value is strongly attached to a sense of place and identity, that is, being able to distinguish among landscapes facilitates attachment to home (or away) landscapes (Antrop 2005). The fact that novel ecosystems are unique expressions of a place therefore means there is the potential for them to be ascribed a heritage value. Historical patterns of biodiversity may be less important to valuing a place as part of our natural heritage than people might assume; in a recent survey, people living in urban Sheffield tended to ascribe values (including heritage) to their natural surroundings on the basis of their perceived 'greenness' rather than biological diversity (Dallimer et al. 2012). People care for specific places, as part of their natural heritage, because

stories about these places are embedded in important cultural narratives (Clayton and Myers 2009). There is no reason in principle that such narratives and the heritage values embedded in them could not attach to novel ecosystems. In fact, there is evidence that they already do.

On the other hand, a less hopeful view is that an influx of non-native species coupled with other prominent changes to a site will prevent novel ecosystems from attaining heritage value. Perhaps some novel ecosystems, like those that develop on vacant blocks of land in the suburbs, will be too ugly for most people to like them or will originate from human activities about which people are not proud (but even this may change with time; see Chapter 35). Managers may be wise to transform such novel ecosystems into places that satisfy the expressed ecological and aesthetic values of the local people (Nassauer 1995) or reflect a narrative with which they can be more comfortable. Additionally, even aesthetically distasteful ecosystems or those with disturbing cultural baggage may still be presented as providing valued services. For example, one of the key roles that even ugly novel ecosystems could play is the provision of habitat for rare and endangered species. Concern about the extinction of species elicits powerful emotive responses from people, more so than the degradation of ecosystems (Maley 1994). This trend is reflected in the policies that govern conservation efforts, most of which are aimed at species rather than ecosystem conservation (Chapter 33).

Clearly, understanding people's attitudes toward novel ecosystems is essential to being able to manage them effectively, that is, to recognize that managers operate within a complex socio-ecological system. In doing so people may evolve from simply reconciling to the presence of novel ecosystems to engaging with them and finding reasons to assign these ecosystems positive value, independent of their novel status. Some novel ecosystems are better positioned than others to be so adopted. For example, some novel ecosystems in urban environments could provide playgrounds for children, thus allowing children to experience biota and ecological processes (i.e. positive values). On the other hand, other novel ecosystems are connected with loss of heritage values and as such may be used to reject the notion that novel ecosystems could be assigned any positive value. One challenge to current and future generations is to recognize novel ecosystems as being embedded in cultural narratives and, in turn, recognize how these narratives influence the

Box 37.2 Heritage values of a novel ecosystem in New Zealand

Prior to human settlement, New Zealand was forested almost totally below the treeline. Polynesians settled in New Zealand about 750 BP and began to burn large areas of forest. Repeated fires resulted in an almost complete removal of the fire-sensitive beech forest in the eastern South Island including seed sources, soil biota and seed dispersers (McGlone and Wilmshurst 1999; Fig. 37.2). In its place, tussock grassland and shrubland communities developed and have been maintained by the farming activities of the European settlers who arrived about 600 years after the Polynesians. Farming meant more frequent fires and heavy grazing by sheep and also by rabbits, which lead to more changes in vegetation composition from tall tus-

socks to short tussocks and invasion by non-native herbaceous and woody species (Duncan et al. 2001). Despite these significant shifts in vegetation composition, the tussock grasslands are considered national treasures for their biodiversity, aesthetic and recreational values (Fig. 37.3). As such, the novel ecosystem has become the goal for management. Currently, grazing sheep are being removed from large areas for the purposes of promoting the values of the tussock grasslands. Ironically, it has become clear that grazing helps to maintain the tussock-dominated state and prevent a switch to a weed-dominated state. The challenge for managers will be to devise a novel management method to maintain the novel ecosystem.

Figure 37.2 Fire-sensitive beech forest in New Zealand. Photograph courtesy of Richard Harris.

Figure 37.3 Tussock grassland in New Zealand. Photograph courtesy of Richard Harris.

extent to which generations of people find such places to be valuable. To the extent novel ecosystems become loci of value by such a means, it is reasonable to expect that social norms will follow and perhaps at this juncture, novel ecosystems will no longer be objected to out-of-hand. There are many academic disciplines that will be able to make valuable contributions to this discussion: science, philosophy, economics, psychology and anthropology, to name a few. Perhaps more impor-

tantly, there is opportunity for humanistic enterprises such as storytelling and the arts to make invaluable contributions.

37.3.3 Hubris

Another persistent concern is that *separation from historical anchors will make it easier to exercise human ambitions*, and that ecosystems will increasingly resemble what humans wish them to be – efficient service providers – rather than what they would be without our intentional intervention. Ambition itself is not an objectionable characteristic of our species, connected as it is with generating many of the advances that help make modern life valued. There is however a sense in which over-reaching our proper domain, or perhaps range of authority, is objectionable. To be over-confident in our ability to take the helm, so to speak, is one way to characterize the vice of hubris. It may seem to some that acknowledging novel ecosystems implies that humanity somehow knows better than 'nature' just how things should or should be allowed to be. For example, if novel ecosystems are valued and managed for the delivery of ecosystem services, rather than the conservation of a historical ecosystem or as habitat for the conservation of particular species, then it may seem as though humans are taking a lead where we shouldn't be.

We provide three considered responses here. First, it should be noted that, by definition, in all cases novel ecosystems deviate from historical conditions consequent of some human activity, activity that has occurred either deliberately or inadvertently, located either on- or offsite (see Chapter 5). In an overwhelming number of cases, the resultant novel ecosystem was not intentionally created or even foreseen. Rather, it was the inadvertent result of human activity – a secondary affect – even where the relevant human activity was not itself inadvertent. In this context we see that it is important to distinguish between discovering and acknowledging extant novel ecosystems on the one hand, and the intentional design and creation of novel systems on the other. Only in cases of the second sort is it obvious how hubris may apply.

Second, rather than connected with their intentional creation, the charge of hubris may be leveled against those who would elect or endorse to intervene in novel ecosystems toward some end other than returning them to a historical condition. Here the idea is that *not pursuing* the restoration of a historical ecosystem, but instead purposefully aiming at some alternative, involves the presumption that we can do something better than to put the system back to how we think it was prior to human interference; *this* presumption exhibits hubris.

Of course, human beings have a long history of intensively managing systems for the delivery of ecosystem services, for example, agricultural and forestry systems. If this is not seen as inherently hubristic, then it is unclear why novel ecosystems that are displaced from historical conditions could not also be managed for ecosystem services without thereby invoking charges of hubris. If policy designates that the land is to be managed to preserve historical conditions, then of course it would be wrong to violate the governing directives which presumably have been justly established. Independent of this wrong-doing, there is nothing inherently wrong, nor specifically hubristic, with managing for alternative ends.

Lastly, in the case of lands which previously had been managed to preserve a historical ecosystem but that now exhibit novelty, one might think that *any type of management at all*, even toward a historical ecosystem, exhibits a form of hubris. Here the idea is that although human activity has resulted in the system deviating from historical conditions, the right thing to do is to now let the 'new nature' run its own course, whatever that may result in, free from human intervention as much as this is possible. Otherwise, we are directing what should become of it and this is a hubristic overreach of our station as defenders or stewards of the wild. It is difficult to know how to respond to this last concern; in this case, novel ecosystems give rise to what will be of persistent concern regarding the purpose of protected 'wild' lands in times of significant and on-going environmental change.

One thing that can be pointed out at this juncture is that worries about hubris are moral concerns about what it is to be a good person or, collectively, what it is for people to be good human beings when faced with questions about how to respond in the face of radical environmental change. Such questions about human virtue (or moral excellence) can be distinguished from other questions that can give rise to a distinct but related set of persistent concerns, that is, questions regarding what kind of value natural ecosystems themselves are thought to possess (Thompson and Bendik-Keymer 2012). Issues of human and other ethical values are taken up directly and at more length in Chapter 31. Still, it is worth considering them briefly here, in connection with some of the persistent concerns they can give rise to.

37.3.4   Intrinsic value

A worry that managing ecosystems as novel fails to respect intrinsic value in nature. It is commonplace in environmental ethics to distinguish between two types of value that might be attributed to parts of the non-human world. Animals, organisms, species or ecosystems may have an instrumental value, that is, value as a means to producing or securing some other good. Alternatively, entities of this sort may be thought to have an intrinsic value, that is, valuable for its own sake, as an end in itself, independent of any instrumental values it may have. In these terms, the last persistent concern about novel ecosystems we discuss pertains to a worry that managing novel ecosystems for something other than a return to a historical ecosystem is a failure to recognize that ecosystems bear an intrinsic value. People may benefit from the services provided by ecosystems that persist in their historical condition, but efforts to preserve them or restore them to this condition are not justified by appeal to that instrumental value. The historical ecosystem is believed by many to be valuable for its own sake, intrinsically. The worry is therefore that managing ecosystems as novel per se represents a failure to respect what is of intrinsic value.

There are several particular forms this concern may take (a more detailed discussion of intrinsic value can be found in Chapter 31). First, if an ecosystem can bear an intrinsic value, it may be thought objectionable that it is managed only for instrumental purposes. The broadest version of this worry will object to the idea that only instrumental properties of the ecosystem are recognized or respected, independent of what or whose ends are thereby promoted. Ecosystems can provide services that benefit not only humans, but other living organisms (individuals and collectives) and functional services that are beneficial to the ecosystem itself, perhaps understood in terms of maintaining or increasing its ecological resilience to changing conditions. Once people give up returning a system to its historical condition, i.e. accepting it as novel, we are thereby acknowledging only its value as instrumental and this is wrong. The narrower version of this instrumentalist concern regarding novel ecosystems is not only about turning away from an ecosystem's intrinsic value but rather that management of a novel ecosystem exclusively for services that benefit human beings is an objectionable form of anthropocentrism or human-centeredness. This concern can be expressed thus: it is wrong to manage novel ecosystems only for the benefit of human beings.

Working backward, we can address these concerns. First, regarding the concern that managing novel ecosystems entails objectionable anthropocentrism,

we are not the first to raise the concern about a management approach that deliberately emphasizes only human well-being. We mentioned the ecosystem-service approach to management in an earlier section of this chapter (Section 37.2.3). The Natural Capital Project is one example of this approach where participating organizations aim to align biodiversity conservation with the delivery of ecosystem services for the promotion of human well-being (www.naturalcapitalproject.org). Researchers have gone to considerable effort to look for evidence of biodiversity conservation being neglected under an ecosystem-services approach to management. Indications are that projects focused on delivering ecosystem services do not necessarily result in trade-offs for biodiversity conservation and, in some cases, win-win scenarios are possible (Goldman and Tallis 2009; Nelson et al. 2009). Moreover, the authors found that an ecosystem-services approach to management engaged new landscapes, stakeholders and funding sources that were not engaged under the traditional approaches to biodiversity conservation. There are reasons to believe intervention can promote multiple objectives simultaneously and so it is a misapprehension to think that, when managing novel ecosystems, managers would be forced to choose between goals that benefit only human beings or those that only affect species conservation, the well-being of non-human animals or ecosystem health. Win-win situations are possible (Rosenzweig 2003; Chapter 39).

Once there is a recognition that valuing and managing novel ecosystems for instrumental reasons does not have to collapse into what many perceive as a morally objectionable form of human-centeredness, we can return to the broader worry about novel ecosystems representing a loss of intrinsic value in nature. This worry will invoke deep and persistent concerns about how people understand value in nature and the nature of our relations with the non-human world.

Novel ecosystems lack the intrinsic value possessed by unaltered, historical ecosystems. Accepting or acknowledging novel ecosystems therefore coincides with accepting or acknowledging the loss of something intrinsically valuable. There are several things that can be said in response to this concern. First, insofar as novel ecosystems are a consequence of anthropogenic environmental changes and not themselves a primary driver of this change, novel ecosystems should not be mistaken for what is causing the loss of historical ecosystems with intrinsic value. Second, addressing this worry with precision and care requires us to draw additional distinctions. What is it exactly about historical ecosystems that make them intrinsically valuable? This question has vexed environmental ethics for decades. On the one hand, the fact that the historical ecosystem has not been significantly influenced by human activity in the past explains why it is intrinsically valuable; intrinsic value here attaches to the causal history of an ecosystem insofar as it does not include human interference. Novel ecosystems, by definition, arise as a result of significant human interference; they therefore cannot bear this intrinsic value. If this is correct, and if people intrinsically value naturalness understood in this way, then novel ecosystems are not the appropriate objects of intrinsic value.

People might find intrinsic value in historical ecosystems for a different reason. A forward-looking consideration – that the structural and functional integrity of the ecosystem is independent of ongoing human intervention and its trajectory of change – is again largely independent of human purpose or design. Sometimes historical ecosystems are thought to be intrinsically valuable in this sense, the sense that they are autonomous, self-directed or 'wild'. If this is the kind of property upon which the intrinsic value of an ecosystem is supposed to depend, then there is no reason in principle why novel ecosystems could not be as intrinsically valuable as historical ecosystems. If this kind of intrinsic value can survive in historical wildlands that are managed to remain in their historical and wild conditions, it is difficult to see a reason why novel ecosystems managed to remain on their own trajectories would not also share in the same kind of value.

While it is possible to conceptually distinguish properties of historical ecosystems that might explain their intrinsic value (i.e. causal autonomy from humans looking backward or directive autonomy from humans looking forward), what many people find intrinsically valuable about historical 'wild' ecosystems is the combination of these properties; this might be called complete autonomy from humans. While novel ecosystems cannot exhibit this total autonomy, in today's heavily humanized world few other ecosystems can either. Anthropogenic global climate change, for example, threatens to put human activities – to some extent – behind any and all future ecological states. Work by environmental historians threatens a mistaken (but commonly held) belief that some 'historical wildlands' do not in fact exhibit the effects of some previous human activity (Glacken 1967; Cronon 1996; Mann 2005; Gammage 2011). In short, it is not at all clear that it is possible to identify some ecosystems as being autonomous from human activity, thus uniquely bearing a kind of intrinsic value in a way that would allow people to categorically separate them from the newly emerging novel ecosystems. This is meant as food for thought, however; it cannot be the last word about this worry, because some people simply *just do* attribute an intrinsic value to historical ecosystems that they do not or will not attribute to the emerging novel ecosystems. This brings us to the last point in the family of persistent concerns regarding intrinsic value and novel ecosystems.

While many people may share the view that natural ecosystems are intrinsically valuable, there is disagreement about the nature of this value: is an ecosystem really valuable itself, independent of people valuing it, or are ecosystems intrinsically valuable only because people value them non-instrumentally? Remember that we are addressing the worry that novel ecosystems somehow pose a threat in reference to the intrinsic value often attributed to historical ecosystems. If the intrinsic value of an ecosystem is objective, thus independent of human acts of evaluation, then it is becoming more and more difficult to identify exactly how historical ecosystems differ in principle from novel ecosystems, rather than in degree. This issue points to a source of persistent concern. On the other hand, if the intrinsic value of ecosystems originates only from human acts of valuation, then again there is nothing in principle that could prevent people from valuing novel ecosystems intrinsically (as was discussed in Section 37.3.2). From this perspective, the fact that people tend to value historical ecosystems for their intrinsic qualities but not novel ecosystems is not logically tied to or justified by some set of properties belonging to one or the other. This is all we will say here, but for a discussion connecting ideas about the intrinsic value of ecosystems with a variety of theoretical orientations in traditional and contemporary environmental philosophy, the reader is referred to Chapter 31.

37.4 CONCLUSIONS

This chapter has focused on some concerns associated with novel ecosystems; the list of concerns we have covered is not exhaustive. Nonetheless, we hope that our readers will agree that some possible worries may simply rest on mistaken beliefs whereas other concerns are more persistent and cannot be dismissed in this way. As authors we share these concerns, and we think it is important to distinguish between the two types. Doing so helps to advance discussions about increased complexity and conflicting values, so that people can start to move forward with the new forms of careful intervention that novel ecosystems demand. Managers and policy-makers worldwide are faced with a dilemma posed by novel ecosystems. While many characteristics of traditional conservation and restoration practice can still be used to guide the management of novel ecosystems, not all of the conventional means and ends of these practices will remain pertinent; it is these changes that give people grounds for concerns.

We acknowledge that change can be difficult, that some changes are avoidable with resolute action and that however well we come to grips with novel ecosystems there will remain lingering concerns including difficult moral and practical challenges. Does it help to articulate the basis of these concerns and open them up for discussion? We think so, and we hope that people are stimulated to keep thinking about this important topic. Novel ecosystems are here to stay; let's make space for an on-going discourse about the ethics of ecosystem intervention in our rapidly changing world.

While arguing that we demilitarize the language used to discuss invasive species, Brendon Larson urged us to consider how invasive species are an expression of ourselves (Larson 2005). We think the same idea could be applied to thinking about and managing novel ecosystems. While it is tempting to ignore them, and it will be challenging to convince dyed-in-the-wool traditional conservationists to do otherwise, there is much to learn and many environmental and social benefits to be derived from the study and management of novel ecosystems. They will prompt us to improve our understanding of how ecosystems work, what we value in them, how they change, how we change them and, consequently, how we change *with* them. An optimist would argue that the study and management of novel ecosystems will require that we road-test new and diverse methods and so generate a better understanding of the consequences of different management interventions. Ultimately, such understanding is likely to enhance our capacity for Earth stewardship and consequently our self-understanding into the future.

ACKNOWLEDGEMENTS

The authors sincerely thank Kristin Hulvey, Mandy Trueman and Keren Raiter for their thoughtful comments on a draft version of this chapter.

REFERENCES

Allison, S.K. (2011) The paradox of invasive species: Do restorationists worry about them too much or too little? in *Invasive and Introduced Plants and Animals: Human Perceptions, Attitudes, and Approaches to Management* (D. Rotherham and R.A. Lambert, eds), Earthscan Press, London, pp. 265–275.

Antrop, M. (2005) Why landscapes of the past are important for the future. *Landscape and Urban Planning*, **70**, 21–34.

Aronson, J., Blignaut, J.N., Milton, S.J. and Clewell, A.F. (2006) Natural capital: the limiting factor. *Ecological Engineering*, **28**, 1–5.

Carroll, S.P. (2011) Conciliation biology: the eco-evolutionary management of permanently invaded biotic systems. *Evolutionary Applications*, **4**, 184–199.

Catford, J.A., Vesk, P.A., Richardson, D.M. and Pyšek, P. (2011) Quantifying levels of biological invasion: towards the objective classification of invaded and invasible ecosystems. *Global Change Biology*, **18**, 44–62.

Chazdon, R.L. (2008) Beyond deforestation: restoring forests and ecosystem services on degraded lands. *Science*, **320**, 1458–1460.

Clayton, S. and Myers, G. (2009) *Conservation Psychology: Understanding and promoting human care for nature.* Wiley-Blackwell, Chichester.

Cronon, W. (1996) The trouble with wilderness: or, getting back to the wrong nature. *Environmental History*, **1**, 7–28.

Dallimer, M., Irvine, K.N., Skinner, A.M.J. et al. (2012) Biodiversity and the feel-good factor: Understanding associations between self-reported human well-being and species richness. *BioScience*, **62**, 47–55.

Davis, M.A., Chew, M.K., Hobbs, R.J. et al. (2011) Don't judge species on their origins. *Nature*, **474**, 153–154.

Duncan, R.P., Webster, R.J. and Jensen, C.A. (2001) Declining plant species richness in the tussock grasslands of Canterbury and Otago, South Island, New Zealand. *New Zealand Journal of Ecology*, **25**, 35–47.

Ehrenfeld, J.G. (2004) The expression of multiple functions in urban forested wetlands. *Wetlands*, **24**, 719–733.

Ehrenfeld, J.G. (2010) Ecosystem consequences of biological invasions. *Annual Review of Ecology, Evolution and Systematics*, **41**, 59–80.

Egler, F.E. (1977) *The Nature of Vegetation, its Management and Mismanagement: An Introduction to Vegetation Science.* FE Egler, Aton Forest, Connecticut.

Gammage, B. (2011) *The Biggest Estate on Earth: How Aborigines Made Australia*. Allen & Unwin, Crows Nest, NSW.

Glacken, C.J. (1967) *Traces on the Rhodian Shore: Nature and Culture in Western Thought from Ancient Times to the End of the Eighteenth Century*. University of California Press, Berkeley.

Goldman, R.L. and Tallis, H. (2009) A critical analysis of ecosystem services as a tool in conservation projects: the possible perils, the promises, and the partnerships. *Annals of the New York Academy of Sciences*, **1162**, 63–78.

Higgs, E.S. (2003) *Nature by Design: People, Natural Process, and Ecological Restoration*. MIT Press, Cambridge, Massachusetts.

Higgs, E.S. (2012) History, novelty and virtue in ecological restoration, in *Ethical Adaptation to Climate Change: Human Virtues of the Future* (A. Thompson and J. Bendik-Keymer, eds), MIT Press, Cambridge, MA.

Higgs, E.S. and Hobbs, R.J. (2010) Wild design: Interventions and ethics in protected areas, in *Beyond Naturalness: Rethinking Park and Wilderness Stewardship in an Era of Rapid Change* (D. Cole and L. Yung, eds), Island Press, Washington, DC, pp. 234–251.

Hobbs, R.J. and Harris, J.A. (2001) Restoration ecology: repairing the earth's ecosystems in the new millennium. *Restoration Ecology*, **9**, 239–246.

Hobbs, R.J. and Suding, K.N. (2009) *New Models for Ecosystem Dynamics and Restoration*. Island Press, Washington, DC.

Hobbs, R.J., Cole, D.N., Yung, L. et al. (2010) Guiding concepts for park and wilderness stewardship in an era of global environmental change. *Frontiers in Ecology and the Environment*, **8**, 483–490.

Hobbs, R.J., Hallett, L.M., Ehrlich, P.R. and Mooney, H.A. (2011) Intervention ecology: applying ecological science in the twenty-first century. *BioScience*, **61**, 442–450.

Homer-Dixon, T. (2009) The newest science. *Alternatives*, **35**, 10–11.

Jackson, S.T. and Williams, J.W. (2004) Modern analogues in quaternary paleoecology: Here today, gone yesterday, gone tomorrow? *Annual Review of Earth Planetary Sciences*, **32**, 495–537.

Jackson, S.T. and Hobbs, R.J. (2009) Ecological restoration in the light of ecological history. *Science*, **325**, 567–569.

Keulartz, J. (2012) The emergence of enlightened anthropocentrism in ecological restoration. *Nature and Culture*, **7**, 48–71.

Kosoy, N. and Corbera, E. (2010) Payments for ecosystem services as commodity fetishism. *Ecological Economics*, **69**, 1228–1236.

Kowarik, I. (2011) Novel urban ecosystems, biodiversity, and conservation. *Environmental Pollution*, **159**, 1974–1983.

Larson, B.M.H. (2005) The war of the roses: demilitarizing invasion biology. *Frontiers in Ecology and the Environment*, **3**, 495–500.

Maley, B. (1994) Ethics and ecosystems. The Centre for Independent Studies, Fyshwick, Australia.

Mann, C.C. (2005) *1491: New Revelations of the Americas before Columbus*. Alfred A. Knopf, New York.

McGlone, M.S. and Wilmshurst, J.M. (1999) Dating initial Maori environmental impact in New Zealand. *Quaternary International*, **59**, 5–16.

Millennium Ecosystem Assessment (2005) *Ecosystems and Human Well-being: Synthesis*. World Resources Institute, Washington, DC.

Morris, L.R., Monaco, T.A. and Sheley, R.L. (2011) Land-use legacies and vegetation recovery 90 years after cultivation in Great Basin sagebrush ecosystems. *Rangeland Ecology and Management*, **64**, 488–497.

Nassauer, J.I. (1995) Messy ecosystems, orderly frames. *Landscape Journal*, **14**, 161–169.

Nassauer, J.I. (2004) Monitoring the success of metropolitan wetland restorations: Cultural sustainability and ecological function. *Wetlands*, **24**, 756–765.

Nelson, E., Mendoza, G., Regetz, J. et al. (2009) Modeling multiple ecosystem services, biodiversity conservation, commodity production, and tradeoffs at landscape scales. *Frontiers in Ecology and the Environment*, **7**, 4–11.

Norton, D.A. (2009) Species invasions and the limits to restoration: learning from the New Zealand experience. *Science*, **325**, 569–571.

O'Neill, J., Holland, A. and Light, A. (2008) *Environmental Values*. Routledge, New York.

Ostrom, E. and Nagendra, H. (2006) Insights on linking forests, trees, and people from the air, on the ground, and in the laboratory. *Proceedings of the National Academy of Sciences USA*, **103**, 19221–19223.

Parks Canada Agency (2005) *Monitoring and Reporting Ecological Integrity in Canada's National Parks. Volume I: Guiding Principles*. Ontario Parks Canada Agency, Ottawa, Canada.

Parks Canada and the Canadian Parks Council (2008) Principles and Guidelines for Ecological Restoration in Canada's Protected Natural Areas. http://www.pc.gc.ca/docs/pc/guide/resteco/index_e.asp. Accessed August 2012.

Pyšek, P., Jarošík, V., Hulme, P.E., Pergl, J., Hejda, M., Schaffner, U. and Vilà, M. (2012) A global assessment of invasive plant impacts on resident species, communities and ecosystems: the interaction of impact measures, invading species' traits and environment. *Global Change Biology*, doi: 10.1111/j.1365-2486.2011.02636.x.

Rejmánek, M. and Richardson, D.M. (1996) What attributes make some plant species more invasive? *Ecology*, **77**, 1655–1661.

Roe Smith, M. (1994) Technological determinism in American culture, in *Does Technology Drive History? The Dilemma of Technological Determinism* (M. Roe Smith and L. Marx, eds), MIT Press, Cambridge, Massachusetts, pp. 1–35.

Roe Smith, M. and Marx, L. (1994) *Does Technology Drive History? The Dilemma of Technological Determinism.* MIT Press, Cambridge, Massachusetts.

Rosenzweig, M.L. (2003) *Win-Win Ecology: How the Earth's Species Can Survive in the Midst of Human Enterprise.* Oxford University Press, New York.

Ruiz-Jaen, M.C. and Aide, T.M. (2005) Restoration success: how is it being measured? *Restoration Ecology,* **13**, 569–577.

Schlaepfer, M.A., Sax, D.F. and Olden, J.D. (2011) The potential conservation value of non-native species. *Conservation Biology,* **25**, 428–437.

Simberloff, D., Alexander, J., Allendorf, F. et al. (2011) Non-natives: 141 scientists object. *Nature,* **475**, 36.

Soulé, M. (1990) The onslaught of alien species, and other challenges in the coming decades. *Conservation Biology,* **4**, 233–239.

Standish, R.J., Sparrow, A.D., Williams, P.A. and Hobbs, R.J. (2009) A state-and-transition model for the recovery of abandoned farmland in New Zealand, in *New Models for Ecosystem Dynamics and Restoration* (R.J. Hobbs and K.N. Suding, eds), Island Press, Washington, pp. 189–205.

Thompson, A. and Bendik-Keymer, J. (eds) (2012) *Ethical Adaptation to Climate Change: Human Virtues of the Future.* MIT Press, Cambridge, Massachusetts.

Vilà, M., Espinar, J.L., Hejda, M., Hulme, P.E., Jarošik, V., Maron, J.L., Pergl, J., Schaffner, U., Sun, Y. and Pyšek, P. (2011) Ecological impacts of invasive alien plants: a meta-analysis of their effects on species, communities and ecosystems. *Ecology Letters,* **14**, 702–708.

Wallington, T.J., Hobbs, R.J. and Moore, S.A. (2005) Implications of current ecological thinking for biodiversity conservation: a review of the salient issues. *Ecology and Society,* **10**, 15.

Woodley, S. (1993) Monitoring and measuring ecological integrity in Canadian National Parks, in *Ecosystem Integrity and the Management of Ecosystems* (S.J. Woodley, G. Francis, and J. Kay, eds), St Lucie Press, Boca Raton, Florida, pp. 155–176.

Worner, S.P. and Gevrey, M. (2006) Modelling global insect pest species assemblages to determine risk of invasion. *Journal of Applied Ecology,* **43**, 858–867.

Zweig, C.L. and Kitchens, W.M. (2010) The Semiglades: The collision of restoration, social values, and the ecosystem concept. *Restoration Ecology,* **18**, 138–142.

Chapter 38

NOVEL URBAN ECOSYSTEMS AND ECOSYSTEM SERVICES

Michael P. Perring[1], Pete Manning[2], Richard J. Hobbs[3], Ariel E. Lugo[4], Cristina E. Ramalho[5] and Rachel J. Standish[6]

[1]Ecosystem Restoration and Intervention Ecology (ERIE) Research Group, School of Plant Biology, University of Western Australia, Australia

[2]School of Agriculture, Food and Rural Development, University of Newcastle, UK

[3]Ecosystem Restoration and Intervention Ecology (ERIE) Research Group, School of Plant Biology, University of Western Australia, Australia

[4]International Institute of Tropical Forestry, USDA Forest Service, Rio Piedras, Puerto Rico

[5]Ecosystem Restoration and Intervention Ecology (ERIE) Research Group, School of Plant Biology, University of Western Australia, Australia

[6]Ecosystem Restoration and Intervention Ecology (ERIE) Research Group, School of Plant Biology, University of Western Australia, Australia

38.1 INTRODUCTION

The Earth's burgeoning human population is becoming increasingly urban: since 2009, more than 50% of the global population are urban dwellers residing in landscapes characterized by high population density and/or high coverage of artificial substrate (UN 2011). As a result, urban areas and the corridors that connect them are the landscapes that most people are most familiar with (Pickett et al. 2001). However, the ecological and evolutionary playground of the urban environment has been somewhat ignored by ecologists until relatively recently, particularly in North America (McDonnell and Pickett 1990) although less so in Europe (Sukopp 2008). With growing interest in the ecology of cities, ecologists are becoming increasingly aware of the sometimes large ecological and evolutionary changes caused by (and feeding back upon) the environmental conditions that characterize urban areas (Shochat et al. 2006; King et al. 2011). The changed conditions in urban environments have led to the creation of novel ecosystems (see Chapters 5 and 6), particularly as some urban ecosystems are founded upon artificially created substrates with almost wholly introduced biotic communities. Urban areas have not featured explicitly in earlier incarnations of the novel ecosystem concept (e.g. Hobbs et al. 2006; but see Lugo 2010; Kowarik 2011), but here we address the ecological novelty that pervades the urban environment.

This chapter includes within its scope a wide variety of ecosystems embedded within and on the edge of urban areas (described later). Further, while the distinction is not always made, the discussion covers both hybrid ecosystems that may have significant novelty or novel elements (see Chapter 6) and novel ecosystems where, in practice, it is impossible to return to hybrid or historical conditions. We note that a range of hybrid and novel ecosystems are found throughout myriad urban environments and suggest ways that they can be managed to promote ecosystem services through enhancing ecological functions.

Awareness of the novel ecosystems created by urbanization provides a number of opportunities. Firstly, and more so than in other less-populated landscapes, it forces ecologists to integrate cultural, social and economic factors with more traditional ecological thinking (Pickett et al. 2001; Grimm et al. 2008). Additionally, a novel ecosystems framework provides the opportunity to change perceptions of the urban environment, particularly in terms of how the urban environment may provide ecosystem services. For example, if the public are aware that novel urban ecosystems can provide valuable ecosystem services, then people may be more willing to accept the presence of altered communities and environments.

In this chapter, we first address the facets of novelty that pervade the urban environment before reviewing the ecosystem services that different components of the urban environment provide. In this discussion we describe the known and potential contribution of novel elements (e.g. soils, plants and animals) to these services, with a particular focus on vegetation. The services we review include biodiversity maintenance, carbon storage, flood regulation, recreation, spiritual fulfilment and education. We discuss how ecosystem service provision could be boosted by novel components but add that experiments are required to test our assertions. Research is also needed to investigate trade-offs among the delivery of ecosystem services in urban areas and to investigate whether they can be successfully implemented in urban areas across the globe and in different socio-cultural, economic and ecological contexts.

38.2 ECOLOGICAL SPACES AND DRIVERS OF NOVELTY IN URBAN ENVIRONMENTS

The urban environment is characterized by ecological spaces that vary in their degree of novelty and land-use history. At one end of the scale are ecosystems that are almost entirely novel e.g. detritus-based systems in sewers and drains, while at the other are large patches of remnant native vegetation that contain very few novel components, often found on the edges particularly where urban areas adjoin native vegetation rather than agricultural areas. Intermediate between these spaces are brownfield sites (i.e. abandoned industrial areas, often with legacies of heavy metal pollution and hydrochemicals); abandoned development sites; transportation and transmission corridors; remnants of former wild land, river corridors or agriculture; and more managed urban spaces such as gardens, parks, corporate grounds, green roofs and even cemeteries (le 38.1), covering both terrestrial and aquatic ecosystems (Bolund and Hunhammar 1999).

The origins of these urban novel ecosystems vary – some arise from benign neglect and subsequent colonization by native and non-native species (e.g. abandoned development or demolition sites with rubble and, in some industrial sites, coal slag; Kowarik 2008; Renforth et al. 2011), others from degradation of native vegetation fragments (e.g. river corridors) and still others from deliberate management (e.g. ornamental parks and gardens). In their original conception, novel ecosystems retained their structure and function without additional human intervention (Hobbs et al. 2006). Thus, sites such as parks, gardens, playing fields and corporate grounds would not classically be considered in the novel ecosystems framework due to their explicit management; we include them in the scope of this chapter however because they include novel elements such as non-native species, changed environmental conditions and overlooked ecological interactions (Sukopp 2008; Goddard et al. 2010). The identification of different types and value of ecological spaces in urban environments is not new (see for example Brady et al. 1979; Sukopp 2008; Kowarik 2011), but at the global scale appreciation of their ecological value is arguably only just emerging (Miller and Hobbs 2002).

Elucidating the origin of novel ecosystems in urban landscapes is important for two reasons. The first is to confirm that the target ecosystem is truly novel. It may be possible to restore degraded systems to some 'original' state whereas this is likely impossible for the entirely novel urban ecosystems that emerge from artificial substrates. The second complementary reason is that defining their degree of naturalness or novelty may influence the management decisions made. It is likely that we would take a more interventionist approach to ecosystems at the more novel end of the spectrum (e.g. brownfield sites with crushed rubble substrates), feeling comfortable to manipulate them to deliver a range of ecosystem services. In systems that are regarded as more 'natural' we may decide that minimal intervention is preferable, thus leaving them in a partially managed or 'wild' state such as Richmond Park in London, where semi-natural acid grassland habitats can be found. Improved ecological understanding of novel urban ecosystems would allow such informed decision making. This will likely be attained by considering perspectives from a range of different frameworks in urban ecology which emphasize the social, cultural and landscape contexts in a hierarchical patch dynamic and temporal framework (Grimm et al. 2000; Alberti et al. 2003; Ramalho and Hobbs 2012).

There are many drivers of novelty in urban areas, but novelty pervades the urban environment essentially because it has been constructed around the needs of humans. Many of its physical and chemical properties differ from non-urban areas including: temperature, precipitation, hydrological regimes, nutrients and pollutants (McDonnell and Pickett 1990; Paul and Meyer 2001; Pickett et al. 2001). These properties partly determine ecological patterns such as species distributions and processes such as primary productivity and mineralization. Commonly observed biophysical changes include the increase of minimum temperature relative to surrounding areas (the 'heat island' effect) and the higher rates of precipitation and atmospheric pollutants in and downwind of cities due to higher concentrations of cloud condensation particles (Pickett et al. 2001).

Interestingly, context is important as to the physical effect of urbanization: in Phoenix, USA, urban areas are actually cooler than surrounding desert in summer because of evaporative cooling from the artificially irrigated urban vegetation (Pickett et al. 2001). Such results have led some authors to suggest that there is a convergence of physical-chemical conditions in urban environments compared to non-urban areas (the 'urban convergence' hypothesis; Pouyat et al. 2006). Hydrology is also strongly modified by human activities

in urban areas with typical changes including increased runoff, more frequent storm flow events and reduced groundwater leading to the incising of urban streams and disconnection of streams from groundwater base flow (Paul and Meyer 2001; Pickett et al. 2001). Transport networks, human activity and preferences also drive novelty in the urban environment by introducing species and fragmenting habitat.

Physical-chemical differences cause changes in the fauna and flora of terrestrial and aquatic urban ecosystems by altering their physiology, phenology and behavior, and by altering the direction of evolutionary adaptation. Responses lead to changes at a community level where alterations to community structure and trophic relationships are seen (McDonnell and Pickett 1990; Shochat et al. 2006). Even small increases in the proportional cover of impervious substrates such as tarmac can lead to dramatic changes in aquatic communities, but the mechanisms behind this are currently unknown (King et al. 2011). Physiological adaptation to the novel urban environment (Partecke et al. 2006) may ultimately change organisms' genotype as well as phenotype (Shochat et al. 2006). Species interactions (e.g. predator–prey relationships) can also be altered by facets of the urban environment. For example, ecological light pollution has led to increased predation on juvenile salmonids by harbor seals (Longcore and Rich 2004).

Changes to species richness and community composition are also caused by urbanization, including nonnative planting in gardens and parks (Goddard et al. 2010) and the resource subsidy of certain species (e.g. rats, pigeons), allowing them to expand their realized niche and out-compete other species (Alberti et al. 2003). In addition human activities change plant dispersal in manifold ways, including through garden throw-outs and soil transport, allowing rapid expansion of some species (Hodkinson and Thompson 1997; Kowarik 2011). These changes led some to suggest that not only is there a convergence in physical and chemical conditions, but also convergence in community composition i.e. biotic homogenization (McKinney 2006). Attempts to identify the factors behind this suggested that the traits of successful urban flora were robust plants of relatively fertile, dry, unshaded, base-rich habitats (Thompson and McCarthy 2008). However, such generalizations regarding biota need to be treated with caution given the heterogeneity of certain aspects of the urban environments across the globe.

The changes to urban conditions e.g. through water additions and fertilizer in managed areas can speed up processes such as net primary production. However, changed conditions may also lead to unexpected alterations of ecosystem process. Pickett et al. (2001) noted that commonly sourced litter decomposed faster in urban areas than in rural areas despite the fact that the urban areas had lower litter fungal biomass and macroinvertebrate abundances compared with the rural areas. These lower biomasses and abundances had initially led the authors to expect slower turnover, particularly since urban litter decomposed slower than non-urban litter under common conditions. However, abundant non-native earthworms in their study area compensated for these other factors (Steinberg et al. 1997). Interestingly, in their urban soil, recalcitrant carbon was higher than passive carbon leading to an expectation of lower nitrogen mineralization. The reverse occurred however, which they likely attributed to methanotrophic microbial community composition (Pickett et al. 2001) although this could also be due to the higher earthworm abundances and/or the heat island effect. These results demonstrate the potential for novel attributes of the urban environment to facilitate unexpected interactions and outcomes for ecological pattern and process.

38.3 CONTRIBUTION OF NOVEL URBAN ECOSYSTEMS TO ECOSYSTEM SERVICES

The novel urban ecosystems outlined earlier are not going to cease being novel, partly by definition and, more importantly, because ongoing disturbances and pressures such as species invasion will assault their communities and help maintain the novel state. Some people may regard these spaces as 'wastelands' with little or no ecological value, particularly those who view an ecosystem's ecological value as being defined by the possession of native biodiversity that is uninfluenced by humans (see Miller and Hobbs 2002; Chapters 31 and 37). However, the myriad different urban novel ecosystems do have ecological value, particularly in terms of ecosystem services but also with respect to outcomes for biodiversity conservation.

In this section, we review a variety of ecosystem services provided by novel urban ecosystems from biophysical regulating and supporting services to cultural services (Millennium Ecosystem Assessment 2005).

Figure 38.1 highlights those services that are particularly supplied by urban areas. Where knowledge is available, we also examine the role of novel elements in delivering these services. We also argue that carefully designed ecosystems may provide opportunities to boost service provision with different novel ecosystems varying in their potential to make these gains (Table 38.1). The context of the ecosystem in question will likely guide such design (Chapter 18); for example, maintenance of native biodiversity may be encouraged in large remnant habitats which still retain many 'natural' elements, whereas the incorporation of novel elements such as non-native trees and shrubs or soil amendments may be more appropriate in abandoned industrial sites. Furthermore, ecosystems incorporating such novel components may be more acceptable to the public than a collection of weedy native species that are often allowed to colonize spontaneously and which may give an impression of dilapidation (Rink 2005). However, such acceptance is heavily context dependent, with people's appreciation of spontaneously developed ecosystems varying greatly and growing in some areas, such as eastern Germany (Keil 2005). Indeed, there is much potential for novel ecosystems to contribute to cultural services; novel urban ecosystems can aid in reconnecting people with their natural environment, help educate people as to the kinds of processes that occur in ecosystems and can also engender a sense of place (Chapter 30).

38.3.1 Carbon sequestration and storage

Vegetation throughout the urban matrix in the different urban novel ecosystems (Table 38.1) has the potential to sequester carbon dioxide (CO_2) from the atmosphere, and increase carbon storage both above- and below-ground. Carbon sequestration is currently regarded as an ecosystem service as it slows the accumulation of greenhouse gases in the atmosphere, mitigating global climate change. Retention of vegetation may also prevent further loss of carbon to the atmosphere through preventing land-use change, as well as promoting benefits such as shading and cooling that reduce energy use and thus further lower emissions and mitigate climate change indirectly. The first question to answer is whether urban areas can store carbon and, if so, where is it found?

Recent research has shown that the capacity for urban areas to store substantial amounts of carbon

Table 38.1 The potential of novel components to bolster ecosystem services in example novel urban ecosystems. The top row in each service line gives the relative value of each service in these environments as currently perceived (the more ticks, the greater the value); support for these assertions can be found in the main text where references to each service are provided. The bottom row for each service hypothesizes the ability of novel components to boost service provision (the more plus signs, the more likely novel components are to boost provision); novel components would typically be non-native species but may also be artificial substrate e.g. when looking to bolster rare native populations or for the sequestration of carbon using biochar. Consideration needs to be given to the potential for dis-services to arise from novel components and other management actions (Pataki et al. 2011), while experiments are required to investigate trade-offs among services as well as the potential contribution of novel components. The table highlights the benefits that already exist in 'wild' remnants and highlights the importance of retaining native vegetation where possible.

	Example novel urban ecosystem		'Wild' remnant	Agricultural remnant	Abandoned industrial	Abandoned residential	Gardens/ parks
Ecosystem Service	Carbon sequestration and storage	Current state	✓✓✓	✓✓	✓	✓	✓✓
		Potential beneficial novel component	–	++	+++	+++	++
	Air quality	Current state	✓✓	✓	–	–	✓✓
		Potential beneficial novel component	–	++	+++	+++	–
	Flood regulation and Water quality	Current state	✓✓✓	–	–	✓	✓✓
		Potential beneficial novel component	–	+++	+++	+++	+
	Spiritual/ psychological/ health	Current state	✓✓✓	✓	–	–	✓✓
		Potential beneficial novel component	–	++	+++	+++	+
	Education/ Recreation	Current state	✓✓✓	✓✓	–	–	✓✓
		Potential beneficial novel component	–	++	+++	+++	+
	Biodiversity maintenance	Current state	✓✓✓	✓✓	–	–	✓✓✓
		Potential beneficial novel component	–	Could use unusual aspects of the urban environment to bolster rare populations elsewhere or test out translocation policies			+

both above- and below-ground has been underestimated (Nowak and Crane 2002; Pouyat et al. 2006; Churkina et al. 2010; Davies et al. 2011; Hutyra et al. 2011). Above-ground, an average stand of urban trees in the United States has been shown to store, on a per-unit-area basis, half the carbon of their forest counterparts (Nowak and Crane 2002). Individual urban trees have four times more carbon than individuals in forest stands because of their typically larger trunk diameter and growth differences due to lower planting density (Nowak and Crane 2002). In Churkina et al.'s (2010) recent study, 10% of the US land carbon stocks were ascribed to urban areas, with a carbon density equal to that found in tropical forests. Twenty percent of this

considerable urban total was stored in vegetation, while soil stored the vast majority (64%). Organic carbon may also be stored in built structures that are often overlooked in carbon accounting (Hutyra et al. 2011), with buildings constituting 5% of the organic carbon stored in urban areas (Churkina et al. 2010). By conducting a thorough on-the-ground survey of both public and private urban space in the city of Leicester, Davies et al. (2011) showed that there may be an order-of-magnitude underestimate in above-ground carbon storage when coarse-scale carbon estimates are used. In the tropical city of San Juan, urban woodlands and forests accumulate large densities of trees and regrow quickly after hurricane passages,

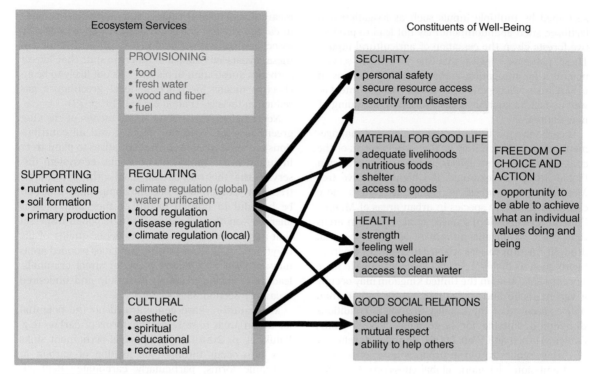

Figure 38.1 The Ecosystem Services framework from the Millennium Ecosystem Assessment (2005) adapted to show particularly important pathways from ecosystem services to human well-being in urban areas (width of arrow indicates hypothesized importance). Grayed text suggests there is little direct impact of urban ecosystems on e.g. food and fresh water provision. The urban ecosystem regulates climate at a local scale, e.g. through the heat island effect. However, it is unlikely to have a large direct bearing on climate regulation at the global scale (Pataki et al. 2011), hence why this is grayed. However, actions to reduce humanity's urban ecological footprint (e.g. by reducing land clearance) would ultimately aid climate regulation at the global scale. Adapted from Millennium Ecosystem Assessment (2005). © 2005 World Resources Institute Island Press, Washington, DC.

suggesting a high capacity for biomass accumulation (Lugo et al. 2011). Such findings question assumptions of limited carbon storage in urban areas, as evidenced in policy documents in the United Kingdom (Davies et al. 2011).

Broad-scale patterns may mask variation among urban areas: in general there is more urban forest cover and greater carbon storage in the urban areas in the northeast of continental North America compared with elsewhere on the continent (Pataki et al. 2006). However, when comparing non-urban and urban areas in the same vicinity, drier and hotter areas tend to store more organic carbon in the urban environment than in surrounding native vegetation, particu-

larly in the soil (Pataki et al. 2006; Pouyat et al. 2006). This is likely a result of the aforementioned urban convergence processes. Management actions (e.g. fertilization, irrigation and species choice) in drier areas result in more productive forest stands that more closely mirror urban stands in cooler, wetter climes compared to the native vegetation that could be found under prevailing, unmanaged conditions (Pouyat et al. 2006). As cities expand over former cultivated lands, the rich formerly agricultural soils coupled to moist climates can also support productive forests and other vegetation on open spaces as has happened in San Juan (Lugo et al. 2011). However, these types of patterns are likely context dependent; where agricultural areas have been sustained by multiple inputs such as irrigation and fertilizer, such abandonment may not lead to productive forests given the cessation of agricultural inputs. These patterns suggest that the ecosystem service potential for mitigating carbon dioxide emissions in urban areas will depend upon the landscape context as well as factors such as species selection, including of non-natives.

If carbon sequestration is the sole focus of management actions, then it may be that novel aspects of the community are best able to boost service provision (Table 38.1). For example, the highly invasive tree *Melaleuca quinquenervia* in Florida sequesters more carbon than native species in urban areas of Miami–Dade and Gainsville (Escobedo et al. 2010). In areas that have significant quantities of made ground, drought-tolerant species adapted to dry, rocky environments such as *Buddleja davidii* (which is native to scree slopes in East Asia) in the United Kingdom may occupy a vacant niche and so sequester more carbon than natives because of their ability to grow in conditions that are unsuitable for large native woody plants (Tallent-Halsell and Watt 2009). However, the efficacy of carbon sequestration in urban areas to offset even local emissions, let alone global emissions, has been called into question (Escobedo et al. 2010; Pataki et al. 2011). In the Florida study, only 1.8–3.4% of local CO_2 emissions were offset by urban tree sequestration; this strategy was therefore considered a moderately effective way of mitigating greenhouse gas emissions compared with other methods such as cleaner and more efficient transportation strategies (Escobedo et al. 2010). Moreover, modeling the effects of putting additional trees on to currently non-forested but potentially available areas did not show significantly improved mitigation potential (Escobedo et al. 2010).

In addition to direct carbon storage and sequestration benefits, there may be indirect benefits of urban vegetation in greenhouse gas reductions. Modeling exercises and empirical investigations suggest that shading effects of trees, and their amelioration of climate/weather extremes, can lead to lower energy usage in surrounding buildings, therefore lowering carbon emissions associated with power generation (Nowak and Crane 2002; Donovan and Butry 2009), although Escobedo et al.'s (2010) study showed only minimal contributions of energy avoidance to overall emissions offsets in subtropical United States. Additionally, the management costs associated with urban trees and their continued maintenance may

mean that some of the direct and indirect gains made in carbon storage and sequestration may be offset by associated emissions (Pataki et al. 2006). Ultimately, these caveats led Pataki and others to state that "direct carbon sequestration in plants . . . is not likely to be an effective means for reaching local greenhouse gas reduction targets" (Pataki et al. 2011; p. 29).

Nevertheless, there is no single means of meeting greenhouse gas reduction targets, and all contributions are to be welcomed. Any initiatives to plant trees in urban areas should also consider ecosystem disservices that they may be associated with. For instance, trees may emit volatile organic compounds that can be harmful to human health or they may possess allergens on the foliage or through pollen dispersal that may also decrease urban residents' quality of life (Lyytimaki et al. 2008). In certain cities wooded areas may also act as a hiding place for street criminals, leading to their perception as unsafe and undesired areas (see Section 38.3.4).

Most authors have concentrated on the potential for urban areas to store organic forms of carbon (e.g. Pataki et al. 2006). However, semi-permanent sinks may also occur through the fixation of carbon in inorganic forms, particularly carbonate (Renforth et al. 2009), as well as more recalcitrant forms of organic carbon (e.g. biochar; Renforth et al. 2011). Investigations in the northeast of England show that more than three times as much carbon can be stored inorganically in urban soils as compared to that found in organic material. This research concentrated on calcium-rich demolition wastes and basic slag in brownfield sites and showed that 97.3% of this carbon was ultimately sequestered from the atmosphere (Renforth et al. 2009). Extrapolating from the figures found in the northeast, the authors suggested that the equivalent of 90% of the emissions associated with cement manufacture could be offset given the likely global distribution of construction and industrial waste, with the potential of other materials (e.g. gypsum) to further contribute to the sequestration of carbon (Renforth et al. 2009).

Such mitigation potential may be further enhanced by management actions and the addition of novel elements to the community (Table 38.1; Renforth et al. 2011). Just as certain species are planted to remediate chemical contamination in old industrial sites because of their traits (e.g. hyperaccumulators and hypertolerators; Kramer 2010), the artificial ground that is characteristic of these sites may also sequester carbon more quickly with an informed selection of species. Non-native species such as *B. davidii* may be more able to grow on these artificial substrates compared to native species, and this ecosystem service could be optimized where their growth is encouraged. Ultimately, these novel components to the flora may be replaced by species more characteristic of any reference community such as has occurred in old fields in Puerto Rico (Lugo 2004), but such assertions need testing experimentally.

38.3.2 Flood mitigation, water quality and stream community recovery

The impervious surface cover of the urban environment leads to highly altered stream hydrodynamics (Paul and Meyer 2001; Pickett et al. 2001) with rapid transfer of water through drainage channels to stream systems, often leading to increased flood severity. For instance, urban landscapes with 50–90% impervious cover can lose 40–83% of rainfall to surface runoff whereas similar rainfall events would only lose 13% in forested ecosystems (Pataki et al. 2011). These and other physical-chemical changes (e.g. pollutants and increased temperature) lead to an 'urban stream' syndrome with concomitant changes in biota (King et al. 2011). Certain urban novel ecosystems (e.g. river corridors, bioretention swales) commonly provide a means of mitigating these effects, providing services such as flood mitigation, improved water quality and pollution reduction from both point and non-point sources as well as recreational value and biodiversity benefits (Ehrenfeld 2000; DeBusk and Wynn 2011; Kazemi et al. 2011; Table 38.1). Vegetated areas intercept precipitation and slow its transfer to drainage channels. Increasing pervious cover could allow infiltration and prevent the rapid surface flow that characterizes impervious areas, as well as mitigating non-point source pollution (Felson and Pickett 2005). Projects to remove pollutants from streams need to bear in mind ecosystem dis-services. For example, trade-offs may exist between nitrate removal and the release of nitrous oxide as a greenhouse gas (Pataki et al. 2011). Pataki et al.'s (2011) recent review suggested that there were actually few studies that have shown water quality changes in urban areas following revegetation. However, studies do show the benefits of biofiltration/bioretention systems (e.g. Read et al. 2008; DeBusk and Wynn 2011).

Novel components may be able to boost multiple services, and not just water quality alone. In New York, Philadelphia and Baltimore (eastern US) floating wetlands have been introduced into river systems. These floating wetlands, created from rafting together plastic bottles that would otherwise litter the shoreline and waterway, provide multiple benefits besides water quality improvements. These benefits include habitat for crabs and other organisms, and cultural services (K. Bowers, pers. comm. 2012; see also http://www.aquabiofilter.com/). Local youth groups participate in the wetlands' creation and thus feel empowered. This latter aspect can be very important to ensuring the project's success and suggests that, when planning ecosystem recovery, techniques that include the community can be usefully considered and included in some cases, as opposed to relying on technical engineered solutions implemented by experts (K. Bowers, pers. comm. 2012).

As with other ecosystem services, the species composition that may best provide water quality and flood regulation services (among others) remains unknown, with research showing large differences among species in their ability to remove pollutants (Read et al. 2008). Importantly, no single species was good at removing all pollutants, suggesting the requirement to have mixed species plantings (Read et al. 2008). It may be that non-native species can provide desired levels of ecosystem functioning and that, in some instances involving artificial substrates and excessive pollution, they may function better than native species. We emphasize that non-natives should not be used without proper consideration of consequences, but there are certain environmental contexts where their use could be considered and not dismissed out of hand because of their origin (see also Chapter 26). More broadly, the protection of native biodiversity could be considered a priority and there is presently not enough evidence to evaluate whether utilizing non-natives in some urban areas harms the protection of such diversity. At a more fundamental level, future economic development pathways could consider how to support native biodiversity rather than degrading habitat in the first place i.e. tackling the causes not the symptoms.

38.3.3 Air quality and noise reduction

Many authors have noted the ability of vegetation in urban areas to improve air quality and therefore human health, with pollution filtration capacity increasing with greater leaf area (Bolund and Hunhammar 1999). However, recent work has cast doubt on the magnitude of this supposed benefit (Pataki et al. 2011), and proposed that there is a limited potential for urban vegetation to improve air quality (Table 38.1). For example, Pataki et al. (2011) suggest that the net effect of planting 1 million trees in New York City would be to add 4.05 hours to the lives of New York City residents. However, there may be indirect benefits of tree cover on air pollution, e.g. through the cooling effect of transpiring vegetation leading to lower ozone levels (Pataki et al. 2011). Vegetation in urban areas can substantially decrease ambient noise levels; for example, in Stockholm, soft lawns decreased noise by as much as 3 dB (A) (Bolund and Hunhammar 1999).

As with carbon storage, the species composition that is best able to provide these benefits remains unknown. It is possible that non-native plant species could improve air quality by being better at absorbing particulates than native forms, as well as being better at reducing noise levels. In some cities this may involve using species with higher leaf areas such as coniferous plants; these can be more sensitive to air pollution however, suggesting a mixture of trees would be better (Bolund and Hunhammar 1999). Again, these assertions require further investigation and experimentation, although mixtures would likely be better for reasons of biodiversity augmentation and maintenance (see Section 38.3.5). Again, such plantings need to be mindful of the requirement to avoid ecosystem dis-services. For example, volatile organic carbons can be emitted by certain species (e.g. *Pinus* species) which may be harmful to human health or increase the risk of fire (Barboni et al. 2011).

38.3.4 Psychological, spiritual and health benefits and educational and recreational opportunities

Novel urban ecosystems may provide a variety of cultural benefits, including psychological and spiritual benefit, educational and recreational opportunities and some direct health benefits. Vegetation in novel urban ecosystems, whether grass or trees, has been related to an improved sense of safety and is preferred by urban dwellers to a non-vegetated state (Kuo et al. 1998). Increased preference and sense of safety occurred regardless of tree density or tree placement.

The results were however based on a survey of people rating different computer generated images, so would require real-world verification. It may also be that such results are context dependent; other work has suggested that vegetation that decreases visibility is not preferred (Nasar and Fisher 1993), with 'semi-natural' areas being associated with a fear of crime (Lyytimaki et al. 2008). Sometimes there is a contradiction in that some people would prefer to have green spaces as their surroundings, yet these same spaces also hold the most fear of crime (Jorgensen et al. 2005). Kuo et al.'s work relates to carefully positioned trees with lower limbs removed, as is typical of urban tree management and maintenance that don't compromise on visibility (Kuo et al. 1998).

Later work by the same research group also found strong evidence for the presence of vegetation being related to decreased property and violent crime in otherwise similar housing blocks (Kuo and Sullivan 2001). The decreased incidence of violent crime with greater vegetative cover was hypothesized to be related to a less mentally fatigued population with a consequent lower likelihood of resorting to violence (Kuo and Sullivan 2001). The decreased incidence of property crime was related to likely increases in the actual and implied surveillance of the area, with a greater use of more vegetated areas compared to more bare areas by the resident population (Kuo and Sullivan 2001).

In addition to work showing the benefits of having vegetation versus not having such areas, recent research in the United Kingdom suggests that people derive greater psychological benefit from more diverse vegetated areas (Fuller et al. 2007). The general public show a good ability to comprehend this diversity, particularly of sessile organisms (Fuller et al. 2007). Interestingly, different components of biodiversity affected different measures of psychological well-being; plant variety tended to be associated with ability to reflect whereas bird variety was associated with participants' emotional attachments. More broadly, the number of different habitats in an area tended to correlate with reflection, with a greater diversity of habitats reinforcing a sense of personal identity (Fuller et al. 2007).

Often urban dwellers may not be aware of the origins of the flora and fauna that they encounter, thus suggesting that the introduction of novel components in urban areas would be relatively uncontroversial and could enhance people's well-being. In some cases this may be through the filling of a vacant niche (as for the

B. davidii example) and as exemplified by non-native plane trees (*Plantanus x acerifolia*) in central London. The latter are some of the only trees able to thrive in an environment with high particulate pollution (Howard 1943) and root compaction; an environment which may be less amenable to woody natives. The presence of plane trees greens this urban space and provides a cultural service among other benefits. Displacement of natives by introduced species may occur in other cases. People may however still derive pleasure from the introduced species, such as grey squirrels in the United Kingdom, while not appreciating other natives (e.g. rats in China). Clearly more research is required to substantiate these findings to different ecological and cultural contexts. Together with research that has shown the psychological benefits of green space of any sort (Kuo et al. 1998; Fuller et al. 2007), this suggests that urban novel ecosystems have the potential to provide emotional solace as well as more physically beneficial manifestations.

In addition to the psychological benefits provided by urban ecosystems, novel or otherwise, they also provide areas for recreation and education (which in turn can bring psychological benefit). Brownfield sites that are generally perceived as 'wasteland' may be used as a playground by children of all ages (Keil 2005). There is even evidence that children play more creatively in areas with vegetation, both trees and other plant materials (Taylor et al. 1998), although in some cases urban wasteland areas may be better managed as an actual playground (Standish et al. 2012). With increasing concern over the 'extinction of experience' (Miller 2005) and given the disconnection of the urban population from 'nature' (Sutherland et al. 2006), urban novel ecosystems also provide an opportunity to educate people about ecological processes (Dearborn and Kark 2010). For example, school groups visit urban ponds in derelict quarries in the UK to learn about food web dynamics and the life cycles of species such as frogs (Braund and Reiss 2004). Such activities can also reinforce the importance of biodiversity underlying ecosystem services and hence our lifestyle; see however Lyytimaki et al. 2008 for an alternative view of the causative chain that our lifestyle dictates what ecosystem services we require and therefore what elements of biodiversity to manage.

Finally, having urban ecosystems may also provide direct health benefits where biodiversity is maintained. For example, Kremen and Ostfeld (2005) showed that Lyme disease risk in the United States decreases with higher diversity vertebrate communities because there is the potential for dilution of disease transmission with a greater range of hosts for the ticks that transmit the disease to humans. Empirical studies confirmed this, with a higher proportion of ticks infected and greater number of infected ticks in smaller fragments where there is also a lower diversity of hosts (Kremen and Ostfeld 2005). Unfortunately, these small fragments are most representative of those which occur in suburban areas of the US and elsewhere. Future urban design could consider a range of remnant sizes for suburban areas. As with the earlier examples, the potential contribution of non-native species is unknown.

38.3.5 Biodiversity restoration and maintenance

Urban areas can be reservoirs of surprisingly high biodiversity (McKinney 2006; Kowarik 2011), with patterns being driven by the socio-economic as well as biophysical context (see McKinney 2006 and Kowarik 2011 for reviews). The maintenance of biodiversity is sometimes regarded as an ecosystem service in its own right (e.g. genetic resources are regarded as a provisioning service), although more commonly biodiversity is regarded as the foundations upon which ecosystem services rely (Millennium Ecosystem Assessment 2005; Mace et al. 2012). Remnant areas, particularly those recently formed and of a larger size, provide a good opportunity to conserve native species populations (McKinney 2006; Ramalho and Hobbs 2012). Managed gardens have the potential to contribute to conservation goals when viewed in a landscape context (Goddard et al. 2010). Moreover, urban woodlands and forests, which range from corridors and patches (both natural and artificial), include in their species complements native, endemic, naturalized as well as introduced species of plants and animals in many novel and traditional combinations (Lugo et al. 2011). Urban streams in San Juan, a tropical city in Puerto Rico, also contain a rich and novel combination of native and introduced aquatic fauna and flora (Lugo et al. 2011). This biotic richness of urban forests and streams has conservation value, as well as providing other ecosystem services as reviewed earlier.

The artificial substrates in urban areas may provide analogues to native systems; such areas consequently harbor an opportunity to bolster populations of threatened species where habitats are under threat, even if they are not native to the area itself. Buildings can substitute as cliff roosting sites, and the lime-rich soils that result from concrete waste can act as habitats for calcareous plants of limestone grasslands (Lundholm 2006). An additional example, which also highlights the difficulties of management decisions, comes from the industrial salt flats and levees of San Francisco Bay. These were artificially constructed out of tidal wetlands but now harbor native species (e.g. Western snowy plover) that are otherwise threatened in their native habitat further south in California, United States (K. Bowers, pers. comm. 2012). Not only do the salt flats provide food in the form of brine shrimps, the non-vegetated levees around them provide ideal breeding grounds (http://www.southbayrestoration.org/). These habitat features have also led to the formation of breeding colonies of American avocets where previously they were winter visitors (Demers et al. 2010). Interestingly, many of these levees and salt pans are now being restored back to tidal wetland. This restoration is potentially threatening breeding populations of desired birds (e.g. Western sandpipers) and thus requires careful management consideration (Warnock and Takekawa 1995). The urban heat island effect and other physical-chemical changes may also allow the testing of ideas surrounding species translocations in an era of environmental change (Standish et al. 2012).

38.4 ECOSYSTEM DIS-SERVICES AND TRADE-OFFS IN NOVEL URBAN ECOSYSTEMS

Although there are many ecosystem services provided by the urban environment, with environments providing many services at the same time (i.e. stacked benefits) which in turn may be boosted by the presence of novel components in the flora and fauna, ecosystem dis-services may also be encountered in the urban environment (Pataki et al. 2011). Our review highlighted some of these potential dis-services such as the presence of volatile organic carbons compromising air quality, or certain species endangering human health through being irritants or toxic. Indeed, a single species may be regarded simultaneously as beneficial and a problem depending upon who is viewing it, where and when. For instance, Lyytimaki et al. (2008) highlight lime trees in Helsinki which were regarded favorably by pedestrians but annoyed others by blocking views,

particularly of traffic. Furthermore, the pedestrians who at one time regarded them favorably later regarded them as a dis-service when an aphid outbreak led to the formation of butyric acid on pavements and a subsequent odor of vomit (Lyytimaki et al. 2008). Also of importance is that ecosystem dis-services may be experienced on a conscious level while the benefits derived from biodiversity may be experienced unconsciously (Lyytimaki et al. 2008). This may be important in how people view managed landscapes in urban environments. It is critical that when assessing urban landscapes both services and dis-services are adequately considered (Pataki et al. 2011).

Trade-offs among services also need to be elucidated. As Escobedo et al. (2010) pointed out in their study, a single invasive tree species (*Melaleuca quinquenervia*) could be regarded as best providing carbon storage in Miami–Dade rather than current vegetation in the urban area. However, such a focus would inevitably compromise the maintenance of native biodiversity if the current vegetation was replaced by the invasive, and may also preclude the psychological benefits of more diverse vegetation. Calculating trade-offs will be aided by comparing measurable quantities that operate at the same scale, particularly for ease of use among policy makers (Wallace 2007). The urban environment provides an opportunity to experimentally test for these trade-offs among different services, both physical and cultural. This will require collaboration between natural and social scientists, including projects that gauge the public perceptions of different community compositions.

Perhaps the largest trade-off in urban environments is that structures associated with shelter and jobs prevent the establishment of the vegetated novel urban ecosystems outlined here. Humanity needs to reconcile competing demands for land, which should ultimately lead to benefits to both humans and other species (Rosenzweig 2003). At the same time, trade-offs will be encountered with regard to where to concentrate management effort. Carefully stated goals and realistic objectives will aid in the focusing of management effort, with the different novel ecosystems likely having alternative optimum strategies (see Box 38.1). Although dis-services do need to be borne in mind, there are sometimes opportunities for win-win scenarios both among services and between goals, such as economic prosperity and security for the family and the restoration and maintenance of biodiversity (Rosenzweig 2003).

Box 38.1 The management framework applied to novel urban ecosystems

The novelty in urban ecological systems is encapsulated by considering the decision tree framework presented in this book (Chapter 18). This framework asks several questions to consider whether a system is novel or not, and suggests management options when a novel ecosystem is present. We show the utility of this framework for the case of an abandoned industrial site used for over a century in the northeast of England, United Kingdom. The case is hypothetical (although based on personal observations) but it illustrates the application of the framework (Fig. 18.1), in particular by highlighting the freedom provided to managers to consider alternative management pathways that embrace the opportunities of novel urban ecosystems.

Question: What type of ecosystem is it?

(1) What is the problem?

Firstly, we have to determine if there is a problem, based on identified goals, that requires intervention. This decision can be informed through an ecosystem assessment to consider the sources of change and to what extent components differ significantly from those that prevailed historically. Recognizing change and determining if such change is undesirable brings in questions of reference conditions (Chapter 18). What is the ecosystem reference for an area with 'made ground', heavy metal contamination and altered hydrology? There may be surrounding ecosystems, likely a mix of natives and non-natives which could serve as a reference site. Such areas are likely heavily modified by human influence in and of themselves, given the situation of a long history of human influence in the northeast of England, although their substrate and vegetation communities will be something of an analog system to refer to. However, the manager may also want to ask whether surrounding systems are the desired target, given the physical and chemical conditions that characterize abandoned industrial sites.

Another question to consider is whether the system will passively recover. The answer depends upon the management goal or target. If the management target is for recovery of native vegetation and associated functions, then the answer will be

no (the changes to abiotic conditions and the distance to seed sources at the very least, given the urban sprawl of the northeast of England, will prevent spontaneous recovery). A target that takes into account the changed abiotic conditions and only desires vegetative cover (say) may allow passive recovery. On balance, the answer to this question will be 'no' and intervention will be required to achieve goals.

(2) Are ecosystem changes reversible via active management?

Again, this question depends upon the management target. In urban areas in general, there are likely to be significant ecological and social barriers to recovery of some surrounding native vegetation, even with active management.

Ecological barriers (see Fig. 18.2) in this particular case include the legacies of past land use (particularly invasive species, loss of soil and pollution), colonization barriers (urban development and fragmentation, loss of a seed bank due to the long-term nature of industry on the site) as well as ongoing shifts in species distribution through climate change and nutrient effects. Even if the target were based on some other reference system (e.g. rare calcareous habitat that the rubble of the industrial site provides an analog to), these ecological barriers are still likely to exist.

Social barriers include the complexity of ownership and multiple stakeholders in urban systems. For example, part of the former industrial site may be earmarked for development while other areas may pass into public ownership. Surrounding land owners will have opinions on the desired development pathway. There will also be significant funding barriers to implementing active management, particularly without clear statement of the management target. Again, on balance, the answer to this question will be 'no'.

Taking the answers from the earlier questions leads us to state that there are both ecological and social barriers to restoring the abandoned industrial site and that ecosystem changes are likely irreversible in practice. This leads us to conclude that it is, and will continue to be, a novel ecosystem. This

Continued

Box 38.1 *Continued*

leads us to consider additional questions about how to manage this novel ecosystem (see Figs 18.1 and 18.3).

(3) What is my management goal?
Once managers are cognizant that restoring to some historical reference is virtually impossible on such an industrial site where social and ecological barriers make ecosystem changes irreversible in practice, more opportunities open up in terms of how to manage the novel ecosystem (Fig. 18.3). In a situation such as this, at least two management goals present themselves: first, to manage for a particular habitat and second, to manage for ecosystem services in a changing environment. Surrounding stakeholders could be made aware of these options and the benefits and costs that are presented by such management pathways. This would then allow a considered management approach to be applied.

One option for management action is using the abiotic conditions to provide habitat for target species and biodiversity. In the UK for instance, calcareous grassland is a rare habitat that is subject to multiple threats, including expansion of intensive agriculture and cessation of traditional management that maintains the grassland (WallisDeVries et al. 2002; Woodcock et al. 2005). The rubble on the abandoned industrial site may actually provide an analog of the abiotic calcareous substrate, and planting native calcareous grassland species could provide an opportunity to bolster these threatened species. In addition, calcareous grassland is usually found further south in England, and providing habitat in the northeast takes account of likely future climate change.

The second option would be to introduce species to bolster ecosystem services. Managers could consult widely and consider which services are most required given the context. For example, if the surroundings of the industrial site were mainly residential, planting could consider amenity value and maintaining residents' sense of safety. Thus, dense plantings of thick shrubs would be discouraged, while groves of trees and grassed areas may be encouraged. Managers could take advantage of the heterogeneity of such abandoned industrial sites, installing water features that could also promote valued native biodiversity; the grassed areas could be planted with native species mixtures rather than monocultures of either non-natives or natives such as *Lolium perenne*. Introductions need not only be plants; bolstering of native fauna could also be considered (e.g. great crested newts). If industry was the predominant land use surrounding the abandoned site, restoration for ecosystem services could consider planting species for sequestering carbon and mitigating costs of surrounding industry through enhanced shading. Managers could also consider planting hyperaccumulators to prevent groundwater contamination. If the potential consequences of each option are unknown, then it would be beneficial to consider adaptive management (see Chapter 18) when carrying out such projects. Alternative management strategies could be implemented and assessed e.g. planting with non-natives that seem adapted to the abandoned industrial conditions (such as *Buddleja davidii*) versus using natives that may or may not need amelioration of abiotic conditions to survive (e.g. *Betula pendula*).

As more and more heavy industrial sites in developed regions of the world become abandoned, more and more reference novel ecosystems are becoming available to inform management. A manager at a site such as this could consider available evidence from other projects (e.g. Kowarik 2005) in considering how best to manage the site. The goals and proposed management actions explored in detail earlier then lead to explicit consideration of the costs and risks of intervention (Fig. 18.3).

(4) Is cost/risk acceptable?
Both of the options outlined earlier carry limited cost and acceptable risk and could therefore proceed. However, it is important to remember this question when implementing any management strategy. If the risks and costs are unacceptable (e.g. planting dense shrubs could compromise public safety beyond a socially acceptable threshold) then options should be reconsidered before implementation. If previously unidentified risks and costs are discovered after implementation, then these outcomes may become undesirable. Any outcomes have the potential to inform similar projects in urban landscapes elsewhere.

38.5 CONCLUSIONS

Despite urban areas constituting only about 4% of the terrestrial land surface (in Davies et al. 2011), their high human population density and its associated demands for food, shelter and economic development has led to and will continue to lead to the formation of novel ecosystems across the globe, from the farthest reaches of the ocean to the most isolated terrestrial areas (Chapter 8). The characterization, spread and future management of these ecosystems is the focus of much of the book, and has been debated in conservation and restoration circles. However, the urban environment that has indirectly created many of the world's novel ecosystems has been overlooked in the novel ecosystems framework (Hobbs et al. 2006, 2009) until recently (Kowarik 2011; Standish et al. 2012). This chapter further addresses this gap and notes that the vast array of ecosystems within the urban environment constitute a novel stage for ecological pattern and process to play out. Rather than bemoaning the loss of environments without human influence, there are multiple benefits to accepting that novel urban environments are here to stay, particularly if ecologists, urban planners, policy makers and conservation managers carefully plan and implement effective management of both hybrid and novel ecosystems in order to provide opportunities for ecosystem service delivery.

ACKNOWLEDGEMENTS

We thank Ingo Kowarik and Keith Bowers for helpful comments on an earlier version of this manuscript.

REFERENCES

Alberti, M., Marzluff, J.M., Shulenberger, E., Bradley, G., Ryan, C. and Zumbrunnen, C. (2003) Integrating humans into ecology: Opportunities and challenges for studying urban ecosystems. *BioScience*, **53**, 1169–1179.

Barboni, T., Cannac, M., Leoni, E. and Chiaramonti, N. (2011) Emission of biogenic volatile organic compounds involved in eruptive fire: implications for the safety of firefighters. *International Journal of Wildland Fire*, **20**, 152–161.

Bolund, P. and Hunhammar, S. (1999) Ecosystem services in urban areas. *Ecological Economics*, **29**, 293–301.

Brady, R.F., Tobias, T., Eagles, P.F.J., Ohrner, R., Micak, J., Veale, B. and Dorney, R.S. (1979) A typology for the urban eco-

system and its relationship to larger biogeographical landscape units. *Urban Ecology*, **4**, 11–28.

Braund, M. and Reiss, M.J. (2004) *Learning Science Outside the Classroom*. Psychology Press, London.

Churkina, G., Brown, D.G. and Keoleian, G. (2010) Carbon stored in human settlements: the conterminous United States. *Global Change Biology*, **16**, 135–143.

Davies, Z.G., Edmondson, J.L., Heinemeyer, A., Leake, J.R. and Gaston, K.J. (2011) Mapping an urban ecosystem service: quantifying above-ground carbon storage at a city-wide scale. *Journal of Applied Ecology*, **48**, 1125–1134.

Dearborn, D.C. and Kark, S. (2010) Motivations for conserving urban biodiversity. *Conservation Biology*, **24**, 432–440.

DeBusk, K.M. and Wynn, T.M. (2011) Storm-water bioretention for runoff quality and quantity mitigation. *Journal of Environmental Engineering*, **137**, 800–808.

Demers, S.A., Takekawa, J.Y., Ackerman, J.T., Warnock, N. and Athearn, N.D. (2010) Space use and habitat selection of migrant and resident American Avocets in San Francisco Bay. *The Condor*, **112**, 511–520.

Donovan, G.H. and Butry, D.T. (2009) The value of shade: Estimating the effect of urban trees on summertime electricity use. *Energy and Buildings*, **41**, 662–668.

Ehrenfeld, J.G. (2000) Evaluating wetlands within an urban context. *Ecological Engineering*, **15**, 253–265.

Escobedo, F., Varela, S., Zhao, M., Wagner, J.E. and Zipperer, W. (2010) Analyzing the efficacy of subtropical urban forests in offsetting carbon emissions from cities. *Environmental Science and Policy*, **13**, 362–372.

Felson, A.J. and Pickett, S.T.A. (2005) Designed experiments: new approaches to studying urban ecosystems. *Frontiers in Ecology and Environment*, **3**, 549–556.

Fuller, R.A., Irvine, K.N., Devine-Wright, P., Warren, P.H. and Gaston, K.J. (2007) Psychological benefits of greenspace increase with biodiversity. *Biology Letters*, **3**, 390–394.

Goddard, M.A., Dougill, A.J. and Benton, T.G. (2010) Scaling up from gardens: biodiversity conservation in urban environments. *Trends in Ecology and Evolution*, **25**, 90–98.

Grimm, N.B., Grove, J.M., Pickett, S.T.A. and Redman, C.L. (2000) Integrated approaches to long-term studies of urban ecological systems. *BioScience*, **50**, 571–584.

Grimm, N.B., Faeth, S.H., Golubiewski, N.E., Redman, C.L., Wu, J., Bai, X. and Briggs, J.M. (2008) Global change and the ecology of cities. *Science*, **319**, 756–760.

Hobbs, R.J., Arico, S., Aronson, J. et al. (2006) Novel ecosystems: theoretical and management aspects of the new ecological world order. *Global Ecology and Biogeography*, **15**, 1–7.

Hobbs, R.J., Higgs, E. and Harris, J.A. (2009) Novel ecosystems: implications for conservation and restoration. *Trends in Ecology and Evolution*, **24**, 599–605.

Hodkinson, D.J. and Thompson, K. (1997) Plant dispersal: the role of man. *Journal of Applied Ecology*, **34**, 1484–1496.

Howard, A.L. (1943) The plane tree. *Nature*, **152**, 421–422.

Hutyra, L.R., Yoon, B. and Alberti, M. (2011) Terrestrial carbon stocks across a gradient of urbanization: a study of the Seattle, WA region. *Global Change Biology*, **17**, 783–797.

Jorgensen, A., Hitchmount, J. and Dunnett, N. (2005) Living in the urban wildwoods: A case study of Birchwood, Warrington New Town, UK in *Wild Urban Woodlands: New Perspectives on Urban Forestry* (I. Kowarik and S. Korner, eds), Springer-Verlag, Berlin, pp. 95–116.

Kazemi, F., Beecham, S. and Gibbs, J. (2011) Streetscape biodiversity and the role of bioretention swales in an Australian urban environment. *Landscape and Urban Planning*, **101**, 139–148.

Keil, A. (2005) Use and perception of post-industrial urban landscapes in the Ruhr, in *Wild Urban Woodlands: New Perspectives on Urban Forestry* (I. Kowarik and S. Korner, eds), Springer-Verlag, Berlin, pp. 117–130.

King, R.S., Baker, M.E., Kazyak, P.F. and Weller, D.E. (2011) How novel is too novel? Stream community thresholds at exceptionally low levels of catchment urbanization. *Ecological Applications*, **21**, 1659–1678.

Kowarik, I. (2005) Wild urban woodlands: towards a conceptual framework, in *Wild urban woodlands: New perspectives for urban forestry*. (I. Kowarik and S. Korner, eds), Springer-Verlag, Berlin, pp. 1–32.

Kowarik, I. (2008) On the role of alien species in urban flora and vegetation, in *Urban Ecology: An International Perspective on the Interaction Between Humans and Nature*. (J.M. Marzluff, E. Shulenberger, W. Endlicher, M. Alberti, G. Bradley, C. Ryan, U. Simon and C. ZumBrunnen, eds), Springer Science + Business Media, New York, pp. 321–338.

Kowarik, I. (2011) Novel urban ecosystems, biodiversity, and conservation. *Environmental Pollution*, **159**, 1974–1983.

Kramer, U. (2010) Metal hyperaccumulation in plants. *Annual Review of Plant Biology*, **61**, 517–534.

Kremen, C. and Ostfeld, R.S. (2005) A call to ecologists: measuring, analyzing, and managing ecosystem services. *Frontiers in Ecology and Environment*, **3**, 540–548.

Kuo, F.E. and Sullivan, W.C. (2001) Environment and crime in the inner city: does vegetation reduce crime? *Environment and Behavior*, **33**, 343–367.

Kuo, F.E., Bacaicoa, M. and Sullivan, W.C. (1998) Transforming inner-city landscapes: trees, sense of safety, and preference. *Environment and Behavior*, **30**, 28–59.

Longcore, T. and Rich, C. (2004) Ecological light pollution. *Frontiers in Ecology and Environment*, **2**, 191–198.

Lugo, A.E. (2004) The outcome of alien tree invasions in Puerto Rico. *Frontiers in Ecology and Environment*, **2**, 265–273.

Lugo, A.E. (2010) Let's not forget the biodiversity of the cities. *Biotropica*, **42**, 576–577.

Lugo, A.E., Ramos Gonzalez, O.M. and Rodriguez Pedraza, C. (2011) The Rio Piedras Watershed and its surrounding environment. USDA Forest Service, Washington DC.

Lundholm, J.T. (2006) How novel are urban ecosystems? *Trends in Ecology and Evolution*, **21**, 659–660.

Lyytimaki, J., Petersen, L.K., Normander, B. and Bezak, P. (2008) Nature as a nuisance? Ecosystem services and disservices to urban lifestyle. *Environmental Sciences*, **5**, 161–172.

Mace, G.M., Norris, K. and Fitter, A.H. (2012) Biodiversity and ecosystem services: a multilayered relationship. *Trends in Ecology and Evolution*, **27**, 19–26.

McDonnell, M.J. and Pickett, S.T.A. (1990) Ecosystem structure and function along urban-rural gradients: an unexploited opportunity for ecology. *Ecology*, **71**, 1232–1237.

McKinney, M.L. (2006) Urbanization as a major cause of biotic homogenization. *Biological Conservation*, **127**, 247–260.

Millennium Ecosystem Assessment (2005) *Ecosystems and Human Well-being: Biodiversity Synthesis*. World Resources Institute, Washington DC.

Miller, J.R. (2005) Biodiversity conservation and the extinction of experience. *Trends in Ecology and Evolution*, **20**, 430–434.

Miller, J.R. and Hobbs, R.J. (2002) Conservation where people live and work. *Conservation Biology*, **16**, 330–337.

Nasar, J. and Fisher, B. (1993) 'Hot spots' of fear and crime: A multi-method investigation. *Journal of Environmental Pyschology*, **13**, 187–206.

Nowak, D.J. and Crane, D.E. (2002) Carbon storage and sequestration by urban trees in the USA. *Environmental Pollution*, **116**, 381–389.

Partecke, J., Schwabl, I. and Gwinner, E. (2006) Stress and the city: urbanization and its effects on the stress physiology in European blackbirds. *Ecology*, **87**, 1945–1952.

Pataki, D.E., Alig, R.J., Fung, A.S., Golubiewski, N.E., Kennedy, C.A., McPherson, E.G., Nowak, D.J., Pouyat, R.V. and Romero Lankao, P. (2006) Urban ecosystems and the North American carbon cycle. *Global Change Biology*, **12**, 1–11.

Pataki, D.E., Carreiro, M.M., Cherrier, J., Grulke, N.E., Jennings, V., Pincetl, S., Pouyat, R.V., Whitlow, T.H. and Zipperer, W.C. (2011) Coupling biogeochemical cycles in urban environments: ecosystem services, green solutions and misconceptions. *Frontiers in Ecology and Environment*, **9**, 27–36.

Paul, M.J. and Meyer, J.L. (2001) Streams in the urban landscape. *Annual Review of Ecology and Systematics*, **32**, 333–365.

Pickett, S.T.A., Cadenasso, M.L., Grove, J.M., Nilon, C.H., Pouyat, R.V., Zipperer, W.C. and Costanza, R. (2001) Urban ecological systems: linking terrestrial ecological, physical, and socioeconomic components of metropolitan areas. *Annual Review of Ecology and Systematics*, **32**, 127–157.

Pouyat, R.V., Yesilonis, I.D. and Nowak, D.J. (2006) Carbon storage by urban soils in the United States. *Journal of Environmental Quality*, **35**, 1566–1575.

Ramalho, C.E. and Hobbs, R.J. (2012) Time for a change: dynamic urban ecology. *Trends in Ecology and Evolution*, **27**, 179–188.

Read, J., Wevill, T., Fletcher, T. and Deletic, A. (2008) Variation among plant species in pollutant removal from stormwater in biofiltration systems. *Water Research*, **42**, 893–902.

Renforth, P., Manning, D.A.C. and Lopez-Capel, E. (2009) Carbonate precipitation in artificial soils as a sink for atmospheric carbon dioxide. *Applied Geochemistry*, **24**, 1757–1764.

Renforth, P., Edmondson, J., Leake, J.R., Gaston, K.J. and Manning, D.A.C. (2011) Designing a carbon capture function into urban soils. *Urban Design and Planning*, **164**, 121–128.

Rink, D. (2005) Surrogate nature or wilderness? Social perceptions and notions of nature in an urban context, in *Wild Urban Woodlands: New Perspectives on Urban Forestry* (I. Kowarik and S. Korner, eds), Springer-Verlag, Berlin, pp. 67–80.

Rosenzweig, M.L. (2003) *Win-Win Ecology. How the Earth's Species Can Survive in the Midst of Human Enterprise.* Oxford University Press, Oxford.

Shochat, E., Warren, P.S., Faeth, S.H., McIntyre, N.E. and Hope, D. (2006) From patterns to emerging processes in mechanistic urban ecology. *Trends in Ecology and Evolution*, **21**, 186–191.

Standish, R.J., Hobbs, R.J. and Miller, J. (2012) Improving city life: options for ecological restoration in urban landscapes and how these might influence interactions between people and nature. *Landscape Ecology*, doi: 10.1007/s10980-012-9752-1.

Steinberg, D.A., Pouyat, R.V., Parmelee, R.W. and Groffman, P.M. (1997) Earthworm abundance and nitrogen mineralization rates along an urban-rural land use gradient. *Soil Biology and Biochemistry*, **29**, 427–430.

Sukopp, H. (2008) On the early history of urban ecology in Europe, in *Urban Ecology: An International Perspective on the Interaction Between Humans and Nature* (J.M. Marzluff, E. Shulenberger, W. Endlicher, M. Alberti, G. Bradley, C. Ryan, U. Simon, and C. ZumBrunnen, eds), Springer Science + Business Media, New York, pp. 79–97.

Sutherland, W.J., Armstrong-Brown, S., Armsworth, P.R. et al. (2006) The identification of 100 ecological questions of high policy relevance in the UK. *Journal of Applied Ecology*, **43**, 617–627.

Tallent-Halsell, N.G. and Watt, M.S. (2009) The invasive *Buddleja davidii* (Butterfly Bush). *The Botanical Review*, **75**, 292–325.

Taylor, A.F., Wiley, A., Kuo, F.E. and Sullivan, W.C. (1998) Growing up in the inner city: Green spaces as places to grow. *Environment and Behavior*, **30**, 3–27.

Thompson, K. and McCarthy, M.A. (2008) Traits of British alien and native urban plants. *Journal of Ecology*, **96**, 853–859.

UN (2011) Population Division of the Department of Economic and Social Affairs of the United Nations Secretariat, World Population Prospects: The 2008 Revision and World Urbanization Prospects: The 2009 Revision, http://esa.un.org/wup2009/unup/. Accessed August 2012.

Wallace, K.J. (2007) Classification of ecosystem services: Problems and solutions. *Biological Conservation*, **139**, 235–246.

WallisDeVries, M.F., Poschlod, P., Willems, J.H. (2002) Challenges for the conservation of calcareous grasslands in northwestern Europe: integrating the requirements of flora and fauna. *Biological Conservation*, **104**, 265–273.

Warnock, S.E. and Takekawa, J.Y. (1995) Habitat preferences of wintering shorebirds in a temporally changing environment: Western sandpipers in the San Francisco Bay estuary. *The Auk*, **112**, 920–930.

Woodcock, B.A., Pywell, R.F., Roy, D.B., Rose, R.J. and Bell, D. (2005) Grazing management of calcareous grasslands and its implications for the conservation of beetle communities. *Biological Conservation*, **125**, 193–202.

Chapter 39

ECOSYSTEM STEWARDSHIP AS A FRAMEWORK FOR CONSERVATION IN A DIRECTIONALLY CHANGING WORLD

Timothy R. Seastedt[1], Katharine N. Suding[2] and F. Stuart Chapin III[3]

[1]Department of Ecology and Evolutionary Biology, University of Colorado, USA
[2]Department of Environmental Science, Policy & Management, University of California, Berkeley, USA
[3]Institute of Arctic Biology, University of Alaska Fairbanks, Fairbanks, USA

"The first law of intelligent tinkering is to save all of the parts."

Aldo Leopold, 1949

39.1 INTRODUCTION

Scientists and conservation advocates have acknowledged the need for actions that contribute to local, regional and global sustainability (Clark and Dickson 2003; Rockström et al. 2009). The sustainability mission involves using the environment to meet the needs of society without jeopardizing resources needed by future generations. Most agree that the challenges to sustainability are immense. Within the next 40 years the human population will increase from seven to nine billion individuals; this growth needs to be accompanied by programs that enhance the quality of life for those individuals. Historically, human population and economic growth initiatives have generally superseded or trumped ecological concerns, but current global-scaled impacts of past and current human activities now dictate that sustainability issues become an essential part of debates on maintaining and improving human welfare (MEA 2005; Rockström et al. 2009). Enlightened societies recognize that human welfare and the welfare of other organisms are linked, but these links range from being obvious to unknown (Raudsepp-Hearne et al. 2010).

All ecosystems on earth show the imprint of human actions. The resulting legacies in new biotic and abiotic characteristics of landscapes have been labeled as 'degraded', 'adventive', 'altered' or, the more value-neutral term, 'novel'(Hobbs et al. 2006). Using the definitions provided in this book (Chapter 6), hybrid and novel ecosystems dominate Earth's biotically active layer that therefore must provide most of the ecosystem services (supporting, provisioning, regulating and cultural services; Chapin 2009) required for sustainability. Similarly, these systems also provide the matrix for future conservation efforts (Kareiva et al. 2011).

That these systems are different or new is undeniable. However, although the ecosystems are different, the living 'pieces' previously found in these areas – the biota – usually persist, albeit in different numbers and combinations than occurred historically. Conservation efforts to preserve biological diversity must therefore find ways to draw upon novel ecosystems to meet the conservation goals of the future. Given that efforts to return these systems to their historical reference point are increasingly problematic in a rapidly changing world, we explore how ecosystem stewardship can contribute to conservation efforts and how accepting and exploiting the reality of novel ecosystems may facilitate these activities.

Ecosystem stewardship is defined as "a strategy to respond to and shape social–ecological systems under conditions of uncertainty and change to sustain the supply and opportunities for use of ecosystem services to support human well-being" (Chapin et al. 2010, p. 241). These services include resources such as food, fuel and water, the capacity of ecosystems to regulate interactions among ecosystems (e.g. movement of water, fire or pollinators among ecosystems), the cultural connections between people and ecosystems and the underlying ecological processes that support these services. Stewardship is seen as a framework for managing these services under the environmental change scenarios predicted for the 21st century (Clark and Dickson 2003; Chapin et al. 2010). The goals of this framework are to sustain required ecosystem services and support human well-being as societies attempt to improve the well-being of disadvantaged segments of an ever-growing human population. This program requires a global perspective to be successful and therefore must address changes that occur at scales ranging from local to global. In many respects, the strategy builds upon the concepts developed for adaptive ecosystem management (e.g. Christensen et al. 1996). In addition to having local, regional and global foci and objectives, what is new about this framework is that it accepts directional environmental change as the reality, and this reality requires strong consideration of a more interventionist strategy to manage activities than has been used in the past. While single environmental drivers such as climate change may not affect the majority of ecological communities in major ways over the next few decades, climate interactions with disturbances such as fires or the ongoing pine beetle epidemic in the forests of western North America are changing ecosystems now. Specifically, the ecological stewardship agenda emphasizes the need to develop resilient communities that can withstand or buffer the impacts of the drivers of environmental change to enhance ecosystem resilience and human well-being. Ecosystem stewardship also explicitly considers humans as a 'keystone species' at regional and global scales, and often at local scales. The stewardship focus broadens the land ethic espoused by Aldo Leopold and recognizes a responsibility to sustain future options for both biota and resources (Chapin et al. 2010). In contrast to a doom-and-gloom message, ecosystem stewardship looks for opportunities and arenas to improve ecosystem services and human welfare.

The relationship between key features of ecosystem stewardship and their implications to conservation and restoration activities is shown in Table 39.1. The stewardship framework relies less on the historical conditions of a community and instead focuses on the dynamics of ecological change and the management options that influence pathways and rates of change. Efforts should be made to maintain properties of the current system that are viewed as desirable by increasing those stabilizing feedbacks that contribute to community persistence. However, the stewardship approach also emphasizes biodiversity from the standpoint of maximizing adaptive capacity – the capacity to control change – instead of focusing on the historical species composition or structure of an ecosystem.

Ecosystem stewardship makes a major departure from classical resource management in stating that expecting and exploiting disturbances such as fires and

Table 39.1 Approaches and goals of ecosystem stewardship differ from traditional resource management. Benefits to biological conservation efforts are identified. From Chapin (2010). Reproduced with permission of Elsevier.

Characteristic	Ecosystem management focus	Ecosystem stewardship focus	Value to conservation
Reference point	Historical condition	Trajectory of change[*]	Resilience focus slows rate of biotic change
Central goal	Ecological integrity	Sustain socio-ecological system and the delivery of ecosystem services	Diversity is a recognized component of services
Approach	Management structure of system	Manage desirable feedbacks	Diversity compatible and contributes to focus
Role of uncertainty	Reduce/minimize uncertainty	(a) Exploit opportunities generated by disturbances (b) Maximize flexibility and have multiple acceptable outcomes	(a) Maximize unassisted migration of biota (b) Emphasis on heterogeneity and promote landscape diversity
Role of research	Researchers inform managers	Researchers and managers partner in programs	Biodiversity explicit management goal
Resources of primary concern	Species composition, structure	Biodiversity, well-being and adaptive capacity	Maintenance of biodiversity explicit

*Ecosystem stewardship might be considered as adaptive ecosystem management for the Anthropocene, i.e. adaptive management that recognizes the significance of directional environmental drivers.

floods is superior to management that attempts to prevent such events. Fire prevention in forests characterized by ground fires can lead to catastrophic fire and potentially ecosystem transformation as in western North America, Australia and other regions of the world (Schoennagel et al. 2004). The economic realities of the current fire control efforts are viewed as unsustainable (Pyne 2004). While we may not want fire, ongoing climatic changes are likely to increase the extent of fire, regardless of the suppression effort applied (Westerling et al. 2011). Further, fire management now emphasizes protection of human-modified landscapes (i.e. the protection and promotion of hybrid or novel ecosystems) in addition to its role as a restoration tool (Schoennagel and Nelson 2011). Given this reality, it is important to rethink where and how to manage for the provision of the ecological services and conservation values provided by both fire-prone and fire-sensitive ecosystems and to manage development in ways that allow adequate protection of life and property.

In a similar fashion, the historical conversion of most of our rivers into channels to maximize water discharge downstream has had catastrophic effects on downstream terrestrial ecosystems, coastal wetlands and coastal shelf ecosystems. As floodplains are reclaimed by these rivers during extreme events, perhaps the best forward-looking conservation and restoration strategy is to allow at least a portion of these floodplains to continue to function in their historical roles. This seems particularly appropriate in those regions where chronically wetter conditions are likely to cause more frequent flood events, such as those seen in the North American midwestern rivers during 2011. Finally, exploiting windows of opportunity in habitat reconstruction following extreme events may provide opportunities for ecosystem reconstruction that benefits biological diversity concerns as well as an opportunity to enhance ecosystem services. Adapting to a new flood regime may involve removal of structures that have channelized rivers as well as adding desirable species believed to be adapted to current and expected future environmental conditions of these areas. A long-term partnership of conservationists with emergency preparedness agencies is not an outrageous idea.

39.2 POTENTIAL DISCONNECTS BETWEEN CONSERVATION AND ECOSYSTEM MANAGEMENT

Environmental scientists and biogeochemists view ecosystems as functional biophysical units that interact with climate and atmospheric chemistry drivers to produce feedbacks affecting the forms and amount of energy, water and other materials that affect future climates. This focus is very different from one concerned with biological diversity. When the species become the focus of concern, the ecosystem blurs to become 'the environment and the species with which they interact'. Historically, such a view was adequate for conservation and restoration purposes. Restore the environment and you restore the appropriate home for the species. However, the current dilemma is that the historical environment cannot be restored to generate a previous biotic configuration or, if it is maintained in some historical state, that state becomes increasingly out of equilibrium with the directionally changing climate and therefore becomes increasingly fragile and non-resilient. Such systems can be viewed as the equivalent of farms, zoos or arboreta, which are sustainable only with constant human intervention. Such habitats have substantial value, but their capacity to provide a conservation umbrella is limited. The environment of the organisms has been changed, perhaps permanently, and therefore the procedures for successful conservation and restoration must address these new realities.

The use of hybrid and novel ecosystems in an ecosystem stewardship agenda has at least two dimensions. One is the identification of the abiotic and biotic changes that are altering the composition of local and regional biotic communities along with assessments of the mechanistic causes of these changes. Such studies include well-researched topics of population and community ecology, as applied to conservation and restoration management. The second dimension of novel ecosystems focuses on management of the stabilizing or amplifying feedbacks that sustain the capacity of the system to provide services over the long term. These dimensions of hybrid and novel ecosystems are linked through feedbacks, yet scientists and advocates often focus on only one or the other. In order to seamlessly merge the science, it is important to identify when ecosystem sustainability initiatives are consistent with biological diversity conservation interests, and whether there are instances when sustainability goals might conflict with these conservation efforts.

39.3 IDENTIFYING CONSERVATION GOALS WITHIN AN ECOSYSTEM STEWARDSHIP AGENDA

39.3.1 The compatibility of conservation and ecosystem services goals

We assume that the conservation agenda for this and future centuries is to preserve biotic diversity. Conservationists may rightly look upon a program that emphasizes the functional properties of ecosystems along with human welfare as one that may not serve the purposes of diverse species preservation. However, the two seemingly different goals – species conservation versus maintaining ecosystem functions and services – can be compatible if the contributions of the species via their functional roles to the characteristics of ecosystems are acknowledged (e.g. Naeem et al. 2009). Although the contribution of rare species to ecosystem services has sometimes been dismissed (c.f. Ridder 2008), redundancy in functional attributes of species becomes important when environmentally driven changes in the relative abundances of species create situations where even rare species are functionally critical. If overgrazing eliminates a common grass, for example, an ecologically similar (but grazing-tolerant) grass may increase in abundance and fulfill many of the functions (e.g. prevention of erosion or invasion of exotic species) of the species that disappeared (Walker et al. 1999). The redundancy in plant diversity allows for grasslands experiencing extreme droughts to maintain some cover and some productivity under such events as the 'dust bowl' that occurred in the last century. Similar scenarios can be constructed for forests experiencing fire and insect outbreaks. If the overarching stewardship goal is to manage for resilience, there are clearly advantages to conserving as much of the resident biodiversity as possible (intelligent tinkering; Leopold 1949) and to adaptively manage over time horizons that allow a range of options to be considered and tested for the preservation of that biological diversity.

39.3.2 Where are the conflicts?

The preservation of biological diversity has historically been in conflict with projects designed (at least in theory) to enhance human welfare, so why won't these conflicts continue? Ecosystem stewardship recognizes that many of the current threats to biological diversity (e.g. pollution, emerging pathogen issues, etc.) are also threats to human welfare. A grayer area, however, is found when attempts to enhance certain specific ecosystem services (e.g. carbon sequestration, increased evapotranspiration, etc.) conflict with those species that do not contribute substantially to those activities and are negatively affected by biotic processes that are used for such efforts. A 'no species left behind' strategy must accommodate those species whose current contributions to ecosystem services appear minimal. Here, interventionist management may again be required to enhance local heterogeneity in ways that can accommodate this diversity. Similarly, a 'no service left behind' strategy might ensure that supporting, provisioning, regulating and cultural services are all considered valuable management goals. One might envision ecosystem services to be maximized at landscape to regional scales, while conservation needs may be achieved by maximizing heterogeneity at smaller spatial scales and emphasizing corridors to facilitate survivorship of metapopulations.

39.3.3 Expanding the conservation focus from local to regional scales

As ecosystems transition to new states and as species abundance is altered due to the suite of directional environmental drivers, conservationists may want to focus on direct and indirect mechanisms to allow for 'unassisted' species redistributions (Hoegh-Guldberg et al. 2008). The conversion of closed-canopy forests to savannas and shrublands, a phenomenon that appears to be becoming more common in semiarid zones, makes life much more difficult for some species but offers opportunities to others that may have historically been rare or found only in adjacent areas. Concurrently, other communities may be tranforming in an opposite direction. For example, much of the eastern grasslands remnants in North America have or will transform to woodlands provided these lands are not used for conventional agriculture purposes (Briggs et al. 2005). If species have corridors that allow for movement, the systems can reform, potentially maintaining most species and ecosystem services (at least at a regional scale). While the debate on assisted migrations is ongoing (McLachlan et al. 2007), natural corridors as well as artificial corridors created deliberately or unintentionally can be viewed as compatible components of both ecosystem stewardship and forward-looking conservation efforts. The new focus recognizes spatial-scale issues that need to be considered now and in the future that previously were ignored for most conservation efforts (e.g. Ervin et al. 2010; Worboys et al. 2010).

This framework has been applied to the National Wildlife Refuge System in the United States to develop a continental-scale conservation strategy (Magness et al. 2011). Across the 540 refuges that comprise this system, the rate of historical temperature changes ranges from negligible to substantial. The sensitivity of each refuge was estimated based on whether it was close to a biome boundary, because species responses to climate change are often associated with biome boundaries (e.g. presence or absence of trees at a forest-grassland boundary; Hampe and Petit 2005). Finally, the capacity of a refuge to adapt to climate change depends on factors that influence opportunities for migration and adjustment (e.g. elevational/latitudinal range within the refuge, road density and proportion of protected lands adjacent to the refuge).

The analysis suggested the following conservation strategy to maximize continental-scale species conservation.

1. Refuges that are not currently exposed to rapid climate change and have low sensitivity (are distant from a biome boundary) and have high adaptive capacity can serve as species refugia. Management need not be intensive and might be focused on addressing non-climatic threats such as invasive species.

2. Refuges that are not currently exposed to rapid climate change and have high sensitivity and/or low adaptive capacity warrant management that mitigates conservation threats using tested conservation methods to maintain or restore historical conditions.

3. Refuges that are exposed to substantial climate change but have high adaptive capacity provide opportunities to learn about processes important to adaptation through natural species adjustments and warrant management of conservation of corridors and other features that facilitate this adjustment.

4. Refuges that are exposed to substantial climate change but have high sensitivity and/or low adaptive capacity may require management that facilitates transformation to novel ecosystem types that support novel species assemblages or that manage refuges as stepping stones for species migration at larger scales.

The first and second approaches are largely reactionary, aimed at maintaining current ecosystem and species composition. The third and fourth require anticipatory efforts that would benefit from adaptive management. Such a framework provides only rough guidance to local managers and must be integrated with local and regional conservation goals. However, it illustrates the value of considering a multi-faceted strategy that views novel ecosystems as potential opportunities under certain circumstances.

39.3.4 Mitigation for climate and atmospheric changes

Another example of this framework is the management of the Tahoe National Forest admininstered by the USDA Forest Service (Joyce et al. 2008; Stephens et al. 2010). In Tahoe National Forest, as with many forest systems in the western United States, drought concerns, amplified by rising temperatures and lowered snowpacks, interact with dense forest structure from fire suppression policies to create severe forest hazards, extreme flood events and new opportunities for invasive species spread. On the Tahoe National Forest, several general principles were recognized as opportunities for stewardship in the face of rapidly changing climate: (1) managing for drought- and heat-tolerant species and ecotypes; (2) reducing the impact of current anthropogenic stressors; (3) managing for diverse successional stages; (4) spreading risks by including buffers and redundancies in natural environments and plantations; and (5) increasing collaboration with interested stakeholders. They considered alternative species mixes and germplasm choice for reforestation, and prioritized sensitive-species management actions at the 'leading edge' of species ranges (likely favorable future habitats) rather than 'trailing edges'. They prioritized projects that were projected to adapt to climate change projections, and put fewer priorities on restoration of systems that had an uncertain future.

Efforts to mitigate vegetation changes due to increased atmospheric deposition of plant-available forms of nitrogen (N) also provide another example of stewardship activities. Chronic increases in N deposition require long-term mechanisms for N reduction or removal. Several management techniques have proven successful to mitigate effects of N deposition on ecosystems: biomass removal; prescribed fire or control of invasive grasses by mowing; selective herbicides; weeding; or domestic animal grazing. However, the efficacy of these techniques varies based on public acceptance and ecosystem type. For instance, 54% of the coastal sage scrub ecosystem in California is estimated to exceed its critical load for nitrogen deposition. Although the most effective large-scale method for controlling annual exotic grasses that invade the coastal sage scrub is burning in the spring before seeds have shattered (Gillespie and Allen 2004), fire has not been used in remnant coastal sage areas because managers are understandably reluctant to burn remnant stands of shrubs. In grasslands, selective grazing on N-rich exotic grasses has been used with success to combat effects of N deposition (Weiss 1999; Marty 2005); however, this same method in coastal sage scrub is less effective because sufficient grass forage only occurs in very wet years. A good management technique (grazing or mowing) has therefore successfully been used to mitigate for N deposition in grasslands. In coastal sage scrub however there is little evidence that management options will be effectively implemented in ecosystems impacted by excess N because they are not technically feasible or cost effective, and in many protected areas site manipulations are prohibited.

39.3.5 Merging researchers with managers

Researchers need to partner with managers on all aspects of resource and conservation management. Resource managers are dealing with current problems and do not have a mandate from stakeholders or policy makers to 'experiment' for future change. Arguably, this has led to a strong emphasis on the removal of current problems such as invasive species. This activity has itself become a self-fulfilling goal and has not always produced desired outcomes (Reid et al. 2009; Kettenring and Adams 2011). Clearly, the purpose of such exercises is forward-looking, i.e. to conserve and enhance the species of concern and interest (Davis et al. 2011; Wardle et al. 2011). Ecosystem stewardship focuses on mantaining and enhancing species that are viewed as important for a variety of ecosystem services, including cultural services, and a common feature is that these species maintain conditions favorable to native communities. Within this framework, species removal would be most appropriate where it maintains native communities with a reasonable expenditure of funds. Scientists should work directly with managers, policy makers and stakeholders and provide them with local interpretations of the significance of directional environmental change drivers on their systems. Right now managers are so overwhelmed in dealing with 'unacceptable alterations or disturbances' that many of the issues associated with novel ecosystems are unlikely to be on their radar screens. 'So many changing variables; so little time' is a common assessment of conservation issues. Solutions require an approach nested within the philosophy that nature is dynamic and resilient (Kareiva et al. 2011).

39.4 CONCLUSIONS

Hybrid and novel ecosystems have a dual role. By default they have become the most widespread terrestrial ecosystems and are therefore important in providing ecological goods and services to society in an era of rapid environmental change. They also provide the most widespread arenas for conservation action. Hybrid and novel ecosystems also provide opportunities for cautious experiments that may foster species conservation in new assemblages that have greater likelihood of persistence under projected environmental conditions. The good news – and it is good news – is that "many orthodox conservation practices . . . will continue to increase species and ecosystem adaptive capacity to climate change" (Dawson et al. 2011, p. 57), and these practices are compatible within an ecosystem stewardship framework. This framework recognizes that there exists a compatibility and mutualism between ecological processes and human needs. Species conservation within the stewardship framework is seen as essential to maximizing options and adaptive capacity. The realities of a rapidly changing world demand that successful conservation activities accommodate uncertainties and exploit the opportunities imposed by directional environmental changes. Both novel ecosystems and interventionist conservation efforts are important and compatible components for achieving this framework.

ACKNOWLEDGEMENTS

We greatly appreciate the opportunity to have participated in the Novel Ecosystems workshop in May of 2011 and been recipients of a number of ideas and perspectives that contributed to this essay. We thank several anonymous reviewers, Janet Prevéy and Dr Erika Zavaleta for contributing ideas and suggesting improvements to an earlier draft of this paper.

REFERENCES

Briggs, J.M., Knapp, A.K., Blair, J.M., Heisler, J.L., Hoch, G.A., Lett, M.S. and McCarron, J.K. (2005) An ecosystem in transition: causes and consequences of the conversion of mesic grassland to shrubland. *BioScience*, **55**, 243–254.

Chapin, F.S. III. (2009) Managing ecosystems sustainabily: the key role of resilience, in *Principles Of Ecosystem Stewardship* (F.S. Chapin III, G.P. Kofinas and C. Folke, eds), Springer, New York, pp. 29–53.

Chapin, F.S. III, Carpenter, S.R., Kofinas, G.P. et al. (2010) Ecosystem stewardship: Sustainability strategies for a rapidly changing planet. *Trends in Ecology and Evolution*, **25**, 241–249.

Christensen, N.L., Bartuska, A.M., Brown, J.H. et al. (1996) The report of the Ecological Society of America Committee on the scientific basis for ecosystem management. *Ecological Applications*, **6**(3), 665–691.

Clark, W.C. and Dickson, N.M. (2003) Sustainability science: The emerging research program. *Proceedings of the National Academy of Sciences*, **100**, 8059– 8061.

Davis, M., Chew, M.K., Hobbs, R.J. et al. (2011) Don't judge species by their origins. *Nature*, **474**, 153–154.

Dawson, T.P., Jackson, S.T., House, J.I., Prentice, I.C. and Mace, G.M. (2011) Beyond predictions: biodiversity conservation in a changing climate. *Science*, **332**, 53–58.

Ervin, J., Mulongoy, K.J., Lawrence, K., Game, E., Sheppard, D., Bridgewater, P., Bennett, G., Gidda, S.B. and Bos, P. (2010) Making Protected Areas Relevant: A guide to integrating protected areas into wider landscapes, seascapes and sectoral plans and strategies. CBD Technical Series No. 44. Montreal, Canada: Convention on Biological Diversity.

Gillespie, I.G. and Allen, E.B. (2004) Fire and competition in a southern California grassland: impacts on the rare forb *Erodium macrophyllum*. *Journal of Applied Ecology*, **41**, 643–652.

Hampe, A. and Petit, R. (2005) Conserving biodiversity under climate change: the rear edge matters. *Ecology Letters*, **8**, 461–467.

Hobbs, R.J., Arico, S., Aronson, J. et al. (2006) Novel ecosystems: theoretical and management aspects of the new ecological world order. *Global Ecology & Biogeography*, **15**, 1–7.

Hoegh-Guldberg, O., Hughes, L., McIntyre, S., Lindenmayer, D.B., Parmesan, C., Possingham, H.P., Thomas, C.D. (2008) Assisted colonization and rapid climate change. *Science*, **321**, 345–346.

Joyce, L.A., Blate, G.M., Littell, S.J., McNulty, S.G., Millar, C.I., Moser, S.C., Neilson, R.P., O'Halloran, K. and Peterson, D.L. (2008) Adaptation options for climate-sensitive ecosystems and resources: national forests. Preliminary Review of Adaptation Options for Climate-Sensitive Ecosystems and Resources (Synthesis and Assessment Product 4.4, US Climate Change Science Program), chapter 3.

Kareiva, P., Lalasz, R. and Marvier, M. (2011) Conservation in the Anthropocene. *Breakthrough Journal*, **2**, http://breakthroughjournal.org/content/issues/issue_2/. Accessed August 2012.

Kettenring, K.M. and Adams, C.R. (2011) Lessons learned from invasive plant control experiments: a systematic review and meta-analysis. *Journal Applied Ecology*, **48**, 970–979.

Leopold, A. (1949) *A Sand County Almanac*. Oxford University Press, Oxford.

Magness, D.R., Morton, J.M., Huettmann, F., Chapin, F.S. III and McGuire, A.D. (2011) A climate-change adaptation framework to reduce continental-scale vulnerability across the US National Wildlife Refuge System. *Ecosphere*, **2**, article 112.

Marty, J.T. (2005) Effects of cattle grazing on diversity in ephemeral wetlands. *Conservation Biology*, **19**, 1626–1632.

McLachlan, J.S., Hellmann, J. and Schwartz, M. (2007) A framework for debate of assisted migration in an era of climate change. *Conservation Biology*, **21**, 297–302.

Millennium Ecosystem Assessment (MEA) (2005) *Ecosystems and Human Well-being: Synthesis*. Island Press, Washington.

Naeem, S., Bunker, D.E., Hector, A., Loreau, M. and Perrings, C. (2009) *Biodiversity, Ecosystem Functioning, and Human Wellbeing: An Ecological and Economic Perspective*. Oxford University Press, NY.

Pyne, S. (2004) *Tending Fire: Coping with America's Wildland Fires*. Island Press, Washington, DC.

Raudsepp-Hearne, C., Peterson, G.D., Tengö, M., Bennett, E.M., Holland, T., Benessaiah, K., MacDonald, G.K. and Pfeifer, L. (2010) Untangling the environmental paradox: Why is human well-being increasing as ecosystem services degrade. *Bioscience*, **60**, 576–589.

Reid, A.M., Morin, L., Downey, P.O., French, K., Virtue, J.G. (2009) Does invasive plant management aid the restoration of natural ecosystems? *Biological Conservation*, **142**, 2342–2349.

Ridder, B. (2008) Question the ecosystem services argument for biodiversity conservation. *Biodiversity Conservation*, **17**, 781–790.

Rockström, J., Steffen, W., Noone, K. et al. (2009) A safe operating space for humanity. *Nature*, **461**, 472–475.

Schoennagel, T. and Nelson, C.R. (2011) Restoration relevance of recent National Fire Plan treatments in forests of the western United States. *Frontiers in Ecology and the Environment*, **9**(5), 271–277.

Schoennagel, T., Veblen, T.T. and Romme, W.H. (2004) The interaction of fire, fuels, and climate across Rocky Mountain forests. *Bioscience*, **54**, 661–676.

Stephens, S.L., Millar, C.I. and Collins, B.M. (2010) Operational approaches to managing forests of the future in Mediterranean regions within a context of changing climates. *Environmental Research Letters*, **5**, doi: 10.1088/1748-9326/5/2/024003.

Walker, B., Kinzig, A. and Langridge, J. (1999) Plant attribute diversity, resilience, and ecosystem function: The nature and significance of dominant and minor species. *Ecosystems*, **2**, 95–113.

Wardle, D.A., Bardgett, R.D., Callaway, R.M. and Van der Putten, W.H. (2011) Terrestrial ecosystem responses to species gains and losses. *Science*, **332**, 1273–1277.

Weiss, S.B. (1999) Cars, cows, and checkerspot butterflies: Nitrogen deposition and management of nutrient-poor grasslands for a threatened species. *Conservation Biology*, **13**, 1476–1486.

Westerling, A.L., Turner, M.G., Smithwick, E.A.H., Romme, W.H. and Ryan, M.G. (2011) Continued warming could transform Greater Yellowstone fire regimes by mid-21st century. *Proceedings of National Academy of Science USA*, **108**(32), 13165–13170.

Worboys, G. L., Francis, W.L. and Lockwood, M. (eds) (2010) *Connectivity Conservation Management: A Global Guide*. Earthscan, London.

Chapter 40

CASE STUDY: NOVEL SOCIO-ECOLOGICAL SYSTEMS IN THE NORTH: POTENTIAL PATHWAYS TOWARD ECOLOGICAL AND SOCIETAL RESILIENCE

F. Stuart Chapin III[1], Martin D. Robards[2], Jill F. Johnstone[3], Trevor C. Lantz[4] and Steven V. Kokelj[5]

[1]Institute of Arctic Biology, University of Alaska Fairbanks, Fairbanks, USA

[2]Arctic Beringia Program Director, Wildlife Conservation Society, New York, USA

[3]Department of Biology, University of Saskatchewan, Canada

[4]School of Environmental Studies, University of Victoria, Canada

[5]Cumulative Impact Monitoring Program, Aboriginal Affairs and Northern Development Canada, Yellowknife, Northwest Territories, Canada

40.1 ARCTIC WARMING AND STEWARDSHIP RESPONSES

The warming that has occurred globally is amplified at high latitudes (ACIA 2005), such that many areas of the Arctic are warming twice as rapidly as the global average (Hinzman et al. 2005; IPCC 2007; Fig. 40.1). Mean annual air temperature in Interior Alaska has increased by 1.3°C during the past 50 years (Shulski and Wendler 2007) and is projected by downscaled climate models to increase by an additional 3–7°C by the end of the 21st century (Walsh et al. 2008; http://www.snap. uaf.edu). Similar increases have occurred across the western Canadian Arctic (Burn and Kokelj 2009).

Rapid Arctic warming is already causing substantial biophysical and ecological change in the north, including

(a)

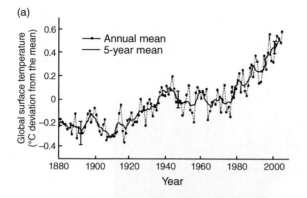

(b)   2005 Surface Temperature Anomaly (°C)

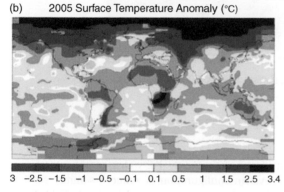

3   −2.5 −1.5   −1   −0.5 −0.1   0.1   0.5   1   1.5   2.5   3.4

Figure 40.1 (a) Time course of the average surface temperature of Earth from 1850 to 2005 (relative to the average temperature for this time period) and (b) geographic pattern of temperature increase. Redrawn with permission from Hansen et al. (2006); © National Academy of Sciences, USA.

diminished summer sea ice, thawing of permafrost, an acceleration of thermokarst, more extensive wildfires, upward movement of altitudinal treeline and outbreak of forest insects, suggesting substantial vulnerability to the changing abiotic conditions of the Arctic (ACIA 2005; IPCC 2007). These changes suggest that Arctic ecosystems may already have crossed tipping points into new system configurations (Lindsay and Zhang 2005).

Given rapid Arctic change, what are the barriers and opportunities that managers and policy-makers face in planning for and responding to change? Many of the changes described earlier, as well as changes that cannot currently be predicted, may be impossible to prevent except at local scales. Permafrost thaw, for example, is sensitive to snow cover and other forms of surface insulation that could be manipulated in

places where permafrost integrity is particularly important. Similarly, fire suppression can reduce fire risk near communities and insects can be locally controlled through use of pesticides or pheromones. However, these short-term efforts to prevent change are management-intensive and, if successful, push the system farther from its balance with key environmental controls and increase the risk of long-term catastrophic change (Holling and Meffe 1996; Simard et al. 2011).

Adaptive ecosystem stewardship is an alternative approach. This involves shaping trajectories of socio-ecological change toward sustaining ecosystem resilience and human well-being within what are often novel ecosystems (Chapin et al. 2009, 2011). This stewardship approach implies an acceptance that species will often occur at different densities, in new assemblages and in new locations, forming novel ecosystems with potentially novel responses to environment and management. Arctic environmental changes are already affecting many of the species and peoples of the Arctic. Below, we describe examples of such novel changes in marine and terrestrial systems.

40.2 EMERGING HIGH-LATITUDE NOVELTY

40.2.1 Marine ecosystems

The Arctic marine systems of the Beaufort, Chukchi and East Siberian Seas have transitioned as a result of climatic forcing to systems where summer sea ice now frequently recedes far north of the shelf break, reaching its seasonal minimum extent over the deep Arctic basin. While historically existing as a complex heterogeneous ice environment throughout much of the Arctic basin in summer (Ray et al. 2010), the areal extent of ice cover in 2010 was 22% below the long-term average for 1979–2009. A suite of amplifying (positive) feedbacks, such as from the reduced albedo of open water and the thinning of sea ice, are likely to increase both the rate of change (Stroeve et al. 2007) and resilience of this new summer state (Comiso et al. 2008). For species that have relied on sea ice over shallow shelf areas or the proximity of sea ice to land, the consequences of this new abiotic system have been dramatic. Pacific walrus are one of these species.

Most female Pacific walrus have traditionally used sea ice as feeding and nursing platforms during

Figure 40.2 Burgeoning walrus herd at haul-out near Chukotka, Russian Federation. Photograph: Maxim Chakilev.

summer. When the sea ice retreated north of the continental shelf and over deep water in summer 2007 (the most extreme retreat since satellite measurements began in 1979), female walruses and their calves could no longer reach the sea floor to feed (Jay et al. 2011). An estimated 3500 female walrus and their calves moved ashore near the villages of Wainwright and Point Lay, Alaska for the first time in recorded history. In neighboring Chukotka (Russian Federation), burgeoning walrus herds at haul-outs have approached 100,000 animals in some places (Fig. 40.2). Since 2007, the large-scale movement of females and their calves has occurred three more times (2009, 2010, 2011), with numbers of walrus on shore in Alaska estimated as high as 50,000 in 2010. A large-scale shift in behavior by female walruses and their calves from using floating sea ice to land-based haul-outs is a novel adaptation to contemporary sea-ice conditions. However, this change forces walrus to concentrate their feeding in a smaller area, increases the risk of trampling of females and calves when animals are spooked and potentially increases their vulnerability to human and non-human predators.

These and similar changes in the ecology of other ice-sensitive species have important implications for conservation of both the ecological and cultural

attributes of the Arctic. For walrus, a transition to an ecosystem configuration of reduced summer sea ice with greater use of terrestrial haul-outs is likely to reduce the carrying capacity for the population as a whole (Robards et al. 2009).

Many coastal indigenous communities in the region have depended nutritionally and culturally on marine mammals such as walrus for millennia. Changing accessibility to walrus, both ecologically (e.g. due to the increased but localized availability of animals while on land) and politically (due to restrictions on access as a consequence of reductions in population size), is likely to have profound cultural implications.

40.2.2 Coastal ecosystems

Combined with more frequent storms and increased sea level, reductions in sea-ice cover (IPCC 2007; Comiso et al. 2008; Sepp and Jaagus 2011) will also have profound effects on terrestrial ecosystems along the Arctic coast. Recent flooding of low-lying Arctic ecosystems in the western Arctic suggest that these impacts have already begun (Borstad et al. 2008; Pisaric et al. 2011; Kokelj et al. 2012). In September of 1999, a storm originating in the Gulf of Alaska tracked across Alaska and the Yukon. Re-intensifying when it reached the Beaufort Sea, it became one of the most severe storms recorded for the outer Mackenzie Delta. Synthetic aperture radar, water survey gauge data and local observations indicate that the surge produced by this storm inundated all low-lying alluvial surfaces within 20–30 km of the coast (Kokelj et al. 2012). The inundation caused by this surge increased mean soil salinity by a factor of 2–20 above the average level measured in unimpacted areas, killing more than 13,000 ha of vegetation in a 129,000 ha portion of the outer Mackenzie delta.

Although this low-lying ecosystem is well adapted to frequent inundation by freshwater (Pearce 1986), several lines of evidence indicate that saline inundation of this magnitude is without precedent in the last 1000 years. Sediment cores from a lake in the outer delta show a rapid (and unprecedented) shift from freshwater to brackish species following the 1999 storm surge event (Pisaric et al. 2011). Geochemical profiles of permafrost, alder ring chronologies and local knowledge of Inuvialuit Hunters also identify the saline storm surge of 1999 as anomalous (Pisaric et al. 2011; Kokelj et al. 2012).

The impact of this event on wildlife habitat has not been examined in detail, but local Inuvialuit hunters have reported shifts in geese and moose abundance in this part of the delta (Kokelj et al. 2012). Ongoing monitoring in the outer delta has shown that low-lying areas adjacent to river channels have shown considerable recovery. However, more than a decade after this event, elevated areas distant from channels of the Mackenzie have shown virtually no recovery (Kokelj et al. 2012; Fig. 40.3). Persistent soil salinities and a lack of recruitment suggest that this disturbance resulted in long-term modifications to this ecosystem (Pisaric et al. 2011).

With the anticipated rise in sea levels and increases in the magnitude and frequency of storms (Manson and Solomon 2007), saline incursions of coastal ecosystems are likely to become more common in the future. Since much of the Arctic is within 100 km of the ocean, the potential for increased coastal inundation represents a critical source of uncertainty regarding ecological change in this biome – one that is critical habitat to the millions of breeding shorebirds and waterfowl that migrate to the Arctic each summer or for the caribou that come to feed and give birth

Figure 40.3 Persistent impacts of the 1999 storm surge on the vegetation of the outer Mackenzie Delta. Photograph: T. Lantz.

to their calves. Consequently, the increased potential for future salinization of low-lying coastal ecosystems must be considered in regional planning and in the assessment and monitoring of the cumulative impacts of development.

40.2.3 Arctic terrestrial ecosystems

Observations across the Low Arctic indicate that other disturbance regimes are also undergoing fundamental changes. Analysis of aerial photographs and satellite imagery show that terrain disturbances associated with thawing permafrost (thermokarst) are becoming more common and expanding more rapidly than in previous decades (Wolfe et al. 2001; Jorgenson et al. 2006; Lantz and Kokelj 2008; Lacelle et al. 2010). As ice-rich landscapes adjust to temperature increases, thermokarst processes have the potential to fundamentally alter hydrological regimes, aquatic food webs and broad-scale vegetation structure.

One of the most conspicuous forms of thermokarst is found in the ice-rich landscapes of northwestern Canada and Alaska. Retrogressive thaw slumping occurs when thawing ground ice creates an exposure. As the ice and sediments thaw, a mud slurry accumu-lates at the base of the headwall and moves downslope as a debris flow. Thawing of the exposed ground ice and sediments causes the headwall to move upslope, causing the disturbance to grow. Thaw slumps impact-ing areas around lakes typically cover 1–2 hectares, while those along stream valleys can exceed 50 hec-tares (Fig. 40.4). Since they are associated with lakes, streams or the coast, they impact both terrestrial and aquatic ecosystems. Slump growth moves large volumes of sediments from the impacted slope to the adjacent water body or floodplain. Suspended sediment and solute concentrations in runoff over freshly thawed soils can be orders of magnitude greater than from the adjacent landscape so that, even if the disturbance is localized, the potential for downstream aquatic impact is significant (Kokelj et al. 2005, 2009a; Thompson et al. 2008; Mesquita et al. 2010; Moquin 2011).

When thaw slumps stabilize, they provide conditions that favor the establishment of unique plant communi-ties. In the upland tundra north of Inuvik in Northwest Territories, canopy-forming tall shrubs (>4m) estab-lish quickly on slump scars and can persist for more than a century (Lantz et al. 2009). Like deciduous forest establishment after severe fire in the subarctic, shrub persistence on stable slumps (and disturbances to tundra) is facilitated by feedbacks between shrub

Figure 40.4 Oblique airphoto showing a thaw slump in the Stony Creek watershed near Fort MacPherson, NWT. Photograph: S.V. Kokelj. © 2012 Her Majesty the Queen in Right of Canada.

growth and abiotic conditions. For example, combined with the concave morphology of the slump scar, the dense vegetation at these sites also helps to create snow drifts that maintain high ground temperatures and deep active layers, and promote nutrient mineralization (Lantz et al. 2009). Modeling studies also indicate that this thermal disturbance can cause stable slumps to re-activate after decades of stability (Kokelj et al. 2009b). Substrates exposed by thaw slumps create unique soil conditions and provide opportunities for rapid colonization and movement of species beyond their present geographic ranges. As the number and size of slumps increases (Lantz and Kokelj 2008), these disturbances will likely have a growing influence on vegetation dynamics in the Low Arctic (Lantz et al. 2009).

40.2.4 Boreal ecosystems

Further inland, the increased extent and severity of wildfire in the boreal forest is also creating conditions that allow ecological novelty to emerge (Chapin et al. 2010). In the past decade, an increase in the number of years with extensive fires has doubled the annual area burned in interior Alaska compared to any decade of the previous 40 years (Kasischke et al. 2010). Warmer and drier summers allow fires to continue burning in late summer when soils are deeply thawed and have lower soil moisture. They therefore burn more deeply, creating a radically different soil environment for seedling establishment. These severe fires have disrupted conditions for black spruce regeneration that sustained black spruce dominance for thousands of years (Johnstone et al. 2010a; Turetsky et al. 2010).

Stabilizing (negative) feedbacks that have sustained the resilience of black spruce forests include biogeochemical processes (thermal insulation by mosses, presence of permafrost, moist soils and low fire severity) that interact with life-history attributes (Fig. 40.5). For example, the thick organic layers that build up during succession in black spruce forests are generally poorly combusted during fire (Dyrness and Norum 1983). These charred organic layers create dark, porous surfaces that are exposed to severe temperature fluctuations and drought stress, reducing seed germination and directly killing young seedlings (Johnstone and Chapin 2006a).

Black spruce trees overcome limitations of poor seedbed quality by producing large quantities of seed from semi-serotinous cones in an aerial seed bank, ensuring that at least some seeds will find microhabitats suitable for recruitment. This advantage disappears when severe fires expose mineral soil, creating high-quality seedbeds that greatly increase seedling establishment and permit even off-site colonizers to successfully recruit at a site. Many of these off-site colonizers grow more rapidly in the mineral soil than black spruce, which has a conservative strategy adapted to the low nutrient availability of cold, organic-dominated soils. In addition, once rapidly growing deciduous trees become established, they alter the plant–soil feedbacks to support a self-sustaining environment of high overstory productivity and low moss growth (Fig. 40.5). The recent increase in mineral soil seedbeds has generated new successional trajectories dominated by deciduous tree seedlings that are expected to be stable through future fire cycles (Johnstone et al. 2010a; Kasischke et al. 2010).

Increases in fire activity that lead to a shortening of the fire return interval can also disrupt stable successional cycles in black spruce forests. Black spruce trees require up to 50 years to reliably produce cones with viable seeds (Viglas 2011). Thus, fires that recur at a frequency of less than 50 years can disrupt regeneration of these forests, leading to the creation of non-forested shrublands in some cases and deciduous-dominated forests in others (Johnstone and Chapin 2006b; Lantz et al. 2010; Brown and Johnstone 2011; Fig. 40.5). Climate may directly interact with fire to drive the creation of alternate forest cover types, as in extremely dry sites, where no tree recruitment may occur after fire (Kasischke et al. 2007; Johnstone et al. 2010b).

These observations of the boreal forest suggest potential shifts in the relative abundance of forest types that currently dominate the Alaskan boreal forest: a decline in abundance of black spruce, which has dominated the lowland landscape and north-facing slopes for the last 6000 years; a potential increase of deciduous forests in former black spruce habitat; and a conversion to grass or shrublands on dry sites (Johnstone et al. 2010b). In the past several millennia, deciduous forests have been largely restricted to south-facing uplands and floodplain corridors and have acted as a stabilizing feedback to fire probability and spread because of their high leaf moisture content and low flammability. As climate warms, however, vegetation

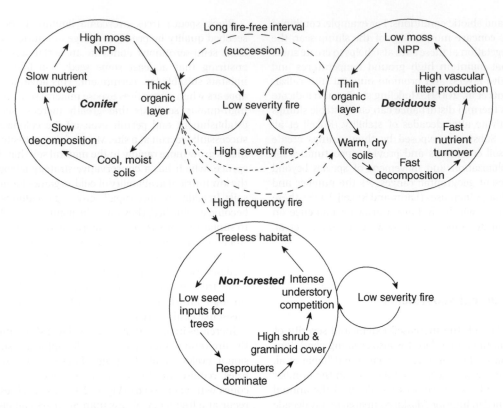

Figure 40.5 A conceptual diagram of successional cycles in boreal forests, representing conifer- and deciduous-dominated forests, as well as non-forested grasslands or shrublands. The resilience of each forest type is maintained by biophysical interactions that strongly favor self-replacement and dominance after fire (inner text within the circles). Shifts between forest cover types are driven by changes to the fire regime expressed as unusual fire events: severe fire, high-frequency fire, or long fire-free intervals that allow prolonged succession (dashed arrows). Modified from Johnstone et al. (2010a).

effects on flammability decline, weakening this stabilizing feedback; the areal extent of fire is therefore projected to continue increasing with climate warming despite the shift to deciduous vegetation (Kasischke et al. 2010; Johnstone et al. 2011).

Hardwoods accumulate less soil organic matter than spruce ecosystems (Van Cleve et al. 1983; Mack et al. 2008), so increased hardwood dominance might reduce carbon sequestration at landscape scales. Increases in the dominance of hardwood forests and non-forested cover caused by a change in fire regime may produce a landscape configuration of Alaska's boreal forests that starts to approach the biome composition that existed in Alaska and eastern Siberia early in the Holocene (Edwards et al. 2005). During

that period, shrublands and deciduous forests dominated the boreal landscape, as conifers had not yet colonized the region from their glacial refugia (Edwards et al. 2005). Thus, although a decrease in conifer forest cover in Alaska may appear novel from the context of current human society, the ecological capacity for novelty may well exist within the range of historical adaptation of current boreal species. Nevertheless, the particular combination of climate, fire regime and forest cover types likely to dominate in the coming century may well represent a unique combination for Alaska's boreal forests and, at the very least, a strong departure from the way these systems have operated over recent millennia. In the tundra north of Inuvik, the continued dominance of tall shrubs on severe tundra fires that burned 40–50 years ago also raise the possibility that an increased frequency of tundra fires will drive changes in vegetation structure in the Low Arctic (Lantz et al. 2010).

40.3 NOVELTY, RESILIENCE AND CONSERVATION

Compounding the types of ecosystem change we describe earlier, some of the most profound changes to Arctic ecosystems are likely to result from the increased accessibility of the Arctic to new people and cultures, leading to a suite of new anthropogenic modifications. Indeed, the human ability to rapidly take advantage of the new opportunities that accompany novel ecosystems may exceed the rate of ecosystem adaptation. Currently, oil and gas development and commercial shipping are rapidly increasing in a more favorable (less ice) physical environment, presenting a new suite of challenges for species such as walrus that may now need to compete with onshore industrial infrastructure as they seek new land-based solutions to their own adaptive needs. However, on land, the shortening of the winter season and thawing of permafrost may work in a counter manner, reducing the opportunity for construction of the ice roads that have been so important to Arctic development activities in regions such as Alaska (Stephenson et al. 2011).

Socio-ecological resilience is the capacity of a system consisting of people and the rest of nature to sustain and shape fundamental system structure and function in the face of perturbations such as climate change. Resilience approaches advocate a shift from *reactive* policies to prevent change to *proactive* policies that endeavor to shape change in a rapidly changing world. Given that future changes are uncertain, resilience places a strong emphasis on building and maintaining a multitude of options that allow flexibility to adapt to change rather than pursuing what might currently seem like the single best option.

There are four basic tenets to building resilience (Chapin et al. 2009):

1. *Sustain the fundamental ecological and social processes that have shaped the current system, but allow enough disturbance so that the system can adjust to change.* For example, reductions in human impacts on the climate system would slow the rates, and perhaps severity, of Arctic change, giving species and ecosystems more time to adapt.

2. *Foster social, economic and ecological diversity to provide a wide range of building blocks and pathways for potential future change.* For example, if sea ice continues to decline, create a portfolio of onshore reserves that partially protect walrus from hunting and minimize the impacts of development activities in the areas likely to host hauled-out walruses.

3. *Experiment with different approaches to provide opportunities to learn what works and what does not.* This could occur through local community-agency initiatives to create novel subsistence arrangements such as ordinances to protect hauled-out walruses or multi-species harvest regulations that allow hunters to target marine or terrestrial game species when they are locally abundant. Assisted migration of native boreal conifers, such as lodgepole pine (Johnstone and Chapin 2003), may provide options for softwood timber production and increased conifer forest resilience under altered fire regimes. Development of a knowledge-sharing network could facilitate widespread learning from such experiments. Learning may also be facilitated through attention to analogous situations. For example, although female Pacific walrus and their calves are transitioning to greater use of terrestrial habitats in Alaska during summer, this is already the norm for both the Atlantic and Laptev walrus populations and for some segments of the Pacific walrus population in Chukotka.

4. *Adapt governance to allow implementation of potential solutions under novel conditions.* For example, although the Marine Mammal Protection Act mandates conservation targets linked to historical population levels for Pacific walrus, a novel ecosystems approach might alter this requirement in favor or one more reflective of changing environmental conditions, allowing greater options at perhaps lesser investment to achieve valuable outcomes (Robards et al. 2009). At a broader scale, creation of bridging mechanisms between pan-Arctic conservation bodies, local, regional, or federal regulatory agencies and hunter-based community monitoring programs would facilitate learning from successes and failures. Similarly, on land, warming-induced increases in wildfire, coastal erosion and thermokarst will likely require exploration of new policies such as those that reconsider the nature of development in vulnerable areas such as the wildland–urban interface or areas at risk to coastal erosion.

Hobbs et al. (2009) regard the types of transitions in the socio-ecological systems that we discuss here as those where traditional conservation and restoration outcomes are unlikely. However, resilience-based stewardship seeks no explicit structural outcome and is therefore at least theoretically compatible with the goal of addressing the novel ecosystems that are emerging in the north. Indeed, it fosters the underlying ecological and social conditions required for conservation or the continued flow of ecosystem services, and opens multiple pathways for potential adaptation to new conditions. Nevertheless, we acknowledge the slow rates of change frequently inherent in the political (and legal) process that would need to support such an approach. Consequently, we posit that one of the biggest challenges of addressing the governance issues associated with novel Arctic ecosystems will be to narrow the gap between policies that enable people to take advantage of novel ecosystem configurations, and the rate at which ecosystems or wildlife can establish their own resilience in the face of change (Robards et al. 2011).

REFERENCES

ACIA (2005) *Arctic Climate Impact Assessment.* Cambridge University Press, Cambridge.

Borstad, G.A., de S. Álvarez, M.M., Hines, J.E. and Dufour, J.-F. (2008) Reduction in vegetation cover at the Anderson River delta, Northwest Territories, identified by Landsat imagery, 1972–2003. Canadian Wildlife Service Technical Report Series **496**.

Brown, C.D. and Johnstone, J.F. (2011) Once burned, twice shy: Repeat fires reduce seed availability and alter substrate constraints on black spruce regeneration. *Forest Ecology and Management,* **266**, 34–41.

Burn, C.R. and Kokelj, S.V. (2009) The environment and permafrost of the Mackenzie Delta area. *Permafrost and Periglacial Processes,* **20**, 83–105.

Chapin, F.S., III, Kofinas, G.P. and Folke, C. (eds) (2009) *Principles of Ecosystem Stewardship: Resilience-Based Natural Resource Management in a Changing World.* Springer, New York.

Chapin, F.S., III, McGuire, A.D., Ruess, R.W. et al. (2010) Resilience of Alaska's boreal forest to climatic change. *Canadian Journal of Forest Research,* **40**, 1360–1370.

Chapin, F.S., III, Power, M.E., Pickett, S.T.A. et al. (2011) Earth stewardship: science for action to sustain the human-earth system. *Ecosphere,* **2**, art89, doi: 10.1890/ES1811-00166.00161.

Comiso, J.C., Parkinson, C.L., Gersten, R. and Stock, L. (2008) Accelerated decline in the arctic sea ice cover. *Geophysical Research Letters,* **35**, doi: 10.1029/2007GL031972.

Dyrness, C.T. and Norum, R.A. (1983) The effects of experimental fires on black spruce forest floors in interior Alaska. *Canadian Journal of Forest Research,* **13**, 879–893.

Edwards, M.E., Brubaker, L.B., Lozhkin, A.V. and Anderson, P.M. (2005) Structurally novel biomes: a response to past warming in Beringia. *Ecology,* **86**, 1696–1703.

Hansen, J., Sato, M., Ruedy, R., Lo, K., Lea, D.W. and Medina-Elizade, M. (2006) Global temperature change. Proceedings of the National Academy of Sciences USA. **103**, 14288–14293.

Hinzman, L.D., Bettez, N.D., Bolton, W.R. et al. (2005) Evidence and implications of recent climate change in northern Alaska and other arctic regions. *Climatic Change,* **72**, 251–298.

Hobbs, R.J., Higgs, E. and Harris, J.A. (2009) Novel ecosystems: implications for conservation and restoration. *Trends in Ecology & Evolution,* **24**, 599–605.

Holling, C.S. and Meffe, G.K. (1996) Command and control and the pathology of natural resource management. *Conservation Biology,* **10**, 328–337.

IPCC (2007) *Climate Change 2007*: The physical science basis, contribution of working group I to the fourth assessment report of the Intergovernmental Panel on Climate Change. Cambridge University Press, Cambridge.

Jay, C.V., Marcot, B.G. and Douglas, D.C. (2011) Projected status of the Pacific walrus (*Odobenus rosmarus divergens*) in the twenty-first century. *Polar Biology,* **34**, 1065–1084.

Johnstone, J.F. and Chapin, F.S. III (2003) Non-equilibrium succession dynamics indicate continued northern migration of lodgepole pine. *Global Change Biology,* **9**, 1401–1409.

Johnstone, J.F. and Chapin, F.S. III (2006a) Effects of burn severity on patterns of post-fire tree recruitment in boreal forests. *Ecosystems,* **9**, 14–31.

Johnstone, J.F. and Chapin, F.S. III (2006b) Fire interval effects on successional trajectory in boreal forests of northwest Canada. *Ecosystems,* **9**, 268–277.

Johnstone, J.F., Chapin, F.S. III, Hollingsworth, T.N., Mack, M.C., Romanovsky, V. and Turetsky, M. (2010a) Fire, climate change, and forest resilience in Interior Alaska. *Canadian Journal of Forest Research,* **40**, 1302–1312.

Johnstone, J.F., Hollingsworth, T.N., Chapin, F.S. III and Mack, M.C. (2010b) Changes in fire regime break the legacy lock on successional trajectories in the Alaskan boreal forest. *Global Change Biology,* **16**, 1281–1295.

Johnstone, J.F., Rupp, T.S., Olsen, M. and Verbyla, D.L. (2011) Modeling impacts of fire severity on successional trajectories and future fire behavior in Alaskan boreal forests. *Landscape Ecology,* **26**, 487–500.

Jorgenson, M.T., Shur, Y.L. and Pullman, E.R. (2006) Abrupt increase in permafrost degradation in Arctic Alaska. *Geophysical Research Letters,* **33**, doi: 10.1029/2005GL024960.

Kasischke, E.S., Bourgeau-Chavez, L.L. and Johnstone, J.F. (2007) Assessing spatial and temporal variations in surface soil moisture in fire-disturbed black spruce forests in

Interior Alaska using space-borne synthetic aperture radar imagery: implications for post-fire tree recruitment. *Remote Sensing of Environment*, **108**, 42–58.

Kasischke, E.S., Verbyla, D.L., Rupp, T.S. et al. (2010) Alaska's changing fire regime: implications for the vulnerability of its boreal forests. *Canadian Journal of Forest Research*, **40**, 1313–1324.

Kokelj, S.V., Jenkins, R.E., Milburn, D., Burn, C.R. and Snow, N. (2005) The influence of thermokarst disturbance on the water quality of small upland lakes, Mackenzie Delta Region, Northwest Territories, Canada. *Permafrost and Periglacial Processes*, **16**, 343–353.

Kokelj, S.V., Lantz, T.C., Kanigan, J., Smith, S.S. and Coutts, R. (2009a) On the origin and polycyclic behaviour of tundra thaw slumps, Mackenzie Delta region, Northwest Territories, Canada. *Permafrost and Periglacial Processes*, **20**, 173–184.

Kokelj, S.V., Zajdlik, B. and Thompson, M.S. (2009b) The impacts of thawing permafrost on the chemistry of lakes across the subarctic boreal-tundra transition, Mackenzie Delta Region, Canada. *Permafrost and Periglacial Processes*, **20**, 185–199.

Kokelj, S.V., Lantz, T.C., Solomon, S., Pisaric, M.F.J., Keith, D. Morse, P. Thienpont, J.R., Smol, J. P. and Esagok, D.(2012) Utilizing multiple sources of knowledge to investigate northern environmental change: Regional ecological impacts of a storm surge in the outer Mackenzie Delta, N.W.T. *Arctic*, **65**, 257–272.

Lacelle, D., Bjornson, J. and Lauriol, B. (2010) Climatic and geomorphic factors affecting contemporary (1950–2004) activity of retrogressive thaw slumps on the Aklavik Plateau, Richardson Mountains, NWT, Canada. *Permafrost and Periglacial Processes*, **21**, 1–15.

Lantz, T.C. and Kokelj, S.V. (2008) Increasing rates of retrogressive thaw slump activity in the Mackenzie Delta region, N.W.T., Canada. *Geophysical Research Letters*, **35**, L06502, doi: 06510.01029/02007GL032433.

Lantz, T.C., Kokelj, S.V., Gergel, S.E. and Henry, G.H.R. (2009) Relative impacts of disturbance and temperature: persistent changes in microenvironment and vegetation in retrogressive thaw slumps. *Global Change Biology*, **15**, 1664–1675.

Lantz, T.C., Gergel, S.E. and Henry, G.H.R. (2010) Response of green alder (*Alnus viridis* subsp *fruticosa*) patch dynamics and plant community composition to fire and regional temperature in north-western Canada. *Journal of Biogeography*, **37**, 1597–1610.

Lindsay, R.W. and Zhang, J. (2005) The thinning of arctic sea ice, 1988–2003: have we passed a tipping point? *Journal of Climate*, **18**, 4879–4894.

Mack, M.C., Treseder, K.K., Manies, K.L., Harden, J.W., Schuur, E.A.G., Vogel, J.G., Randerson, J.T. and Chapin, F.S. III (2008) Recovery of aboveground plant biomass and productivity after fire in mesic and dry black spruce forests of Interior Alaska. *Ecosystems*, **11**, 209–225.

Manson, G.K. and Solomon, S.M. (2007) Past and future forcing of Beaufort Sea coastal change. *Atmosphere-Ocean*, **45**, 107–122.

Mesquita, P.S., Wrona, F.J. and Prowse, T.D. (2010) Effects of retrogressive permafrost thaw slumping on sediment chemistry and submerged macrophytes in Arctic tundra lakes. *Freshwater Biology*, **55**, 2347–2358.

Moquin, P. (2011) Effects of shoreline retrogressive thermokarst slumping on the productivity and food web structure of upland Arctic lakes: an experimental approach. MSc thesis. University of Victoria, Victoria, BC.

Pearce, C.M. (1986) The distribution and ecology of the shoreline vegetation on the Mackenzie Delta, NWT. PhD Dissertation. University of Calgary, Calgary.

Pisaric, M.F.J., Thienpont, J.R., Kokelj, S.V., Nesbitt, H., Lantz, T.C., Solomon, S. and Smol, J.P. (2011) Impacts of a recent storm surge on an Arctic delta ecosystem examined in the context of the last millennium. *Proceedings of the National Academy of Sciences, USA*, **108**, 8960–8965.

Ray, G.C., Overland, J.E. and Hufford, G.L. (2010) Seascape as an organizing principle for evaluating walrus and seal sea-ice habitat in Beringia. *Geophysical Research Letters*, **37**, L20504.

Robards, M.D., Burns, J.J., Meek, C.L. and Watson, A. (2009) Limitations of the optimum sustainable population or potential biological removal approaches for conservation management of marine mammals: Pacific walrus case study. *Journal of Environmental Management*, **91**, 57–66.

Robards, M.D., Schoon, M.L., Meek, C.L. and Engle, N.L. (2011) The importance of social drivers in the resilient provision of ecosystem services. *Global Environmental Change*, **21**, 522–529.

Sepp, M. and Jaagus, J. (2011) Changes in the activity and tracks of Arctic cyclones. *Climatic Change*, **105**, 577–595.

Shulski, M. and Wendler, G. (2007) *The Climate of Alaska*. University of Alaska Press, Fairbanks.

Simard, M., Romme, W.H., Griffin, J.M. and Turner, M.G. (2011) Do mountain pine beetle outbreaks change the probability of active crown fire in lodgepole pine forests? *Ecological Monographs*, **81**, 3–24.

Stephenson, S.R., Smith, L.C. and Agnew, J.A. (2011) Divergent long-term trajectories of human access to the Arctic. *Nature Climate Change Letters*, **1**, 156–160.

Stroeve, J., Holland, M.M., Meier, W., Scambos, T. and Serreze, M.C. (2007) Arctic sea ice decline: faster than forecast. *Geophysical Research Letters*, **34**, L09501.

Thompson, M., Kokelj, S., Wrona, F. and Prowse, T. (2008) The impact of sediments derived from thawing permafrost on tundra lake water chemistry: an experimental approach, in *Ninth International Conference on Permafrost*, Institute of Northern Engineering, University of Alaska Fairbanks, Fairbanks, pp. 1763–1768.

Turetsky, M.R., Mack, M.C., Hollingsworth, T.N. and Harden, J.W. (2010) The role of mosses in ecosystem succession and function in Alaska's boreal forest. *Canadian Journal of Forest Research*, **40**, 1237–1264.

Van Cleve, K., Oliver, L., Schlentner, R., Viereck, L.A. and Dyrness, C.T. (1983) Productivity and nutrient cycling in taiga forest ecosystems. *Canadian Journal of Forest Research*, **13**, 747–766.

Viglas, J. (2011) Age effects on seed productivity in northern black spruce (*Picea mariana*). MSc thesis, Department of Biology, University of Saskatchewan, Saskatoon, Saskatchewan, Canada.

Walsh, J.E., Chapman, W.L., Romanovsky, V., Christensen, J.H. and Stendel, M. (2008) Global climate model performance over Alaska and Greenland. *Journal of Climate*, **21**, 6156–6174.

Wolfe, S.A., Kotler, E. and Dallimore, S.R. (2001) Surficial characteristics and the distribution of thaw landforms (1970 to 1999), Shingle Point to Kay Point, Yukon Territory. Geological Survey of Canada, Open File 4088. Geological Survey of Canada, Ottawa.

Chapter 41

PERSPECTIVE: IS EVERYTHING A NOVEL ECOSYSTEM? IF SO, DO WE NEED THE CONCEPT?

Emma Marris[1], Joseph Mascaro[2] and Erle C. Ellis[3]

[1]Columbia, Missouri, USA

[2]Department of Global Ecology, Carnegie Institution for Science, Stanford, California, Stanford, California, USA

[3]Geography & Environmental Systems, University of Maryland, USA

In Chapters 5 and 6 of this volume the authors wrestled with the precise definition of novel ecosystems. They found that while they could fashion a rough-and-ready discrete definition, with which one could categorize any ecosystem as novel or not novel, a more accurate representation of reality called for a continuous space, with areas exhibiting degrees of novelty. That a continuum best models the concept of novel ecosystems – that its boundaries are not clear lines but gray areas – raises the possibility that a novel ecosystem may not be a distinct entity. We explore that possibility here. In doing so, we find that many instinctively appealing definitions, upon careful consideration, turn out to apply to the whole planet.

If a novel ecosystem must be changed from its historical range of species composition or ecosystem processes, then there are very few natural landscapes that would not qualify as a novel ecosystem. No ecosystem on Earth is the same as it was 12,000 years ago and most have changed much more recently (Jackson 2006; see also Chapter 7); change is an inherent prop-erty of ecosystems. Human influence is omnipresent, from land-use change to climate change (Ellis 2011).

The degree to which a specific piece of land is under active management or subject to human intelligence – another potential criteria for a novel ecosystem – is clearly a continuous variable. Lands which we might be tempted to disqualify from the category 'novel ecosystem' on the basis of some nominal amount of management may be much more self-directed than such hyper-managed spaces as an Iowa cornfield. Grazing land may be used only for the occasional grazing of livestock, which may or may not closely replicate the grazing regime of displaced herbivores. What of the person who camps in the novel forest? Is she 'using' or 'managing' the ecosystem as she roasts weenies over her fire? This vagueness works the other way as well. 'Unmanaged' spaces are seldom if ever really uninflu-enced by human activities. Areas never used directly for agriculture or settlements may still be used for hunting or spangled with species transported by humans. Even a forest that is never visited by humans may burn or not burn in the dry season due to large-scale fire prevention efforts, regional land use and so on. No ecosystem is an island, even islands. It is clear that the seeming absence of local human influence is an illusion on a planet undergoing rapid anthropogenic change. There might be some meaningful threshold beyond which an ecosystem becomes a novel ecosystem, but it is almost impossible to determine what that threshold should be. By these criteria, then, no systems are 0% novel. Is every organism, landscape or water body on Earth now part of a novel ecosystem?

Before we go on, we must stop to acknowledge that not everyone agrees that concepts have to be mutually exclusive or watertight in order to describe real entities or in order to be useful. Philosophers of science, in particular, have accepted vagueness and pluralistic definitions for scientific concepts. In the debate over the species concept, given the endless problems biologists and philosophers face pursuing a universal definition, some reject the reality of the category of species but many others accept cluster concepts, a plurality of definitions or are prepared to wait for a better, tighter definition to arrive some day in the future (Ereshefsky 2010).

In addition, we could conceive of the novel ecosystem concept as a tool. The phrase "boundary object" (Kueffer 2011) refers to a concept that is used by different groups in subtly different ways, "at the same time powerful enough to structure conceptual thinking and 'soft' or adaptive enough to allow flexible use by different groups of people".

Despite these perfectly good options for reconciling ourselves to the imperfect definitions of novel ecosystems so far proposed, we maintain that the tendency of the category to take over the entire planet is instructive. Perhaps it is the case that simplistic definitions tend to categorize the whole Earth as a novel ecosystem because the whole Earth *is* a novel ecosystem: a creation of anthropogenic change under varying levels of day-to-day management, a global garden with some corners gone feral and others planted in neat rows (Ellis 2011; Marris 2011; Steffen et al. 2011). For the purpose of this chapter, we will assume that everything is a novel ecosystem. Given this assumption, is the concept still useful?

'Novel ecosystems' is by no means the first descriptor of landscapes to face accusations of meaninglessness. Many have pointed out that if human beings are considered natural, then so are all their works and one finds that 'nature' has expanded to cover the entire universe. 'Wilderness' has in recent decades come under fire for describing a mirage: a land that has never been touched by human hands (Cronon 1996; Sörlin 2011). Yet 'nature' and 'wilderness' are still widely used terms. We want some terms to differentiate places not simply out of a joy of categorization, but because we value different places differently. Conservationists have used 'nature' and 'wilderness' to describe places they like: places to be protected and revered.

The proponents of the 'novel ecosystem' concept have a similar motivation. They seek to rebrand lands currently described by ecologists as 'degraded' or, less formally, as 'trash' so that some such lands can come to be valued. Or, as Mascaro et al. write in Chapter 5 on the origins of a novel ecosystem, "It is important that we appreciate that we are living in a changing world and that societal restoration norms of systems with historical species composition may not be a suitable or even possible future intervention target." With such an aim, it is perhaps no problem that the concept 'novel ecosystem' may be as meaninglessly broad as 'nature' when rigorously examined. Its purpose is not to describe certain places as they are, but to color our emotional reaction to certain places in order to make us see possibilities were we formerly only saw failure. Thus the concept may well be useful as a cultural idea or, to put it more bluntly, as a propaganda tool.

If that is the case, then a day should come when the term is no longer needed. Imagine a summer field season many years from now where young ecologists hike through forests and scramble up steep mountain meadows, identifying plants and animals, measuring leaf litter and tree diameters at breast height and honestly enjoying the beauty of the scene, despite the fact that all the species around them are far outside their historical ranges. With all of the ecosystems of their experience having similarly high degrees of change, would these future ecologists think of the landscape that surrounds them as a 'novel ecosystem' or just an ecosystem?

That day may be long in coming. Despite a broad realization in ecology that all ecosystems are dynamic and, by now, at least somewhat anthropogenic, conservationists and the public at large still cling to the comforting vision of the single historically correct timeless wilderness paradise. In many introductory environmental science courses, *A Sand County Almanac* is assigned as the gospel of conservation. Beautiful and elegantly written, it is hard to argue against its status as the greatest work on the subject; on the finer points of ecology, however, is it woefully outdated. Leopold notes that "Paleontology offers abundant evidence that wilderness maintained itself for immensely long periods; that its component species were rarely lost, neither did they get out of hand". He also describes wilderness as "the most perfect norm", "a base datum of normality; a picture of how healthy land maintains itself as an *organism*" (emphasis added).

At the time of publication in 1949, these statements were in step with prevailing ecology theory. Clementsian succession and the organismal 'climax' view of ecosystems reigned, and the recent past ecologically was viewed as stoic and rarely changing. Over the next half century, however, a steady stream of discoveries reversed these views. The super-organism model collapsed with a new understanding of how species moved during periods of environmental change (Davis 1981; Jackson 2006), and accounts of one-sided and pre-human 'biological invasions' surfaced (Vermeij 2005). It seemed that component species were often lost, and frequently got out of hand.

Lest it strike you as terribly unfair to argue that a 60-year-old text is, well, old, consider the modern dialogue in conservation and ecological restoration, which is just beginning to pound idly at its Clementsian shackles. Editorial boards are being dragged kicking and screaming into even minute concessions to historical efficacy. One recent high-profile review took the tellingly timid step of arguing that restoration projects might benefit from relaxing strict adherence to the historical *relative abundances* of native species ('restoration through reassembly'; Funk et al. 2008). This study envisions restoration projects only within the outlines of historical composition, but allows the relative abundances of natives to deviate from historical levels in an effort to produce a stable system. Despite its very minor relaxations of the strict goal of historical fidelity, one colleague of the second author called this study "splashy", as though the field of Restoration Ecology was conceding that it cannot produce a perfectly historically accurate ecosystem. Another paper suggested that some ecological restorations might benefit from non-native species, which in many instances help recover basic ecosystem functions such as productivity and nutrient cycling (Ewel and Putz 2004). A different colleague felt this paper should not have been published. Fredrick Clements would doubtless find each of these papers abhorrent. But why do some modern ecologists feel the same way?

And then there are practical matters. Ultimately, accepting that all ecosystems are novel means chucking out a single shared goal – historical fidelity – that was supposed to have optimally provided everything we might have wanted out of an ecosystem, from beauty to resilience to function to services to wildness. When that goal is no longer tenable, or is reduced to a minor category of lands managed for museum-style pedagogic values or as shelters for precarious endangered species, then we, collectively, have to decide on management goals for every piece of land. Specific ecosystem management strategies (which can broadly be divided into the categories of protection, restoration or development) will now ideally be decided on within a specific landscape in proportion to their ability to offer local, regional and global stakeholders the opportunity to add to the values and services of multifunctional landscapes. Decisions should hinge on factors such as the human system context (economics, culture, institutions) within which a particular ecosystem was embedded, its suitability (relative productivity per unit development and maintenance cost) for providing these values and the opportunity costs induced by not using the ecosystem for some other use.

Hashing out all these uses will not be easy, as any veteran of stakeholder processes knows. If the day comes when historical fidelity is usurped as the default goal for every landscape, we can imagine grizzled old ecologists reminiscing about a time when conservationists, at least, were all on the same page: *put it back the way it used to be*. By then, however, society will have realized that this is an option for only a few boutique restoration jobs. Everyone will be looking forward instead, trying to balance all the potential values to be gained from every acre. Restoration, which began as a retrospective enterprise, will have become wholly prospective.

Modern restorationists will doubtless view this as a pessimistic future, but we believe that intensive use will not always win the day if values such as biodiversity, recreation and beauty still have adherents. There will also be options on the table previously not considered. One relatively new value that will have to fight it out with all the others will be 'undirected succession', or the value of letting an area 'go wild' and seeing what becomes of it. Here will be our strongholds of Nature with a capital N, our new wildernesses. Leopold said that "wilderness is a resource that can shrink but not grow". However, the novel ecosystems concept helps describe a future in which a new and expanding wilderness is indeed possible and even likely. The unfolding of natural processes – whatever the level of human interference – seems to be the engine of nature.

Until this day, when 'novel ecosystems' have disappeared into 'nature' because the set of landscapes and seascapes that qualify as novel ecosystems is ultimately seen to exactly match the set of landscapes and seascapes that qualify as nature, we believe the term still has utility. In science, the novel ecosystems framework can help embolden restoration ecologists to use new tools and new goals, and shake off the yoke of history. It can also be a useful descriptor for a certain kind of landscape, a shorthand that lacks the negative connotation of terms such as 'trash ecosystem'. In policy, the concept can create a space for incentivizing the protection of spaces currently regarded as trash. Appropriately deployed, it may even be able to create monetary value for huge swathes of the undeveloped Earth. In the culture more broadly, the concept *may* be useful to teach a certain affluent outdoorsy demographic not to automatically despise and undervalue ecosystems altered by past human activity (and thus spiral into a black hole of despair from which no checks to conservation organizations can escape). But, whether 'novel ecosystem' is the right brand for this education process is up for debate. The term has a jargon feel, and doesn't tap into terms which already inspire warm emotions in the outdoorsy public. Brands which include connotations of wilderness, such as 'the new wild', might be more appealing. After all, 'historically correct' ecosystems are no longer wild, as they require ever more intensive management as they are prevented from adapting to changing conditions. Novel ecosystems are the new center of undirected evolution and wildness.

It is interesting to speculate who will grow out of the novel ecosystem concept first. Will restorationists (by then, presumably called 'ecosystem designers') learn to manage each landscape for forward-looking goals without regard to historical composition? Will policymakers come to value all ecosystems according to metrics divorced from historical composition? Will the public learn to love weeds?

Change among professional ecologists and restorationists could happen in a teaching environment; over time, historical systems visited by students will look less and less like they did to their mentors. Novelty will lose visible context to people, at which point we will simply be living in a new normal of rapidly changing ecosystems. In such a normal, what is the point of talking about categorical designations?

We have argued that the novel ecosystem concept is a useful transitional concept, even if rigorous attempts at definition cause it to balloon to unhelpful proportions. There is a counterargument that using the novel ecosystems concept even as a bridge for just a few decades is dangerous, as the existence of the category implies the existence of its opposite. This reinforces the notion that there is or can be some static, pristine alternative to novel ecosystems and thus perpetuates the very myths of ecosystem stasis and 'untouched wilderness' it seeks to uproot. If there is a 'novel' ecosystem, then there must be an 'old' or 'normal' or 'unchanged' ecosystem out there somewhere, right?

Whatever the past, today we live in a biosphere governed by human legacy and design. Humans have altered everything, and there is no going back. Our hope is that the novel ecosystems concept will help us understand this reality. Time will tell how useful we find the term, and for how long, as we learn to be conscious managers to our rapidly changing planet.

REFERENCES

Cronon, W. (1996) The trouble with wilderness: Or, getting back to the wrong nature. *Environmental History*, **1**, 7–28.

Davis, M.B. (1981) Quaternary history and the stability of forest communities, in *Forest Succession: Concepts and Applications* (D.C. West, H.H. Shugart and D.B. Botkin, eds), Springer-Verlag, New York, pp. 134–153.

Ellis, E.C. (2011) Anthropogenic transformation of the terrestrial biosphere. *Proceedings of the Royal Society A: Mathematical, Physical and Engineering Science*, **369**, 1010–1035.

Ereshefsky, M. (2010) *Species. The Stanford Encyclopedia of Philosophy*. Spring 2010 Edition (E.N. Zalta, ed.), http://plato.stanford.edu/archives/spr2010/entries/species/. Accessed August 2012.

Ewel, J.J. and Putz, F.E. (2004) A place for alien species in ecosystem restoration. *Frontiers in Ecology and the Environment*, **2**, 354–360.

Funk, J., Cleland, E., Suding, K. and Zavaleta, E. (2008) Restoration through reassembly: Plant traits and invasion resistance. *Trends in Ecology and Evolution*, **23**, 695–703.

Jackson, S.T. (2006) Vegetation, environment, and time: The origination and termination of ecosystems. *Journal of Vegetation Science*, **17**, 549–557.

Kueffer, C. (2011) Novel ecosystem – a useful boundary object? Unpublished manuscript, Institute of Integrative Biology, ETH, Zurich.

Leopold, A. (1949) *A Sand County Almanac*. Oxford University Press, Oxford.

Marris, E. (2011) *Rambunctious Garden: Saving Nature in a Post-Wild World*. Bloomsbury USA.

Sörlin, S. (2011) The contemporaneity of environmental history: Negotiating scholarship, useful history, and the new human condition. *Journal of Contemporary History*, **46**, 610–630.

Steffen, W., Grinevald, J., Crutzen, P. and Mcneill, J. (2011) The Anthropocene: conceptual and historical perspectives. *Philosophical Transactions of the Royal Society A: Mathematical, Physical and Engineering Sciences*, **369**, 842–867.

Vermeij, G.J. (2005) Invasion as expectation: a historical fact of life, in *Species Invasions: Insights into Ecology, Evolution, and Biogeography* (D.F. Sax, J.J. Stachowicz and S.D. Gaines, eds), Sinauer Associates, Inc. Publishers, Sunderland, Massachusetts, pp. 315–340.

Part VII

Synthesis and Conclusions

Chapter 42

WHAT DO WE KNOW ABOUT, AND WHAT DO WE DO ABOUT, NOVEL ECOSYSTEMS?

Richard J. Hobbs[1], Eric S. Higgs[2] and Carol M. Hall[2]

[1]Ecosystem Restoration and Intervention Ecology (ERIE) Research Group, School of Plant Biology, University of Western Australia, Australia

[2]School of Environmental Studies, University of Victoria, Canada

"That's what conservation management will be like in the future – laced with irony. Old mines will be saved for bats, and pine plantations kept for endangered cockatoos. Experts will argue about the ecological value of weeds. There will be less clarity of purpose (do we recreate the past, preserve the present, or usher in the future?) and more potential for misguided actions. Intervention, after all, is more difficult than a hands-off approach. Expect to see plenty of blunders made, as reserves are burned when they should not be, and cows are let in for the wrong reasons. Conservation is intervention, and intervention isn't easy."

(Low 2002, p. 301)

42.1 NEW NATURES

Low's quote neatly encapsulates many of the issues that have been raised and discussed in chapters throughout the book. Its relevance also indicates that we have not really progressed far in the decade since Low's controversial and forward-looking book *The New Nature* was published. Few of the issues covered have been resolved effectively. If anything, they loom larger as pressing concerns for current and future generations of people interested and involved in managing the planet and its ecosystems for the well-being of both humans and other species.

We hope that the material contained in the chapters, case studies and perspectives in this book provide a basis for constructive discussion of these issues and some clues for ways forward in dealing with them. As indicated in Chapter 1, we did not set out to arrive at an agreed-upon set of answers and recommendations. Indeed, clear differences of opinion remain among authors even after the extensive process of discussion surrounding the various chapters. A standard academic approach is to mull over and dissect these differences (with focus on ongoing theoretical debate in the literature and areas for future research). However, a more pragmatic approach is to continue to examine

Novel Ecosystems: Intervening in the New Ecological World Order, First Edition. Edited by Richard J. Hobbs, Eric S. Higgs, and Carol M. Hall.
© 2013 John Wiley & Sons, Ltd. Published 2013 by John Wiley & Sons, Ltd.

important differences while at the same time acknowledging key areas of commonality that offer a path forward. Our emphasis on the common perspectives that are prevalent in the book has allowed a move beyond simply debating what novel ecosystems are to considering how we might effectively intervene in such systems, within the broader context of ecosystem management and policy. In this final chapter we reflect on some of the ongoing points of discussion and key themes emerging in the book and offer a few last thoughts (for now) on how we start to apply these ideas to intervening in a new ecological world order.

42.2 WHAT ARE NOVEL ECOSYSTEMS?

The book contains a variety of different perspectives on what novel ecosystems are, how they can be characterized and how extensive they are. An extreme view is that, since ecosystems are always in a state of flux if viewed on the appropriate timescales, all systems are ipso facto novel. The observation that human influence is pervasive across the planet further feeds this line of argument. However, it is not clear that this argument actually helps. We know that we are changing many aspects of the Earth system more rapidly now than ever before in human history. These modifications are acting synergistically and in unpredictable ways to cause ecosystem change (Steffen et al. 2004; Chapin et al. 2008). By recognizing that ecosystems are being pushed into new configurations that do not necessarily resemble what was there in the past, we are forced to confront the dynamism of nature in a more urgent and cohesive way. This includes identifying or adapting tools and approaches for determining if, when and how to intervene, and requires a good deal more than simply concluding that everything is novel.

Having said that, what then does novelty mean? What characterizes a novel ecosystem? Can a novel ecosystem be defined in a way that is both useful and general? Such questions (as discussed in Chapter 6) created perhaps the most heated debate both at the 2011 workshop (see Chapter 1) and during the book-writing process: definitions inevitably provoke strong opinions and attachments to particular wordings and emphases. One definition of novel ecosystems is presented in Chapter 6; this is our attempt at a working characterization.

Towards arriving at this definition, disagreements surfaced around various aspects of novel ecosystems;

although some strong differences remain the debate has been constructive. One such aspect is the place of history or historical systems in considering novel ecosystems. Historical conditions are frequently referred to as a baseline or reference state for assessing novelty (i.e. a departure from baseline or reference state). 'Historical' is however often used in a vague and unreferenced way and can refer to anything in the past, contributing to confusion and a sense by some that history is no longer relevant. But how that past is defined makes all the difference. As Steve Jackson reminds us (Chapter 7) and as noted earlier, ecosystems are in constant flux and hence all ecosystems are novel at some timescales. Novelty, like beauty, is in the eye of the beholder. This means that the criteria and time–space scales being used to talk about novelty should always be carefully defined.

The role of historical knowledge in informing decisions around novel ecosystems, as with decisions on ecosystem management generally, is increasingly being dismissed as irrelevant as rapid environmental change pushes systems into new configurations that have diminishing connection with past conditions. On the other hand, some argue that historical configurations still yield the best information on how systems work and provide guidance and limits to management endeavors. As Jackson and Hobbs (2009) point out, at this juncture we need to embrace both perspectives and try to find a resolution to the apparent contradiction this presents. Novel ecosystems go to the heart of this contradiction and provide an important test-bed for ideas and approaches.

If novelty is measured against some reference, how should that reference be determined? This question is broached in several places in the book (e.g. Chapters 3, 18, 19 and 24), and during writing much debate surrounded the question of whether historical references were valid or possible and whether modern analogs or current reference states were more appropriate. All of these alternatives have both advantages and limitations, with questions of temporal or spatial uncertainties. While new approaches may be needed, current best practices that are based on years of successes and failures have much to offer and should be considered first, especially in the face of uncertainty over the current state and trajectory of the system. In all cases, it is important to recognize that reference ecosystems, however determined, are not meant to be a rigid template towards which the system should be managed or restored. Rather, they act to provide guidance and need to be considered in tandem with current and projected conditions and dynamics.

An important aspect that can lead to confusion is that phrases such as 'historical condition' or 'reference state' should really be viewed as shorthand for a more dynamic perspective; recognizing that dynamism is an inherent property of all ecosystems, past, present and future states should be considered more as trajectories of change. A key question in attempting to resolve these issues is to consider when novelty actually matters – at what stage does change and degree of difference from a past or present condition trigger the need for management intervention?

This question leads to another key point of debate that revolved around the question of whether novelty is a continuum or whether there are there clear breakpoints where it is evident that one system is novel and another is not. Several 2011 workshop participants strongly suggested that novelty occurred along a continuum, while others posited that taking a categorical approach – being able to classify systems as novel or not novel – had greater management utility. Central to this discussion was the question of whether clear thresholds exist in the state space illustrated in Figure 3.2, as put forward by Hobbs et al. (2009). The idea of threshold dynamics has proven very useful in considering ecosystem management choices and techniques (e.g. Hobbs and Harris 2001; Suding and Hobbs 2009), indicating that the use of thresholds in a framework for managing novel ecosystems is appropriate.

This was the predominant line of thought carried through the book in chapters dealing with management interventions (Chapters 3 and 18). Nevertheless, our ability to observe, measure and diagnose ecological thresholds and hence to make decisions effectively remain painfully limited in many instances, as illustrated by the challenges of navigating management decisions (e.g. in the Colorado Front Range; Chapter 20). As already pointed out, this issue is not restricted to discussions on novel ecosystems, but is a live topic of urgent debate and research in applied ecology in general (Andersen et al. 2008; Samhouri et al. 2010; Bestelmeyer et al. 2011). Certainly, many ecosystem characteristics that might be measured to gauge novelty are continuous rather than discrete. To some extent the debate on whether novelty should be considered a continuum or not mirrors extensive debate in ecology over, for instance, whether ecological communities can be considered discrete or continuous (Whittaker 1975; McIntosh 1985). The answer is, of course, that they can be both, depending on perspective, purpose and mode of measurement. Physics seems to have survived effectively with the dual notion of light as particles and as waves. Similarly, for some purposes it will be useful to consider a continuum of measurements and in others to seek dichotomies to make maps, aid management decisions and so on.

Clear black and white distinctions between 'novel' and 'not novel' may not always be possible and shades of gray are the more likely scenario. As Low (2002) observes, "Like computers, humans are probably wired to think in opposites – yes and no, good and bad, winners and losers, nature and culture, natural and artificial. The world isn't divided up like this, but our minds like to see it that way." Indeed, as pointed out in Chapter 30, exploring novel ecosystems ideas can actually help us to confront and tackle in more nuanced ways the simple binaries that permeate conservation discourse. This tendency toward simple binaries has been evident in editing the book, where often any ecosystem that has 'novel elements' or a degree of 'novelty' is labeled as novel, when in fact many of those described are more likely to be a hybrid system. These hybrid systems are situated most solidly in the gray zone between historical and novel ecosystems, where we find some of the most wicked issues.

A further topic for debate is the relative irreversibility of changes depicted as threshold crossings. A core characteristic of novel ecosystems discussed in many chapters is their persistence or relative 'stability', with an underlying assumption that the system has crossed a threshold (ecological or social) that is, for all intents and purposes, irreversible. But how realistic is this assumption? Irreversibility is a big concept, and one that might not hold water in all situations. The example of restoring the Hole in the Donut in the Everglades (Chapter 2) shows clearly that, with enough effort, even massive changes can be to some extent reversed. Three important observations follow from this. Firstly, we must take care when assuming that any given ecosystem state change is 'irreversible'; throwing enough resources at it may well result in at least partial reversal. Secondly, on the other hand, many changes may be irreversible in a practical sense because resources, institutional will, policy settings and other social factors conspire to make it virtually impossible to implement management interventions that could force a reversal. At the same time, while social barriers can be very difficult to alter and can result in crossing of thresholds, these should not be trotted out as an excuse to avoid action (Chapters 6 and 18). Finally, the idea that we know how to reverse ecosystem changes and direct the system effectively may, in many cases, be a fine example of misguided hubris.

42.3 DO WE NEED NEW MEASURES, TOOLS AND APPROACHES FOR NOVEL ECOSYSTEMS?

Are novel ecosystems a new class of ecosystem that require new ways of thinking and doing things, or are they simply a subset of the types of ecosystem that are already studied and understood to varying degrees and managed on the basis of that current understanding? This is a tricky question to answer for a number of reasons. On one hand, since novel ecosystems comprise the same basic elements and processes as any ecosystem (i.e. an abiotic stage with biotic players and processes such as nutrient cycling, trophic interactions and the like), we should theoretically not need any new science to understand them or tools to manage them. On the other hand, so little attention has been paid to novel ecosystems until recently that there is not a well-developed body of literature describing how they work.

As highlighted in Chapter 24, measuring novelty is really about measuring ecological difference. There are many well-established techniques for doing this, together with ongoing development of measurement techniques, statistical methods and so on. What to actually measure remains a key question, and it will be important in questions of novel ecosystems to focus clearly on what might be relevant and reliable to measure rather than on individual preference, fad or availability of measuring capability (Lindenmayer and Likens 2010). In novel ecosystems, perhaps more than anywhere else, the dynamics of new species mixtures and new abiotic settings have the capacity to result in ecological surprises. The capacity to predict and detect such surprises remains remarkably low (Lindenmayer et al. 2010), and the monitoring and measurement of novel ecosystems might provide an ideal vehicle for improving this capacity.

Finding useful measures to apply to novel ecosystems is particularly important in attempts to map the past, current and future extent of novel ecosystems, as discussed in Chapters 8 and 9. While global mapping (see Chapter 8) has some utility, it is currently limited because it has to rely on characteristics that can be measured remotely. An important element of many novel ecosystems is a shift in species composition, and this is not always easily detectable from space. Finer-scale observations made with on-ground sampling may sometimes, or often, be the only way to detect changing compositions, especially in non-canopy species. Hence, broad-brush estimates of the extent of novel ecosystems based on remotely sensed data represent a partial view of the situation. Again, this may come down to definitional issues; as long as it is made clear what exactly is being represented in such maps, then potential confusion should be minimized.

One perspective highlighted in several chapters is that novel ecosystems do not necessarily demand the development of entirely new tools and approaches, but that they may in fact open the door to the opportunity for utilizing existing management techniques in novel ways (see Chapters 3 and 18). The key will be in finding innovative ways of solving apparently intractable problems, for example, perhaps not hitting a key driver of change square on but taking a roundabout route to alleviating its impacts (see Box 20.1). An example of this is using cows to remove unwanted growth of non-native grasses in serpentine grassland caused by elevated nitrogen deposition in California (Chapters 3 and 5; Weiss 1999). It does not solve the N deposition problem but it does provide at least a temporary solution to aggressive grass growth and loss of native plant diversity.

Some lessons learned from managing island ecosystems where novelty is widespread offer a window into whether there is a need for new tools and approaches (see Chapter 4). Based on experience managing these ecosystems, the authors of Chapter 4 suggest an approach that considers factors such as assessment of the need to intervene, the potential barriers to intervention and feasibility of intervention. The authors also point to the importance of gaining stakeholder support, assessing the probability of success or failure before intervening, trying to both avoid and plan for unintended consequences and of working within local constraints while looking for local opportunities.

Interestingly, these suggestions apply equally well to the management of any ecosystem, novel or not. In a world of rapid and pervasive change where there are frequent surprises and the need for more explicit trade-offs, these lessons do suggest that certain approaches, such as taking into account multiple scales and integrating social factors, carry greater importance to pay attention to and learn from in the attempt to get it 'right'. We note however that these features, and others including adaptive management, are regularly recognized as important but still only rarely incorporated effectively in ecosystem management.

Further, the recognition that novel ecosystems are an increasing part of the mix of systems to be managed perhaps more than anything requires a change in the overall approach to how management goals are set and which techniques and outcomes are deemed important. This book (Chapters 3 and 18) provides a framework to start working through the potential morass that lies at the heart of many of the concerns surrounding novel ecosystems. How do we make decisions about if and how to intervene and what goals we manage for when it is no longer possible to manage for historical conditions? However, the morass is wider than just novel ecosystems and results from the steadily increasing recognition that traditional methods and goals are not always appropriate in today's rapidly changing world. The fields of conservation biology and restoration ecology are, in their modern forms, only a few decades old, and recent conservation gains in terms of protected areas and the like have been hard won. It is therefore easy to see the perceived threat presented by now starting to ponder on the likelihood that many of the tenets of these young fields and the practice and policy developing from them are possibly no longer appropriate and almost certainly no longer sufficient.

42.4 SOCIAL, ETHICAL AND POLICY DIMENSIONS OF NOVEL ECOSYSTEMS

Clearly, as with ecosystem management more broadly, many management decisions have a strong values component embedded in them. The concept of novel ecosystems can invoke strong reactions, and much of this arises from the challenge that novel systems appear to present to existing dearly held values and norms concerning nature and humanity's relationship to it. Reciprocally, it is often these same values and norms that are behind important drivers of change or barriers to the effective management of altered ecosystems, emphasizing the need to consider the social and ecological aspects of the issue in tandem. Numerous chapters (e.g. 18, 21 and 39) highlight aspects of novel ecosystems over and above the straightforward understanding and management of the biophysical realm.

Questions of value are far from straightforward (and contribute significantly to the potential morass surrounding the idea of novel ecosystems). Many of the concerns about novel ecosystems can be addressed by clarifying misunderstandings, whereas others are more persistent and less likely to be malleable in the face of more facts and figures (see Chapter 37). However, there is a tendency to think that public perception of novel ecosystems will be unflinchingly negative, and this idea is clearly challenged by examples indicating that people find value in altered ecosystems. For instance, the shale bings in central Scotland have moved relatively rapidly from being ugly blots on the landscape to treasured features with cultural and biodiversity value (Chapter 35). Is this simply a case of insidious 'shifting baselines' (Papworth et al. 2009), where people begin to value the ecosystems they encounter currently while not recognizing the altered state of these systems and the loss of past values? Or is it more a positive admiration for the systems that have developed as a result of the interplay of environment and human culture and industry?

The shale bings of Scotland offer a striking modern example of how the products of human endeavors inevitably become part of the overall landscape fabric and contribute to its processes, support of fauna and flora and the human experience (also see Cei Balast example, Chapter 32). This has been occurring throughout human history. Some view human impacts on ecosystems as inevitably negative – a degrading influence. However, where humans have been present in ecosystems for long periods of time, the dependence of ecosystem structure and function on human presence and activity can be large. This is clearly obvious in Europe, where valued ecosystems and biodiversity depend on human management activities. Where traditional management ceases the system changes in sometimes undesirable ways, for instance in Mediterranean areas where the decline of traditional management has led to increasing shrub and tree cover in once open landscapes and to increased wildfire incidence. Human impacts on ecosystems in the new world are still a topic of much debate, but it is becoming increasingly clear that such impacts often shaped the systems we know and value today.

The only difference (but a very big difference) between historical human impacts and those in evidence today is the pervasiveness and rapidity of change. Novel ecosystems thus loom larger in humanity's perception of 'nature'. Further, the recognition that human endeavors are embedded in the very fabric of our landscapes pushes us to consider these social factors not just in tandem with ecological factors but as interwoven elements that affect each other in multiple and often unexpected ways (Turner 2005).

In considering how we value novel ecosystems, it is important to distinguish between two differing aspects of novel ecosystems. First, there is the need to accept and even embrace novel ecosystems that have emerged and continue to develop as a result of organisms responding to altered conditions (albeit given that the alterations themselves are human-induced). But second is the desire and motivation to create new systems on purpose. The ideas of 'designer ecosystems' and ecosystem engineering have been mooted for some time (MacMahon and Holl 2001; Mitsch and Jørgensen 2003). However, recent proposals within conservation biology that aim to deal with climate change and other large environmental concerns include relatively radical suggestions involving moving species around, either to extend their range (as in managed relocation or assisted migration; Hoegh-Guldberg et al. 2008; Richardson et al. 2009) or to place them in environments to fulfill or restore specific functions (as in Pleistocene rewilding; Donlan et al. 2005, 2006). These proposals have spurred vigorous ongoing debate (much more so than the concept of novel ecosystems).

As an example, a recent proposal raises the possibility of importing elephants to Australia to assist in management of plant invasion and fire regimes (Bowman 2012). While the idea has been roundly criticized, it is a logical extension of the idea that we can create new systems and use species for specific purposes, even if they do not 'belong' in an area from a biogeographic perspective. Bringing elephants to Australia would undoubtedly help create an entirely novel system, with highly unpredictable consequences. So, do such proposals court hubris and represent a step too far along the gradient of human intervention in ecosystems? Are we heading towards designing ecosystems that contain only attractive or useful species? Are we at risk of adopting a technological mindset in how we manage ecosystems? Is the new approach 'anything goes'? On the other hand, at what point is it no longer reasonable to attempt to return to historical conditions? The Hole in the Donut in the Everglades also mentioned earlier is an example at the other end of the spectrum where park managers took extraordinary measures in attempt to return a novel ecosystem to historical conditions, despite the cost (efforts required in moving more than 3 million cubic meters of soil) and evidence that the novel ecosystem provided some valued functions (see

Chapter 2). As a less extreme example, consider the case from the Mascarene Islands (described in Chapter 4) where recently extinct endemic giant tortoises were replaced with extant tortoises from other islands in the region. As a result, important ecosystem functions such as herbivory and seed dispersal likely to benefit native plants (see also Fig. 4.6) were re-instated. In this case, management intervention is consistent with the framework presented in this book (see Chapter 3) where the emphasis shifts from managing for specific species or conditions to managing for (historical) function.

There are again clearly ecological dimensions to the earlier questions, but ultimately societal value judgments have an important role in deciding what is and is not acceptable and, in each case, where we draw the line. Elsewhere, Higgs and Hobbs (2010) have called for a measured approach to intervention decisions that involves recognition that some element of 'design' is increasingly inevitable and that such design can be carried out within a framework that limits the potential for overstepping the mark. Similarly, work from other disciplines such as that around 'virtue ethics' (see Chapter 31) can help guide goal setting and 'good' management. Well-designed policy also can provide a mechanism to help navigate trade-offs and difficult decisions around prioritizing goals (see Chapter 33) and avoiding perverse outcomes (Lindenmayer et al. 2012). A comprehensive approach to policy is needed that simultaneously recognizes novelty and the dynamic nature of ecosystems and takes measures to slow land-use conversion and climate change, the main drivers of ecosystem change. Further, we need to re-shape the policy process itself (and related institutions) to be more dynamic and flexible (Chapter 33) in the face of uncertainty and rapid socio-ecological change.

Digging around social, ethical and policy issues exposes the ever-present 'slippery slope' that concerns many people (Chapter 37). By starting to recognize the existence of novel ecosystems, to think about how to manage them for goals detached from historical conditions and even to suggest that such systems can have values, some fear that these steps will result in a rapid slide further into the morass and away from traditional concepts and methods. This is a real concern and one that should not be dismissed lightly. However, is it better to refrain from recognizing and discussing these issues for fear of precipitating a slide, or should we embark on the discussion and at least aim to stand at the top of the slide with our eyes open?

42.5 LOVING OUR MONSTERS

In a chapter in a recent book entitled *Love Your Monsters – Postenvironmentalism and the Anthropocene*, Bruno Latour (2011) examines humanity's relationship with technology through the story of Frankenstein: "Dr. Frankenstein's crime was not that he invented a creature through some combination of hubris and high technology, but rather that he abandoned the creature to itself. When Dr. Frankenstein meets his creation on a glacier in the Alps, the monster claims that it was not born a monster, but that it became a criminal only after being left alone by his horrified creator, who fled the laboratory once the horrible thing twitched to life. "Remember, I am thy creature," the monster protests, "I ought to be thy Adam; but I am rather the fallen angel, whom thou drivest from joy for no misdeed . . . I was benevolent and good; misery made me a fiend. Make me happy, and I shall again be virtuous."

Novel ecosystems can also be viewed as monsters – created either deliberately or inadvertently through human activity, though not necessarily sustained by human intervention thereafter. But following Latour's analogy, we can either lament our crime in creating these monsters or we can perpetuate a crime of ignoring and abandoning them. Unlike Frankenstein's monster, whose future was sealed by events, novel ecosystems can have multiple pathways to the future and there are choices in terms of whether humanity should intervene to maintain such systems or direct them to a different trajectory.

The issues around novel ecosystems are part of a broader dialog about humanity's changing relationship with nature. All ecosystems – including novel ones – will continue to change into the future, perhaps at unprecedented rates. The challenge is to find a path through the complex and pervasive issues that need to be tackled in the quest to nurture and maintain human populations and the world's ecosystems and species. We hope that this book contributes in some small way to this quest.

REFERENCES

Andersen, T., Carstensen, J., Hernandez-Garcia, E. and Duarte, C.M. (2008) Ecological thresholds and regime shifts: approaches to identification. *Trends in Ecology and Evolution*, **24**, 49–57.

Bestelmeyer, B.T., Goolsby, D.P. and Archer, S.R. (2011) Spatial perspectives in state-and-transition models: a missing link to land management? *Journal of Applied Ecology*, **48**, 746–757.

Bowman, D. (2012) Bring elephants to Australia? *Nature*, **482**, 30.

Chapin, F.S., Randerson, J.T., McGuire, A.D., Foley, J.A. and Field, C.B. (2008) Changing feedbacks in the climate-biosphere system. *Frontiers in Ecology and the Environment*, **6**, 313–320.

Donlan, C.J., Berger, J., Bock, C.E. et al. (2006) Pleistocene rewilding: An optimistic agenda for twenty-first century conservation. *American Naturalist*, **168**, 661–681.

Donlan, J., Greene, H.W., Berger, J. et al. (2005) Re-wilding North America. *Nature*, **436**, 913–914.

Higgs, E.S. and Hobbs, R.J. (2010) Wild design: Principles to guide interventions in protected areas, in *Beyond Naturalness: Rethinking Park and Wilderness Stewardship in an Era of Rapid Change* (D. Cole and L. Yung, eds), Island Press, Washington, pp. 234–251.

Hobbs, R.J. and Harris, J.A. (2001) Restoration ecology: Repairing the Earth's ecosystems in the new millennium. *Restoration Ecology*, **9**, 239–246.

Hobbs, R.J., Higgs, E. and Harris, J.A. (2009) Novel ecosystems: implications for conservation and restoration. *Trends in Ecology and Evolution*, **24**, 599–605.

Hoegh-Guldberg, O., Hughes, L., McIntyre, S., Lindenmayer, D.B., Parmesan, C., Possingham, H. and Thomas, C.D. (2008) Assisted colonization and rapid climate change. *Science*, **321**, 345–346.

Jackson, S.T. and Hobbs, R.J. (2009) Ecological restoration in the light of ecological history. *Science*, **325**, 567–569.

Latour, B. (2011) Love your monsters, in *Love Your Monsters – Postenvironmentalism and the Anthropocene* (M. Shellenberger and T. Nordhaus, eds), The Breakthrough Institute, Amazon Digital Services.

Lindenmayer, D.B. and Likens, G.E. (2010) Improving ecological monitoring. *Trends in Ecology & Evolution*, **25**, 200–201.

Lindenmayer, D.B., Likens, G.E., Krebs, C.J. and Hobbs, R.J. (2010) Improved probability of detection of ecological "surprises". *Proceedings of the National Academy of Sciences*, **107**, 21957–21962.

Lindenmayer, D.B., Hulvey, K.B., Hobbs, R.J. et al. (2012) Avoiding bio-perversity from carbon sequestration solutions. *Conservation Letters*, **5**, 28–36.

Low, T. (2002) *The New Nature: Winners and Losers in Wild Australia*. Viking, Camberwell, Victoria.

MacMahon, J.A. and Holl, K.D. (2001) Ecological restoration: A key to conservation biology's future? in *Conservation Biology: Research Priorities For The Next Decade* (M.E. Soulé, and G.H. Orians, eds), Island Press, Washington, DC, pp. 245–269.

McIntosh, R.P. (1985) *The Background of Ecology: Concepts and Theory*. Cambridge University Press, Cambridge.

Mitsch, W.J. and Jørgensen, S.E. (2003) Ecological engineering: A field whose time has come. *Ecological Engineering*, **20**, 363–377.

Papworth, S.K., Rist, J., Coad, L. and Milner-Gulland, E.J. (2009) Evidence for shifting baseline syndrome in conservation. *Conservation Letters*, **2**, 93–100.

Richardson, D.M., Hellmann, J.J., McLachlan, J.S. et al. (2009) Multidimensional evaluation of managed relocation. *Proceedings of the National Academy of Sciences*, **106**, 9721–9724.

Samhouri, J.F., Levin, P.S. and Ainsworth, C.H. (2010) Identifying thresholds for ecosystem-based management. *PLoS ONE*, **5**, e8907.

Steffen, W., Sanderson, A., Tyson, P.D. et al. (2004) *Global Change and the Earth System: A Planet Under Pressure*. Springer-Verlag, Berlin, Heidelberg, New York.

Suding, K.N. and Hobbs, R.J. (2009) Threshold models in restoration and conservation: A developing framework. *Trends in Ecology and Evolution*, **24**, 271–279.

Turner, N.J. (2005) *The Earth's Blanket: Traditional Teachings for Sustainable Living*. Douglas & McIntyre Ltd., Vancouver, British Columbia.

Weiss, S.B. (1999) Cars, cows, and checkerspot butterflies: nitrogen deposition and management of nutrient-poor grasslands for a threatened species. *Conservation Biology*, **13**, 1476–1486.

Whittaker, R.H. (1975) *Communities and Ecosystems*, 2nd edition. MacMillan, New York.

INDEX

Note: page numbers followed by b, f, or t refer to Boxes, Figures or Tables

Novel Ecosystems: Intervening in the New Ecological World Order, First Edition. Edited by Richard J. Hobbs, Eric S. Higgs, and Carol M. Hall.
© 2013 John Wiley & Sons, Ltd. Published 2013 by John Wiley & Sons, Ltd.

Printed and bound by CPI Group (UK) Ltd, Croydon, CR0 4YY

27/10/2024

14580355-0004